Novel Psychoactive Substances: Classification, Pharmacology and Toxicology

This page intentionally left blank

Novel Psychoactive Substances: Classification, Pharmacology and Toxicology

Edited by

PAUL I. DARGAN MB, BS, FRCPE, FACMT, FRCP, FAACT
Consultant Physician and Clinical Toxicologist, Clinical Director and Reader in Toxicology
London, UK

DAVID M. WOOD MD, FRCP, FACMT, FBPharmacolS
Consultant Physician and Clinical Toxicologist, Senior Lecturer
London, UK

AMSTERDAM • BOSTON • HEIDELBERG • LONDON • NEW YORK • OXFORD
PARIS • SAN DIEGO • SAN FRANCISCO • SINGAPORE • SYDNEY • TOKYO
Academic Press is an imprint of Elsevier

Academic Press is an imprint of Elsevier
32 Jamestown Road, London NW1 7BY, UK
225 Wyman Street, Waltham, MA 02451, USA
525 B Street, Suite 1800, San Diego, CA 92101-4495, USA

British Library Cataloguing-in-Publication Data
A catalogue record for this book is available from the British Library

Library of Congress Cataloging-in-Publication Data
A catalog record for this book is available from the Library of Congress

ISBN: 978-0-12-415816-0

For information on all Academic Press publications
visit our website at elsevierdirect.com

Typeset by MPS Limited, Chennai, India
www.adi-mps.com

Contents

Section 2
ANALYTICAL TECHNIQUES

6. Analytical Techniques for the Detection of Novel Psychoactive Substances and Their Metabolites

FRANK T. PETERS AND MARKUS R. MEYER

Section 3
INDIVIDUAL NOVEL PSYCHOACTIVE SUBSTANCES

7. Synthetic Amphetamine Derivatives

JEFF LAPOINT, PAUL I. DARGAN AND ROBERT S. HOFFMAN

8. 1-Benzylpiperazine and other Piperazine-based Derivatives

PAUL GEE AND LEO SCHEP

9. Mephedrone

DAVID M. WOOD AND PAUL I. DARGAN

10. Pipradrol and Pipradrol Derivatives

MICHAEL W. WHITE AND JOHN R.H. ARCHER

11. Aminoindane Analogues

SIMON D. BRANDT, ROBIN A. BRAITHWAITE,
MICHAEL EVANS-BROWN AND ANDREW T. KICMAN

12.   Ketamine

QI LI, WAI MAN CHAN, JOHN A. RUDD, CHUN MEI WANG,
PHOEBE Y.H. LAM, MARIA SEN MUN WAI, DAVID M. WOOD,
PAUL I. DARGAN AND DAVID T. YEW

13.   Synthetic Cannabinoid Receptor Agonists

VOLKER AUWÄRTER, PAUL I. DARGAN AND DAVID M. WOOD

14.   Natural Product (Fungal and Herbal) Novel Psychoactive Substances

SIMON GIBBONS AND WARUNYA ARUNOTAYANUN

15.   Tryptamines

SHAUN L. GREENE

16.   Benzofurans and Benzodifurans

SHAUN L. GREENE

17.   Miscellaneous Compounds

SHAUN L. GREENE

This page intentionally left blank

Forewords

Over the last decade the world has gone through a period of extensive change driven by the twin engines of globalisation and technological advancement. We now live on a smaller and more joined up planet in which geographical boundaries have been transcended by the creation of a virtual space that facilitates the rapid flow of goods, services and ideas. The way people work, play, and socialise has been transformed in ways that would have been hard to imagine even a decade ago. Profound economic, social and political changes have occurred, bringing with them both positive and negative consequences. Amongst these, the growing dislocation of commerce and communication from physical jurisdictions means that the power of nation states to control or regulate these activities has been considerably weakened.

As so many aspects of our modern life have been transformed by these developments it is not surprising that they are also having an impact on the consumption and availability of psychoactive substances. With time and hindsight, the changes we are now seeing in drug use and the drug market may come to be viewed as predictable developments. They have however been experienced as rapid and unexpected changes that have taken our existing policies and practices by surprise. In terms of both public health and social control the paradigm on which contemporary responses to drug use has been historically based has become increasingly challenged by the emergence of a wide range of new and novel substances. They have been bought to market in sophisticated new ways, and with a speed that only serves to highlight the slow pace at which countermeasures can be mustered. Within a short period of time this has become a global phenomenon, albeit one with products specifically tailored to be attractive to different consumer groups. The power of modern search engines has allowed the back-catalogue of the chemical and pharmacological research industry to be exploited to find obscure chemicals with proven actions. The Internet has become both a vehicle for the diffusion of innovation in this area, and increasingly now also a new marketplace where existing regulatory models are failing to find traction. Products sometimes containing complex cocktails of chemicals have been developed and marketed using attractive packaging that appeals so well to its intended consumers that it has an influence that can even be seen leaking out into wider design trends. The end result of all this is that an increasing number of young people are experimenting with chemicals about which we simply have no knowledge of the acute or chronic risks from either human or animal studies. The long-term implications of this are unclear but worrisome. Sufficient experience has already been accumulated however to know that for some of these substances at least the potential for them to cause severe acute adverse consequences is considerable.

Developments in this area have also highlighted the inadequacies of our current conceptual framework for supporting scientific discourse and responses. We are struggling with terminology, lacking in appropriate analytical tools for identification and finding that many of our current monitoring approaches are simply not fit for purpose when faced by consumers who are using 'brands' that contain unknown and often complicated mixtures of obscure chemicals that can change over time. Self reports that are the main stay of many illicit-drug studies are of little value when the user may be ignorant of the substance(s) that they are consuming, or refer to a brand name. Equally, medical services are increasingly faced by having to respond to individuals with acute toxicity with a history of consumption of substances of unknown provenance and content.

The pace of change in this area has been so rapid that the information available for guiding policies and practice has struggled to keep up. Clinicians are handicapped in responding to this new reality in drug use by a severe knowledge deficit and currently lack the tools needed to ground their practice in hard evidence. This book is therefore both timely and important. Paul Dargan and David Wood bring an unparalleled vision and understanding to this topic that comes from their pioneering work as researchers and clinicians who have followed this phenomenon closely from its inception. As scientists they guide us in this publication through the pharmacological and toxicological issues that are critical to understanding the implications of developments in this area. But they also do this from the perspective of clinicians who have unparalleled experience of working at the 'coalface' of clinical practice in responding to drug induced emergencies occurring in the London club scene. Their work has justifiably earned them a world-wide reputation and they have exploited this to assemble a global cast of leading experts to steer us through this complex topic. This multi-authored publication provides a comprehensive review of what we know about the detection, pharmacology and toxicology of novel psychoactive substances. It also provides the reader with the concepts and framework necessary to understand this emergent area. Individual classes of novel psychoactive substances that have recently emerged onto the recreational drug scene are described, accompanied by a review of the pharmacology and acute and chronic toxicity. This information is much needed and has never previously been brought together in such a comprehensive and authoritative form.

We can only conclude by noting how happy we are to endorse this ground breaking publication, which offers the much needed foundation to support an understanding of this complex and emergent area. It not only provides the reader with the conceptual tools necessary to understand this phenomenon, but also imparts an understanding of critical issues related to the pharmacology and identification of novel psychoactive substances. Most importantly it also includes a state of the art review of the clinical implications of the consumption of new and novel psychoactive substances in respect to their acute and chronic toxicology. Put simply, this book is essential reading and a valuable reference source for any scientist, clinician, policy maker or law enforcement professional wanting to become more familiar with what is probably the most important contemporary development in the drugs field.

Paul Griffiths MSc
Scientific Director, EMCDDA

Wolfgang Götz MSc
Director, EMCDDA , European Monitoring
Centre for Drugs and Drug Addiction
Lisbon, Portugal

The market for synthetic drugs has always been characterised by a large variety of substances. Historically, novel (new) psychoactive substances (NPS) have always appeared in this market. However, in recent years, the pace at which these substances have emerged has increased considerably. In 2011, 49 NPS were reported to the early warning system of the European Monitoring Centre on Drugs and Drug Addiction. This means that one NPS was reported almost every week. And this monitoring system only covers some 30 European countries, out of the more than 200 countries and territories in the world.

At the global level, nobody knows the exact number of NPS that are currently circulating in the market today. However, preliminary figures from a survey on NPS carried out by the United Nations Office on Drugs and Crime suggest that more than 100 different NPS were available on the market in 2012.

As the number of NPS increases, so do the potential health problems that some of these substances entail. Often, their appearance is accompanied by an increase in reported health problems, emergency room visits and even fatalities. National and local authorities face numerous challenges when trying to address the issue. Drug analysis laboratories struggle to identify the active ingredients of the substances, as conventional analytical techniques methods often do not provide reliable results. Health care professionals in particular find it difficult to provide proper care as there are few resources on the pharmacology and toxicology of these substances when they first emerge.

The present publication narrows this information gap and is the first book to provide a wide spectrum of facts on everything one has always wanted to know about NPS, including information on their chemical structure, their availability, supply and their epidemiology of use. It will be of great use to the increasing number of people requiring reliable scientific information on this issue.

Early warning systems on NPS have been pivotal in national efforts to identify and monitor them. At present such a system exists only for a number of European States. However, there is a worldwide dimension to the emergence of these substances that deserves our attention. Many NPS have spread to all corners of the globe – synthetic cannabinoid receptor agonists ('Spice'), for example, have been reported from Africa, the Americas, Asia, Europe and Oceania. Like cocaine or heroin, NPS are a truly international phenomenon – often manufactured in one region of the world, packaged and sold to users in another – and therefore warrant an international response.

Sharing information and best practices and using our collective knowledge to develop common strategies are some of the elements that should be part of an effective action plan to address the challenge. Monitoring the issue will provide useful information that will assist in making evidence-based decisions that respond to the rapid changes that encompass the supply and demand of NPS.

Beate Hammond BA, MBA

Beate Hammond manages the Global Synthetics Monitoring: Analyses, Reporting and Trends (SMART) Programme of the United Nations Office on Drugs and Crime. The views expressed in this preface are those of the author and do not necessarily reflect the views of the United Nations.

This page intentionally left blank

List of Contributors

John R.H. Archer Clinical Toxicology, Guy's and St Thomas' NHS Foundation Trust, King's Health Partners and King's College London, London, UK

Warunya Arunotayanun Department of Pharmaceutical and Biological Chemistry, UCL School of Pharmacy, London, UK

Volker Auwärter Institute of Forensic Medicine, Forensic Toxicology, University Medical Center Freiburg, Freiburg, Germany

Robin A. Braithwaite Drug Control Centre, Department of Forensic Science, Division of Analytical and Environmental Sciences, King's College London, London, UK

Simon D. Brandt School of Pharmacy and Biomolecular Sciences, Liverpool John Moores University, Liverpool, UK

Paul I. Dargan Clinical Toxicology, Guy's and St Thomas' NHS Foundation Trust, King's Health Partners and King's College London, London, UK

Michael J. Evans-Brown Centre for Public Health, Liverpool John Moores University, Liverpool, UK

Ana Gallegos European Monitoring Centre for Drugs and Drug Addiction (EMCDDA), Lisbon, Portugal

Fiona M. Garlich University of Calgary, Calgary, Canada

Paul Gee Christchurch Hospital, University of Otago, Christchurch, New Zealand

Simon Gibbons Department of Pharmaceutical and Biological Chemistry, UCL School of Pharmacy, London, UK

Shaun L. Greene Victorian Poisons Information Centre, Melbourne, Australia

Robert S. Hoffman Department of Emergency Medicine, New York University School of Medicine, New York City Poison Control Center, Bellevue Hospital Center, New York, NY

Padraigin Kenny European Monitoring Centre for Drugs and Drug Addiction (EMCDDA), Lisbon, Portugal

Andrew T. Kicman Drug Control Centre, Department of Forensic Science, Division of Analytical and Environmental Sciences, King's College London, London, UK

Leslie A. King Former part-time advisor to the European Monitoring Centre for Drugs and Drug Addiction (EMCDDA), Lisbon, Portugal

Phoebe Y.H. Lam Department of Oncology, Cancer Research UK/Medical Research, Council Gray Institute for Radiation Oncology and Biology, University of Oxford, Oxford, UK

Jeff Lapoint Emergency Medicine, Medical Toxicology, Southern California Permanente Medical Group San Diego, CA

Qi Li Department of Psychiatry, The University of Hong Kong, Hong Kong

Wai Man Chan School of Biomedical Sciences, The Chinese University of Hong Kong, Hong Kong

James McVeigh Centre for Public Health, Liverpool John Moores University, Liverpool, UK

Fiona Measham Professor of Criminology, School of Applied Social Sciences, Durham University, Durham, UK

Chun Mei Wang School of Biomedical Sciences, The Chinese University of Hong Kong, Hong Kong

Markus R. Meyer Department of Experimental and Clinical Toxicology, Institute of Experimental and Clinical Pharmacology and Toxicology, Saarland University, Homburg (Saar), Germany

Jane Mounteney European Monitoring Centre for Drugs and Drug Addiction (EMCDDA), Lisbon, Portugal

Maria Sen Mun Wai School of Biomedical Sciences, The Chinese University of Hong Kong, Hong Kong

Frank T. Peters Institute of Forensic Medicine, University Hospital Jena, Jena, Germany

John A. Rudd School of Biomedical Sciences, The Chinese University of Hong Kong, Hong Kong

Leo Schep National Poisons Centre, Department of Preventive and Social Medicine, University of Otago, Dunedin, New Zealand

Roumen Sedefov European Monitoring Centre for Drugs and Drug Addiction (EMCDDA), Lisbon, Portugal

Silas W. Smith Department of Emergency Medicine, New York University School of Medicine, New York, NY

Harry Sumnall Centre for Public Health, Liverpool John Moores University, Liverpool, UK

Michael W. White Formerly of the Forensic Science Service Ltd, London, UK

David M. Wood Clinical Toxicology, Guy's and St Thomas' NHS Foundation Trust, King's Health Partners and King's College London, London, UK

David T. Yew School of Biomedical Sciences, The Chinese University of Hong Kong, Hong Kong

BACKGROUND

This page intentionally left blank

Legal Classification of Novel Psychoactive Substances
An International Comparison

Leslie A. King

Former part-time advisor to the European Monitoring Centre for Drugs and Drug Addiction (EMCDDA), Lisbon, Portugal

INTRODUCTION

Historical Background

Limited drug control began in the early years of the 20th century, following the Shanghai Opium Commission in 1909 and the League of Nations Conventions of 1925 and 1931. These early controls were largely restricted to traditional plant products (e.g. opium, cannabis, cocaine) and semi-synthetics such as heroin. To a great extent, the drug legislation of most countries now originates from the precepts of the United Nations (UN) Treaties, namely the Single Convention of 1961 and the UN 1971 Convention on Psychotropic Substances. The schedules of the two UN Conventions comprise mostly traditional drugs [1,2] and, as discussed later, apart from a few phenethylamines, do not include any examples of the more recent drug groups. The organisation of chemical entities into various schedules in the UN Conventions is partly based on whether the substances have any therapeutic value and partly on the risk of harm associated with their use. However, national legislatures have often incorporated the UN scheduling scheme as a basis for determining penalties associated with various offences such as possession, supply, production, importation etc. A notable exception to this rule is the United Kingdom (UK). In the UK the schedules of the Misuse of Drugs Regulations 2001 [3] largely reflect the UN classification, but the separate Misuse of Drug Act, 1971 sets out the same substances (known as controlled drugs) in three Classes (A, B and C). In other words, the Regulations set out what should be done, i.e. their use within a clinical context, while the Act sets out what should not be done.

Novel Psychoactive Substances.
DOI: http://dx.doi.org/10.1016/B978-0-12-415816-0.00001-8

The term 'Novel Psychoactive Substance' is the latest in a series of expressions to describe a relatively recent phenomenon. A few miscellaneous phenethylamines, such as STP (2,5-dimethoxy-4-methamphetamine) and its bromine analogue DOB (bromo-STP; 4-bromo-2,5-dimethoxyamphetamine) had been misused in the United States (US) since at least the early-1960s; in the UK an illicit tablet was found to contain STP [4] in 1969. However, it was the appearance in the US, during the early 1980s, of derivatives of the narcotic analgesics fentanyl and α-prodine (where desmethylprodine is the reverse ester of pethidine/meperidine) that gave rise to major concerns. In particular, two of the substituted fentanyls (α-methylfentanyl and 3-methylfentanyl) were typically several hundred times more potent analgesics than morphine. Not surprisingly, these high potencies led to many accidental, often fatal, overdoses. The α-prodine series caused a major public health issue when it was found that a by-product of clandestine synthesis (MPTP; 1-methyl-4-phenyl-1,2,5,6-tetrahydropyridine) produced a rapid and irreversible chemically-induced Parkinsonism. These events [5] led to the coining of the term 'designer drugs', which were defined as:

> Analogues, or chemical cousins, of controlled substances that are designed to produce effects similar to the controlled substances they mimic.

Following the publication of the book *PIHKAL* [6] ('Phenethylamines I have known and loved') in 1991, large numbers of mostly ring-substituted phenethylamines began to appear in Europe. These raised questions about possible health risks and the problems that could arise if such substances were arbitrarily controlled in some Member States, but not in others. It was agreed that progress could be made by sharing information and by establishing a risk-assessment procedure and a mechanism for their eventual control across the European Union (EU). This led, in 1997, to the 'Joint action concerning the information exchange, risk assessment and control of new synthetic drugs'. These 'new synthetic drugs' were defined as those that had a limited therapeutic value and were not at that time listed in the 1971 United Nations (UN) Convention on Psychotropic Substances, yet posed as serious a threat to public health as the substances listed in Schedules I and II of that Convention. The 'Joint action' was superseded by an EU Council Decision of 2005 [7,8], leading to a more comprehensive and robust system for monitoring what then became known as 'New Psychoactive Substances'. The Council Decision introduced procedures for risk assessment and EU-wide control in appropriate cases. The definition of these substances is:

> Narcotic or psychotropic drugs that are not scheduled under the United Nations 1961 or 1971 Conventions, but which may pose a threat to public health comparable to scheduled substances.

The words 'new' and 'novel' refer to the fact that these substances are newly-misused, but some of them had been first synthesised many years ago. In the meantime, other expressions have appeared to describe the phenomenon, including 'legal highs', 'research chemicals', 'party pills' and specific phrases such as 'plant food' or 'bath salts' (often used to describe white powders) and 'incense' as a euphemism for smoking mixtures containing synthetic cannabinoid receptor agonists (cannabimimetics). The term novel psychoactive substance is used to refer to all of these in this textbook.

Legal Concerns

The appearance of novel substances has continued to cause problems for drug control authorities in many countries. Following the lead of the UN Treaties, it has been an accepted part of drug legislation that a substance should only be brought under control (scheduled) if it can be shown to be harmful, either to individuals, to society or both. And therein lies the central difficulty: almost nothing is known about the pharmacology of many new substances or their potential for abuse. Some were developed by academic laboratories or the pharmaceutical industry as potential medicines, but never succeeded to market authorisation. The synthesis and basic chemical properties of these 'failed pharmaceuticals' will often have been described in the scientific or patent literature, yet apart from *in vitro* studies and occasional limited animal testing, their pharmacodynamic and pharmacokinetic properties and metabolic fate in humans usually remain largely unexplored. Other substances are closer to the original definition of a designer drug; in other words they have been deliberately created as entirely novel compounds by clandestine laboratories and synthesised by analogy with better-known substances. Their properties have never been published and even the most basic information is lacking; what little we do know comes from occasional fatal poisonings in humans and clinical observations of intoxicated patients. Anecdotal reports from users, such as may be found on Internet 'chat rooms', must be treated with caution since the exact identity of the substances concerned may be unknown, often being described by street terms or product names, the composition of which often changes with time.

In the UK, the Misuse of Drugs Act, 1971 provides greater room for manoeuvre [9]. Thus, there is no strict requirement to demonstrate actual harm, provided that the substance concerned might have the potential for harm. In Section 1(2) of the Act, which sets out the duties of the Advisory Council on the Misuse of Drugs (ACMD), there is a definition of what constitutes a controlled drug. Thus, the Council should:

> ... keep under review the situation in the United Kingdom with respect to drugs which are being or appear to them likely to be misused and of which the misuse is having or appears to them capable of having harmful effects sufficient to constitute a social problem...

This flexibility has allowed the UK to introduce a wide range of generic controls. These are described in more detail later, but an inevitable consequence is that an essentially infinite group of substances will be subsumed where, for most, information is unavailable, and is never likely to become available. And it is quite certain that some substances will not only be harmless, but will have no physiological effect of any kind.

While the basic properties of new substances could be investigated by relatively inexpensive research programs, perhaps using *in vitro* receptor binding, metabolic studies and other methods, governments often wish to act at an early stage of misuse. There is a belief that it is better to control a substance because of the severe consequences of permitting open sale of a substance that later turns out to be harmful. On the other hand, restricting a substance that is later shown to be harmless has far fewer negative consequences. The problem is made worse by the number of compounds involved and the rapid replacement of controlled substances by non-controlled analogues. Thus even those substances that remain uncontrolled often have a short lifetime on the illicit market. Furthermore, reliable population surveys and information on prevalence may not become available until a substance is well-established, assuming it ever is.

There is a general view in many countries that existing drug law is inadequate to deal with new substances, and that better solutions are needed. A recent review commissioned by the UK Drug Policy Commission [10] has elaborated on the problems and opportunities for restricting new substances. This is just one of many reviews in the past few years that have scrutinised drug policy in the most general sense. For example, the UK has seen numerous reports that have been specifically targeted at the working of the Misuse of Drugs Act, 1971 [11–14]. During 2012, two separate Committees of the UK Parliament undertook reviews of drug policy. The All-Party Parliamentary Group on Drug Policy Reform was specifically focused on novel substances, while one of the terms of reference of the Home Affairs Select Committee was 'the availability of "legal highs" and the challenges associated with adapting the legal framework to deal with new substances'. The reports from both Committees were published in late 2012 [15, 16].

Increasingly, questions are now asked almost daily and at an international level about whether drug prohibition, and particularly prohibition of possession, is the right course of action in the modern world. This wider debate is largely driven by attitudes to established drugs of misuse, but it cannot be entirely separated from prohibitions surrounding new substances. Although it is beyond the scope of this chapter to examine them in detail, questions are also being asked about whether scheduling substances under the criminal law has any impact on usage, whether 'drugs of misuse' might have benefits to the user, to what extent penalties should reflect the harm caused to individuals and society and whether some controls do more harm than good. These issues will be explored in more depth in Chapter 5, 'Social issues in the use of novel psychoactive substances', of this book Finally, there is the question of whether law enforcement agencies consider drug misuse, and particularly misuse of new substances, a priority issue, particularly at a time when police budgets are being reduced [17].

The absence of appropriate information on the properties of new substances has led legislatures around the world to look for new ways of restricting their supply. At its most basic, a novel drug might be considered as just another chemical entity. It is quite normal for chemical retailers to restrict supply of their products to *bona fide* companies and research establishments. And in all countries, legislation exists to control certain chemical entities such as is required by the UN 1988 Convention Against Illicit Traffic in Narcotic Drugs and Psychotropic Substances [18] for drug precursors and the UN 1997 Chemical Weapons Convention [19] for chemical weapons and their precursors. In the UK, the Poisons Act 1972 and the Poisons Rules 1982 [20] require that the sale of some poisonous chemicals is only possible through pharmacies or registered sellers. The list includes, for example, certain 'organophosphorus' compounds and other pesticides, salts of arsenic, barium and mercury, mineral acids, nicotine, paraquat and formic acid. That Act was designed to guard against the misuse by accident, inadvertence or criminal design of non-medicinal poisons to which the public need to have access. At another level, consumer protection legislation exists to guard against the harmful effects of products, and to ensure that they are properly labelled. Alternatively, novel substances could be classified as medicinal products. Established and relatively safe medicinal products might be on open sale or available through a pharmacy, but others can only be obtained under medical supervision. In practice, countries that classify one of these novelties as a medicinal product do not issue product licences for them. Since most of the new substances in question are manufactured in the Far East, the option is open for other countries to introduce import controls. Some legislatures have introduced entirely new controls, and a few case studies will be described. A recent report by Reuter [21] considered options for regulating new drugs and discussed the experiences in the US and Europe with four substances (1-benzylpiperazine [BZP], 'Spice' [mixtures of synthetic cannabinoid receptor agonists and probably inert vegetable matter intended for smoking], mephedrone and naphyrone [NRG-1]).

Chemical and Pharmacological Classification

As will be seen from the above discussion, novel substances are mostly synthetic compounds. This was clearly the focus of the 1997

EU Joint Action, and is reflected in what has been seen world-wide over the past few decades. However, the 2005 EU Council Decision broadened the scope to include, for example, herbal products (see Chapter 14, 'Natural product (fungal and herbal) novel psychoactive substances') and even medicines. Only a few plants/fungi or their extracted products have been reported since 2005; they include *Salvia divinorum*, which contains the hallucinogen salvinorin-A, *Piper methysticum* (kava kava; active principals kawain and related substances), *Tabernanthe iboga* (ibogaine), *Areca catechu* (betel nut; active principal arecoline) and *Mitragyna speciosa* (kratom; active principles mitragynine and 7-hydroxymitragynine). There are many other plant extracts that might be added to this list, some of which contain established scheduled drugs, for example, *Diplopterys cabrerana*, *Psychotria viridis* and *Mimosa hostilis* (N,N-dimethyltryptamine), *Catha edulis* (khat; cathinone), and *Psilocybe semilanceata* and other 'magic' mushrooms (psilocin and psilocybin). However, most of these 'non-synthetics' pose particular legal problems and are rarely amenable to an all-encompassing control regime. Many legislatures are reluctant to specify herbal materials beyond the traditional products (e.g. cannabis, coca leaf and opium), because of the botanical, taxonomic and physical difficulties that can arise in their identification.

It is sometimes useful to consider solvents and gases, such as nitrous oxide (laughing gas), alkyl (now mostly isopropyl) nitrite (poppers), aliphatic hydrocarbons (e.g. cigarette lighter fuel) and aromatic hydrocarbons (e.g. adhesive solvents) within the group of novel psychoactive substances. But solvents and gases, particularly when used by inhalation (contrast ingestion of the solvent GBL, gamma-butyrolactone) pose problems with analysis and proof of possession by virtue of their high vapour pressures. That said, some volatile substances are amenable to restrictions on supply. In the UK, the Intoxicating Substances (Supply) Act 1985 [22] makes it an offence for a retailer to sell solvents to anyone under the age of 18, knowing that they are being purchased to be abused. It does not make it illegal to buy or own solvents. The Cigarette Lighter Refill (Safety) Regulations 1999 [23] – an amendment to the Consumer Protection Act 1987 – makes it illegal to supply gas cigarette lighter refills to anyone under the age of 18. Furthermore, European Directive 2005/59/EC of 26th October 2005 prohibits the placing on the market, for sale to the general public, the substance toluene and adhesives and spray paints containing in excess of 0.1% toluene [24]. Nitrous oxide has clinical use as an anaesthetic, but is also a commercially-available foaming agent for dairy cream, where restrictions on the small pressurised containers would be difficult to enforce. Attempts in the UK to classify alkyl nitrites (other than the once clinically-useful amyl nitrite) as medicinal products have so far been unsuccessful.

Active pharmaceutical ingredients and medicinal products in general represent a further group that can fall under the heading of 'novel psychoactive substances'. It is not usually the established use of such products that is the cause for concern, but rather their unlicensed consumption, often in pharmaceutical forms or routes of administration that differ from those authorised. Examples here include dextromethorphan (DXM), a common antitussive when used in small quantities (e.g. 10 mg), which is alleged to produce psychoactive effects when 100–200 mg are ingested. Illicit tablets containing large amounts of DXM are now rarely seen. Ketamine (Chapter 12), when in the form of injection ampoules, is an established licensed medicinal product for use as an analgesic and anaesthetic, but tablets and white powders are unlicensed products that may be ingested or snorted. Other pharmaceutical ingredients that have been notified to the European Monitoring Centre for Drugs and

Drug Addiction (EMCDDA) under the terms of the Early Warning System include phenazepam, pyrazolam and etizolam (none of which is included with other benzodiazepines in the UN 1971 Convention), pregabalin, benzydamine, glaucine and GHB (gamma-hydroxybutyrate). It is appropriate in this context to mention misuse of licensed cognitive enhancers, the most common example of which is modafinil. Apart from the US, this licensed medicine is not commonly subject to drug legislation, but is widely misused, and is available through similar Internet channels as other novel substances. The European Medicines Agency announced in November 2011 that the use of modafinil should be restricted to the treatment of narcolepsy. The review by the Agency's Committee for Medicinal Products for Human Use (CHMP) was initiated because of a number of safety concerns, relating to psychiatric disorders, skin and subcutaneous tissue reactions as well as significant off-label use and potential for abuse [25].

As noted earlier, the detailed pharmacological properties of many novel substances are unknown, but in terms of general effects it is clear that users seek out substances which are primarily central nervous system (CNS) stimulants like amphetamine or behave as entactogens and empathogens like 3,4-methylenedioxymethamphetamine (MDMA). The synthetic cannabinoid receptor agonists are often smoked as substitutes for cannabis, but hallucinogens are less common, while novel narcotic analgesics are now rare.

TRADITIONAL CONTROL MECHANISMS

Novel psychoactive substances are not an entirely new phenomenon. What is new is the rate at which they have appeared on illicit drug markets in numerous countries over the last few years. Although the concept of a designer drug was first recognised and defined 30 years ago [5], the commonly-heard term 'legal high' is more recent. In a world where new substances did not appear too often it is unsurprising that the simplest method of control would be to name them as individual chemical entities or plant products. This is known as 'specific listing'. However, the UK and a few other countries such as Ireland and New Zealand recognised some years ago that controlling a chemically-defined group of substances might be more efficient: a process known as generic control. Meanwhile, an alternative approach, known as 'analogue control' first appeared in the US legislation in 1986. The administrative procedures involved in adding a substance to national drug laws show considerable variation. For example, they may require approval of Parliament, the Government or simply a Minister. Depending on which process occurs, the speed of control varies from a few weeks to many months. Detailed information on the methods used in individual countries of the European Union, the substances concerned and the penalties for specific offences can be found in the European Legal Database on Drugs [26] and the review by Hughes and Blidaru [27].

In the following paragraphs, different methods of drug control are described, but it should be recognised that they are not all mutually exclusive. In other words, both specific and generic methods might be subject to temporary control measures, and generic and analogue control can be used concurrently. In addition, some substances might be listed in the drug control as well as other legislation but, at least in the UK, the drug legislation (i.e. Misuse of Drugs Act, 1971) takes precedence. To a large extent, the focus is on the primary objects of concern, i.e. synthetic compounds.

Specific Listing

Specific listing, that is to say the individual listing of substances by their chemical names, has the advantage, in principle, that there is no ambiguity about whether or not a substance is covered by the legislation. In other words, it satisfies the legal principle of certainty in criminal law. The major drawback of specific listing is that, when new substances arise in quick succession, the legislative process of adding them one-by-one can prove increasingly burdensome. However, even within the UN treaties, the concept of specific listing has been partly compromised as it became necessary to deal with certain derivatives of scheduled substances. For example, the esters and the ethers of morphine first came under international control through the Geneva Convention of 1931 [28]. This was further extended by the United Nations 1961 Convention to refer to all substances in Schedule I, and was designed to prevent the production of non-scheduled substances that had a similar effect to, or could easily be converted into, scheduled drugs. Thus heroin (diacetylmorphine) and codeine (3-methylmorphine) remain as named substances, but without the modification, other esters and other ethers of morphine would have had a similar misuse potential, yet are chemically distinct from heroin and codeine respectively. Likewise, salts of scheduled substances are now treated in the same way as the parent compound. Ignoring salts, esters and ethers, there are over 100 substances named in each of the two UN Conventions. However, by 2011 the individual countries of the EU had controlled, in total, over 600 named substances [26].

Generic Definitions

The UK was the first country to introduce generic controls. The essence of a generic definition is that it starts with a core molecular structure. This may not in itself be controlled or even liable to misuse, but the definition goes on to set out particular substituent groups at specified positions in that core molecule that do lead to controlled substances. In 1964 [29],

an attempt was made to group a large number of CNS stimulants into a single definition. The Drugs (Prevention of Misuse) Act contained the definition (with certain named exceptions):

> Any synthetic compound structurally derived from either α-methylphenethylamine or β-methylphenethylamine by substitution in the side chain, or by ring closure therein, or by both such substitution and such closure…

However, although this did indeed include compounds such as phentermine, methylphenidate and other prescription anorectics common in those days, it soon became clear that a refined interpretation included many drugs that were not stimulants [30]. It was even argued that some barbiturates such as phenobarbitone were also captured. Difficulties then arose with interpretation when multiple bonds were present in the side chain or substitution by oxidation occurred in the side chain. This generic control was repealed in 1970. Following this early failure, it would be some years before generic control of phenethylamines again entered the legislation. But this time (1977), the focus was on ring-substituted phenethylamines; this was, more robust and was be followed by generic controls for many other groups. Table 1.1 lists the groups for which generic control now exists in the UK under the Misuse of Drugs Act, 1971, showing their year of introduction and classification. The penalties for offences involving controlled drugs decrease in the order A > B > C. It will be seen that recently-classified new substances are either in Class B or Class C.

Despite the UK having over 30 years' experience of operating generic controls, numerous arguments against them or perceived difficulties continue to be raised. These include:

- **They would hinder the development in the pharmaceutical industry of novel compounds for legitimate clinical use.** This has not been a problem in the UK. Even if the pharmaceutical industry did wish to develop substances that were covered by generic

TABLE 1.1 Chemical Groups for which Generic Controls Operate under the UK Misuse of Drugs Act, 1971, Showing their Year of Introduction and Classification

Group	Year	Class
Anabolic steroids	1996	C
Barbiturates	1984	B
Cannabinoid agonists	2009	B
Cannabinols	1971	B
Cathinones	2010	B
Naphthylpyrovalerone and related compounds	2010	B
Ecgonine derivatives	1971	A
Fentanyls	1986	A
Lysergide and derivatives of lysergamide	1971	A
Pentavalent derivatives of morphine	1971	A
Pethidines	1986	A
Phenethylamines	1977	A
Phenyl- and benzylpiperazines	2009	C
Pipradrol derivatives	2012	B
Tryptamines	1977	A

controls, it would be a simple matter to either issue licences or modify the legislation.

- **Control of chemical groups may cover substances with a range of different pharmacological effects and some with no effects whatsoever.** Because the Act relies on the concept of actual or potential social harm, rather than the specific pharmacological or toxicological properties of a controlled drug, no great difficulty arises from the introduction of generic control. This would be more of concern in those jurisdictions (and the UN itself) where there is an *a priori* need to review the pharmacological and toxicological

properties of every substance considered for control. It is quite certain that amongst the essentially infinite number of generically defined substances there will be compounds that have little abuse potential and some may have no physiological effect of any sort. Without these effects, a substance will not be marketed by the pharmaceutical industry and neither will it be produced as a misusable drug. However, it cannot be denied that this blurs the principle that penalties associated with a drug offence should correlate with the harmful properties of that drug.

- **Useful medicines and other substances will be inadvertently controlled.** Provided that the definitions of included substances are sufficiently rigorous, this should rarely happen. In the generic definition of phenethylamines (see later), a specific exclusion was made for methoxyphenamine (*o*-methoxy-*N*-methamphetamine), the active pharmaceutical ingredient in now obsolete proprietary bronchodilators, for example Orthoxine®.

- **Generic controls will be difficult to comprehend.** One of the most complex definitions in the Misuse of Drugs Act, 1971 involves ring-substituted phenethylamines, but in the past 30 years many tens of thousands of witness statements, involving the identification of MDMA in seized samples, have been submitted in evidence by UK forensic science laboratories. These statements have incorporated the definition without any apparent problems. Nevertheless, it is still perceived as a weakness that certain common substances, e.g. MDMA, mephedrone, are not named specifically, but rather are hidden within a definition that may be accessible only to forensic chemists.

As an example of the complexity of generic controls, consider the definition of

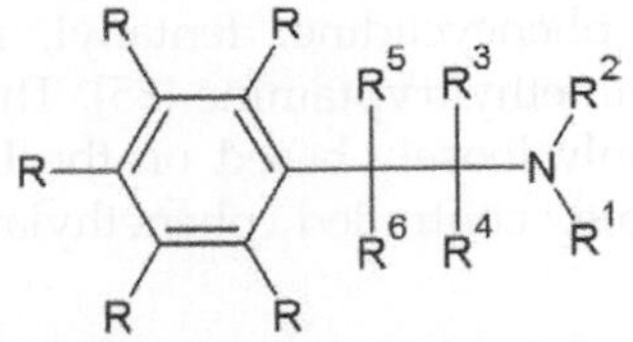

FIGURE 1.1 Generalised structure of a phenethylamine showing substitutions in the ring and side-chain.

ring-substituted phenethylamines introduced in 1977:

> any compound (not being methoxyphenamine or a compound for the time being specified in subparagraph (a) above) structurally derived from phenethylamine, an N-alkylphenethylamine, α-methylphenethylamine, an N-alkyl-α-methylphenethylamine, α-ethylphenethylamine, or an N-alkyl-α-ethylphenethylamine by substitution in the ring to any extent with alkyl, alkoxy, alkylenedioxy or halide substituents, whether or not further substituted in the ring by one or more other univalent substituents.

This can be illustrated by the structural diagram in Figure 1.1.

To meet the above definition, the following criteria must be satisfied:

R^1 = H or alkyl
$R^2 = R^3 = R^5 = R^6$ = H
R^4 = H, methyl or ethyl
R = alkyl, alkoxy, alkylenedioxy or halogen (either singly or in any combination) with or without any other substitution in the ring.

The focus of this rather daunting definition is ring-substitution in amphetamine-like molecules. The reasoning behind this is that the attachment of other atoms (especially oxygen, sulfur or halogen) to one or more of the carbon atoms (commonly the 2-,4- or 5-positions) in the aromatic ring of phenethylamine leads to major changes in pharmacological properties. Whilst amphetamine and many of its side-chain isomers and other simple derivatives (e.g. methamphetamine, methcathinone and benzphetamine) are all CNS stimulants, suitable

substitution in the ring can create hallucinogens (e.g. mescaline) or empathogenic/entactogenic agents that may or may not retain some stimulant activity.

Despite some apparent difficulties with generic controls, it cannot be denied that they represent efficient ways of capturing a large group of substances. For example, when the book PIHKAL [6] was published in 1991, almost 80% of the substances shown in the principal monographs were covered by the above definition of ring-substituted phenethylamines. Of the 50 ring-substituted phenethylamines notified to EMCDDA since 1997, only a few fall outside the definition – typically those with more complex N-substituents or multi-ring systems such as the 'FLY' series (e.g. 2C-B-FLY and bromodragonFLY), where the phenyl ring bears two fused furanyl rings.

As of late August 2012, 40 cathinone derivatives had been reported (Chapter 9, Mephedrone), yet almost all are subsumed by the generic definition [31,32], the only notable exceptions again being those with anomalous N-substitution, e.g. N-benzyl-substituted compounds. Cannabimimetic activity is found in a diverse group of compounds with multiple sites for substitution; the generic definitions for synthetic cannabinoid receptor agonists [33] capture less than half of the ca. 65 substances reported by late August 2012 (Chapter 13, Synthetic Cannabinoid Receptor Agonists). However, it is quite conceivable that the existing generic definitions could be modified.

Table 1.2 shows the structural and broad pharmacological classification of the 252 substances reported to EMCDDA between 1997 and late August 2012. As will be seen, most are the subject of generic definitions in the UK legislation (viz. phenethylamines, tryptamines, piperazines, cathinones, synthetic cannabinoid receptor agonists, pipradrol derivatives), although not all substances within each group are necessarily subsumed by the respective definitions. Around two-thirds are

TABLE 1.2 Structural Classification of the 252 Substances Reported to EMCDDA Between 1997 and Late August 2012

Group	%	Pharmacology
Phenethylamines	20	Mostly CNS stimulants
Tryptamines	11	Hallucinogens
Piperazines	5	CNS stimulants
Cathinones [a]	16	CNS stimulants
Cannabimimetics	26	CB_1 receptor agonists
Pipradrol derivatives	2	CNS stimulants
Miscellaneous	20	Mostly CNS stimulants

[a]includes naphthylpyrovalerone and related compounds.

CNS stimulants. There are a number in the miscellaneous group that could be brought under generic control if the need arose. Thus the 2-aminoindans [34] are currently represented on the European Database on New Drugs (EDND) by three examples: 2-aminoindan itself (2-AI); 5, 6-methylenedioxy-2-aminoindan (MDAI); and 5-iodo-2-aminoindan (5-IA). However, several other members of this group (e.g. 5,6-methylenedioxy-N-methyl-2-aminoindane (MDMAI) and 5-methoxy-6-methyl-2-aminoindane (MMAI) have been described in the scientific literature and could be potentially new drugs (see Chapter 11). Smaller miscellaneous groups are represented by: 1) ketamine (see Chapter 12) and its analogues such as methoxetamine; 2) the related group of phencyclidine analogues, i.e.: 1-[1-(4-methoxyphenyl)-cyclohexyl]-piperidine [4-MeO-PCP] and 3-methoxyeticyclidine [3-MeO-PCE]; and finally 3) the positional isomers (i.e. 4- and 6-) of 5-(2-aminopropyl)benzofuran (5-APB).

In New Zealand, the legislation has generic definitions for derivatives of amphetamine,

pethidine, phencyclidine, fentanyl, methaqualone and dimethyltryptamine [35]. These definitions are only loosely based on the UK model. For example, controlled phenethylamines are defined as:

Amphetamine analogues, in which the 1-amino-2-phenylethane nucleus carries any of the following radicals, either alone or in combination:

(a) 1 or 2 alkyl radicals, each with up to 6 carbon atoms, attached to the nitrogen atom:
(b) 1 or 2 methyl radicals, or an ethyl radical, attached to the carbon atom adjacent to the nitrogen atom:
(c) a hydroxy radical, attached to the carbon atom adjacent to the benzene ring:
(d) any combination of up to 5 alkyl radicals and/or alkoxy radicals and/or alkylamino radicals (each with up to 6 carbon atoms, including cyclic radicals) and/or halogen radicals and/or nitro radicals and/or amino radicals, attached to the benzene ring.

A number of other countries have adopted the generic system based on the UK model. The Republic of Ireland introduced generic controls at an early stage for many of the groups shown in Table 1.2. In late 2011, Switzerland introduced legislation to capture novel substances, including a number of generic definitions (see Switzerland Schedule 'e' below). In Denmark, a new drug strategy – Kampen mod Narko II [36] – was introduced in October 2010. This is expected to lead to an amendment to the Euphoriants Act that will incorporate generic definitions, and is expected to come into force soon [37]. The intention is to introduce a staged system of group definitions that is expected to include synthetic cannabinoid receptor agonists, phenethylamines, cathinones and tryptamines. On 10 August 2011, Lithuania [26] added generic definitions for cathinones and synthetic cannabinoid receptor agonists to its legislation (Amendment of the Order of the Minister of Health of the Republic of Lithuania No V-776). In Hong Kong, under the Dangerous Drugs Ordinance (Amendment of First Schedule) Order 2011, generic control now extends to piperazines, cathinones and synthetic cannabinoid receptor agonists [38]. In January 2009, the Austrian Government used a decree under the Pharmaceutical Law to declare that 'smoking mixes containing JWH-018' are prohibited from being imported or marketed. In March 2009, this was extended to include CP-47,497 and its homologues (a generic concept) and HU-210 [39]. These provisions were further amended in May 2011 when the Austrian Government brought a wide range of synthetic cannabinoid receptor agonists under generic control based on the UK model. In December 2011, the Austrian Government introduced yet further amendments, this time controlling a wide range of derivatives of cannabimimetics, phenethylamines, cathinones, isocathinones, aminoindans, tryptamines, 1-phenyl- and 1-benzyl-piperazines, arylcyclohexylamines and diphenylmethylpiperidines [26]. In 2011, the Italian Government enacted a rather broad control on 'derivatives of 3-phenylacetylindole and 3-(1-naphthoyl)indole' [26].

Alongside the structure-substitution generic model, many legislatures had to deal with a different type of generic control, namely the problem caused by isomers, and specifically stereoisomers. In the UK, the first such modifications predated the current (1971) legislation. In 1998, following a proposal from the Spanish Government, the World Health Organisation (WHO) considered extending control of substances listed in the UN 1971 Convention to 'isomers, esters, ethers and analogues'. However, WHO considered that the changes might have a negative impact on legitimate industry, and they were rejected. It was also stated that control of analogues would contradict its mandate of evaluating individual substances. In addition, the proposed control of isomers, as opposed to stereoisomers, was widely regarded as being too vague. However, the UN 1971 Convention, but not the UN 1961 Convention, was later modified to allow control of stereoisomers.

The Analogue Approach

The USA was the first country to adopt analogue controls. These are much broader than the generic system. The Controlled Substances Analogue Enforcement Act 1986 (sometimes called the Federal Analogue Act) [40] defines analogues in the following way:

> Controlled substance analogue means a substance –
>
> (i) the chemical structure of which is substantially similar to the chemical structure of a controlled substance in schedule I or II; and
> (ii) which has a stimulant, depressant, or hallucinogenic effect on the central nervous system that is substantially similar to or greater than the stimulant, depressant, or hallucinogenic effect on the central nervous system of a controlled substance in schedule I or II; or
> (iii) with respect to a particular person, a substance which such person represents or intends to have a stimulant, depressant, or hallucinogenic effect on the central nervous system substantially similar to or greater than the stimulant, depressant, or hallucinogenic effect of a controlled substance in schedule I or II.

In an appeal heard in 1996 (United States v. Allen McKinney), the Federal Analogue Act was deemed not to be constitutionally vague [41]. The case concerned sale of aminorex (5-phenyl-4,5-dihydro-1,3-oxazol-2-amine) before it became explicitly controlled, and the sale of phenethylamine as a substitute for methamphetamine. Some of the limits of what was meant by 'substantially similar' were argued in the case of United States v. Damon S. Forbes et al. in 1992 [42], where it was decided that α-ethyltryptamine (AET; Fig. 1.2) was not an analogue of either N,N-dimethyltryptamine (DMT; Fig. 1.3) or N,N-diethyltryptamine (DET; Fig. 1.4).

The reasons for this decision were that: AET is a primary amine, but DMT and DET are tertiary amines; AET cannot be synthesised from DMT or DET; the effects of AET are not substantially similar to those of DMT or DET. By contrast, it has been accepted that

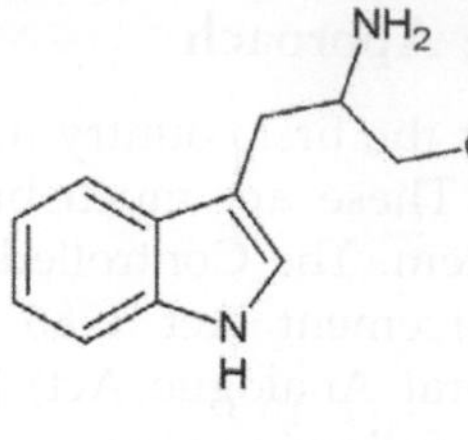

FIGURE 1.2   The structure of α-ethyltryptamine (AET).

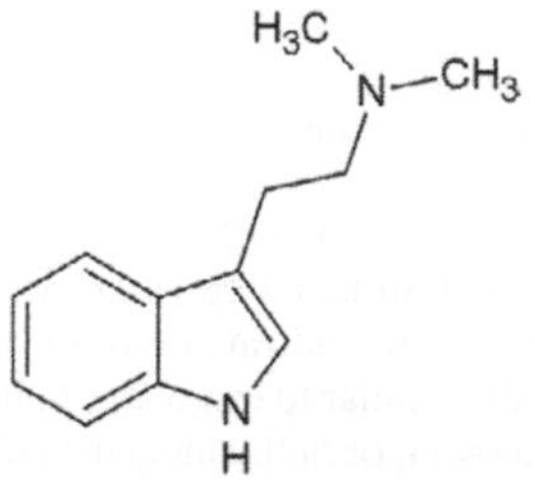

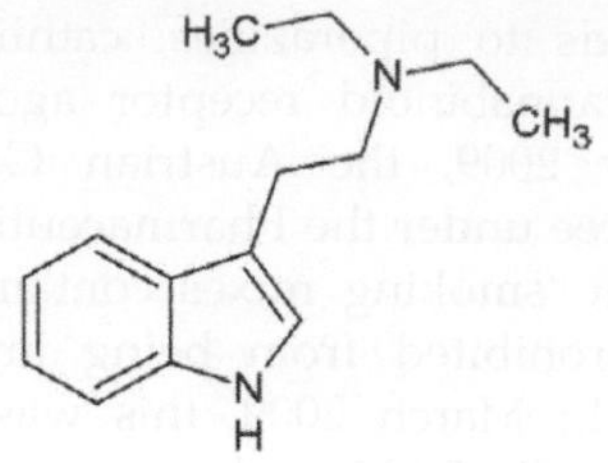

FIGURE 1.4   The structure of *N,N*-diethyltryptamine (DET).

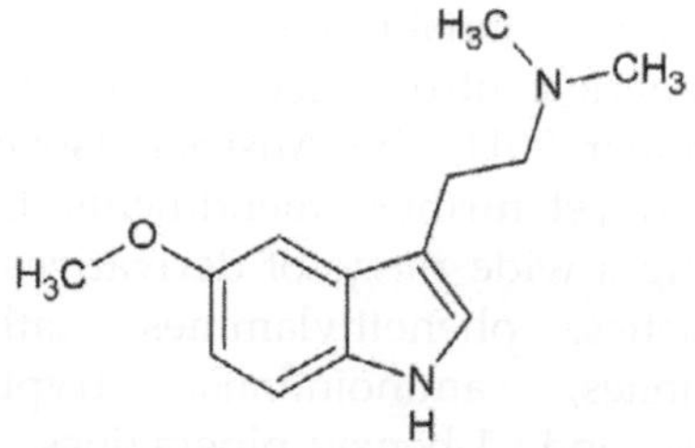

FIGURE 1.3   The structure of *N,N*-dimethyltryptamine (DMT).

FIGURE 1.5   The structure of 5-methoxy-*N,N*-dimethyltryptamine (5-MeO-DMT).

5-methoxy-DMT (5-MeO-DMT; Fig. 1.5) is an analogue of DMT even though it cannot readily be synthesised from it.

A further example is provided by United States v. T.W. Washam [43] where it was determined that 1,4-butanediol (1,4-BD) was substantially similar to GHB. Nevertheless, there is a view in Europe that analogue controls are less satisfactory from a legal viewpoint. Whereas, with explicit listing of substances in a schedule or even a generic definition, the status of a substance is clear from the outset; the use of analogue legislation requires that a court process should determine whether the substance is or is not controlled. It has been argued that such a retrospective process undermines the right of a defendant to know from the outset whether an offence has been committed. Case-by-case decisions on whether a substance is or is not an analogue might be seen as cumbersome, requiring as they do expert chemical and pharmacological testimony on each occasion, but from a US perspective, it appears that the Controlled Substances Analogue Enforcement Act was

successful in curtailing the proliferation of an earlier generation of designer drugs. The US government prosecuted a substantial number of individuals for the manufacture and distribution of analogues of MDMA, amphetamine, pethidine (meperidine), fentanyl and others. However, the Act may no longer be fit for purpose. In the last 2 years, and as discussed later, the US has sidestepped the option of analogue control by placing a number of new substances such as cathinone derivatives and synthetic cannabinoid receptor agonists under temporary drugs legislation.

In New Zealand, the Misuse of Drugs Act [35] includes the definition of a 'Controlled Drug Analogue' as 'any substance, such as the substances specified or described in Part VII of the Third Schedule to this Act, that has a structure substantially similar to that of any controlled drug; …'. The definition goes on to exclude any substance listed elsewhere in the Misuse of Drugs Act as well as pharmacy-only medicines, restricted and prescription medicines. To a certain extent, this was inspired by the US analogue controls. The application of the analogue provisions is not limited to the families of substances listed in Part VII of the Third Schedule (i.e. amphetamine, pethidine, phencyclidine, fentanyl, methaqualone and dimethyltryptamine). But, the definition of what constitutes 'substantially similar' is a potentially arguable issue for substances other than those six categories, and thus far there is minimal case law to clarify this. In Australia, in 2007, the Queensland Government [44] introduced a similar definition of an analogue, i.e. '…structurally similar and has a similar pharmacological effect to a dangerous drug…'. Only a few European countries have introduced analogue legislation. Luxembourg has controlled 'CP-47,497, JWH-018, HU-210 and other synthetic agonists of cannabinoid receptors or synthetic cannabimimetics' [26]. This is a limited use of the analogue system since it is based solely on pharmacological activity without reference to chemical structure. Analogue controls with a restricted scope also operate in Malta and Latvia, while broader controls have been implemented in Bulgaria and Norway [26].

The ACMD has suggested that the UK Government should consider analogue legislation [45]. It could be used in conjunction with generic controls in situations where a set of related substances are not sufficiently similar to merit a concise generic definition. An example might be to consider 4-fluorotropacocaine and dimethocaine as analogues of cocaine. Because the structures have common features, yet are rather diverse, this group would be less easy to control generically.

A comprehensive critique of the Federal Analogue Act, and by implication other analogue controls, has been provided by Kau [46]. In addition to the constitutional validity of retrospective control noted above, Kau pointed out several main problems, namely: the difficulty of determining what is meant by 'substantially similar'; that no court has ever given guidelines on what is 'not substantially similar';

that decisions can degenerate into a 'battle of experts', which are founded more on opinion than scientific evidence; decisions about which analogue is a controlled substance may not be binding on other Courts and the related possibility that different Courts might come to different conclusions about the same chemical entity. Another fact emerges when the US case law is examined: most of it is quite old. In a presentation to the Home Office in 2010, the US Drug Enforcement Administration (DEA) claimed that the Analogue Act was an 'imperfect law', and recommended that the UK should not adopt a similar approach [47]. However, in a 2011 report on novel psychoactive substances, the ACMD [45] proposed a means of avoiding some of the problems of analogue control by suggesting that it should be the task of a statutory agency to determine what qualifies as a controlled analogue. This could still lead to problems if the decisions of that agency were to be challenged in a criminal trial. Furthermore, the process might be seen as lacking legal certainty [48].

Finally, it is clear that some new substances will be beyond the current scope of the analogue definitions. The US Courts have interpreted the separate parts of the analogue definition as being additive. In other words, in the above definition, paragraphs (i) *and* (ii), namely a substantially similar chemical structure and a substantially similar pharmacology, or (i) *and* (iii), namely a substantially similar chemical structure and the representation of a substantially similar pharmacology must apply. From this we can conclude that salvinorin-A the active principle of the hallucinogenic herb *Salvia divinorum*, being chemically distinct from any other controlled substance, would immediately fail the test. The same applies to the active constituents in many other herbal materials such as kawain, mitragynine, arecoline and ibogaine. The plant products containing these alkaloids have all been reported to EMCDDA in the past few years as 'new psychoactive substances'

[49]. It should also be recognised that analogue control is likely to impact on legitimate pharmaceutical research and development. Although this criticism is sometimes levelled at generic controls (see above), history has shown that no serious problems arise since it is open to all to determine *a priori* if a new compound is covered by a generic definition. With analogue control no such surety exists. Finally, although a few other countries (e.g. Canada, New Zealand and some Australian States) adopted analogue control in the 1980s based on the US model, in all cases the legislation was rarely used.

In 2011, the US Senate [50] started to debate 'The Synthetic Drug Control Act'. Amongst other provisions, the proposed Act would see a set of controls on synthetic cannabinoid receptor agonists based on modified versions of the original UK definitions, but which would include a residual analogue test. Thus to qualify for control, a substance must not only fall within the generic definition, but must also show cannabimimetic, i.e. cannabinoid agonist activity. It is not yet clear how the US Courts would apply this proposed legislation. Not all reported synthetic cannabinoid receptor agonists have been described in the literature, and for many their receptor affinity constants (K_i values) have not been published. Even where K_i values are available, these do not in themselves uniquely identify an agonist as opposed to an antagonist. As of late August 2012, 'The Synthetic Drug Control Act' had not received Presidential approval.

Almost all countries that are signatories to the UN Conventions have adopted specific listing, but Table 1.3 shows examples of countries which use generic or analogue control in addition to specific control. Apart from the UK, Ireland and New Zealand, the use of generic control in other countries is much more recent and is mostly restricted to cathinone derivatives and synthetic cannabinoid receptor agonists. Furthermore, these latter controls are usually based on the original UK definitions, albeit with

TABLE 1.3 Examples of Countries which Use Generic or Analogue Control in Addition to Specific Control

Country	Analogue Control	Generic Control
Australia (certain states)	Yes	No
Austria	No	Yes
Bulgaria	Yes	No
Canada	Yes	No
Cyprus	No	Yes
Hong Kong	No	Yes (limited)
Hungary	No	Yes
Ireland	No	Yes
Latvia	Yes (limited)	No
Luxembourg	Yes (limited)	No
Malta	Yes (limited)	No
New Zealand	Yes	Yes
Norway	Yes (limited)	No
Slovakia	No	Yes
United Kingdom	No	Yes
United States	Yes	No

modifications in some cases. Further details of the generic and analogue controls in European countries are provided by EMCDDA [26].

RECENT DEVELOPMENTS

As mentioned earlier, drug control usually requires that the substance to be controlled has been shown to be harmful. Since this is not always possible to demonstrate, one solution would simply be to allow the substance to remain outside legal control. Yet for many governments that course of action seems unacceptable; there is a belief that the precautionary principle should be invoked. In other words, if there is doubt then the safest or 'failsafe' option is to introduce some form of control. This is sometimes guided by the proposition that if harms are unknown, then if the substance cannot be shown to be safe it should be restricted. In reality, demonstrating that a substance is safe can be an impossible task. The precautionary principle has its origins in environmental protection and food safety, and became a means of protecting the public from the activities of commercial and industrial concerns. As discussed by Nutt [51], the precautionary principle when applied to drug control suffers from a number of weaknesses. These include: the risk of causing more harm by criminalising users than is caused by the substance itself; distorting markets; entrenchment of a moral attitude to drug use; and encouraging other drug use. The European Commission counselled against the overuse of the principle, and has stated that:

> the implementation of an approach based on the precautionary principle should start with a scientific evaluation, as complete as possible, and where possible, identifying at each stage the degree of scientific uncertainty [52].

Alongside this is the separate, but related question: 'how should a State treat a substance that is known to have a very low level of harm?' One answer to this question is described below, namely the case of 1-(3-trifluoromethyl-phenyl)piperazine (TFMPP) in the US.

The following paragraphs describe some of the alternative approaches that have been, or are being, addressed to resolve these problems. As noted earlier, some countries have also recently adopted generic legislation (see Table 1.3).

Temporary Listing

In its reaction to the designer drugs of the day, the first response of the US government, in 1984, was to introduce emergency scheduling provisions. Emergency scheduling, otherwise known as temporary control, was a scheme whereby a substance could be added to Schedule 1 of the Controlled Substances Act 1970 [53] for a period of one year. The conditions that had to be satisfied were that the substance presented an imminent hazard to the public safety, and that it wasn't already listed in another Schedule of the Act. This temporary measure could be extended by six months provided, by then, procedures had been initiated to control the substance permanently. There was still a requirement on the authorities to provide some evaluation of the abuse potential of the substance, even if these had to be inferred from structure–activity relationships and comparison with similar compounds. Within a few years, it was recognised that, whilst a valuable tool, emergency scheduling was not in itself enough to limit the illicit manufacture of designer drugs. This need for a more proactive stance gave rise to the Controlled Substances Analogue Enforcement Act (see above). To a certain extent, analogue control reduced the need for emergency action, but temporary scheduling is still used in the US. One example included the listing of BZP and TFMPP in 2002. Subsequently, BZP was made subject to permanent control (Schedule I) while TFMPP was removed from control because of a lack of evidence of harmful properties. In 2010, the DEA published plans [54] to control, for a limited period of one year, five synthetic cannabinoid receptor agonists: JWH-018, JWH-073, JWH-200, CP-47,497 and the C8 homologue of CP-47,497. In late 2011, the DEA announced its intention [55] to place three cathinone derivatives under temporary control: 4-methylmethcathinone [mephedrone], 3,4-methylenedioxy-methylmethcathinone [methylone] and 3,4-methylenedioxypyrovalerone [MDPV].

Following the first identification of synthetic cannabinoid receptor agonists in Germany in late 2008, the German government acted to bring several (CP-47,497 and its C6, C8 and C9 homologues as well as JWH-018) under temporary control for a period of one year. Together with JWH-019 and JWH-073, those compounds were incorporated into the substantive legislation (Betäubungsmittelgesetz) from January 2010 [56]. There are plans to bring further compounds under control in early 2012 including JWH-200, JWH-250, JWH-015, JWH-081, JWH-122, JWH-007, JWH-203, JWH-210, JWH-251, 1-adamantyl(1-pentyl-1H-indol-3-yl)methanone, AM 694 and RCS-4. In the Netherlands a similar 1-year temporary control also exists [57]. In these countries, the penalties associated with substances under temporary control apply to all offences including possession. However, temporary (one year) control in Hungary excludes a possession offence [26].

Meanwhile, the UK has decided to adopt a similar approach. Temporary Class Drug Orders are set out in Section 151 of the Police Reform and Social Responsibility Act, which came into force on 15 November 2011 [58]. Following consultation with ACMD, new substances may be added to a new Class under the Misuse of Drugs Act, 1971 for a period of one year. There will be no possession offence, but in some circumstances law enforcement officers will have powers to seize and destroy a 'Temporary Class Drug'. Unlike the arrangements in the US where the DEA, as an executive agency, can determine which substances are added to the temporary list, in the UK this will require Parliamentary approval. But as with those other emergency scheduling procedures, there must be a decision at the end of the 1-year period on whether the substance should be substantively controlled or removed from the Act entirely. In early 2012, methoxetamine, which has a chemical structure similar to that of ketamine, became the first substance to be added to the list of Temporary Class Drugs [59].

In New Zealand, Temporary Class Drug Notices have been in operation since late 2011. To date, they have been used to control a number of synthetic cannabinoid receptor agonists [60].

New Zealand: Class D

In New Zealand, a somewhat different approach to emergency legislation was originally used [61,62]. From around 2000, BZP was sold as a 'safer alternative to methamphetamine', but without restriction from either the Misuse of Drugs Act 1975 or the Medicines Act 1981. Dosage units, known as 'party pills' were widely available, and often contained TFMPP, which in combination was thought to mimic the effects of MDMA. In 2004, the Expert Advisory Committee on Drugs (EACD) stated that 'After considering the evidence, … there is no current schedule under the Misuse of Drugs Act 1975 under which BZP could reasonably be placed'. The Ministry of Health therefore created a new schedule of 'Restricted Substances', informally referred to as 'Class D', with BZP as the first example of this new class of substance. Unlike the new Class created in the UK (see above), this was not necessarily seen as a temporary measure. From 2005, 'Restricted Substances' attracted no penalty for possession, but were regulated through control of manufacture, advertising and sale, rather than prohibition.

Subsequently, a number of studies were published which indicated that BZP did pose some health risks (these are discussed in detail in Chapter 8). The EACD therefore issued a follow-up report in 2006 based on this new evidence. Their advice was that BZP and the related compounds *m*-chlorophenylpiperazine (*m*CPP), TFMPP, *p*-fluorophenylpiperazine (*p*FPP), methylbenzylpiperazine (MBZP) and methoxyphenylpiperazine (MeOPP) should be moved to Class C1, the same classification as, for example, cannabis. This came into effect as the Misuse of Drugs (Classification of BZP) Amendment Act 2008. Based on general household surveys, past year usage of BZP in New Zealand declined from 15% in 2006 to 3% in 2009, while last month usage fell from 5% to 1% [63]. There are now no drugs in the 'Restricted Substances' classification, although it remains available for future use.

UK: Importation Controls

Insofar as most legal highs are imported, often from countries in the Far East, then one means of restricting their supply would be to prohibit their importation. In the UK, this power was first used in 2010 for the substance desoxypipradrol, (2-DPMP) [64]. This is a designer drug based on the now obsolete anorectic drug pipradrol, a Schedule IV substance in the UN 1971 Convention (Section 3, Chapter 4). The Import of Goods (Control) Order 1954 bans the importation of all goods except those permitted to be imported under licence. In practice, most goods can be imported freely by an 'Open General Import Licence' except those listed in the schedule. The Government amended that scheduled list by including 2-DPMP thereby prohibiting its importation. In 2011, phenazepam was likewise made the subject of an import ban [65]. In both cases, and shortly after the bans had been announced, the ACMD recommended that 2-DPMP and a group of generically-defined analogues [66] as well as phenazepam should become controlled drugs under the Misuse of Drugs Act, 1971. Because of immediate concerns about the harmful properties of 2-DPMP and related compounds, and the fact that the necessary amendment of the Misuse of Drugs Act, 1971 might take some time to enact, the ACMD later recommended in November 2011 that those related substances, i.e. diphenylprolinol (D2PM) and diphenylmethylpyrrolidine, should be added to the list of substances prohibited at importation [67]. The pipradrol derivatives were added to the Misuse of Drugs Act, 1971 as Class B substances in June 2012 [68].

Ireland: Criminal Justice (Psychoactive Substances) Act 2010

This Act, which came into force in August 2010, was designed specifically to deal with the problem caused by novel substances, and stands as a piece of legislation quite separate from the existing Misuse of Drugs Act 1977 of the Republic of Ireland [69]. It makes it a criminal offence, with a maximum penalty of five years in prison, to advertise, sell or supply, for human consumption, psychoactive substances not specifically controlled under existing legislation. They are defined as 'substances which have the capacity to stimulate or depress the central nervous system, resulting in hallucinations, dependence or significant changes to motor function, thinking or behavior.' The Act excludes medicinal and food products, animal remedies, alcohol and tobacco. There is no personal possession offence. This Act may be seen as a form of 'reduced analogue' control, i.e. there is a requirement to show psychoactivity, which by general implication means a pharmacological effect substantially similar to that of existing novel substances, but there is no requirement for the test substance to be substantially similar in a chemical-structural sense.

Although there has so far been no formal evaluation of the 2010 Act, it does appear to have restricted the supply of new substances; the number of 'head shops' in the Republic of Ireland had fallen dramatically by late 2011 [70]. Test purchases indicate that they are now only supplying paraphernalia, cannabis seeds and hydroponic equipment, but not 'substances'. It is less clear if the Act will have any impact on Internet sales. It is also uncertain how the law courts will deal with the definition of psychoactivity and how this will be objectively determined for substances where such information is currently lacking.

Japan: Non-authorised Pharmaceuticals

As with many other countries, Japan has experienced a wide availability of new substances. Because of their unknown harms, it has likewise been unable to incorporate them into the Narcotics and Psychotropics Control Law. These novel substances are formally described as 'non-authorised pharmaceuticals' and informally as 'dappo drugs' [71]. In June 2006, the Pharmaceuticals Affairs Law was modified to introduce the category of 'designated substances', where there is a general prohibition on importation, manufacture and distribution. By mid-2011, this included 13 tryptamine derivatives, 24 phenethylamines, 7 cathinones, 4 piperazines, 10 synthetic cannabinoid receptor agonists and 6 alkyl nitrites. However, certain other substances are controlled under the Narcotics and Psychotropics Control Law, including methylone and ketamine, the phenethylamines 2C-I, 2C-T-2, 2C-T-4, 2C-T-7, TMA-2, and MBDB (all of which had previously been risk-assessed by EMCDDA [72]), the piperazines BZP, TFMPP and *m*-CPP and the tryptamines AMT and 5-MeO-DIPT. Like many countries, Japan is equally reluctant to classify psychoactive plants. In the absence of more research, possibly involving DNA analysis, the authorities cite: lack of knowledge of their active components and pharmacology; the fact that many species of plants have the same psychoactive components; and the widespread distribution of such plants.

Switzerland: Schedule 'e'

In Switzerland, new legislation came into force on 1 December 2011. It contains a list (schedule e), which includes so-called 'research chemicals' and compounds with assumed psychotropic effects. In the first instance, there are 52 named substances as well as generic definitions for cathinones, naphthylpyrovalerones and related compounds, and five groups of synthetic cannabinoid receptor agonists, all of which are based on the UK system [73]. The legislation also makes provision for a fast-track scheduling procedure.

Poland, Romania: Substitute Drugs

A new law entered into force in Poland on 27 November 2010 [74–76]. It was adapted from the Act on Counteracting Drug Addiction and is enforced by the state sanitary inspectorate. Designed to eliminate the open sale of psychoactive substances not controlled under drug laws, this law was prompted by the large number of 'head shops' that existed in Poland. In October 2010, over a thousand such premises had been closed following inspections by police and state sanitary inspectors. As with the 2010 Psychoactive Substances Act in the Republic of Ireland (see above), the legislation penalises suppliers rather than users. And again, there is no requirement on the authorities to show that the banned substances are harmful. The new law prohibits the manufacture, advertising and introduction of 'substitute drugs' into circulation. In some respects the law may be considered as a broad form of analogue control since it defines these designer drugs as any type of substitute product that could be used as a narcotic or for the purpose of 'getting high'. In 2011, the Romanian Government announced its intention to introduce similar controls [77].

Finland: Intoxication and Harm

In Finland, following concerns about emerging substances, the Narcotics Act of 2008 was modified in 2011 to allow for the inclusion of a formal risk assessment process. The definition of drugs has been modified to include the statement: 'substances used for the purpose of intoxication that are harmful to health.' These intoxicating properties and harms are to be evaluated by the Finnish Medicines Agency together with police, customs and the National Institute for Welfare and Health [78].

Medicines Legislation

The common EU definition of a medicinal product is set out in Article 1 of Directive 2001/83/EC:

(a) Any substance or combination of substances presented as having properties for treating or preventing disease in human beings; or

(b) Any substance or combination of substances which may be used in or administered to human beings either with a view to restoring, correcting or modifying physiological functions by exerting a pharmacological, immunological or metabolic action, or to making a medical diagnosis.'

A detailed account of the meaning of the term 'medicinal product' within the context of EU legislation has been provided by Rogers [79]. However, individual countries differ not only in how they interpret this definition, but in their readiness to use medicines legislation rather than the national drug control laws, where penalties are usually much more severe. For example, in 2009, Austria brought certain synthetic cannabinoid receptor agonists within the Pharmaceutical Law [26]. This was to avoid criminalising users, and had the effect that supply of 'Spice' products soon ceased in Austria. The Netherlands has been active in wishing to classify novel substances under the medicines law; it may be seen as a way of avoiding criminalising users that would otherwise occur under the Dutch Opium Law.

Evans-Brown et al. [80] have argued that 'legal highs' should be regulated as medicinal products. In the UK, the Medicines and Healthcare Products Regulatory Agency (MHRA) did define BZP as a medicinal product [81] and actively prosecuted some suppliers. However, when mephedrone (4-methylmethcathinone) became widespread in the UK after late 2009, the MHRA decided that mephedrone could not be treated as a medicinal product [82]. The MHRA referred to the above definition, noting that mephedrone was commonly sold as 'bath salts', or 'plant food' or was otherwise labelled as 'not for human consumption'. It therefore decided that mephedrone failed the first limb of the definition. Since at the time, there was essentially no information in the published scientific literature concerning the pharmacological properties of mephedrone, it was probably concluded that the second limb of the definition could not apply either. The European

Court has made it clear that the pharmacological properties of a product must be demonstrated by national medicines agencies if a substance is to qualify as a medicinal product, and that the onus in cases of classification is on the medicines agencies to prove that a product has such an effect, not for the supplier to show that it does not. Since most of the new substances recently encountered are never advertised overtly for human consumption, and are hardly mentioned in the established literature, then they would presumably fall into the same category as mephedrone, i.e. they could not be classed as medicinal products. Unless the definition of a medicinal product is revised, then this situation is unlikely to change. Hughes and Winstock [83] have discussed the control of novel substances based on medicines legislation and other forms of marketing regulation. As discussed by Winstock and Ramsey [84], an unintended consequence of medicines legislation is that distributors cannot disclose the true purpose of their product without risking prosecution.

The Australian Government proposed in late 2011 that synthetic cannabinoid receptor agonists and certain other new substances should be included in the Therapeutic Goods Act, 1989 [85].

Consumer Protection Legislation

Consumer protection legislation could limit the number of vendors entitled to supply certain substances and require that those vendors demonstrate that their product meets particular standards on product safety. These could include age restrictions on sales, requirements that they are sold with information on dosage levels and side-effects, and controls on marketing and packaging. Civil or criminal sanctions could be applied for breaches of the legislation. Several recent reports have come to the conclusion that consumer protection legislation could provide a useful method of controlling the supply of novel substances. These included the Demos report for the UK Drug Policy Commission [10], and the ACMD [45]. In Sweden, the Ordinance on Prohibition of Certain Goods Dangerous to Health [86] lists substances that are not otherwise classified as narcotics. This includes gamma-butyrolactone (GBL) and 1,4-butanediol (1,4-BD) (where permits are provided for legitimate use) as well as a number of synthetic cannabinoid receptor agonists.

In 2011, the New Zealand Law Commission published a report on regulating drugs including a review of the New Zealand Misuse of Drugs Act [87]. A major impetus for that report was the emergence of a rapidly evolving market in new psychoactive substances. Despite the fact that a 'Restricted Substances' category is available (and was used for BZP – see earlier), and that analogue controls are also part of the existing Act, it was concluded that a new regime of drug regulation was required, which would replace both of those mechanisms. The Commission proposed a form of consumer protection with elements of the 'Restricted Substances' regime. Thus there would be restrictions on the sale of novel substances to persons under the age of 18, restrictions on advertising and restrictions on where they could be sold. An independent regulator would be established to determine applications from suppliers. If the regulator decided that a substance was so harmful that it should not be approved then that substance would be added to the prohibited drugs list. The system would not cover solvent misuse, where it was thought the existing Hazardous Substances and New Organisms Act 1996 should be adequate. In July 2012, the New Zealand Government announced that it would adopt the recommendations of the Law Commission [88]. These new regulations would place the burden on the 'legal highs' industry to prove that its products were low risk, where the approval process would be similar to that for medicinal products.

In the UK, the control of certain volatile solvents was mentioned earlier; the Cigarette Lighter Refill (Safety) Regulations 1999 being an example of specific consumer protection. In Italy, 'Spice' products were confiscated because they were not labelled in Italian [26]. There are otherwise few examples of the successful use of this approach, but if it is felt that existing consumer protection legislation is insufficient then it could be amended.

International-level Initiatives

From the foregoing discussion, it is clear that the problem of whether and how to control new substances is being tackled in many different ways by different countries. The substances concerned are available everywhere, often via Internet retail sites, and there is a need for more co-ordinated action. Yet, despite the early lead taken in drug control by the WHO and UN agencies, the current world-wide concern with new substances has not been adequately reflected by these international bodies. The 34th meeting of the WHO Expert Committee on Drug Dependence (ECDD) took place in 2006. Following a long gap, the 35th meeting was not held until June 2012 [89]. The following substances were listed for review: ketamine; dextromethorphan; tapentadol; N-benzylpiperazine (BZP); 1-(3-trifluoromethyl-phenyl)piperazine (TFMPP); 1-(3-chlorophenyl) piperazine (mCPP); 1-(4-methoxyphenyl)piperazine (MeOPP); 1-(3,4-methylenedioxybenzyl)piperazine (MDBP); Gamma-butyrolactone (GHB) and 1,4-butanediol (1,4-BD).

From 2006 to late August 2012, almost 200 new substances were reported to EMCDDA as part of the EU-wide early warning system, and it is probable that more have been found if other countries are included. However, it is unlikely that the ECDD would have the resources to provide detailed reviews of this large group of compounds, even if the information were available. Similarly, EMCDDA can only carry out risk assessments on one substance at a time. Between 1997, when the monitoring system began, and late August 2012, over 250 new substances were reported, yet to date it has assessed only eleven in the same time period, including BZP (2006) and mephedrone (2010). Of the original 11 substances risk-assessed, eight were recommended for EU-wide control. The remaining three are controlled in the UK, and one of them (GHB) is now under international control (Schedule IV of the UN 1971 Convention). Since then, the EMCDDA has also undertaken risk assessments of 4-MA (2012) and 5IT (2013).

Other Control Options and Future Developments

Alcohol and tobacco products are each the subject of specific controls, such that the law defines where, when and to whom such products may be sold, and sets up the concept of licensed premises. Despite the fact that these two products may be far more harmful than controlled drugs [90, 91], the licensed sale of new products would probably be unacceptable to many Governments if only because it would be argued that their harms are unknown.

Some scope might exist for a more comprehensive control of chemical substances. This was first put forward by the Royal Society of Arts in 2007 in a review of the UK Misuse of Drugs Act, 1971 [14]. This could bring together various pieces of currently overlapping legislation, for example, control of precursors, poisons, medicines, chemical weapons, solvents, other hazardous materials as well as including new psychoactive substances [92] into a single 'Harmful Substances Act'.

There is the possibility of modifying current legislation to allow classification to depend on the type of offence [92]. At present, substances are placed in different schedules largely as a means of determining a gradation of penalties. While it is not appropriate here to discuss the merits or otherwise of decriminalisation or even legalisation of drugs, offence-dependent classification could provide a means of, for example, decriminalising possession of a limited group of substances, while retaining more severe penalties for activities such as production, importation and supply. This need not represent a radical step since, as described earlier, there are examples from several countries where simple possession of new substances is not a criminal offence.

The procedures for monitoring new substances at an international level, including the European Early Warning System [7,8] are described in (Chapter 2). A number of weaknesses have been recognised in the 2005 Council Decision. These include: the problem of confirming psychoactivity when no scientific data may be available; the exclusion of substances used to manufacture medicinal products, where there was no clear distinction between an active pharmaceutical ingredient and a precursor such as *m*CPP; the limitations posed by restricting risk-assessments to one substance at a time; the lack of time and resources to carry out even limited pharmacological and toxicological testing; and the lack of flexibility in control options. In late 2011, the European Commission announced that it would strengthen its anti-drugs policy [93] particularly with reference to new substances, and include a review [94] of the current Council Decision.

The UK Government has been open to new ideas in drugs control, and the examples of generic definitions, import controls and temporary legislation were discussed earlier. Nevertheless, in the minds of many, there is still some way to go in updating the law to recognise the realities of the modern world.

REFERENCES

[1] United Nations. Single convention on narcotic drugs, Available: <http://www.unodc.org/pdf/convention_1961_en.pdf>; 1961 [accessed 10.05.13].

[2] United Nations International Narcotics Control Board. List of psychotropic substances under international control. Available: <http://www.incb.org/documents/Psychotropics/green_lists/Green_list_ENG_2010_53991_with_logo.pdf>; 1971 [accessed 10.05.13].

[3] United Kingdom Government. The misuse of drugs regulations. Available: <http://www.legislation.gov.uk/uksi/2001/3998/contents/made>; 2001 [accessed 01.03.13].

[4] Phillips GF, Mesley RJ. Examination of the hallucinogen 2,5,dimethoxy-4-methylamphetamine. J Pharm Pharmacol 1969;21:9–17.

[5] Baum RM. New variety of street drugs poses growing problem. Chem Eng News 1985;63:7–16.

[6] Shulgin A, Shulgin A. PIHKAL: a chemical love story. Berkeley, California: Transform Press; 1991.

[7] European Commission. Council decision 2005/387/JHA on the information exchange, risk-assessment and control of new psychoactive substances. Official journal of the European union, L 127/32. Available: <http://www.emcdda.europa.eu/html.cfm/index5173EN.html?pluginMethod=eldd.showlegaltextdetail&id = 3301&lang=en&T=2>; 2005 [accessed 01.03.13].

[8] King LA, Sedefov R. Early-warning system on new psychoactive substances: operating guidelines. Lisbon: EMCDDA; 2007. Available: <http://www.emcdda.europa.eu/html.cfm/index52448EN.html> [accessed 01.03.13].

[9] United Kingdom Government. The misuse of drugs act. Available: <http://www.legislation.gov.uk/ukpga/1971/38/contents>; 1971 [accessed 01.03.13].

[10] Birdwell J, Chapman J, Singleton N. Taking drugs seriously: a demos and UK drug policy commission report on legal highs. London: Demos; 2011. Available: <http://www.ukdpc.org.uk/resources/Taking_Drugs_Seriously.pdf> [accessed 01.03.13].

[11] The Police Foundation. Drugs and the law: report of the independent inquiry into the misuse of drugs act 1971. London; 2000.

[12] House of Commons Home Affairs Committee. The government's drugs policy: is it working? third report of session 2001–02, vol. 1: Report and proceedings of the committee. London: The Stationery Office. Available: <http://www.parliament.the-stationery-office.com/pa/cm200102/cmselect/cmhaff/318/31804.htm>; 2002 [accessed 01.03.13].

[13] House of Commons Science and Technology Committee. Drug classification: making a hash of it? fifth report of session 2006–06. Available: <http://www.tdpf.org.uk/Drug%20classification.pdf>; 2006 [accessed 01.03.13].

[14] Royal Society of Arts. Drugs: facing facts. The report of the RSA commission on illegal drugs, communities and public policy. London. Available: <http://www.thersa.org/projects/past-projects/drugs-commission/drugs-report>; 2007 [accessed 01.03.13].

[15] The All-Party Parliamentary Group on Drug Policy Reform. Towards a safer drug policy: challenges and opportunities arising from 'legal highs' Available: <https://docs.google.com/file/d/0B0c_8hkDJu0DODg3UXpfa2U0SFk/edit?usp=sharing&pli=1>; 2012 [accessed 10.05.13].

[16] House of Commons Home Affairs Committee. Ninth Report. Drugs: Breaking the cycle. Available: <http://www.publications.parliament.uk/pa/cm201213/cmselect/cmhaff/184/18402.htm>; 2012 [accessed 10.05.13].

[17] Beck H. Drug enforcement in an age of austerity: key findings from a survey of police forces in England. London: UK Drug Policy Commission; 2011. Available: <http://www.ukdpc.org.uk/resources/Drug_related_enforcement.pdf>.

[18] United Nations. Convention against illicit traffic in narcotic drugs and psychotropic substances. Available: <http://www.unodc.org/pdf/convention_1988_en.pdf>; 1988.

[19] United Nations. Convention on the prohibition of the development, production, Stockpiling and use of chemical weapons and on their destruction.Available: <http://www.un.org/disarmament/WMD/Chemical/>; 1997.

[20] United Kingdom Government. Poisons act. Available: <http://www.legislation.gov.uk/ukpga/1972/66>; 1972.

[21] Reuter P. Available: <http://www.ukdpc.org.uk/resources/Reuter_Legal_highs_report.pdf> Options for regulating new psychoactive drugs: a review of recent experiences. London: UK Drug Policy Commission; 2011.

[22] United Kingdom Government. Intoxicating substances (Supply) Act. Available: <http://www.legislation.gov.uk/ukpga/1985/26/contents>; 1985.

[23] United Kingdom Government. The cigarette lighter refill (safety) regulations. Available: <http://www.legislation.gov.uk/uksi/1999/1844/contents/made>; 1999.

[24] The European Parliament and the Council of the European Union. Directive 2005/59/EC. Available: <http://eur-lex.europa.eu/LexUriServ/LexUriServ.do?uri=OJ:L:2005:309:0013:0014:EN:PDF>; 2005.

[25] European Medicines Agency. European Medicines Agency recommends restricting the use of modafinil. Available: <http://www.ema.europa.eu/ema/index.jsp?curl=pages/news_and_events/news/2010/07/news_detail_001061.jsp&murl=menus/news_and_events/news_and_events.jsp&mid=WC0b01ac058004d5c1>; 2011.

[26] EMCDDA European Legal Database on drugs, Lisbon. Available: <http://www.emcdda.europa.eu/eldd>; 2011.

[27] Hughes B, Blidaru T. Legal responses to new psychoactive substances in Europe. Lisbon: European Legal Database on Drugs, EMCDDA; 2009.

Available: <http://www.emcdda.europa.eu//html.cfm//index5175EN.html?>

[28] League of Nations Convention for Limiting the Manufacture and Regulating the Distribution of Narcotic Drugs. League of Nations Treaty Series, CXXIX, No. 3219, Geneva; 1931.

[29] King LA. The misuse of drugs act: a guide for forensic scientists. London: Royal Society of Chemistry; 2003. Available: <http://www.rsc.org/ebooks/archive/free/BK9780854046256/BK9780854046256-FP001.pdf>

[30] Phillips GF. Controlling drugs of abuse. Chem Br 1972;8(3):123–30.

[31] Advisory Council on the Misuse of Drugs. Consideration of the cathinones. Available: <http://www.namsdl.org/documents/ACMDCathinonesReport.pdf>; 2010.

[32] Advisory Council on the Misuse of Drugs. Consideration of the naphthylpyrovalerone analogues and related compounds. Available: <http://www.drugslibrary.stir.ac.uk/documents/naphyrone-report.pdf>; 2010.

[33] Advisory Council on the Misuse of Drugs. Consideration of the major cannabinoid agonists. Available: <http://www.namsdl.org/documents/ACMDMajorCannabinoidReport.pdf>; 2009.

[34] Sainsbury PD, Kicman AT, Archer RP, King LA, Braithwaite R. Aminoindanes – the next wave of 'legal highs'? Drug Test Anal 2011;3(7–8):479–82.

[35] New Zealand Government. Misuse of Drugs Act 1975. Available: <http://www.legislation.govt.nz/act/public/1975/0116/37.0/DLM436101.html>; 1975.

[36] Danish Government. Kampen mod Narko II – Handlingsplan mod Narkotikamisbrug. Available: <http://www.im.dk/Aktuelt/Publikationer/Publikationer/narkoplanII.aspx>; 2010.

[37] Pallavicini B, Europol. Personal communication 2011.

[38] Government of Hong Kong Legislative Council Brief. Dangerous Drugs Ordinance (Chapter 134) (Amendment of First Schedule) Order. Available: <http://www.legco.gov.hk/yr10-11/english/subleg/brief/7_brf.pdf>; 2011.

[39] Hughes B. May 2011 Legal responses to new psychoactive substances in eu: theory and practice – typologies, characteristics and speed. Presented at the first international multidisciplinary forum on new drugs. Lisbon: EMCDDA; 2011.

[40] United States Government. The controlled substances analogue enforcement act. Available: <http://en.wikipedia.org/wiki/Federal_Analog_Act>; 1986.

[41] United States v. Allen McKinney. Available: <http://law.justia.com/cases/federal/appellate-courts/F3/79/105/555999/>; 1995.

[42] United States v. Damon S. Forbes. Available: <http://www.erowid.org/psychoactives/law/cases/federal/federal_analog1.shtml>; 1992.

[43] United States v. T.W. Washam. Available: <http://caselaw.findlaw.com/us-8th-circuit/1262827.html>; 2002.

[44] Office of the Queensland Parliamentary Counsel. Drugs Misuse Amendment Bill 2007. Available: <http://www.legislation.qld.gov.au/Bills/52PDF/2007/DrugsMisuseAB07Exp.pdf>; 2007.

[45] Advisory Council on the Misuse of Drugs. Consideration of the novel psychoactive substances ('Legal Highs'). Available: <http://www.homeoffice.gov.uk/publications/agencies-public-bodies/acmd1/acmdnps2011?view=Binary>; 2011 [accessed 01.03.13].

[46] Kau G. Flashback to the federal analog act of 1986: mixing rules and standards in the cauldron. Univ PA Law Rev 2008;156:1077–115. Available: <http://www.law.upenn.edu/journals/lawreview/articles/volume156/issue4/Kau156U.Pa.L.Rev.1077(2008).pdf> [accessed 01.03.13].

[47] Wong L, Dormont D, Matz, HJ. United states controlled substance analogue act: legal and scientific overview of an imperfect law. Presented to ACMD, 2010; 7 July 2010.

[48] King LA, Nutt D, Singleton N, Howard R. Analogue controls: an imperfect law, UK drug policy commission and independent scientific committee on drugs. Available: <http://www.ukdpc.org.uk/wp-content/uploads/Analogue-control-19.06.12.pdf>; 2012 [accessed 01.03.13].

[49] EMCDDA. EMCDDA-Europol 2010 Annual report on the implementation of Council Decision 2005/387/JHA. (See also earlier reports in this series). Available: <http://www.emcdda.europa.eu/publications/searchresults?action=list&type=PUBLICATIONS&SERIES_PUB=a104>; 2011 [accessed 01.03.13].

[50] United States Senate. The synthetic drug control act. Available: <http://thomas.loc.gov/cgi-bin/query/z?c112:H.R.1254.IH>; 2011 [accessed 01.03.13].

[51] Nutt D. Precaution or perversion: eight harms of the precautionary principle. Available: <http://profdavidnutt.wordpress.com/2010/06/>; 2010 [accessed 01.03.13].

[52] European Commission. Communication from the commission of 2 february 2000 on the precautionary principle. Available: http://europa.eu/legislation_summaries/consumers/consumer_safety/l32042_en.htm>; 2000 [accessed 01.03.13].

[53] United States Government. Controlled Substances Act. Available: <http://en.wikipedia.org/wiki/Controlled_Substances_Act>; 1970 [accessed 01.03.13].

[54] United States Government Schedules of controlled substances: temporary placement of five synthetic cannabinoids into schedule I. Microgram Bull 2010;44(1):1–5. Available: <http://www.justice.gov/dea/programs/forensicsci/microgram/mg2011/mg0111.pdf> [accessed 01.03.13].

[55] United States Government. Schedules of controlled substances: temporary placement of three synthetic cathinones into schedule I. Available: <http://www.federalregister.gov/articles/2011/09/08/2011-23012/schedules-of-controlled-substances-temporary-placement-of-three-synthetic-cathinones-into-schedule-i>; 2011 [accessed 01.03.13].

[56] German Government. Betäubungsmittelgesetz. Available: <http://de.wikipedia.org/wiki/Bet%C3%A4ubungsmittelgesetz_(Deutschland)>; 2011.

[57] Netherlands Government. Opium law. Available: <http://en.wikipedia.org/wiki/Opium_Law>; 2011 [accessed 01.03.13].

[58] United Kingdom Government. The police reform and social responsibility act 2011 (Commencement No. 1) Order 2011. Available: <http://www.legislation.gov.uk/uksi/2011/2515/contents/made>; 2011 [accessed 01.03.13].

[59] Home Office. A change to the misuse of drugs act 1971: control of methoxetamine under a temporary class drug order. Available: <http://www.homeoffice.gov.uk/about-us/corporate-publications-strategy/home-office-circulars/circulars-2012/008-2012/>; 2012 [accessed 01.03.13].

[60] New Zealand Ministry of Health. Temporary class drug notices. Available: <http://www.health.govt.nz/our-work/mental-health-and-addictions/drug-policy/temporary-class-drug-notices>; 2001 [accessed 01.03.13].

[61] Bassindale T. Benzylpiperazine: the New Zealand legal perspective. Drug Test Anal 2011;3(7–8):428.

[62] Bowden M, Trevorrow P. BZP and New Zealand's alternative approach to prohibition. Drug Test Anal 2011;3(7–8):426.

[63] Wilkins C. The impact of the prohibition of Benzylpiperazines (BZP) in New Zealand. Presented at the first international multidisciplinary forum on new drugs, EMCDDA, Lisbon, May 2011 2011.

[64] United Kingdom Government. Import ban of ivory wave drug 2-DPMP introduced. Available: <http://www.homeoffice.gov.uk/media-centre/press-releases/ivory-wave>; 2010 [accessed 01.03.13].

[65] Home Office. Phenazepam. Available: <http://www.homeoffice.gov.uk/drugs/drug-law/phenazepam/>; 2011 [accessed 01.03.13].

[66] Advisory Council on the Misuse of Drugs. Consideration of desoxypipradrol (2-DPMP) and related pipradrol compounds. Available: <http://www.homeoffice.gov.uk/publications/agencies-public-bodies/acmd1/desoxypipradrol-report?view=Binary>; 2011 [accessed 01.03.13].

[67] Advisory Council on the Misuse of Drugs. Further advice on diphenylprolinol (D2PM) and diphenylmethylpyrrolidine. Available: <http://www.

homeoffice.gov.uk/publications/agencies-public-bodies/acmd1/acmd-d2pm?view=Binary>; 2011 [accessed 01.03.13].

[68] Home Office. A change to the misuse of drugs Act 1971: control of pipradrol-related compounds and phenazepam. Available: <http://www.homeoffice.gov.uk/about-us/corporate-publications-strategy/home-office-circulars/circulars-2012/014-2012/>; 2012 [accessed 01.03.13].

[69] Government of Ireland. Criminal justice (Psychoactive Substances) Act. Available: <http://www.irishstatutebook.ie/2010/en/act/pub/0022/index.html>; 2010 [accessed 01.03.13].

[70] Long J, Connolly J. Report on new psychoactive substances and the outlets supplying them. Drugnet Ireland 2011;39:9–10. Available: <http://www.drugsandalcohol.ie/16160/> [accessed 01.03.13].

[71] Kikura-Hanajiri R. Drug control in Japan – designated substances – update. Presented at the first international multidisciplinary forum on new drugs, EMCDDA, Lisbon, May 2011; 2011.

[72] EMCDDA. Risk assessments. Available: <http://www.emcdda.europa.eu/publications/searchresults?action=list&type=PUBLICATIONS&SERIES_PUB=w12>; 2011 [accessed 01.03.13].

[73] Joos M. Swissmedic, Schweizerisches Heilmittelinstitut, Switzerland. Personal communication 2011.

[74] Kolasinska M. The online market of 'legal highs' in Poland. ReDNet News, 5, 2–3. Available: <https://www.rednetproject.eu/documents/newsletters/Rednet_News_AUG11.pdf>; 2011 [accessed 01.03.13].

[75] Hughes B, Malczewski A. Poland passes new law to control 'head shops' and 'legal highs'. Drugnet Europe online 73, EMCDDA. Available: <http://www.emcdda.europa.eu/publications/drugnet/online/2011/73/article12>; 2011 [accessed 01.03.13].

[76] Kapka-Skrzypczak L, Kulpa P, Sawicki K, Cyranka M, Wojtyła A, Kruszewski M. Legal highs – legal aspects and legislative solutions. Ann Agric Environ Med 2011;18(2):304–9.

[77] Simionov V. Romanian Harm Reduction Network. Personal communication 2011.

[78] Hughes B. Finland and the UK respond to emerging substances. EMCDDA: Drugnet Europe; 2011:76. Available: <http://www.emcdda.europa.eu/publications/drugnet/76> [accessed 01.03.13].

[79] Rogers S. Medicines legislation Feldschreiber P, editor. The law and regulation of medicines. Oxford: Oxford University Press; 2008. p. 23–72.

[80] Evans-Brown M, Bellis MA, McVeigh J. 'Legal highs' should be regulated as medicinal products. BMJ 2011;342:501.

[81] Medicines and Healthcare Products Regulatory Agency. Press release: benzylpiperazine (PEP) pills are

dangerous and illegal. Available: <http://www.mhra.gov.uk/NewsCentre/Pressreleases/CON2030603>; 2007.[accessed 01.03.13].

[82] Jason-Lloyd L. Mephedrone. Criminal law and justice weekly 29 May 2010, Available: <http://criminal-lawandjustice.co.uk/index.php?/Analysis/mephedrone.html>; 2010.

[83] Hughes B, Winstock AR. Controlling new drugs under marketing regulations. Addiction 2011 doi:10.1111/j.1360-0443.2011.03620.x. [accessed 01.03.13].

[84] Winstock AR, Ramsey JD. Legal highs and the challenges for policy makers. Addiction 2010;105(10):1685–7.

[85] Australian Government. Final decisions and reasons for decisions by delegates of the secretary to the department of health and ageing. Available: <http://www.tga.gov.au/pdf/scheduling/scheduling-decisions-1202-final.pdf>; 2011 [accessed 01.03.13].

[86] Reitox National Focal Point, Sweden. National report (2009 data) to the EMCDDA; New development, trends and in-depth information on selected issues. Available: <http://www.emcdda.europa.eu/attachements.cfm/att_142550_EN_SE-NR2010.pdf>; 2010 [accessed 01.03.13].

[87] New Zealand Law Commission. Controlling and regulating drugs: A review of the Misuse of Drugs Act 1975, Report 122. Wellington, New Zealand. Available: <http://www.lawcom.govt.nz/sites/default/files/publications/2011/05/part_1_report_-_controlling_and_regulating_drugs.pdf>; 2011 [accessed 01.03.13].

[88] Topnews. NZ Govt seeks proof of safety from legal high manufacturers. Available: <http://topnews.net.nz/content/223502-nz-govt-seeks-proof-safety-legal-high-manufacturers>; 2012 [accessed 01.03.13].

[89] World Health Organisation. Thirty-fifth meeting of the expert committee on drug dependence. Available: <http://www.who.int/medicines/areas/quality_safety/35thecddmeet/en/index.html>; 2012 [accessed 01.03.13].

[90] Nutt DJ, King LA, Saulsbury W, Blakemore C. Developing a rational scale for assessing the risks of drugs of potential misuse. Lancet 2007;369:1047–53.

[91] Nutt DJ, King LA, Phillips LD. Drug harms in the UK: a multicriteria decision analysis. Lancet 2010;376:1558–65.

[92] King LA. Forensic chemistry of substance misuse: a guide to drug control. London: Royal Society of Chemistry; 2009. Available: <http://www.rsc.org/shop/books/2009/9780854041787.asp>.

[93] European Commission. European Commission seeks stronger EU response to fight dangerous new synthetic drugs. Press release, 25 October.Available: <http://europa.eu/rapid/pressReleasesAction.do?reference=IP/11/1236&type=HTML>; 2011 [accessed 01.03.13].

[94] European Commission. Report from the Commission on the assessment of the functioning of Council Decision 2005/387/JHA on the information exchange, risk assessment and control of new psychoactive substances. Available: <http://ec.europa.eu/justice/policies/drugs/docs/com_2011_430_en.pdf>; 2011 [accessed 1.03.13].

This page intentionally left blank

Monitoring Novel Psychoactive Substances
A Global Perspective

Roumen Sedefov, Ana Gallegos, Jane Mounteney and Padraigin Kenny

European Monitoring Centre for Drugs and Drug Addiction (EMCDDA), Lisbon, Portugal

INTRODUCTION

The emergence of novel psychoactive substances that are not controlled under existing drug laws is not a new phenomenon. However, over the past few years, the accelerating pace of globalisation and the innovation it has brought has allowed the unprecedented growth in both the number and availability of these substances. In the European Union (EU), 24 novel psychoactive substances were identified for the first time in 2009, 41 in 2010, and 49 in 2011. Currently, more than 250 substances are monitored by the EU Early warning system on new psychoactive substances [1,2]. Many of the novel substances that appear on the market have been previously described in the scientific and patent literature, or are structural modifications of these substances. In most cases it appears that entrepreneurs have searched this literature for suitable substances, apparently paying particularly close attention to those that have the potential to mimic the effects of well-known controlled drugs such as amphetamine, 3,4-methylenedioxy-N-methylamphetamine (MDMA) or cocaine. In other cases novel substances may emerge from the diversion and misuse of medicines (such as ketamine or pregabalin). While to a lesser degree, as discussed in Chapter 14, plant-based substances may also emerge (such as kratom, kava kava and *Salvia divinorium*).

Until approximately the mid-2000s, most of the novel substances that emerged in the EU were produced from chemical precursors in clandestine laboratories and distributed through the same channels used by the illicit drug market. Such substances became known as 'designer drugs' [3–6]. A classic example of this phenomenon is the emergence of 'ecstasy' (MDMA) [3,4,7], which rapidly established itself on the drug market in the United States

Novel Psychoactive Substances.
DOI: http://dx.doi.org/10.1016/B978-0-12-415816-0.00002-X

in the mid-to-late 1980s [3,4,8] (although often called a 'designer drug', MDMA was initially sold openly by mail order and in bars in Texas, for example) and the European market in the early 1990s [9,10]. More recently, a second diverse group of largely synthetic substances have appeared, known as 'legal highs', which take advantage of cheap organic chemical synthesis in emerging economies such as China, along with expedited cargo and courier services. Using sophisticated marketing techniques these substances are sold through the Internet and bricks and mortar shops (often known as 'head' or 'smart' shops). However, there is also some overlap with the illicit drug market and they may also be sold by street-level drug dealers. The supply of novel psychoactive substances is covered in more detail in Chapter 3.

Undoubtedly, the licit nature of these substances will increase their attractiveness to producers, distributors, retailers, established drug users and even individuals who typically do not use controlled (illicit) drugs. At least initially, however, many of these substances are used only by a small number of people, often with a specialised knowledge or interest in them ('innovators' or 'psychonauts'). However, over time, broader interest in some of these substances may develop and they may gain a foothold in the market, leading to more widespread diffusion. Initially this includes diffusing to groups such as club-goers ('early adopters') and, in some cases, eventually to sections of the broader population ('later adopters'). Mephedrone (4-methylmethcathinone), a synthetic cathinone derivative, and 'Spice', herbal smoking mixtures that contain synthetic cannabinoid receptor agonists, provide examples of such patterns of diffusion. On occasion, often without the knowledge of the user (and sometimes without the knowledge of producers/dealers), novel psychoactive substances have been used as substitute for active ingredients in illicit drugs. These include

meta-chlorophenylpiperazine (*m*CPP), *para*-methoxyamphetamine (PMA) or *para*-methoxymethylamphetamine (PMMA) in tablets sold as ecstasy [11] and the use of 4-methylamphetamine in powder or pastes sold as amphetamine. In some cases, users may be at risk of harm caused by these novel substances [12].

Those new substances that do diffuse beyond small groups – such as gamma-hydroxybutyrate (GHB) [13–17] in the 2000s – may pose significant social and health risks to society. In order to take appropriate measures to minimise these harms, a range of stakeholders, including policy makers, practitioners, and researchers need access to timely evidence-based and authoritative information on these substances and trends in their use. It is here that drug information systems play a critical role in detecting, identifying, and monitoring such substances, as well as helping to inform the responses that are likely to be required. This chapter provides an overview of some of the most important monitoring systems in place at the national, regional and international level. The EU Early warning system on new psychoactive substances is described in some depth, alongside an overview of the risk assessment process and an accompanying case study on the synthetic cathinone derivative mephedrone.

Attempts to monitor novel psychoactive substances increasingly struggle to keep pace with a sophisticated, highly innovative and fast moving market where entrepreneurs actively seek out new substances and marketing strategies [18]. For this reason it has become necessary to develop new ways of detecting, identifying and monitoring novel substances. The final section of the chapter presents a number of the more promising novel approaches, including wastewater analysis, leading-edge indicators (such as Internet monitoring and the use of hospital emergency data), alongside a review of findings from some of the first European studies on the prevalence of use of novel psychoactive substances.

DEFINITIONS

Over the last 20 years, a variety of terms and definitions have been used for novel psychoactive substances that emerge on the market and are not under international control [19]. Here we briefly review those relating to the experience in the EU, while recognising that other definitions and terms are used in different settings.

In the EU, a new synthetic drug [20] has been defined as one that had a limited therapeutic value, and is not listed under the 1971 United Nations Convention on Psychotropic Substances [21], but which poses a comparably serious threat to public health as those substances listed in Schedules I and II to that Convention. In this context, the term 'new' is not intended to refer to newly invented, but rather to a 'newly available' or a 'newly misused' substance. In practice, most 'new' drugs were first described in the scientific and patent literature many years ago as part of legitimate research and development but have not been widely available or used. For practical purposes, a 'designer drug' is probably best thought of as a psychoactive substance produced from chemical precursors in a clandestine laboratory, which, by slight modification of the chemical structure, has been intentionally designed to mimic the properties of known psychoactive substances, and which is not under international control.

Currently in the EU a new psychoactive substance is defined as a new narcotic drug or a new psychotropic drug in pure form or in a preparation, that has not been scheduled under: 1) the 1961 United Nations Single Convention on Narcotic Drugs [22], and that may pose a threat to public health comparable to the substances listed in Schedule I, II or IV; or, 2) the 1971 United Nations Convention on Psychotropic Substances [21], and that may pose a threat to public health comparable to the substances listed in Schedule I, II, III or IV. This definition is used by the Council Decision 2005/387/JHA, which is the legal instrument that establishes the basis for the information exchange mechanism (known as the EU Early warning system), the risk-assessment and control of new psychoactive substances at the level of the EU [23].

More recently, the concept of 'legal highs' has been used as an umbrella term for unregulated novel psychoactive substances, or products claiming to contain them, which are intended to mimic the effects of controlled drugs. The term encompasses a wide range of synthetic and/or plant derived substances and products. These may be marketed as 'legal highs' (emphasising 'legality'), 'herbal highs' (stressing the natural/plant origin), as well as 'research chemicals' and 'party pills'. These products are often deliberately misbranded in an effort to disguise the fact that they are intended for use in humans and to hide the identity of the active substance.

So-called 'legal highs' are usually sold via the Internet or in bricks and mortar shops (often known as 'head' or 'smart' shops). In some cases they are intentionally mislabelled with regard to their intended use (e.g. labelled as 'not for human consumption', 'plant food', 'bath salts', 'room odourisers') and the active substances that they contain. This 'legal highs' market can be distinguished from other drug markets by the speed at which suppliers circumvent drug controls by offering new alternatives to restricted products.

THE IDENTIFICATION, RISK ASSESSMENT AND MONITORING OF NOVEL PSYCHOACTIVE SUBSTANCES

Novel psychoactive substances appearing on the drugs market in Europe have historically belonged to a small number of chemical families, with the phenethylamines and tryptamines accounting for the majority of reports to the EU Early warning system (EU EWS). In the past decade, however, increasing numbers of novel psychoactive substances from an expanding range of chemical families have been reported (see Fig. 2.1). The identification of

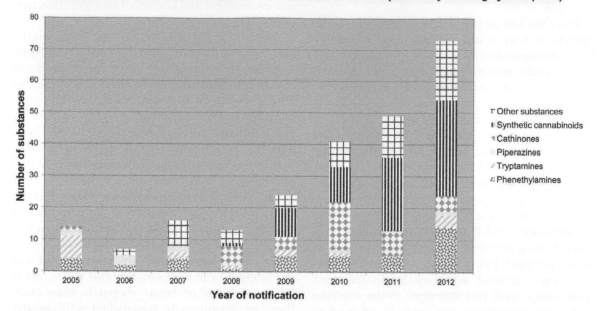

New psychoactive substances notified for the first time to the European Early warning system (EWS)

FIGURE 2.1 **Number and main groups of new psychoactive substances identified by the European Union Early warning system [24].** Phenethylamines encompass a wide range of natural or synthetic substances that may exhibit stimulant, entactogenic or hallucinogenic effects. Examples include the synthetic substances amphetamine, methamphetamine and MDMA (3,4-methylenedioxy-methamphetamine), and mescaline. Tryptamines include a number of substances that have predominantly hallucinogenic effects. The main representatives are the naturally occurring compounds dimethyltryptamine (DMT), psilocin and psilocybin (found in hallucinogenic mushrooms) as well as the semi-synthetic lysergic acid diethylamide (LSD). Piperazines are best represented by two synthetic substances – *m*CPP (1-(3-chlorophenyl)piperazine) and BZP (1-benzylpiperazine), both of which are central nervous system stimulants. The principal active components in khat (*Catha edulis* Forsk.) are cathinone and cathine (norpseudoephedrine) which have stimulant effects [25]. The main cathinone derivatives are the semi-synthetic methcathinone and the synthetic compounds mephedrone, methylone and MDPV (3,4-methylenedioxypyrovalerone). Synthetic cannabinoid receptor agonists are functionally similar to delta-9-tetrahydrocannabinol (THC), the active principle of cannabis. Like THC, they bind to the same cannabinoid receptors in the brain and have hallucinogenic, sedative and depressant effects. More correctly designated as synthetic cannabinoid receptor agonists, although often referred to simply as synthetic cannabinoids, many of the substances are not structurally related to the 'classical' cannabinoids, i.e. compounds, like THC. They have been detected in herbal smoking mixtures such as 'Spice' [26,27].

novel psychoactive substances is a specialised task, primarily associated with forensic and toxicological analysis. At the time of writing in 2012, about one novel psychoactive substance is identified in one of the countries of the EU every week.

Other substances reported to the EU EWS include various plant-derived and synthetic psychoactive substances (e.g. indanes, benzodifuranyls, narcotic analgesics, synthetic cocaine derivatives, ketamine and phencyclidine derivatives), which do not strictly belong to any of the previous families. Also included here are a small number of medicines as well as designer medicines (derivatives of medicines). Further information on the drug families mentioned here can be found on the EMCDDA Drug profiles webpages [28].

Assessing the risks and harms related to novel psychoactive substances is high up on the political agenda. Recent efforts to index risks and harms associated with novel, as well as known psychoactive substances, have been undertaken using expert panels based on the Delphi method and scoring schedules [29,30]. These studies provided a framework for the development of the EU operating guidelines for risk assessment presented later in this chapter [31]. Most of the problems highlighted by the studies undertaken with established drugs (such as heroin, cocaine and MDMA) are exacerbated when attempting to assess the risks and harm associated with novel psychoactive substances. Here the available information is even more limited; pharmacological, clinical and epidemiological studies rarely exist, and reliance on anecdotal and soft data is often a necessity. Wherever possible, the risks associated with use of novel psychoactive substances tend to be judged in relation to other known substances.

The monitoring of novel psychoactive substances may take different forms. Once identified, a new substance can be tracked, for example by the EU EWS to ensure it does not constitute a threat and that prevalence levels remain negligible or low. More intensive surveillance methods are appropriate on the rare occasion that a new substance diffuses more widely, and succeeds in finding a foothold in the illicit drug market or is linked potential public health risks and harm. As trends develop over time, routine epidemiological information, such as that from general population and school-based surveys, can be used to monitor the diffusion of these substances. With the knowledge that new trends in drug use tend to emerge within restricted social groups and geographical settings, more focused qualitative and specialised information sources can make a valuable addition to routine data [32].

THE EU EWS INFORMATION EXCHANGE

Following the emergence of MDMA (ecstasy) and other synthetic drugs on the European drug market in the early 1990s [9,10], a consensus was reached on the need for an EU-wide response. This led to the establishment of the EU Early warning system (EU EWS) in 1997 [20,33]. At this time, as noted earlier, most new drugs were produced in clandestine laboratories and distributed through the same channels used by the illicit drug market [34,35]. There was also a concern that the new EU mechanism should not impact on human and animal health by unnecessarily restricting the availability of substances with therapeutic value or lead to a large number of obscure chemicals becoming controlled under drug laws. Between 1997 and 2005, the number of drugs reported through the mechanism was relatively low [20]. Many of these quickly disappeared, but those that appeared to pose a significant threat underwent risk assessment [36], and, as a result, some were subjected to control measures by EU countries [37].

In 2005, the original early warning system was strengthened under the provisions of the Council Decision 2005/387/JHA [23]. This rapid-response mechanism has three stages: i) information exchange (known as the European Union Early warning system); ii) risk assessment; and iii) control [38]. The EMCDDA, in collaboration with its network of national early warning systems, plays a central role in detecting novel psychoactive substances, assessing their characteristics and, if necessary, paving the way for eventual control measures.

The EWS operates in real time and provides a rapid channel for dissemination and awareness-raising. It collects, appraises and disseminates information on new substances and products that contain them. Once a new substance is identified, it is logged in the EWS database (the EMCDDA European database on new drugs; EDND) [1] and information is then collated and shared on its manufacture, traffic and use. The EMCDDA, Europol (the European Police Office) and the European Medicines Agency (EMA) all contribute to the system. If the EMCDDA and Europol consider that information collected on a substance merits an active follow-up, a Joint Report is prepared and presented to the European institutions. The analysis provided in this report is used by EU decision makers to determine if a formal risk assessment is required.

In addition to monitoring novel psychoactive substances, the EWS assists in the identification, monitoring and exchange of information on emerging trends for known illicit drugs (e.g. the recent heroin drought experienced by some European countries) [39], and on possible public health-related problems (e.g. anthrax outbreaks). The system also allows the collection and exchange of information on misused psychoactive medicines as well as suspected adverse reactions related to these that are reported under the EU pharmacovigilance system [40].

Risk Assessment

The Council Decision 2005/387/JHA also provides for an assessment of the risks associated with new substances. The risk assessment component is an important instrument to support decision-making on novel psychoactive substances at EU level, adding value to national actions in this area. Formal risk assessments are rare, with only two conducted from 2005 until the writing of this chapter in 2012: 1-benzylpiperazine (BZP) in 2007 [41], and 4-methylmethcathinone (mephedrone) in 2010 [42].

Risk assessments are conducted by the EMCDDA Scientific Committee; however, the decision on whether to subject the substance to control measures across the EU countries is a political one. Should this option be taken, then EU countries have to introduce legislation to control the substance within one year.

In 2009, the EMCDDA published risk assessment operating guidelines [31], whose principal aim is to provide a sound methodological and procedural basis for carrying out a risk assessment, including providing a conceptual framework for consideration of risk. A risk assessment takes into account all factors which according to the UN Conventions (1961 [22] or 1971 [21]) would warrant placing a substance under international control.

The risk assessment process reviews the possible health and social risks of the substance and the implications of placing it under control. In general, the scientific knowledge on a novel psychoactive substance will accumulate over time and as experience with the substance develops. In the interim, risk assessments will have to be based on a broad range of available evidence, the quality of which needs to be appraised. Data reliability and relevance need to be assessed and weighed separately. For example, unpublished recent data may be considered to have a lower formal quality, but still may be considered relevant.

An assessment of the risk–benefit ratio of a novel psychoactive substance is also needed. Factors including whether the substance has legitimate uses, such as potential therapeutic benefits, industrial use or other economic value may be taken into account. Indeed, substances with a known therapeutic value or those that are used to manufacture medicinal products may be exempted from risk assessment. At the risk assessment stage, the prevalence of use of a new substance will usually be low. Here the majority of the available information comes from anecdotal reports, forensic and toxicology laboratories and law enforcement agencies. Triangulation of ethno-epidemiological methods are required to assess the extent of use among limited user groups and an expert judgement needed on the likelihood that use of a novel psychoactive substance will spread. Reference to similar known substances in evaluating the possible risks of a new substance can be helpful.

The concept of risk includes both the element of probability that some harm may occur (usually defined as 'risk') and the degree of seriousness of such a harm (usually defined as 'hazard') [31]. Substance-related risks can originate from several sources and it is vital to clarify their type and origin. The risk assessment conceptual framework differentiates between a) sources from which substance hazards emanate, and b) types of hazardous effects that may be caused by substance use. Acknowledging the problems of interaction between different domains of hazard and harm, the risk assessment process has adopted a semi-quantitative expert judgement approach drawing broadly on the above categories.

Case Study: Mephedrone Risk Assessment

In 2010, increased evidence of the use and availability of the synthetic cathinone mephedrone [43] within the EU countries prompted decision makers to request that the EMCDDA Scientific Committee assess the health and social risks of the drug. The main findings of the risk assessment are summarised below [42].

At the time of the risk assessment, there were no published formal studies that had examined the effects of mephedrone in humans or animals. Users reported psychological and behavioural effects similar to known stimulants, such as improved mental function, euphoria, decreased hostility, general stimulation and mild sexual stimulation [44]. They also reported a number of adverse effects such as sweating, palpitations, chest pain, nasal irritation, bruxism (teeth grinding), nausea, tachycardia, agitation and paranoia. In terms of acute toxicity, evidence suggests that patients typically present with sympathomimetic features (dilated pupils, agitation, tachycardia, and hypertension); severe clinical features including seizures, significant hypertension, and arrhythmias were also reported in a small number of cases [45]. Whilst withdrawal symptoms were not problematic for most users, there were

reports of users experiencing strong cravings for mephedrone [42]. In some cases these were reported to be stronger than those experienced with other stimulants [42]. Mephedrone was detected in a number of post-mortem blood and/or urine toxicology screening at the time of the risk assessment, but only two deaths had been documented with mephedrone assumed to be the sole cause (one in the UK and one in Sweden) [42].

In the absence of representative studies, prevalence rates for mephedrone proved difficult to estimate. Non-representative studies reported lifetime use of mephedrone at around 40% among readers of *Mixmag*, the dance music and clubbing magazine published in the UK [44], 20% amongst a group of Scottish students [46] and 40% for a focus group of school children from Northern Ireland [47]. Qualitative reports identified mephedrone use in other countries but provided no information on consumption levels [42]. The evidence suggested that mephedrone was being combined with alcohol, cannabis, cocaine, and ecstasy [42].

Consumption of mephedrone was reported in a number of sub-populations, including among night clubbers, students, 'psychonauts', opiate injectors and gay men [42]. Users were reported to be aged between their late teens and late twenties, and predominantly male [42], although both younger and older users were identified in studies from the UK. Mephedrone was reported to be used in a similar way to cocaine or ecstasy in nightlife settings, with recreational weekend use a commonly reported pattern. A limited number of users appeared to progress to daily use [42].

The analysis of seised and purchased mephedrone showed quality and purity to be as generally high [42]. Sources for purchase of mephedrone included the Internet, head shops and street-level drug dealers [42,48]. Internet sites tended to market mephedrone as a 'research chemical', 'plant food' or 'bath salts', in order to take advantage of legal grey areas in consumer protection and marketing regulations. Internet monitoring found the number of sites selling mephedrone increased substantially between December 2009 and March 2010, but decreased immediately after the UK controlled the drug in April 2010 [42].

Seizures of mephedrone were reported by 22 EU countries as well as two other countries that report to the EMCDDA. The largest single seizure of more than 130 kg occurred in the Netherlands in October 2009 [42]. Mephedrone had also been detected through pill testing schemes or ad-hoc test purchases in six countries [42]. Some seizures of mephedrone tablets had logo imprints indicating they were being sold to users as 'ecstasy'. Several countries reported information on legal production and distribution from Asia and in particular from China, with European suppliers responsible for final packaging prior to sale. Seizures of tableting machines used for mephedrone processing also suggest the involvement of organised crime in the preparation of the drug for sale on the illicit market [42].

The risk assessment report concluded with a summary of the main findings regarding the health and social risks of mephedrone. The limited scientific evidence available for drawing conclusions and need to exercise care when interpreting the findings of the risk assessment exercise was reiterated. The report concluded with consideration of the possible consequences of controlling mephedrone: a decision to control this drug has the potential to reduce availability and use of the drug; however, it was also acknowledged that control measures could create an illicit market with the associated risk of criminal activity.

Following the risk assessment policy makers decided that EU countries should take the necessary measures, in accordance with their national law, to submit mephedrone to control measures and criminal penalties, as provided for under their obligations under the United Nations Convention on Psychotropic Substances, 1971 [49]. Although 15 countries had already controlled the substance by the time of this decision, the requirement for all EU countries to control the drug enabled better cooperation between judicial authorities and law enforcement agencies across the EU.

National European EWS Networks

The EU EWS is a multidisciplinary network consisting of the national early warning systems of the 27 EU countries, Croatia, Turkey, and Norway. At the heart of these systems is data on the identification of novel substances from forensic and toxicological laboratory networks. To varying degrees, these data are supplemented by information drawn from sources that include, health and care systems, medicine agencies, law enforcement agencies, key informants (such as users, organisers of music festivals, owners and staff of clubs), the media and the Internet. Overall, such an approach allows the collection, assessment and rapid reporting of information on the appearance of new substances found at national level to the EMCDDA [50]. The organisation and functioning of the national EWS is a national responsibility. While these systems have developed to meet national needs, they draw on a common format and guidelines to provide information to the EU EWS.

Presented below are two examples of national EWS, the UK and France, as well as the Nordic Network for the Current Situation of Drugs (NADiS) that is a collaborative effort at regional level. In addition, the Drug Information Monitoring System (DIMS), which is at the core of the Dutch EWS, is discussed later in relation to pill testing.

The UK EWS

Prior to the introduction of the EU EWS in 1997, the UK had set up and managed an informal early warning system since the mid-1990s.

Consisting of representatives from forensic science laboratories in Europe, its primary function was to circulate information and analytical data on the new synthetic substances that started to appear in the early 1990s. Since 1997 the primary role of the UK EWS has been to report information to the EMCDDA, as well as disseminate data and information received from EMCDDA to the members of the UK EWS. The network has a broad membership, including government departments, law enforcement agencies, healthcare agencies and academics, forensic science laboratories and toxicology laboratories. Whilst most data comes from the latter two, academic departments also play a significant role, particularly in relation to analysis of amnesty bin and Internet test purchase samples [48,51–55]. The UK EWS also benefits from important clinical input allowing collection of unique data from a specialist centre in London on the clinical patterns of toxicity associated with NPS. The UK EWS operates as an independent information provider and is not directly involved in the risk assessment of new substances nor in policy matters; risk assessment and advice to the UK government are undertaken by the UK Advisory Council on the Misuse of Drugs (ACMD).

More recently, on behalf of Home Office, the UK EWS has been involved in providing information to support the assessment of new substances that may be subject to a temporary control in the UK. The detection and identification of novel substances in the UK has been improved recently with the establishment of the Forensic Early Warning System (FEWS). This is done by: analysing samples suspected to contain novel substances that have been collected from a range of sources including online and bricks and mortar shops, music festivals and law enforcement agencies; creating chemical reference standards to help with the identification of novel substances; working with the UK Border Agency in order to better detect and detain suspicious packages; and, rapidly identifying novel substances in emergency cases

involving serious toxicity (including fatalities). Analytical data and related contextual information (such as information on packaging) on novel substances identified by FEWS is provided to the UK EWS, which, in turn, is formally reported to the EU EWS [56].

The French EWS

The French EWS, known as the Système d'Identification National des Toxiques et Substances (SINTES) was established in 1999 within the French Monitoring Centre for Drugs and Drug Addiction (Observatoire Français des Drogues et des Toxicomanies; OFDT) [57]. At that time the authorities were concerned that the growing popularity of the 'techno' dance music movement would lead to the emergence of new substances. Initially, a network of local partners was formed, with a network of monitoring centres set up in 2001 covering the major French cities.

SINTES focuses on the identification of substances and products on the illicit market and ensures real-time transmission of information. It is supported by the Recent Trends and New Drugs (Tendances Récentes et Nouvelles Drogues; TREND) surveillance system [58], which provides more in-depth information on use, including a description of users, substances used, use patterns, and, where possible, consequences in populations with a high prevalence of use. TREND also monitors local markets and micro-trafficking (i.e. the availability, accessibility, price and substances on the market at local level).

In 2006, the scope of the surveillance activities of SINTES, which had previously been limited to synthetic substances, was extended to include all types of illicit drugs. Coverage was also extended to include all regions of the country. At the same time it was divided into an 'observation' function and a 'monitoring' function, based on two different data collection approaches. The annual 'observational' surveys of SINTES provide information on the composition of a substance in circulation on the basis of simultaneous collection of drug samples for laboratory analysis, as well as a questionnaire from drug users. They supplement the data resulting from seizures by providing a snapshot of the composition of substances used or potentially available to users.

The Nordic Countries: Nordic-NADiS

The Network for the Current Situation of Drugs (NADiS) was set up as a national network in Sweden in 2000. In 2003, Denmark joined the network followed by Norway in 2004, Finland in 2006, and Iceland in 2011, thus establishing a broader cross-border collaboration (the national EWS for Denmark, Finland, Sweden and Norway report individually to the EU EWS). The extended network, known as the Network for the Current Situation of Drugs in Nordic countries (Nordic-NADiS), is an expert system that involves key national institutions with competence in the field of new psychoactive substances, including the police, medical care and health institutes. The purpose is to enhance collaboration and increase knowledge on novel substances, the way in which they are used, and their possible medical use, ultimately, with the aim of possible regulation at the national level. The work of the network is operationalised through a web-based information exchange platform (NADiS web).

INTERNATIONAL, REGIONAL AND NATIONAL DRUG MONITORING SYSTEMS

Over recent decades, a number of specialised drug information systems outside of the EU have developed the capacity to monitor new and emerging trends in drug consumption. A number of other regional and national models,

as well as some at the international level, will be presented in the following section.

National Drug Monitoring Systems
USA

The United States Drug Enforcement Administration (DEA) (Department of Justice) [59] is responsible for drug scheduling and control. As such it receives reports on new drugs from poison control centres, hospitals and law enforcement. Where necessary it has the authority to temporarily control a substance under emergency scheduling. The Office of Diversion Control maintains a public list on drugs and chemicals of concern, including novel psychoactive substances. The list includes brief summaries of licit uses, chemistry, pharmacology, user population, illicit distribution and control status. The National Forensic Laboratory Information System (NFLIS) [60] is a programme of the DEA that systematically collects drug identification results and associated information from drug cases which have been submitted to and analysed by federal, state and local forensic laboratories. They publish annual reports which provide insight into which known drugs are most used in non-medical circumstances.

SENTRY was an Internet-based programme run by the DEA, which primarily collected information on synthetic drugs of abuse in order to identify new trends at an early stage. Although SENTRY was primarily focussed on drugs produced via a chemical process, it also monitored prescription drugs, over-the-counter medication, chemicals involved in the manufacturing of synthetic drugs and botanical substances and extracts. Registered users accessed SENTRY through a secure URL or via the National Drug Intelligence Centre (NDIC) [61] site and submitted information which was then verified by the NDIC to ensure accuracy and reliability of information. This qualitative/anecdotal information was then combined with additional data collected by the NDIC to produce a DrugAlert Watch (when a pattern of synthetic drug-related activity is first identified) and/or a DrugAlert Warning (when a trend is first detected). At the warning level, agencies could respond to the trend with formal action. For example, law enforcement agencies could identify and target the supply source for local distributors, or treatment providers could begin formal studies into short- and long-term physical and psychological effects of abuse of the drug. The SENTRY project ended in November 2011, and the NDIC closed in June 2012.

Established in 1976 by the National Institute on Drug Abuse (NIDA), the Community Epidemiology Work Group (CEWG) [62] is a network of researchers from the USA and selected third countries which meet to discuss the current epidemiology of drug abuse. It meets biannually to provide on-going community-level public health surveillance of drug use. Multiple sources of information are used to indicate characteristics of drug abuse trends and different types of drug abusers. Admissions to drug abuse treatment programs, drug-involved emergency department reports, seizure data, drug-related deaths reports, arrestee urinalysis or other toxicology results, surveys of drug use and the poison control centre data are used by several or most of the CEWG area representatives. Following the CEWG meeting a report is produced annually based on the presentations of the representatives. Placing a particular focus on new substances, the CEWG held a special session at the June 2010 CEWG meeting, 'New Drugs: United States and International Perspectives'. In June 2012, NIDA and EMCDDA co-organised the 'New and emerging psychoactive substances: Second interdisciplinary forum' [63], which gathered over 300 participants from 72 countries. This conference, which built on the First international multidisciplinary forum on new drugs organised by the EMCDDA in Lisbon in May 2011 [64], focused on new and emerging synthetic and natural drugs, such as synthetic cannabinoid receptor agonists (e.g. 'Spice') and stimulants (e.g. cathinones).

The CEWG has often been the first to report emerging drug trends in the USA. These include: the abuse of the sedative-hypnotic, flunitrazepam, in 1992; the rise of ecstasy and club drugs in 1996 and their decline in 2001; and the occurrence of fentanyl-contaminated heroin in 2006 [65].

The Substance Abuse and Mental Health Services Administration (SAMHSA) [66] is responsible for reducing the impact of substance use and mental illness in communities in the USA. As part of this role it conducts and publishes a number of drug-related surveys and operates reporting systems, such as the Drug Abuse Warning Network (DAWN) [67]. The DAWN Emergency Department Data System provides national and local-area estimates of drug-related department visits, and drug-related mortality. Among its objectives, it seeks to identify substances associated with drug abuse episodes as well as to detect new drug abuse entities and combinations in order to provide data for national, state and local drug abuse policy and programme planning. The DAWN network consists of information provided on emergency department (ED) episodes related to drug abuse and information from medical examiners on drug-related deaths from their affiliated organisations. Since 2003, a DAWN case has a very expansive definition such that any ED visit related to recent drug use or where drug use is implicated is included. Over-the-counter medications, dietary supplements, psychoactive inhalants (such as 'poppers') and alcohol (alone or in combination) are included in the data collection. The cases are categorised according to the substance involved: illicit drugs, alcohol or the non-medical use of pharmaceuticals. Annual reports on the drug-related emergency department admittance rates are made, as well as periodic reports on topics and trends of interest such as drug-related suicide attempts and underage drinking. One critique of the system is that it may not provide an up-to-date and accurate picture of overall drug use nationally, as only episodes where the patient admits to taking drugs or the hospital staff can identify drug use can be included. Because of the large number of DAWN cases, those involving the use of novel psychoactive substances make up the minority. For example, in 2008 'other illicit drugs', which includes MDMA, GHB, flunitrazepam, ketamine, PCP (phencyclidine), LSD, and other hallucinogens and psychoactive inhalants, were each involved in less than 4% of ED visits involving illicit drugs [68].

Australia

The National Drugs and Alcohol Research Centre (NDARC) [69] in Australia is responsible for two national systems tasked with providing an early warning of new drug trends. These are the Ecstasy and Related Drugs Reporting System (EDRS) [70] and the Illicit Drug Reporting System (IDRS) [71]. EDRS is a national monitoring system for ecstasy and related drugs intended to serve as a strategic early warning system in identifying emerging trends of local and national interest in the markets for these drugs. It is managed by different research institutions in each Australian state or territory. The methods employed in this system include interviews with regular ecstasy users, interviews with key experts and analysis of indicator data related to ecstasy and other related drugs. This includes monitoring the price, purity, availability and patterns of use of ecstasy, methamphetamine, cocaine, ketamine, GHB, MDA (3,4-methylenedioxyamphetamine) and LSD. In 2010, mephedrone use was found to be relatively prevalent (17% in the last 6 months) among regular ecstasy users. This system is sensitive to trends and provides data in a timely manner. Its sister system, IDRS, monitors drug use trends across Australia for heroin, amphetamines, cocaine and cannabis.

New Zealand

The Illicit Drug Monitoring System (IDMS) [72] was established in 2005 and is run by the Social and Health Outcomes Research and Evaluation (SHORE) at Massey University. The aim of the IDMS is to provide a brief 'snapshot' of illicit drug use in New Zealand. Typically, it consists of a survey conducted among drug users about purity, price and potency of a range of drugs, as well as the appearance of any new substances in the market. The 2010 IDMS sample included frequent methamphetamine, ecstasy and injecting drug users [73]. The interviews with frequent drug users were then combined with secondary data sources of drug use, such as drug seizure statistics, telephone calls to the Alcohol and Drug Helpline, and admissions to drug treatment services. The report provides a picture of current issues, including the rise and fall of BZP usage, which was a widely used legal drug in New Zealand during the early to mid-2000s until it was controlled in April 2008 [74]. While the 2008 IDMS report found a significant reduction in the availability and use of BZP following its control [75], the 2009 IDMS showed some recovery in its availability, although it is yet to match the levels observed in the years pre-control [76].

South Africa

The South African Community Epidemiology Network on Drug Use (SACENDU) [77], established in 1996, is a network of researchers, practitioners and policy makers from all areas of South Africa. Members of SACENDU meet every six months to provide community-level public health surveillance of alcohol and other drug (AOD) use trends and associated consequences through the presentation and discussion of quantitative and qualitative research data. Demographic information on patients, primary substance of abuse, mode of administration, and frequency of use are reported from six substance abuse treatment centre sites across South Africa. SACENDU reports regularly on the nature and pattern of AOD use, emerging trends, risk factors associated with AOD use, characteristics of vulnerable populations, and consequences of AOD use in South Africa [78].

Although this network does not focus specifically on novel psychoactive substances, within the reports there is scope for detection of these substances. In December 2010, methcathinone was noted in most sites, with one particular region reporting that 4% of patients had 'CAT' (a street term used for methcathinone) as a primary or secondary drug of abuse [79].

The United Nations System

World Health Organisation (WHO)

The World Health Organisation (WHO) is a specialised agency of the United Nations. Through its Expert Committee on Drug Dependence (ECDD) [80], it conducts the medical, scientific and public health evaluation of psychoactive substances in order to inform the decisions made by the United Nations Commission on Narcotic Drugs (CND) [81] on whether to control a substance under the United Nations drug control conventions.

In 2010, the World Health Organisation adopted a revision of their guidelines for the review of psychoactive substances for international control [82]. The guidelines ensure that the review process is of clear methodology, transparent, and based on scientific and public-health related principles. The guidelines detail the procedure for preparing a critical review, including how it is decided if a pre-review or critical review will be held, how to prepare the review reports and the criteria on which the ECDD should base their judgement. The ECDD is to first consider the applicability of the 1961 Convention [22] – that is, whether the substance shows similar abuse liability profile and dependence-producing properties to drugs already controlled under this convention – and if not, whether the 1971 Convention [21] is

applicable. For all substances that are reviewed, a summary assessment giving a description of the ECDD's findings should include the extent or likelihood of abuse, the degree of seriousness of the public health and social problem, the degree of usefulness of the substance in medical therapy, and advice on the control measures that would be appropriate. The guidelines also cover the meeting, membership and functions of the ECDD and its collaboration with other organisations and experts for the decision-making process.

United Nations Global SMART Programme

The Global SMART (Synthetics Monitoring: Analyses, Reporting and Trends) Programme was launched in September 2008 as a project of the United Nations Office on Drugs and Crime (UNODC) [83]. The main aim of the programme is to improve the quality of information on synthetic drugs. This includes information on the patterns of trafficking and use, as well as to increase the information exchange between the participating countries on illicit synthetic drugs in order to design effective policy and programme interventions. It further aims to improve methods for detecting and reporting emerging trends within these countries.

The SMART programme holds national training and review sessions in participating countries, facilitates on-line data collection, verifies data, as well as analyses of country situation reports. It also carries out regional assessments, for example on the patterns and trends of amphetamine type stimulants in East and South East Asia [84].

The Global SMART programme also provides regular updates on patterns and trends in the global synthetic drug situation [83]. It reports information in several categories, such as significant or unusual drug or precursor seizures, new locations or methods for clandestine manufacture, new trafficking groups or routes, changes in legislation to combat synthetic drugs, as well as information on emerging drugs or user groups and health implications related to their use [85].

Inter-American Observatory on Drugs

The Inter-American Observatory on Drugs (OID) [86] is a pan-American surveillance system created in 2000 and run by the Inter-American Drug Abuse Control Commission (CICAD) in partnership with national statistics and information focal points (national observatories on drugs). It works with 21 Latin American and Caribbean drug councils to develop and supervise standardised drug use prevalence (SIDUC) studies and publish an annual compendium of supply-side drug data [87]. Internet based software is used to collect and report data on drug-related arrests, crop eradication, destruction of drug laboratories, seizures of drugs and chemicals and other law enforcement data in the member countries. This surveillance system has the potential to serve as an early warning system on the appearance of new substances, new methods of using and manufacturing substances and changing trafficking patterns.

SOURCES FOR DETECTING AND MONITORING NOVEL PSYCHOACTIVE SUBSTANCES

This section reviews a number of data collection approaches and methodologies which have potential to enhance detection, understanding, and monitoring of novel psychoactive substances. Forensic science remains the key method for identifying such substances, whether through development, of, and facilitating access to, reference samples, test purchasing, pill testing, wastewater analysis or new approaches to predict the psychoactivity and toxicity of novel psychoactive substances. In terms of settings, data from hospital emergency departments are able to provide interesting insights into both patterns of use of novel psychoactive substances and associated acute health problems. The use of Internet monitoring, with repeat cross-sectional snapshots can provide insights into different aspects of the online marketplace for new

substances, including highlighting the range of substances/products available to the public. Finally, there are now a limited but growing number of European epidemiological surveys and studies reporting on the prevalence of use of novel psychoactive substances in the general population or particular sub-populations.

Forensic Science

Forensic science represents a core component in systems monitoring novel psychoactive substances. The first step in the identification of any novel psychoactive substance involves determining the chemical structure of the compound by means of chromatography (e.g. medium pressure liquid chromatography [MPLC], high-performance liquid chromatography [HPLC]) as well as spectrometric techniques (e.g. mass spectrometry [MS], nuclear magnetic resonance [NMR]) or both (e.g. gas chromatography coupled with mass spectrometry [GC-MS]). This information is provided by forensic chemistry and forensic toxicology laboratories.

Among the most pressing challenges hampering development of forensic responses are the practical difficulties of identifying novel substances. In this dynamic marketplace, neither buyer nor seller may be accurately aware of what substances, or mixtures of substances, are being sold or consumed [48]. Some countries have begun test purchase projects, but this information is not routinely collected or shared. Moreover, the availability of reference materials (substances) is of critical importance if forensic and toxicology laboratories are to identify novel psychoactive substances, especially in the case of those substances for which limited scientific literature is available.

Pill Testing

Forensic testing and identification of substances in pills and powders is the cornerstone of some of national early warning systems in Europe, notably in the Netherlands, Austria,

France and Spain. The most well-known initiative here is the Drugs Information and Monitoring System (DIMS) [88], which originally started as a local initiative in Amsterdam to give drug users the opportunity to determine the composition of their drugs. In 1992 DIMS was rolled-out nationally, and since the 1990s it has played a central role in the Dutch early warning system. DIMS is a network of cooperating institutions which consists of the coordinating and steering centre at the Trimbos Institute, which is the Dutch research institute for mental health and addiction, 30 'test-offices', and numerous anonymous drug users.

DIMS is tasked with the daily monitoring, surveillance and the acute assessment of risks of novel psychoactive substances and is based on a system of information exchange. Drug users have their drugs analysed in order to know the exact composition and in return provide information about observed effects, use and market. Signals pointing to the appearance of new drugs on the market can thus be picked up by the system. The main focus for DIMS is to identify the compounds of (synthetic) drugs, to describe prevalence of drugs in the market and trends in drug use, and to identify health risks for drug-users. Apart from its monitoring function, DIMS also has an important surveillance task. When a drug with clear health risks appears on the market, the DIMS-bureau initiates and co-ordinates a national warning campaign (known as a 'Red Alert') to prevent and reduce health risks. To determine the acute risks for the Red Alert procedure a network of different experts is consulted. Field workers, policymakers and scientists participate in these warning campaigns [89], following an established protocol.

Wastewater Analysis and Pooled Urine Analysis

Wastewater analysis (sewage epidemiology) is a rapidly developing scientific discipline with the potential for monitoring population level trends in illicit drug consumption [90,91]. Advances in analytical chemistry have made it possible to identify urinary excretion of illicit drugs and their main metabolites in wastewater at very low concentrations. This is comparable to taking a much diluted urine sample from an entire community (rather than from an individual user). With certain assumptions, it is possible to back-calculate from the amount of the drug and/or metabolite(s) in the wastewater to an estimate of the amount of a drug consumed in a community [92]. While early research focused on identifying cocaine and its metabolites in wastewater, recent studies have produced estimates on levels of cannabis, amphetamine, methamphetamine, heroin and methadone. The identification of less commonly used drugs, such as ketamine and novel psychoactive substances may also be possible. This area of work is developing in a multidisciplinary fashion, with important contributions from, among others, analytical chemistry, physiology, biochemistry, sewage engineering and conventional drug epidemiology. At the top of the research agenda is the development of a consensus on sampling methods and tools, as well as the establishment of a code of good practice for the field. While wastewater appears promising for the identification and monitoring of community and general population drug use (particularly with established drugs), pooled urine analysis is a more sensitive technique being used for the identification of novel psychoactive substances in sub-groups where early adoption of such substances is often seen [93]. Collection of samples from sewage-treatment plants for the detection of NPS is problematic as often little is known about factors such as stability or metabolism by bacteria. Pooled urine collection avoids these problems by collecting urine samples closer to users. This technique utilises samples of urine collected from portable toilets, for example those used at music festivals, placed outside nightclubs or used in city-centres at weekends or public holidays to prevent public urination in the streets.

Computational Modelling

The possibility of predicting the mode of action of novel compounds for which scarce scientific literature is available can be extremely valuable [94]. The potential use of inexpensive structure-activity relationships, molecular classification models as well as bioactivity models for the prediction of the properties of novel substances (i.e. toxicity, pharmacology, psychoactivity, etc.) without the need for conducting experimental studies in animals or humans is currently being explored [95]. The chemogenomics principle, which assumes that structurally similar targets share similar ligands [96], allows some degree of prediction of the effect of a chemical compound on a large set of receptors. The combination of machine learning techniques with bioactivity databases enables the prediction of the mode of action and effects of novel compounds.

However, caution is required when inferring effects of substances based on structure-activity relationships, for example when making predictions about the activity and potency of novel cathinone analogues by analogy to the structure-activity relationships derived from amphetamine-related substances [97].

Hospital Emergencies Data

With the exception of the Drug Abuse Warning Network (DAWN) system in the USA [67], there are limited data that are systematically collected and published internationally on acute toxicity associated with recreational drug use. Hospital coding systems are typically based around the ICD-10 system (International Statistical Classification of Diseases and Related Health Problems, 10th Revision), which does not include the majority of recreational drugs and in most countries will only capture hospital admissions, and not patients who are discharged

directly from the emergency department [98]. Studies have also shown that cases may be coded according to the presenting clinical feature rather than based on the drug causing the presentation (e.g. cocaine-related chest pain being coded as chest pain with no coding relating to cocaine) [99]. Acute recreational drug toxicity is a significant clinical issue with the potential for significant morbidity and mortality.

A recent feasibility study, undertaken in the UK and Spain, explored methods for collecting emergency data in two different units in busy nightlife areas. The two centres involved in the study (London, UK and Palma, Mallorca) designed a tool to collect data prospectively on all cases presenting to their emergency departments with acute recreational drug toxicity in June and July 2009 [100]. The study demonstrated that it is feasible to collect and collate data to help detect and allow differences in acute recreational drug toxicity in different EU countries to be examined. Analysis is made on types of recreational drugs used, place of drug use, country of origin of individuals presenting, and patterns of acute toxicity seen. Establishing the epidemiology of acute recreational drug toxicity across EU countries and the patterns of toxicity seen is important in determining the harm associated with recreational drug use. The European Drug Emergencies Network (EuroDEN) project is a European Commission funded project developing techniques to collect health emergency data from countries around Europe to look at seasonal and geographical trends in the acute harms associated with the use of novel psychoactive drugs. Currently, club/night-time economy specific guidelines for improving prehospital care of recreational drug users adapted to a European context are under development.

Internet Monitoring

The use of the Internet to market and sell novel psychoactive substances poses an important challenge to policy-makers. The Internet offers direct access to the world of the drug user through online forums and chat rooms, as well as to online shops selling 'legal' alternatives to controlled substances. This market has grown dramatically over the last few years and now includes a wide range of different types of substances: from plants to herbal mixtures, to pills and research chemicals. Online shops often specialise in particular drug-related products; some selling drug paraphernalia, some specialising in hallucinogenic mushrooms or 'party pills', while others market a wide range of herbal, semi-synthetic and synthetic substances. Illustrations of the growing importance of the Internet can be witnessed with the 'Spice' phenomenon which was largely Internet based, the sale of 'legal highs' and the recent mephedrone phenomenon [26,27,42].

Monitoring the Internet not only improves the ability to identify and track new and emerging trends, but can also contribute to and complement the rapid information exchange of information and risk assessment on novel psychoactive substances. Here, Internet monitoring has the potential to act as a confirming source, and also as a trigger in the identification of potentially threatening new substances. The nature of the Internet, which is intrinsically unstable and multilingual, presents a number of monitoring challenges, in particular the development of comparable methods for searching and methods for assessing the validity of the information.

To date there have been relatively few published research studies that have examined the novel psychoactive substances available on the Internet. Schifano and Deluca [101] in the context of the 'Psychonaut' research project [102,103], presented a European overview on the availability of online information on psychoactive substances, analysing websites presenting information on the consumption, manufacture, and sale of psychoactive substances. In 2005, the EMCDDA initiated its European Perspectives on Drugs (E-POD)

project, which aimed to provide practical experience for the development of a European system to detect, track, and understand emerging trends. Since 2005, the EMCDDA have conducted a number of studies, including studies on magic mushrooms (2006) [104], GHB (gamma-hydroxybutyrate) and GBL (gammabutyrolactone) (2007) [105], 'legal highs' (2008–2011) [26,106], 'Spice' products (2009) [26] and mephedrone (2010, 2011), in the process developing an Internet 'snapshot' monitoring methodology (see Box 2.1). EMCDDA snapshots of online drug shops have become increasingly pivotal in understanding a rapidly adapting Internet-based drug market. They provide information on the working methods of online retailers, the way that they respond to users' demands, changes in the law and other supply issues. The 2011 'legal highs' snapshot undertaken in July in 20 languages identified a total of 631 online shops that were supplying products to European consumers (see Table 2.1) [107].

Surveys and Studies on the Prevalence of Use of Novel Psychoactive Substances

In Europe, there are a limited number of surveys that are capable of reporting on the prevalence of novel psychoactive substances in the general population or particular sub-populations. As the use of novel psychoactive substances has only recently emerged as a social and cultural phenomenon, their inclusion in national drug surveys is a relatively recent development. As well as being scarce, prevalence data on novel psychoactive substances has many limitations and associated methodological issues. These include: a lack of common definitions, non-representative samples, small sample size and as a result often overall weak data. Problems may be confounded by the rapidly changing legal situation at a country level and the variable content of the drugs which limits the reliability of users self-report of the drugs they are/have been using.

BOX 2.1

CORE COMPONENTS OF THE EMCDDA INTERNET SNAPSHOT METHODOLOGY

Scope

- Online websites (retailers and wholesale) easily accessible to a random Internet user interested in buying psychoactive substances
- Targeting and addressing an EU audience.

Exclusion Criteria:

- Websites selling only paraphernalia or seeds or non-psychoactive mushrooms
- Websites not shipping psychoactive substances to an EU country
- Discussion forums and/or drug-related chat rooms, social networking sites or tools such as Skype, Messenger, Facebook, Twitter, etc.

Identification of Killer String:

Select the search string to achieve the maximum coverage. Identify 'killer' term or a 'killer' combination of terms of each search ('killer' = more relevant hits on the search performed).

Sampling to Exhaustion:

Mandatory look at the first 100 links and after 101, 'sample to exhaustion' (ceasing when 20 successive links are irrelevant).

Use of Multiple Search Engines:

Coverage and performance (in term of accuracy of search) is enhanced with the use of Metacrawler.com + Google(.national) + 1 additional specific national search engine (the most relevant by languages).

Broad Coverage of EU Languages:

Maximised number of EU languages used for searching.

Common Reporting Template:

Search results reported the same way in different languages.

The 2011 Eurobarometer survey, which examined the attitude of young people towards drugs, interviewed more than 12000 people (aged 15–24) across EU countries. In most countries the level of use was not more than 5% [108]. However, in the UK, Latvia and Poland, use was close to 10%, whilst in Ireland 16% reported use of a new substance. Of those young people who had used a new substance, 54% indicated that they had been offered them by friends, compared to 37% who had been offered them during a party or in a pub and 33% who had bought them in a specialised

shop, e.g. a head/smart shop. Interestingly, the Internet was the least common source for these substances; only 7% of interviewees indicated that they had purchased these substances on the Internet.

At national level, there are general population and/or school-based surveys from the UK, Poland, Ireland, Spain, New Zealand and the USA which provide an indication of the prevalence of use of new drugs or 'legal highs'. The 2011 British Crime Survey reported that for those aged 16–24, mephedrone use (4.4%) was at a similar level of use as powder cocaine

TABLE 2.1 Most Frequently Identified New Psychoactive Substances/'Legal Highs' Available for Online Sale and Prices in 2011 Internet Snapshot

New Psychoactive Substance/ 'Legal High'	Number of Online Shops		Nature	Price for 10 g (EUR)
	July 2011	January 2011		
Kratom	128	92	Natural	6–15
Salvia	110	72	Natural	6–12
Hallucinogenic mushrooms	72	44	Natural	10–14
MDAI (aminoindane)	61	45	Synthetic	100–110
Methoxetamine (arylcyclohexylamine)	58	14	Synthetic	145–195
6-APB (benzofuran)	49	35	Synthetic	230–260
4-MEC (cathinone)	32	11	Synthetic	120–200
MDPV (cathinone)	32	25	Synthetic	115–239
Cactus	30	17	Natural	20–40 (plant)
Methiopropamine (thiophene)	28	5	Synthetic	115–130
5-IAI (aminoindane)	27	25	Synthetic	95–120
Dimethocaine (benzoate)	27	22	Synthetic	85–150
Methylone (cathinone)	26	17	Synthetic	76–130
5-APB (benzofuran)	23	6	Synthetic	250–330
AM-2201 (cannabinoid)	22	1	Synthetic	180–210
JWH-018 (cannabinoid)	20	5	Natural	200–230
JWH-250 (cannabinoid)	19	4	Natural	110–195
AMT (tryptamine)	19	13	Synthetic	230–460
MDAT (aminotetralin)	18	22	Synthetic	110–130
Mephedrone (cathinone)	18	23	Synthetic	120–200
JWH-122 (cannabinoid)	17	4	Synthetic	50–55/200–240
Ayahuasca (active principle DMT)	17	10	Natural	15–30 (kit)
4-FA (phenethylamine)	17	2	Synthetic	120–200
3,4-DMMC (cathinone)	16	3	Synthetic	90–200
Hawaiian baby woodrose (active principle lysergamides)	16	10	Natural	4–8 (10 seeds)
GBL (GHB precursor)	15	12	Synthetic	35–45 (½ litre)

(4.4%; the second most used drug amongst young people) [109]. Among those aged from 16 to 59, last year use of mephedrone was at a similar level as ecstasy use (1.4%; the third most used drug within this age group). In 2012, the percentages for mephedrone use had fallen to 3.3% and to 1.1%, respectively [110].

Mephedrone and 'legal highs' were included for the first time in the drug prevalence survey of households in Ireland and Northern Ireland conducted in 2010/11 [111]. In Northern Ireland, the prevalence of use for respondents aged between 15 and 64 was around 2% for lifetime and 1% for last year use for both mephedrone and 'legal highs'. The lifetime prevalence of use rate for people aged 15–24 was 6% for both mephedrone and 'legal highs'. In the Republic of Ireland, with regard to drug use by adults in the year prior to the survey, new psychoactive substances (4%) were the second most frequently reported drugs after cannabis (6%). Among younger adults aged 15–24 years, 10% reported last year use of novel psychoactive substances.

In Spain, the National Survey on Drug Use in Students aged 14–18 years (2010) found that 3.5% of students had consumed one or more of the new or emerging drugs at some time in their life, and 2.5% had consumed during the previous year and 1.3% last month [112]. The substances most often consumed were magic mushrooms, 'Spice' and ketamine; however, the overall prevalence levels were very low, confirming the sporadic nature and experimental use of these substances among students in this age group. A 2009 study from Frankfurt, Germany also explored use of 'Spice' amongst students, reporting that a total of 7% of the 15–18 year-olds in Frankfurt reported experience with smoking mixtures.

A 2008 Polish study among students found that just under 4% had used 'legal highs' at least once in their life, while a follow-up study in 2010 reported an increase to 11%. The use of 'legal highs' during the previous 12 months

was reported by around 3% of students in 2008, and increased to 7% in 2010. This period saw a proliferation of smart shops in Poland. An additional Polish general population study conducted in both 2009 and 2010, provided further insights into consumption levels of 'legal highs', in particular the drop in use after closure of the Polish smart shop network in 2010. In this year, 3% of the respondents admitted using 'legal highs' at least once, which was a reduction from 6% in 2009. Just fewer than 2% had used them in the last 12 months, compared with 5% in 2009.

A small number of studies in countries outside of the EU have attempted to ascertain prevalence of use of 'legal highs'. In 2011, for the first time, the US Monitoring the Future annual school survey reported on prevalence of synthetic cannabinoids use among young people [113]. Among 12th graders, past-year use of these synthetic cannabinoids (Spice and K2) was found to be just over 11%. The popularity of BZP (1-benzylpiperazine) in New Zealand led to a number of studies on this substance and its effects. A 2006 national household study of people aged 13–45 years found that 20% of New Zealanders had ever tried BZP and 15% had used it in the previous year [114]. At the time the survey was conducted, BZP-party pills were the fourth most commonly used recreational substance in the country behind alcohol, tobacco and cannabis [115].

A number of sources, most commonly Internet surveys, provide insight into the use of new drugs by specific populations. As these are largely self-selected convenience samples, the findings do not reflect the use of these substances in the wider population. As previously noted, the 2010 Internet survey among readers of *Mixmag*, the dance music and clubbing magazine published in the UK, found lifetime use of mephedrone at around 40% (33% last month use) [44]. In 2011, a similar survey found that lifetime use among respondents was 61%, while use in the last month was 25%

[116]. A survey conducted in South London in gay-friendly nightclubs found that mephedrone had the highest prevalence of last month use (53.2%) and use on the night of the survey (41.0%) [117]. Another Internet survey has reported on use of 'legal highs' in Ireland [118]. Similar to findings among *Mixmag* readers, high levels of mephedrone use (66%) were reported by survey respondents, and 18% had used BZP party pills. Interestingly, more individuals had tried 'smoking blends' (presumably containing synthetic cannabinoids) than any other type of novel psychoactive substance. Among the drugs of herbal origin, *Salvia divinorum* was most popular, being sampled by almost a quarter of all respondents in both the Irish and UK studies. A survey in the first half of 2011 among Internet users looking into the use of new synthetic drugs ('legal highs') was conducted among respondents aged 15–34 in the Czech Republic, with 4.5% reporting lifetime use of a new synthetic drug [119].

attempts by simply moving on to alternative non-controlled substances. To date, most their interest has been focused on providing novel substances that target the recreational stimulant and cannabis markets. However, new substances are now emerging that appear to target more chronic problematic drug users. A further worrying development is the interest in exploring the potential of psychoactive medicines as a basis for new non-controlled substances.

The combined use and triangulation of routine epidemiological indicators, qualitative research and leading-edge indicators are likely to improve the chance of obtain a holistic picture of new trends at the local, national, regional and international level. There is a need to respond rapidly to novel substances and related trends that are identified through early warning systems, while simultaneously fine tuning other monitoring instruments to track diffusion, increased consumption, as well as evidence of new health and social risks.

CONCLUSIONS

In the last few years, record numbers of novel psychoactive substances have been identified globally. The availability of cheap organic chemical synthesis has been a catalyst here, with chemical companies, predominantly in emerging economies such as China, working with entrepreneurs who have developed sophisticated marketing and distribution approaches. Products are sold on the Internet or specialist bricks and mortar head shops as well as by street-level drug dealers. Some overlap and interaction exists with the illicit drugs market.

It is likely that synthetic substances will continue to dominate the phenomenon of novel psychoactive substances. Entrepreneurs, including chemists, are currently researching hundreds of potential psychoactive substances and will respond rapidly to any control

REFERENCES

[1] EMCDDA. Action on new drugs, <http://www.emcdda.europa.eu/activities/action-on-new-drugs> [accessed 04.03.13].

[2] EMCDDA/Europol. EMCDDA–Europol 2011 Annual Report on the implementation of Council Decision 2005/387/JHA. Lisbon. Available: <http://www.emcdda.europa.eu/publications/implementation-reports/2011>; 2011 [accessed 04.03.13].

[3] Climko RP, Roehrich H, Sweeney DR, Al-Razi J. Ecstasy: a review of MDMA and MDA. Int J Psychiatry Med 1986;16:359–72.

[4] Beck J, Morgan PA. Designer drug confusion: a focus on MDMA. J Drug Educ 1986;16:287–302.

[5] Baum RM. New variety of street drugs poses growing problem. Chem Eng News 1985;63:7–16.

[6] Henderson GL. Designer drugs: past history and future prospects. J Forensic Sci 1988;33:569–75.

[7] Shulgin AT. The background and chemistry of MDMA. J Psychoactive Drugs 1986;18:291–304.

[8] Renfroe CL. MDMA on the street: analysis anonymous®. J Psychoactive Drugs 1986;18:363–9.

[9] Davies JB, Ditton J. The 1990s: decade of the stimulants?. Br J Addict 1990;85:811–3.

[10] Holland J. The history of MDMA Holland J, editor. Ecstasy: the complete guide. Rochester VT: Park Street Press; 2001. p. 11–20.

[11] Bossong MG, Brunt TM, Van Dijk JP, Rigter SM, Hoek J, Goldschmidt HMJ, et al. mCPP: an undesired addition to the ecstasy market. J Psychopharmacol 2010;24:1395–401.

[12] Sedefov R, Brandt SD, Evans-Brown M, Sumnall HR, Cunningham A, Gallegos A, et al. PMMA in 'ecstasy' and 'legal highs'. Br Med J BMJ 2011;341:c3564. (published 8 August 2011). Rapid Response to Brandt, et al. (2010) BMJ.

[13] Williams H, Taylor R, Roberts M. Gamma-hydroxybutyrate (GHB): a new drug of misuse. Ir Med J 1998;91:56–7.

[14] EMCDDA Report on the risk assessment of GHB in the framework of the joint action on new synthetic drugs, European monitoring centre for drugs and drug addiction. Luxembourg: Office for Official Publications of the European Communities; 2002. Available: <http://www.emcdda.europa.eu/html.cfm/index33345EN.html>.

[15] Degenhardt L, Copeland J, Dillon P. Recent trends in the use of 'club drugs': an Australian review. Subst Use Misuse 2005;40:1241–56.

[16] Maxwell JC. Party drugs: properties, prevalence, patterns, and problems. Subst Use Misuse 2005;40:1203–40.

[17] Drasbek KR, Christensen J, Jensen K. Gamma-hydroxybutyrate–a drug of abuse. Acta Neurol Scand 2006;114:145–56.

[18] EMCDDA Responding to new psychoactive substances. Drugs in focus 22, European monitoring centre for drugs and drug addiction. Luxembourg: Publications Office of the European Union; 2011. Available: <http://www.emcdda.europa.eu/publications/drugs-in-focus/responding-to-new-psychoactive-substances> [accessed 04.03.13].

[19] Daughton CG. Illicit Drugs: contaminants in the environment and utility in forensic epidemiology. Rev Environ Contam Toxicol 2011;210:59–110.

[20] Council of the European Union. Joint Action of 16 June 1997 concerning the information exchange, risk assessment and the control of new synthetic drugs (97/396/JHA). Official Journal L 167, 25/6/1997, 1–3. Available: <http://eur-lex.europa.eu/LexUriServ/LexUriServ.do?uri=CELEX:31997F0396:EN:HTML>; 1997 [accessed 04.03.13].

[21] UN. Convention on psychotropic substances, United Nations. Available : <http://www.unodc.org/pdf/convention_1971_en.pdf>; 1971.

[22] UN. Single convention on narcotic drugs, as amended by the 1972 Protocol amending the single convention on narcotic drugs, 1961, United Nations. Online: <http://www.unodc.org/pdf/convention_1961_en.pdf>; 1961.

[23] Council Decision 2005/387/JHA of 10 May 2005 on the information exchange, risk assessment and control of new psychoactive substances. Official Journal L 127, 20/5/2005, 32–37. Available: <http://eur-lex.europa.eu/LexUriServ/LexUriServ.do?uri = CELEX:32005D0387:EN:HTML> [accessed 04.03.13].

[24] EMCDDA Annual report 2012, the state of the drugs problem in Europe. Luxembourg: Publications Office of the European Union; 2012. Available: <http://www.emcdda.europa.eu/publications/annual-report/2012> [accessed 04.03.13].

[25] Griffiths P, Lopez D, Sedefov R, Gallegos A, Hughes B, Noor A, et al. Khat use and monitoring drug use in Europe: the current situation and issues for the future. J Ethnopharmacol 2010;132:578–83.

[26] Sedefov R, Gallegos A, King A, Lopez D, Auwärter V, Hughes B, et al. Understanding the 'Spice' phenomenon, European monitoring centre for drugs and drug addiction. : Office for Official Publications of the European Communities; 2009. Available: <http://www.emcdda.europa.eu/publications/thematic-papers/spice> [accessed 04.03.13].

[27] Griffiths P, Sedefov R, Gallegos A, Lopez D. How globalisation and market innovation challenge how we think about and respond to drug use: 'spice' a case study. Addiction 2010;105:951–3.

[28] EMCDDA drug profiles. Available: <http://www.emcdda.europa.eu/publications/drug-profiles> [accessed 04.03.13].

[29] van Amsterdam JG, Best W, Opperhuizen A, de Wolff FA. Evaluation of a procedure to assess the adverse effects of illicit drugs. Regul Toxicol Pharmacol 2004;39:1–4.

[30] Nutt D, King L, Saulsbury W, Blakemore C. Development of a rational scale to assess the harm of drugs of potential misuse. Lancet 2007;369:1047–53.

[31] EMCDDA Risk assessment of new psychoactive substances: operating guidelines, European monitoring centre for drugs and drug addiction. Luxembourg: The Publications Office of the European Union; 2010. Available: <http://www.emcdda.europa.eu/html.cfm/index100978EN.html> [accessed 04.03.13].

[32] Griffiths P, Vingoe L, Hunt N, Mounteney J, Hartnoll R. Drug information systems, early warning, and new drug trends: can drug monitoring systems become more sensitive to emerging trends in drug consumption?. Subst Use Misuse 2000;35:811–44.

[33] Sedefov R, Gallegos A. The European early warning system: responding to novel and emerging recreational drugs. Clin Toxicol 2011;49:200.

[34] Commission of the European Communities. Communication from the commission to the council and the european parliament on the control of new synthetic drugs 'designer drugs' (COM(97) 249 final). Office for

Official Publications of the European Communities; 1997.

[35] Soine WH. Clandestine drug synthesis. Med Res Rev 1986;6:41–74.

[36] EMCDDA. Action on new drugs, Risk assessments. Available: <http://www.emcdda.europa.eu/html.cfm/index16776EN.html> [accessed 04.03.13].

[37] EMCDDA. Action on new drugs, Control measures. Available: <http://www.emcdda.europa.eu/html.cfm/index16783EN.html> [accessed 04.03.13].

[38] EMCDDA Early warning system on new psychoactive substances – operating guidelines, European monitoring centre for drugs and drug addiction. Luxembourg: Office for Official Publications of the European Communities; 2007. Available: <http://www.emcdda.europa.eu/html.cfm/index52448EN.html> [accessed 04.03.13].

[39] Griffiths P, Mounteney J, Laniel L. Understanding changes in heroin availability in Europe over time: emerging evidence for a slide, a squeeze. Addiction 2012;107:1539–40.

[40] Regulation (EU) No 1235/2010 of the European Parliament and of the Council of 15 December 2010 amending, as regards pharmacovigilance of medicinal products for human use, Regulation (EC) No 726/2004 laying down Community procedures for the authorisation and supervision of medicinal products for human and veterinary use and establishing a European Medicines Agency, and Regulation (EC) No 1394/2007 on advanced therapy medicinal products. Official Journal L 348/1, 31/12/2010, 1–16.

[41] EMCDDA Report on the risk assessment of BZP in the framework of the Council decision on new psychoactive substances, European monitoring centre for drugs and drug addiction. Luxembourg: Office for Official Publications of the European Communities; 2009. Available: <http://www.emcdda.europa.eu/publications/risk-assessments/bzp> [accessed 04.03.13].

[42] EMCDDA Report on the risk assessment of mephedrone in the framework of the Council Decision on new psychoactive substances, European monitoring centre for drugs and drug addiction. Luxembourg: Publications Office of the European Union; 2011. Available: <http://www.emcdda.europa.eu/html.cfm/index116639EN.html> [accessed 04.03.13].

[43] Joint report on the new psychoactive substance 4-methylmethcathinone (mephedrone). 8145/10 CORDROGUE 36/SAN 68.

[44] Dick D, Torrance C. Mixmag drugs survey. Mixmag 2010;225:44–53.

[45] Wood DM, Davies S, Puchnarewicz M, Button J, Archer R, Ovaska H, et al. Recreational use of mephedrone (4-Methylmethcathinone, 4-MMC) with associated sympathomimetic toxicity. J Med Toxicol 2010;6:327–30.

[46] Dargan PI, Albert S, Wood DM. Mephedrone use and associated adverse effects in school and college/university students before the UK legislation change. Q J Med 2010;103:875–9.

[47] Meehan C. Doctoral study on adolescent drug taking in Northern Ireland. : University of Ulster; 2010.

[48] Davies S, Wood DM, Smith G, Button J, Ramsey J, Archer R, et al. Purchasing 'legal highs' on the Internet – is there consistency in what you get? Q J Med 2010;103:489–93.

[49] Council Decision 2010/759/EU of 2 December 2010 on submitting 4-methylmethcathinone (mephedrone) to control measures. Official Journal L 322, 8/12/2010, p. 44.

[50] EMCDDA Early warning system – national profiles, European monitoring centre for drugs and drug addiction. Luxembourg: Publications Office of the European Union; 2012. Available: <http://www.emcdda.europa.eu/thematic-papers/ews>.

[51] Ramsey JD, Butcher MA, Murphy MF, Lee T, Johnston A, Holt DW, et al. A new method to monitor drugs at dance venues. BMJ 2001;323:603.

[52] Kenyon SL, Ramsey JD, Lee T, Johnston A, Holt DW. Analysis for identification in amnesty bin samples from dance venues. Ther Drug Monit 2005;27:793–8.

[53] Ramsey J, Dargan PI, Smyllie M, Davies S, Button J, Holt DW, et al. Buying 'legal' recreational drugs does not mean that you are not breaking the law. Q J Med 2010;103:777–83.

[54] Wood DM, Panayi P, Davies S, Huggett D, Collignon U, Ramsey J, et al. Analysis of recreational drug samples obtained from patients presenting to a busy inner–city emergency department: a pilot study adding to knowledge on local recreational drug use. Emerg Med J 2011;28:11–13.

[55] Dargan PI, Hudson S, Ramsey J, Wood DM. The impact of changes in UK classification of the synthetic cannabinoid receptor agonists in 'Spice'. Int J Drug Policy 2011;22:274–7.

[56] Home Office Annual report on the Home office forensic early warning system (FEWS). A system to identify new psychoactive substances in the UK. Home Office; 2012.

[57] Observatoire français des drogues et des toxicomanies (OFDT). Available : <http://www.ofdt.fr> [accessed 04.03.13]

[58] Cadet-Taïrou A, Gandilhon M, Lahaie E, Chalumeau M, Coquelin A, Toufik A, et al. Drogues et usages de drogues en France. État des lieux et tendances récentes 2007–09. St Denis: Neuvième édition du rapport national du dispositif TREND, OFDT; 2010. p. 281.

[59] Drug enforcement administration home – Department of justice. Available: <www.justice.gov/dea/> [accessed 04.03.13].

[60] National forensic laboratory information system (NFLIS). Available: <www.deadiversion.usdoj.gov/nflis/index.html> [accessed 04.03.13].

[61] United states of america department of justice, national drug intelligence center (NDIC) (closed on 15 June 2012, archives available: <http://www.justice.gov/archive/ndic/> [accessed 04.03.13].

[62] National institute on drug abuse (NIDA), the Community epidemiology work group (CEWG). Available: <http://www.drugabuse.gov/about-nida/organization/workgroups-interest-groups-consortia/community-epidemiology-work-group-cewg> [accessed 04.03.13].

[63] NIDA International Forum, Available: <http://international.drugabuse.gov/meetings/international-forum>; 2012 [accessed 04.03.13].

[64] EMCDDA First international multidisciplinary forum on new drugs, Available: <http://www.emcdda.europa.eu/events/2011/new-drugs-forum>; 2011 [accessed 04.03.13].

[65] Hall JN. Community-based surveillance of drug use provides adds valuable perspectives on trends. Available: <http://www.cicad.oas.org/oid/NEW/Information/Observer/08_01/community_nets.asp>; 2008 [accessed 04.03.13].

[66] Substance abuse and mental health services administration (SAMHSA). Available: <http://www.samhsa.gov/> [accessed 04.03.13].

[67] Drug abuse warning network (DAWN). Available: <http://dawninfo.samhsa.gov/>.

[68] Substance abuse and mental health services administration, center for behavioral health statistics and quality. Drug abuse warning network, 2008: National estimates of drug-related emergency department visits. HHS Publication No. SMA 11-4618. Rockville, MD; 2011.

[69] National drugs and alcohol research centre (NDARC). Available: <http://ndarc.med.unsw.edu.au/> [accessed 04.03.13].

[70] The ecstasy and related drugs reporting system (EDRS). Available: <http://ndarc.med.unsw.edu.au/project/ecstasy-and-related-drugs-reporting-system-edrs>.

[71] The illicit drug reporting system (IDRS). Available: <http://ndarc.med.unsw.edu.au/project/illicit-drug-reporting-system-idrs> [accessed 04.03.13].

[72] Illicit drug monitoring system (IDMS) Available: <http://www.shore.ac.nz/massey/learning/departments/centres-research/shore/projects/illicit-drug-monitoring-system.cfm>.

[73] Wilkins C, Sweetsur P, Smart B, Warne C, Griffiths R. Recent trends in illegal drug use in New Zealand, 2006-2011: findings from the 2006, 2007, 2008, 2009, 2010 and 2011 illicit drug monitoring system (IDMS). Social and Health Outcomes Research and Evaluation (SHORE), Massey University; 2011. Available: <http://www.shore.ac.nz/massey/fms/Colleges/College of Humanities and Social Sciences/Shore/reports/IDMS 2011 Final Report.pdf>.

[74] Wilkins C, Sweetsur P. The impact of the prohibition of benzylpiperazine (BZP) 'legal highs' on the prevalence of BZP, new legal highs and other drug use in New Zealand. Drug Alcohol Depend 2013;127:72–80.

[75] Wilkins C, Griffiths R, Sweetsur P. Recent trends in illegal drug use in New Zealand, 2006-2008: findings from the 2006, 2007 and 2008 illicit drug monitoring system (IDMS). Auckland: Social and Health Outcomes Research and Evaluation, School of Public Health, Massey University; 2009. Available: <http://www.shore.ac.nz/projects/2008IDMS Report.pdf>.

[76] Wilkins C, Griffiths R, Sweetsur P. Recent trends in illegal drug use in New Zealand, 2006-2009: findings from the 2006, 2007, 2008 and 2009 illicit drug monitoring system (IDMS). Auckland: Social and Health Outcomes Research and Evaluation, School of Public Health, Massey University; 2010. Available: <http://www.shore.ac.nz/projects/Final2009IDMS Report.pdf>.

[77] South african community epidemiology network on drug use (SACENDU): <http://www.sahealthinfo.org/admodule/sacendu.htm> [accessed 04.03.13].

[78] South african community epidemiology network on drug use (SACENDU). Monitoring Alcohol and Drug Abuse Treatment Admissions in South Africa. Available: <http://www.sahealthinfo.org/admodule/sacendu/SACENDUFullReport11a.pdf>; 2012 [accessed 04.03.13].

[79] SACENDU. Alcohol and drug abuse trends: January – June 2010 (Phase 28). Available: <http://www.sahealthinfo.org/admodule/sacendu/updatedec2010.pdf>; 2010 [accessed 04.03.13].

[80] World health organization (WHO), Expert committee on drug dependence (ECDD): <http://www.who.int/medicines/areas/quality_safety/ECDD/en/> [accessed 04.03.13].

[81] United nations commission on narcotic drugs (CND). Available: <http://www.unodc.org/unodc/en/commissions/CND/> [accessed 04.03.13].

[82] WHO Guidance on the WHO review of psychoactive substances for international control. Geneva: World Health Organisation; 2010. Available: <http://www.who.int/medicines/areas/quality_safety/GLS_WHORev_PsychoactSubst_IntC_2010.pdf> [accessed 04.03.13].

[83] UN ODC The global SMART programme (Synthetics monitoring: analyses, reporting and trends) Available: <http://www.unodc.org/unodc/en/scientists/smart.html> [accessed 04.03.13].

[84] UNODC. Patterns and trends of amphetamine-type stimulants and other drugs – Asia and the Pacific. Available: <http://www.unodc.org/documents/scientific/Asia_and_the_Pacific_2011_Regional_ATS_Report.pdf>; 2011 [accessed 04.03.13].

[85] UN ODC. Global smart update 2012. United nations office on drugs and crime 7, March 2012. Available: <http://www.unodc.org/documents/scientific/Global_SMART_Update_7_web.pdf>; 2012 [accessed 04.03.13].

[86] The inter-american observatory on drugs. Available: <http://www.cicad.oas.org/OID/MainPage/About%20Us.htm> [accessed 04.03.13].

[87] 2011 Annual report of the inter-american drug abuse control commission to the forty-second regular session of the General Assembly of the Organization of American States, 2011. Available: <http://www.cicad.oas.org/apps/Document.aspx?Id=1440> [accessed 04.03.13].

[88] Brunt TM, Niesink RJ. The drug information and monitoring system (DIMS) in the Netherlands: implementation, results, and international comparison. Drug Test Anal 2011;3:621–34.

[89] Keijsers L, Bossong MG, Waarlo AJ. Participatory evaluation of a Dutch warning campaign for substance-users. Health Risk Soc 2008;10:283–95.

[90] EMCDDA Assessing illicit drugs in wastewater. EMCDDA Insights 9, European monitoring centre for drugs and drug addiction. Luxembourg: Publications Office of the European Union; 2008. Available : <http://www.emcdda.europa.eu/publications/insights/wastewater> [accessed 04.03.13].

[91] Thomas KV, Bijlsma L, Castiglioni A, Covaci A, Emke E, Grabic R, et al. Comparing illicit drug use in 19 European cities through sewage analysis. Sci Total Environ 2012;432:432–9.

[92] Zuccato E, Chiabrando C, Castiglioni S, Bagnati R, Fanelli R. Estimating community drug use by sewage analysis. Environ Health Perspect 2008;116:1027–32.

[93] Archer JR, Hudson S, Wood DM, Dargan PI, Analysis of urine from pooled urinals - a novel method for the detection of novel pyschoactive substances. Current Drug Abuse Reviews 2013, In press.

[94] Cronin MT, Jaworska JS, Walker JD, Comber MH, Watts JS, Worth AP, et al. Use of QSARs in international decision-making frameworks to predict health effects of chemical substances. Environ Health Perspect 2003;111:1391–401.

[95] Mohd-Fauzi F, Bender A. Computational analysis of the possibility of Ostarine eliciting psychoactive effects. : University of Cambridge; 2012.

[96] Bender A, Glen RC. Molecular similarity: a key technique in molecular informatics. Org Biomol Chem 2004;2:3204–18.

[97] Dal Cason TA, Young R, Glennon RA. Cathinone: an investigation of several N-alkyl and methylenedioxy-substituted analogs. Pharmacol Biochem Behav 1997;58:1109–16.

[98] Wood DM, Conran P, Dargan PI. ICD-10 coding: poor identification of recreational drug presentations to a large emergency department. Emerg Med J 2011;28:387–9.

[99] Shah AD, Wood DM, Dargan PI. Survey of ICD-10 coding of hospital admissions in the UK due to recreational drug toxicity. QJM 2011;104:779–84.

[100] EMCDDA. Hospital and Emergency services data. EMCDDA contract code: CT.08.EPI.042.1.0, Final report. 2009.

[101] Schifano F, Deluca P, Agosti L, Psychonaut 2002 Research Group New trends in the cyber and street market of recreational drugs? The case of 2C-T-7 ('Blue Mystic'). J Psychopharmacol 2005;19:675–9.

[102] Psychonaut Web Mapping Research Group Psychonaut web mapping project: final report. London UK: Institute of Psychiatry, King's College London; 2010. Available: <http://www.psychonaut-project.eu/technical.php>.

[103] Davey Z, Schifano F, Corazza O, Deluca P, on behalf of the Psychonaut Web Mapping Group e-Psychonauts: conducting research in online drug forum communities. J Ment Health 2012;21:386–94.

[104] Hillebrand J, Olszewski D, Sedefov R. Hallucinogenic mushrooms: an emerging trend case study. European Monitoring Centre for Drugs and Drug Addiction; 2006. Available: <http://www.emcdda.europa.eu/publications/thematic-papers/mushrooms> [accessed 04.03.13].

[105] Hillebrand J, Olszewski D, Sedefov R. GHB and its precursor GBL: an emerging trend case study. European Monitoring Centre for Drugs and Drug Addiction; 2008. Available : <http://www.emcdda.europa.eu/publications/thematic-papers/ghb> [accessed 04.03.13].

[106] Hillebrand J, Olszewski D, Sedefov R. Legal highs on the internet. Subst Use Misuse 2010;45:330–40.

[107] EMCDDA Online sales of new psychoactive substances/'legal highs': summary of results from the 2011 multilingual snapshots: European Monitoring Centre for Drugs and Drug Addiction; 2011. Available: <http://www.emcdda.europa.eu/publications/scientific-studies/2011/snapshot> [accessed 04.03.13].

[108] Gallup Organisation. Youth attitudes on drugs, Flash Eurobarometer 330. 2011. Available : <http://ec.europa.eu/public_opinion/flash/fl_330_en.pdf>.

[109] Smith K, Flatley J. Drug misuse declared: findings from the 2010/11. British Crime Survey of England and Wales: Statistical Bulletin, UK Home Office; 2011.

[110] Anonymous Drug misuse declared: findings from the 2011/12 crime survey for England and Wales. Statistical Bulletin, UK Home Office; 2012.

[111] NACD PHIRB Drug use in ireland and northern ireland: first results from the 2010/11 drug prevalence survey: National Advisory Committee on Drugs and Public Health Information and Research Branch; 2011. Available: <http://www.nacd.ie/index.php/publications/29-drug-use-in-ireland-and-northern-ireland-first-results-from-the-20102011-drug-prevalance-survey-bulletin-1.html>.

[112] PNSD. Emerging drugs. Comisión clínica de la delegación del gobierno para el plan nacional sobre Drogas. 2011. Available: <http://www.pnsd.msc.es/Categoria2/publica/pdf/DROGAS_EMERGENTES_ingles_WEB.pdf>.

[113] NIDA's 2011 Monitoring the future survey. Available: <http://www.monitoringthefuture.org> [accessed 04.03.13].

[114] Wilkins C, Girling M, Sweetsur P, Huckle T, Huakau J. Legal party pill use in New Zealand: prevalence of use, availability, health harms and 'gateway effects' of benzylpiperazine (BZP) and triflourophenylmethylpiperazine (TFMPP). : Centre for social and health outcomes research and evaluation (SHORE), Massey University; 2006.

[115] Wilkins C, Griffiths R, Sweetsur P. Recent trends in illegal drug use in New Zealand, 2006–2009. : Illicit Drug Monitoring System (IDMS); 2010. July 2010.

[116] Winstock A. The 2011 Mixmag drugs survey. Mixmag 2011:49–59.

[117] Wood DM, Hunter L, Measham F, Dargan PI. Limited use of novel psychoactive substances in South London nightclubs. QJM 2012 doi:10.1093/qjmed/hcs107.

[118] NACD An overview of new psychoactive substances and the outlets supplying them. : National Advisory Committee on Drugs; 2011. Available: <http://www.nacd.ie/index.php/publications/38-an-overview-of-new-psychoactive-substances-and-the-outlets-supplying-them.html>.

[119] 2011 National report (2010 data) to the EMCDDA by the Reitox national focal point. The Czech Republic.

Availability and Supply of Novel Psychoactive Substances

Silas W. Smith[*],[†] *and Fiona M. Garlich*[**],[‡]

[*]Department of Emergency Medicine, New York University School of Medicine, New York, NY
[†]New York City Poison Control Center, New York, NY
[**]University of Calgary, Calgary, Canada
[‡]Poison and Drug Information Service of Alberta, Calgary, Canada

BACKGROUND

Numerous factors hinder the adequate characterisation of the availability and supply of novel psychoactive substances (NPS). Diverse substances – botanicals; synthetic cannabinoid receptor agonists (SCRAs); cathinone analogues; pyrrolidinophenones; substituted phenylethylamines; 'FLYs;' piperazines; aminoindanes; and analogues of tryptamines, arylcyclohexylamines, opioids, cocaine, and gamma-hydroxybutyrate – compete in a shifting market space, defying generalisations. 'Supply side' analysis is complicated by ongoing market entries, including compounds, suppliers, distributors and vendors. Recent years have witnessed an exceptional surge of entirely new psychoactive substances. Online platforms for NPS transactions have flourished, complimented by increasing product diversity [1,2].

Evolving usage patterns in various nationalities or communities ('demand side'), coupled with assorted market stressors, ensure a constantly shifting landscape. NPS compete with other established abused substances for selection, as well as for available interventions and treatment resources. User experiences, perceptions of harm and purchasing sources are rapidly disseminated via various social media. The legal, administrative and regulatory environments continue to evolve at local, state, national and transnational levels, often in a reactionary manner. Medical or forensic analysis, characterisation, detection and measurement in patient biological specimens or product samples are complicated by deficiencies in widely available, validated methodologies or established reference standards.

Further adding to the difficulties of NPS market assessment is the fact that much NPS development, production and introduction is unlinked from traditional avenues and pressures. Marijuana, cocaine, heroin, khat, peyote and psilocybin mushrooms (as well as nicotine and grain or fruit sources for ethanol production) require some degree of cultivation or rendering. Unlike ketamine, or prescribed opioids, benzodiazepines, or amphetamines analogues, numerous NPS do not possess the legal exposure derived from diversion from legitimate sources. Similar to amphetamine-type stimulants (ATS) and synthetic opioids, many NPS are completely divorced from botanical origins, requiring only knowledge, materials, standard laboratory equipment and a relatively brief production cycle. Additionally, NPS remain largely free from the national and international efforts targeting precursor chemicals and ATS and non-pharmaceutical opioid reagents (e.g., The Combat Methamphetamine Epidemic Act of 2005, 1988 United Nations Convention Against Illicit Traffic in Narcotic Drugs and Psychoactive Substances, Drug Enforcement Administration (DEA) fentanyl chemical precursor scheduling, etc.) [3–5].

Classical data sources struggle to address the NPS hyperkinetic market. Inherent delays in the conduct of research, (peer) review and production limit journal publications, books and reports. The surreptitious 'shadow economy' surrounding NPS and the absence of accessible, complete, recorded sources preclude a conventional 'market research report' analysis. Traditional databases assessing usage and demand often do not include NPS or update recent introductions with adequate frequency. For example, through July 2012, usage statistics for kratom (*Mitragyna speciosa*), SCRAs, cathinones, piperazines, or ring-substituted phenylethylamines were unavailable in recent US Drug Abuse Warning Network (DAWN), National Survey on Drug Use and Health, or the Youth Risk Behavior Surveillance Systems databases [6–8]. Drugs of abuse classifications often maintain broad, historical categories inapplicable to NPS. The 2012 DAWN report of drug-related emergency department visits, summarising 2010 data, continued to employ such traditional categorisations [7]. SCRA data were unaddressed until the 2011 Monitoring The Future study (published 2012); cathinones

and other NPS were absent [9]. DEA drug seizures, mostly recently available for 2010, reflect only cocaine, heroin, marijuana, methamphetamine and 'hallucinogens' [10]. Available European (EMCDDA) seizure data, mostly recently from 2009, are similarly constrained by category limitations [11]. While the US National Poison Data System (NPDS) identified emerging trends for SCRAs and cathinones in 2009, formal publication (non-blogs or press releases) occurred in 2011 [12]. Other flexible data systems and organisations have fallen victim to economic constraints. The US National Drug Intelligence Center ceased operation of its Synthetic Drug Early Warning and Response System (SENTRY) on November 1, 2011 and itself folded on June 15, 2012.

Established US government initiatives which purchase or assess illicits to determine supply, street value and purity (e.g. the US Heroin-, Cocaine- and Methamphetamine Domestic Monitoring Programs) often omit NPS. Governmental public educational resources, which could be analysed for access patterns or queries, have also disregarded them. For example, despite the well-established US presence of SCRAs and cathinones [13], as of 1st July 2012, they were absent from the US National Institute on Drug Abuse's 'Drugs of Abuse' and Substance Abuse and Mental Health Services Administration's 'Substances' websites [14,15].

Linguistic diversity further challenges research. Obfuscation to circumvent potential legal restrictions has created a myriad of NPS descriptions – 'legal high,' 'research chemical,' 'synthetic,' 'designer drug,' 'herbal,' 'incense,' 'natural,' 'spice,' 'organic,' 'smoking blend,' 'potpourri,' 'traditional medicine,' 'bath salt,' 'party pill,' 'plant food,' 'room odouriser,' 'jewellery cleaner,' 'glass cleaner,' 'pond water cleaner,' 'toilet bowl cleaner,' 'natural stain remover,' 'insecticide,' 'insect repellent,' 'novelty collector's item,' 'collectible,' and other products 'for use in laboratory experiments' or 'not for human consumption.' 'Brand' names for SCRAs and cathinones, as well as chemical common names and acronyms (2-DPMP, desoxypipradrol; 4-FMC, 4-fluoromethcathinone; DMC, dimethocaine; etc.) number in the hundreds. Product mixtures (intentional or unintentional), patient co-ingestants or drug 'cocktails,' and adulterants or contaminants all obscure data gathering.

While non-traditional sources – news articles, Internet sources (discussion forums, blogs, 'information' sites, sellers), drug seizures and case reports indicative of NPS sources – can provide potential information, they lack definitive, stringently evaluable data. With these difficulties in mind, we attempt to provide an overview of NPS availability and supply.

NOVEL PSYCHOACTIVE SUBSTANCE MARKET

What is the potential or apparent NPS market? In one study, two suggested methods were applied to estimate a national illicit market using reported consumption and reported expenditure [16]. The first evaluated demographic groups and adjusted for usage rates (per group), frequency of use, quantity per use and cost per substance unit to calculate total expenditure. The second required expenditure data for a sampled population, subsequently appropriately extrapolated to a larger population of interest. The aforementioned data limitations, including the lack of assessments of type, frequency or amount of NPS use, as well as deficits in price and purity datasets, preclude effective use of these established methodologies. The 'unknown unknowns' (unrecognised NPS, markets, or demographics) further limit precision.

Certain demographic data, sales of previously legal NPS and seizures can provide a preliminary analytic basis. UNODC 2012 World Drug Report annual user estimates for the major drug markets – citing 2010 data – were

14.3–52.5 million for ATS, 13.2–19.5 million for cocaine, 10.5–28.1 million for 'ecstasy', 119–224 million for cannabis, 13.0–21.0 million for opiates and 26.4–36.1 million for opioids [17]. The DEA estimate of the US domestic illicit cannabis market is $35 billion [18]. These represent potential areas for displacement by NPS such as SCRAs, cathinones, piperazines and others. SCRA users are usually young and demographic displacement is evident in studies of current undergraduate students, with usage rates of 8% or higher [19–21]. Similar studies implicate significant increased *Salvia divinorum* (salvia) market penetration in educational settings (exceeding 4%), overall 18–25 year-old demographics (exceeding 6%) and in pre-existing substance abusers (exceeding 20% depending on the substance) [22–24]. Cathinones demonstrated extensive market infiltration in certain populations. Pre-control, mephedrone (4-methylmethcathinone) achieved a 20.3% prior use rate in UK college/university students [25] and significant use in the club market [26].

When legal in New Zealand, 1-benzylpiperazine (BZP) market data included: a market penetration of almost 50% among 20–24 year-olds (ever-users); an excess of 120 brands; estimated yearly sales of NZ$24–50 million and 1 800 000 doses (150 000/month); and a median user consumption of 400 mg per occasion [27–29]. UK BZP sales were reportedly 2 million doses from 2002 to 2006, approaching 3600 daily dose shipments by one supplier [30]. The estimated size of UK NPS market had already reached £10 million in 2006 [31]. A US-based NPS trade association estimates a market with annual taxable revenues of $4–6 billion, based on self-reported members' NPS sales statistics [32]. A single manufacturer selling 41 000 SCRA packets monthly recovered a six-month profit of $500 000 on revenues of $2.5 million [33].

The potential scope and sophistication of Internet drug operations was revealed in April 2005 by 'Operation Cyber Chase,' the DEA's first investigation of no-prescription websites.

It disrupted a multi-national organisation with *monthly* distributions of 2.5 million dosage units of controlled substances, 100 websites, 40 bank accounts, with seizures of $8.5 million, 10 million dose units of controlled pharmaceuticals and 231 pounds of ketamine [34]. Several months later 'Operation CybeRx' targeted an illegal Internet-based enterprise with $50 000 in *daily* profits.

Drug seizure and interdiction efforts also confirm NPS market capture. BZP rose rapidly from less than 10 mentions in 2006 to become the ninth most frequently identified drug in 2009 in the US National Forensic Laboratory Information System (NFLIS) [35]. While a decline occurred in 2010, mirroring European trends reported by the UK Forensic Science Service due to cathinone displacement, BZP still accounted for over 4% of stimulant reports (equivalent to amphetamine) [36,37]. NFLIS also documented the rapid increase and geographic variation of SCRA and cathinone seizures in the USA, which now number in the thousands [13]. Cathinones have dispersed throughout the USA since 2009. In 2011, the DEA noted mephedrone in 32 states and MDPV (3,4-methylenedioxypyrovalerone) in 34 states and NPDS documented extensive national cathinone exposures [12,38,39]. European nations have reported confiscation of mCPP tablets since 2005 (Hungary), BZP since 2006 (Malta), bromo-dragonfly blotters and cathinones since 2007 (Norway) and SCRAs ('Spice') since 2009 (Estonia, Finland) [11].

NPS ORIGINS AND MARKET ENTRY

Many NPS have conceptual roots in legitimate scientific efforts exploring and developing neuroactive or neuropharmacological substances. Potential chemicals, analogues and published synthesis pathways are widely available simply by perusing recent or prior medicinal chemistry journals or medical literature, patent applications, catalogues of compounds never commercially marketed, or the Internet. Some SCRAs were procured from academic and industry research originally aimed at elucidating cannabinoid receptor properties and development of novel anti-inflammatories and analgesics. Recognition of the prevalent natural cannabinoids in *Cannabis* species led to synthesis of tetrahydrocannabinol analogues such as Hebrew University's HU-210 in the 1960s [19]. Pfizer's cyclohexylphenol series (CP-47,497 and others) were developed later in the 1980s for potential antidepressant properties. Research into the structure, functionality and receptor bioactivity of endogenous and synthetic cannabinoids by John W. Huffman extended the classes of agents active at the CB_1 and CB_2 receptors [40,41]. Appropriation of published research methods by underground chemists lead to the synthesis and mass-distribution of JWH-series and other SCRAs in 'Spice' blends [19,42].

David E. Nichols and colleagues have published abundantly on the synthesis and potential use of a variety of psychoactive chemicals. Research into the structure-activity relationship of hallucinogens, with the goal of furthering their legitimate use as psychotherapy adjuncts and psychiatric medications, resulted in the development of numerous compounds. The 'FLYs' such as 'bromo-dragonFLY,' 1-(8-bromobenzo[1,2-b;4,5-b']difuran-4-yl)-2-aminopropane and '2-CB-FLY,' 2-(8-Bromo-2,3,6,7-tetrahydrofuro[2,3–f][1]benzofuran-4-yl)ethanamine, were synthesised in the 1990s at Purdue as a research chemicals probing rat serotonin receptors [43,44]. The aminoindaines (5-IAI, 5-iodo-2-aminoindane; MDAI, 5,6-methylenedioxy-2-aminoindane; etc.) were also explored as part of their research work aimed at understanding 3,4-methylenedioxymethamphetamine (MDMA), particularly in drug discrimination studies [45]. Their similarity to known hallucinogens such as MDMA and lysergic acid diethylamide (LSD), as well as their serotonin and dopamine receptor activities and behavioural effects in animals are extensively documented. Amateur chemists exploited Nichols' work in the late 1990s, using published methods as a synthesis 'roadmap' [46,47]. MTA (4-methyl-thioamphetamine) was subsequently distributed in 'flatliners' tablets, which were associated with several fatalities [47,48]. Bromo-dragonFLY appeared as a drug of abuse in Scandinavia around 2006 and has been associated with significant toxicity in both Scandinavia and the UK [49,50]. The aminoindaines are potential 'next wave' NPS [45,51].

BZP was originally evaluated as a potential antidepressant in the 1970s, but retained an abuse liability similar to amphetamine and was not pursued for this indication [52]. BZP subsequently entered the market, particularly early in the last decade in New Zealand, as a legal 'party pill' until legislation changes in 2008 [29]. Another piperazine derivative, meta-chlorophenylpiperazine (mCPP), was used extensively to evaluate serotonin function in vitro, in animal and in human psychiatric research dating back to the 1980s [53]. mCPP entered the Dutch Internet and retail ('head shop') market as an MDMA mimetic in 2004 [54]. mCPP market consolidation has increased to the point where visually distinguishing mCPP-containing tablets from regular MDMA is impossible [55].

Other NPS followed a pattern of re-introduction and promotion as new 'designer drugs' despite knowledge of their existence and synthesis pathways for years. This is analogous to the 1980s eruption in the use of MDMA, first created in 1912 by Merck Darmstadt while attempting to produce an appetite suppressant [56]. Of the three synthetic cathinones recently temporarily scheduled by the DEA, mephedrone synthesis has been known since 1929, MDPV synthesis was described in a Boehringer Ingelheim 1969 patent (USP 3,478,050) for use as a CNS stimulant and methylone was patented in 1996 (PCT/US06/09603) through research in anti-depressant and anti-Parkinson agents [57–59]. Methylone surfaced in the Dutch market

as early as 2004 [54] and continues to supplant MDEA (3,4-methylene-dioxyethylamphetamine) and MDA (3,4-methylene-dioxyamphetamine) [60]. Pipradrol was explored as 'a new stimulant drug' in the mid-1950s [61], and other pipradrol class agents such as desoxypipradrol (Ciba USP 2,820,038) were patented later in that decade. They later reappeared after cathinone control measures. Methylhexanamine (dimethylamylamine, DMAA) was trademarked in 1948 as Forthane by Eli Lilly and Company and patented in 1971 for gingival hypertrophy (USP 3,574,859). It emerged as an NPS in Ireland following 2010 legislative action [62] and has been most recently banned as a 'party pill' ingredient in New Zealand [63].

NPS development and manufacturing efforts are aided by those with scientific training who develop or scan research literature for novel structures or derivatives [46]. Substituted phenylethylamines and tryptamines were explicitly described in the 1990s by synthetic chemist Alexander Shulgin in the books *PiHKAL* and *TiHKAL*, which maintain an online presence [64,65]. The first volume of *The Shulgin Index*, cataloguing psychedelic phenethylamines, amphetamines, phenylpiperazines and others, which includes synthesis and pharmacological properties, was released in 2011 [66]. Methoxetamine (MXE), an arylcyclohexylamine derivative, allegedly was conceived by an underground chemist trained in biochemistry and neuropharmacology, aided by his familiarity with ketamine [67]. Leading compounds can be further expanded using a variety of well-established medicinal chemistry methods. For example, ongoing analysis of Internet purveyed compounds identified novel halogenated cathinone analogues (2-, 3- and 4-fluoromethcathinones) in 2009 [68]. This diversification thwarts legislative or administrative control efforts, as alternative products are already available, awaiting release in response to regulatory environments [18,46]. This was evident in the speed of SCRA replacement products following their initial ban in Germany [19].

It also contributes to market volatility, as products never tested or without established use in humans are widely distributed, essentially experimenting 'on the fly' in human populations, with potentially unpredictable results in toxicity (and popularity). The NPS developmental model has driven the number of new psychoactive substances reported to the EMCDDA Early Warning System to extraordinary levels – from 13 substances in 2008, 24 in 2009, 41 in 2010, to 49 in 2011 [69,70]. Similar diverse NPS introduction of tryptamines, phenethylamines, cathinones, piperazines, SCRAs and psychotropic plants (including voacanga alkaloids) has occurred in Asian markets [71].

NPS MARKET STRUCTURE

Supply Chain: Manufacturing and Distribution

The supply chain of novel psychoactive substances comprises an intricate web that joins research chemists, underground labs, large-scale industrial manufacturers and a vast, complex marketing and distribution network. Unlike opium poppy or coca cultivation, ATS and most NPS production are geographically unlimited; all areas of the world participate with increasing frequency [17]. As detailed above, drug development and manufacturing efforts may be spearheaded by entrepreneurs who hire synthetic chemists or pharmacologists to aid in the development of novel structures to accentuate or mimic known drugs effects [46]. Initial synthesis often occurs in clandestine, but well-funded laboratories (initially primarily in Europe, but also in the USA and elsewhere) [46]. For example, US Internet supply of 2C-T-7 [2,5-dimethoxy-4-(n)-propylthiophenethylamine] was provided by an Indiana company, which itself was supplied by a clandestine laboratory in Las Vegas, Nevada [72]. However, despite domestic production of some

substances, mass production may require large-scale manufacturing facilities that are not economically or legally feasible in Europe or the USA. Thus, China and neighbouring countries in South East Asia have emerged as a primary source of NPS [46,67,73].

The involvement of Chinese chemical manufacturing plants in the NPS supply chain takes several forms. Chinese plants may synthesise and ship the final product, or they may ship intermediary chemicals to facilities where final synthesis, packaging and distribution occur [33,46,67,74]. The former model appears prevalent. Web-based export directories facilitate acquisition of potential raw or finished products by suppliers and distributors (and users). In June 2011, 3800 Chinese laboratories were selling JWH products online [33]. Many offered custom orders and mass-production capacities. One claimed monthly shipment capabilities of 5000 kg of JWH-019. Individual chemists or entrepreneurs in Europe or the USA may contract with Chinese chemical plants to synthesise desired substances at relatively low cost, which are then imported via multiple mechanisms in sacks or drums [73]. An SCRA manufacturer's model is illustrative: it hired a chemical laboratory to formulate cannabinoid 'recipes' and a Chinese manufacturer to synthesise the substances [33]. These 'special additives' were imported, dissolved, spread over vegetation on large drying trays and then measured, packaged and sealed into branded foil packets. Packets were delivered via UPS to stores and wholesalers, who in-turn re-distributed the packets or sold them online. Custom batches containing different substances were prepared, depending on state-specific SCRA legal designations. In 2011 a myriad of synthetic drug distributors were estimated in the USA, many with distinct brands [33,59].

Shipment labelling varies greatly. Chemical contents are sometimes accurately disclosed, but many shipments are unlabelled or mischaracterised [73]. The identification of miscellaneous research chemicals, incorrectly labelled powders and substances of evolving legal designation challenges border agencies attempting to control the flow of illicit substances. US Customs and Border Protection has encountered mephedrone, methylone, MDPV and 4-MEC (4-methyl-N-ethylcathinone) and other cathinones such as butylone, fluoromethcathinone and dimethylcathinone, originating primarily from China and India [59]. Most cathinone seizures in New Zealand in 2009 originated from the UK or directly from China [75]. Supply and distribution of botanical NPS such as salvia and kratom are aided by the additional flexibility of multiple primary cultivation sites and supplemental user self-propagation [76].

Sale of NPS

Once synthesised, either abroad or in domestic laboratories, NPS are distributed to the end user, often via a complex chain of second- and third-party resellers. Ultimately, users obtain NPS via three primary vending modes: online, via website distributors; retail vendors with physical storefronts; and non-retail vendors such as family, friends, associates, or dealers who distribute their wares at concerts, in clubs, or on the street. The interactions of buyers, sellers and resellers may result in intermingled distribution networks.

Internet

The Internet provides a key exchange platform for NPS information and increasingly, for NPS distribution. Numerous websites, chat-rooms, forums and instant messaging resources are devoted to drug-related information and experiences [77–82]. Web-based NPS sources influence a range of drug-use behaviours, including the use of new drugs and drug combinations, modifications of preferred drugs and cessation [77,81]. Indeed, the Internet is the primary source of illicit drug information for 15–24 year-olds [83]. 70% of American 12–17

year-olds visit social networking sites daily, a practice associated with a greater ease in obtaining illicit drugs and controlled prescription drugs [84]. Erowid [85], Dancesafe [86], Multidisciplinary Association for Psychedelic Studies [87], Bluelight [88], Lycaeum [89], The Shroomery [90] and other websites have figured prominently in disseminating NPS information [78,79].

The Internet has evolved from a static information source to a dynamic, interactive means of illicit drug acquisition and dispersal [91]. In retrospect, Dutch companies had used the Internet to sell cannabis seeds and derivatives since 1996 [92]. In 1997, the International Narcotics Control Board (INCB) noted that the Internet was providing an exchange forum for advice on illegal drug use and manufacture [93]. Online sales of diverted pharmaceuticals expanded greatly following the first online pharmacy opening in the UK in 1999 [94]. By 2001, the INCB identified the Internet as a growing source of drug trafficking and it had become the most widely used mechanism for expanding synthetic drug production in some countries [92]. Scheduled and prescription psychiatric medications, analgesics and CNS stimulants became extensively available online throughout the ensuing decade [95,96]. The DEA currently recognises Internet substance purchases for non-medical indications as an entrenched drug consumer practice [97].

The NPS Internet market space now reflects the modern, demand-based approach describing legitimate corporate enterprise – providing solutions, information, value and access ('SIVA') [98]. Effective NPS transnational marketing, sales and distribution have all been facilitated by the Internet [51,70,92,99]. The burgeoning demand for 'legal highs' resulted in an enormous increase in psychoactive substance online transactions [74]. By removing the requirement for personal interactions with drug dealers, the Internet emerged and succeeded as an alternative avenue for NPS acquisition [1,2,51].

Internet purchase capability was reported by some as a significant factor that led users to consume 'legal highs' [100]. The relative anonymity, personal safety and novelty of purchasing psychoactive substances appealed to many. Despite methodological limitations, a 2010 survey detailed that the vast majority (92%) of legal high products were obtained online [100]. However, a subsequent, more global survey suggested a lower rate (24%) of overall Internet drug sourcing among users [101].

The Internet now significantly substantiates NPS sales and distribution [102]. Specific and actionable information on how to obtain a vast range of hallucinogens has been available online for over a decade [82]. In 2005–6, 'salvia divinorum' employed as a search term yielded websites offering sale, distribution, or links to sites that did so 58% of the time [103]. In 2010, 33% of one group of European mephedrone users reported web purchase, although this decreased to 1% following legislative restrictions the following year [26,100]. In 2011, 38% of worldwide 'Spice' users purchased online [21]. Of those consuming (still) legal highs in 2012, 45% bought online, compared to 22.5% dealer purchases [104]. Wide online availability of bromo-dragonFLY, kratom and psychoactive plant products is established [71,105,106]. In Japan, multiple psychoactive substances, known as 'dap' or 'ibo drugs' are easily procurable online [71,105]. Prior to 2007, tryptamine derivatives dominated this market. When these became designated substances, they were replaced by cathinones, phenylethylalamines, piperazines and SCRAs, all sold over the Internet [71].

The vast size and ever-changing content of the Internet precludes continuous monitoring of drug-related websites. Thus, sequential 'snapshots' are often used to provide insight into NPS availability and trends. In 2008, the EMCDDA investigated online availability of 'legal high' drug retailers in Europe [2]. 69 EU-based online retailers were identified, the vast majority of which were located in the UK (52%) and the Netherlands (37%). More than 500 products were offered, the most common of which were salvia, kratom, Hawaiian baby woodrose, magic mushrooms, 'party pills' and 'smoking blends' (SCRAs). In 2009, 39 online legal high retailers based in the UK or selling to the UK public were identified [30] with 346 unique products advertised: an average of 16 items per site, similar to those reported by the EMCDDA. Later investigations revealed that online vendors providing EU sales and shipments increased from 170 online shops in January 2010 to 631 in July 2011, with the number of US-based sites tripling [1]. Kratom, salvia, hallucinogenic mushrooms, MDAI, MXE, 5- and 6-APB [5- and 6-(2-aminopropyl) benzofuran], 4-MEC, MDPV, mescaline, methioporpamine, 5-IAI and dimethocaine led the evolving NPS Internet market, with SCRAs and Hawaiian baby woodrose less widely available. 2011 snapshots revealed that the UK and the USA were the predominant online vendors' apparent country of origin, followed by the Netherlands, Germany, the Czech Republic and Hungary [102]. The primary transaction and interface language was English (83%). Despite the high percentage of American online vendors, many focus their sales efforts on the EU, listing prices in Euros or British pounds. This may reflect European consumers' relative comfort with purchasing drugs online, compared with Americans [46].

Sites may be straightforward about NPS vending and intent, or may imitate other enterprises, such as 'gardening sites' with images of foliage, despite selling 'plant food' products such cathinones, aminoindanes or botanical NPS [107]. Purchase and delivery practices advertised online are designed to be easy and discrete, although some online shops invoke destination restrictions. Vendor shipping policies promote circumspect packaging designed to avoid suspicion [30]. Some vendors include same-day or even more rapid delivery options (e.g. within 90 minutes to a London address for £95) [74].

Online NPS availability is variable, frequently-changing and highly dependent on market pressures, evolving legislation and the demographic characteristics of certain population sectors. For example, the online sale of mephedrone in Europe showed dramatic growth in 2008–2010. In March 2008 less than 10 online mephedrone vendors existed. By June, new sites appeared weekly [74]. The percentage of mephedrone-using UK students reporting Internet purchase prior to the April 2010 UK control legislation was 10.7% [25]. As legislation shifted, online mephedrone sourcing fell from 33% to 1%, and street-level sales dramatically increased [26]. New products quickly filled the online void: other cathinone analogues, aminoindanes and MXE [1].

In addition to online retailers with websites directly purveying NPS, the Internet supports several additional market spaces. NPS sales have extended to social networking sites, though the extent of market penetration is difficult to completely assess. NPS vendors are reported to increasingly exploit Twitter® and Facebook® [108]. Social media's safety and efficiency are appreciated by dealers due to consistent and persistent buyer access and discreet buyer selection. Conversely, law enforcement officials have accessed social media to monitor and disrupt drug-dealing rings, which discuss and sometimes advertise their drugs in various posts [109]. Allegedly, online media sites unintentionally fuelled NPS sales because web advertising programs linked press reports containing 'keyword' mentions of NPS to vendors via automatically generated ads; this provided an instant mechanism for potential NPS acquisition and market expansion [110]. Despite prohibitions by administrators, dealers broker NPS through popular online auction sites, which can facilitate the sale of numerous abusable substances [111]. Auction sites (such as eBay®) were reported to provide the primary US sale

and purchase market for poppy pods (crushed and brewed into tea), until the practice was halted in 2009 [112]. Users report ongoing auction purchases of nitrous oxide, smoking blends, 2-C series, MDPV and mephedrone [113]. The authors uncovered current auction site vendor advertisements for multiple NPS 'research chemicals,' with promised next-day local delivery and international shipping [114]. Classified advertising also offers the opportunity to 'hide in plain sight.' The authors also found innumerable examples of international, national and local vendors presenting NPS in 'Home/Garden,' 'Health/Beauty,' and even 'Traditional medicine' sections. Vendors may also provide raw chemical components, associated production or usage products (packaging materials, scales, etc.), or links to sellers. Additionally, Internet vendors may employ 'hidden' or 'underground websites', such as 'Silk Road,' accessible only through anonymous 'Tor' (the onion router) networks. These transmit traffic through privately built pathways in international volunteer server networks in order to obscure traffic analysis and detection. Transactions occur via virtual 'Bitcoin' 'cryptocurrency,' to further obscure identities and operations [115].

Retail Vendors

While sales and distribution via the Internet is common, many NPS sales are via retail, 'brick and mortar' establishments. 'Head shops,' 'smoke shops,' and 'smart shops,' stores that cater to a drug-using market by selling paraphernalia and counter-culture consumer products, are a common source of 'legal highs.' While terminology overlaps, 'head shop' generally refers to stores selling smoking paraphernalia (pipes, bongs, rolling paper, etc.), while 'smart shops' sell herbals and psychoactive substances. Smart shops are more common in the Netherlands, where psychoactive mushrooms are sold legally [116]. Gas stations and convenience stores, which often sell 'energy' products alongside snacks and soft drinks, have also become an NPS outlet, especially in the USA and New Zealand. In Japan, many 'dap' or 'ibo' drugs are purchased in video stores [71]. Head shops are popular among young people, where the lack of age restriction and the ability to make purchases without a credit card makes psychoactive substances easily available [117].

Novel psychoactive substances surfaced in smart shops in the late 1990s. It appears that many drugs debuted in the Dutch market and were subsequently dispersed throughout Europe. For example, the emergence of synthetic phenylethylamines was observed first with 2C-B (4-bromo-2,5-dimethoxyphenethylamine) in 1994, 2C-T-2 [2-[4-(ethylthio)-2,5-dimethoxyphenyl]ethanamine] and 2C-T-7 in 1997 and a branded version of 2C-T-7 ('Blue Mystic') in 2000 [118]. 2C-T-7 later appeared in Germany [119]. BZP had become increasingly popular in New Zealand since its arrival in 1999, where it was sold extensively in retail stores [120]. European piperazine introduction occurred around 2000 [121]. Reports of synthetic cathinone (e.g., mephedrone) use began occurring in 2009 in Europe, while the first cases of exposures to "bath salts" were reported to US Poison Centers in 2010 [122]. The introduction of NPS into the US retail market has tended to lag behind Europe.

Assessment of the NPS retail market is difficult due to a paucity of published data regarding locations and sales figures for head shops and other stores in this category. While the number of shops selling NPS is unknown, there are likely to be hundreds or thousands of retail vendors in each country. In 2006, 200 head shops filled the 'high street' of Bournemouth alone, a city of less than 200 000 [31]. In the USA, 2446 head shops were listed in one directory [123], and 2423 were listed in another [124]. In Poland in 2010, 3500 head shops were raided following new legislation outlawing the sale of 'legal highs,' resulting in the closure of 1200 [125]. Closure of head shops in Poland

resulted in increased online sales [1]. A police inventory indicated that at their peak in early 2010, there were 113 head shops in Ireland, with weekly shop openings in certain regions [126]. The number of shops did decrease markedly to less than 10 following the ban of piperazine derivatives, SCRAs and cathinones [126]. However, similar to the Internet evolution of available products in response to governmental action, head shops quickly replaced outlawed compounds with dimethylcathinone, naphyrone, fluorotropacocaine, desoxypipradrol and dimethylamylamine [62].

Users provide further estimates of retail utilisation. In one online survey of 168 SCRA users, 87% reported purchasing it from retail vendors [21]. In another study, 13.9% of urine samples from patients at Ireland's Drug Treatment Centre Board were positive for head shop products (mephedrone, methylone and BZP) [127]. 7% of young people attending adolescent addiction services centres in Dublin reported using head shop products as part of a multidrug cocktail [128]. Annual surveys of UK club-goers in 2011 reported mephedrone purchase from head shops 5% of the time [26]. In the USA, where online purchasing of drugs of abuse is not as commonplace as in Europe, the primary source of cathinones were small, independent stores, such as gas stations and head shops [122].

NPS sold by retail vendors are extensive, variable by region and subject to rapid change with legal regulations. However, some trends are identifiable. In the USA, products available from retail vendors are primarily SCRAs, cathinones and salvia. Mephedrone, methylone and BZP were widely available at head shops in the UK and Europe under a variety of creative brand names [127]. An analysis of 29 'natural' and 'herbal high' products seized from head shops by police in Poland in 2009 revealed a mixture of JWH products and cannabicyclohexanol. In New Zealand, the primary substance abused was BZP, which was commonly sold in retail stores [28]. There, it was often supplied in packs with 'recovery pills' containing 5-hydroxytryptophan. Numerous case reports and case series report clinical toxicity from novel substances of abuse in which head shops or smart shops were the substance source [122,129–132]. Other attempts at identifying and analysing active agents in emerging NPS has resulted from the methodical purchase of products from local head shops [118,133,134].

Street Level Drug Dealers

While the Internet and head shops provide the source for the majority of novel psychoactive substances, a significant number are purchased from dealers who broker sales on the street, in clubs and in schools. Drugs are also acquired from friends, classmates or family members who obtain NPS from a variety of means and then subsequently distribute them. Concerts and music festivals also provide a venue for NPS sale and distribution [30,135]. Toxicity has been associated with 'Foxy' (5-MeO-DIPT) traced to street purchase [136], and with the sale of BZP in night clubs [137].

UK mephedrone sales illustrate the impact of non-retail, non-Internet vendors and the adaptability of the market to legislation changes. Prior to mephedrone regulation, 48% of UK student users acquired mephedrone from street level dealers and the vast majority reported facile acquisition [25]. Common mephedrone sources for club-goers were friends (38%) and dealers (24%) [26]. Following the ban, the use of dealers to obtain mephedrone increased dramatically (58%), as opposed to friends (41%), the Internet or head shops (<1% each). Telephone survey results were similar, with a significant increase in dealer-originated mephedrone [138]. While reported use of mephedrone decreased somewhat after the legislation, it was still widely consumed and readily available in the UK on the street, as 38% reported that mephedrone was easily or very easily accessible after the ban [26].

In the USA, NPS have appeared on the street-level drug market, as evidenced by regional seizures. Large-scale trafficking networks have developed across state borders. An April 2011 case associated with the sale of mephedrone and 4-MEC resulted in multiple arrests and a 25 kg drug seizure (estimated street value, $525 000) [139]. Organised criminal groups are now engaged in international cathinone trafficking and distribution [76].

PRODUCTS

Pricing of NPS

NPS pricing is associated with supply, demand, vendors, competitors and market dynamics, such as regional legal designations. In 2009, the average price per 'legal high' product in the UK was £9.68 (range £1.75–£54.99) [30]. Compared with traditionally controlled drugs such as cocaine and MDMA, 'legal highs' are often cheaper, in addition to being readily accessible. In contrast to cocaine, typically priced at €50–70 ($69–97) per gram in Europe [46], the average online cost of MDAI, MXE, 6-ABP ('Benzo fury'), mephedrone and JWH-250 were €10.50, €17, €24.50, €16 and €15 per gram, respectively [1]. This is similar, if not cheaper, to the cost of MDMA, which typically sold at €2.80 (UK) to €15.90 (Italy) per pill, or approximately €11–64 per gram [140]. Plant products such as salvia and kratom are significantly cheaper at €0.60–1.50 per gram.

Many users report turning to 'legal highs' as a less expensive way to get high and some credit the global recession as an impetus for NPS popularity [141]. UK students reported using mephedrone because it was cheaper than alcohol [142]. In 2011, UK street drug prices were reported to be £17 per gram of mephedrone, less than half the price of cocaine [143]. Typical mephedrone maximum purchase amounts were about 5 grams (20% of respondents), but not infrequently exceeded 5 grams (43%) [144]. NPS pricing also exhibits temporal changes. In 2011, a 3-gram packet of 'Spice' SCRA cost from €12 to €18, a decrease from €20 to €30 in 2009 [102]. The average reported cathinone purchase price in the UK increased from £10–£12 to £16–20 following regulation [26,138]. NPS transactions also reflect quantity discounts observed in legitimate enterprises. Vendors provide scaled pricing for larger purchases, similar to that seen with other illicit [37,145]. Table 3.1 provides some reported NPS pricing. Simple Internet queries permit users to promptly accomplish current 'price discovery' and comparisons. Unfortunately, many young adults are excessively swayed by price and lack both adequate Internet health literacy skills and health decision-making capacity to appropriately evaluate Internet drug vendors [154].

Transaction Mechanisms

A variety of payment mechanisms are supported. Credit or debit card purchase is common for online products [30]. Online money transfers, traditional bank transfers, checks, postal money orders and digital currencies are also accepted [114,115,155].

Marketing

Aggressive marketing has furthered NPS usage. This includes websites encouraging use, complete with portals to purchase sites [103]. Media reports appear to have inadvertently promoted the NPS abuse both indirectly (by sparking interest in so-called 'legal alternative' drugs) and directly, as detailed above via linked advertising [91,156].

Branding and Packaging

The concept of branding is a commercial market strategy to command higher prices and establish customer loyalty. This practice is previously well established at the wholesale and retail level in the traditional illicit trade. Cocaine bricks and heroin point-of-sale packages have contained a variety of stamps or markings; marijuana 'designer varieties' have acquired brand labels or 'strain names;' and ecstasy pills have been assigned a variety of logos, colours and names [145,157,158]. Unique packaging or brand names serve to differentiate products from competitors and to associate certain drugs with a recognised purity, potency, or desirable special qualities [145,159]. Logos confer significant preferences among established designer drug users and positive messages and apparent safety are also key elements to effective diffusion of new drug use [160,161]. The NPS trade mirrors these market tactics. Affordability and permissive legal designations encourage further access [162]. These characteristics are promoted either on product labels or by vendors' advertising content. NPS are frequently sold in colourful packets, professionally designed to attract and arouse interest [156]. Many brand names remind consumers of illegal street drugs or their effects [162]. Other NPS brands have appropriated and evoked names and graphic designs from common, nontoxic food products – e.g. knock-offs of Cadbury's® 'Flake' (containing methiopropamine, MDAI, caffeine, benzocaine and lidocaine) and Kellogg's® 'Special K' (containing MXE) [153]. Labelled mixtures of 'natural' blended herbs may be included to connote safety; pill formulations may be designed to appear as pharmaceuticals. Where branding is the least mature with some emerging NPS, products may simply be supplied in zip-lock plastic bags with little descriptive information [133]. NPS have incorporated established branding 'line extension' strategies for product diversity, thus, 'Spice Silver,' 'Spice Gold,' and 'Spice Diamond;' 'Chill zone original,' 'Chill zone cherry,' and 'Chill zone mint;' and 'Exclusive original,' 'Exclusive cherry,' and 'Exclusive mint;' etc [134,156]. The range of

TABLE 3.1 Reported Pricing of Some Novel Psychoactive Substances

Class	Substance	Price	Unit	Year reported	Source	Ref
Aminoindanes	5-IAI	£26.99	2g	2010	Internet	[165]
		€9.50–12	1g	2011	Internet	[1]
	MDAI	£14.50–15	1g	2010	Internet	[165]
		€10–11	1g	2011	Internet	[1]
Arylcyclohexylamines	MXE	€14.50–19.50	1g	2011	Internet	[1]
Benzofurans	5-APB	€25–33	1g	2011	Internet	[1]
	6-APB Benzo fury)	£16	2 pellets	2010	Internet	[165]
		€23–26	1g	2011	Internet	[1]
Synthetic cannabinoid receptor agonists	General/ smoking mixtures	€9–12	1g	2011	Unspecified	[175]
		€27–36	3g	2011	Unspecified	[175]
		€26–30	3g	2011	Internet, Head shops	[140]
	Homemade K2	US$0.44–160	3g	2010	Internet	[163]
	K2/Spice	US$24–75	3g	2010	Internet	[163]
	Spice Diamond	€8–10	1g	2008	Internet	[2]
	Spice Gold	€5.30–10.30	1g	2008	Internet	[2]
	AM-2201	€18–21	1g	2011	Internet	[1]
	JWH 018	US$33–120	3g	2010	Internet	[163]
		€20–23	1g	2011	Internet	[1]
	JWH-122	€5–5.50	1g	2011	Internet	[1]
	JWH-250	€11–19.5	1g	2011	Internet	[1]
Cathinones	Generic	US$20	Package	2011	Store	[147]
		US$25–75	Packet	2011	Unspecified	[59]
		US$60–70	1g	2011	Unspecified	[59]
		€18–25	1g	2011	Internet, Head shops	[140]
	Mephedrone	£5.50	1 tablet/ capsule	2009	Internet	[99]
		€10–15	1g	2010	Internet, Head shops	[140]
		£12.20	Not specified	2010 pre-UK ban)	Unspecified	[26]

(Continued)

TABLE 3.1 (Continued)

Class	Substance	Price	Unit	Year reported	Source	Ref
		£19.30	Not specified	2010 post-UK ban)	Unspecified	[26]
		€12–20	1 g	2011	Internet	[1]
		US$40	1 g	2011	Distributor	[59]
	4-MEC	€12–20	1 g	2011	Internet	[1]
	MDPV	€11.50–23.90	1 g	2011	Internet	[1]
		€7.60–13	1 g	2011	Internet	[1]
	Methylone	€9–20	1 g	2011	Internet	[1]
	3,4-DMMC	£9–20	1 g	2010	Internet	[1]
	Naphyrone	£12.50	1 g	2010 pre-UK ban)	Unspecified	[149]
		£2500	1000 g	2010 pre-UK ban)	Unspecified	[149]
	NRG-3	£27.99	2 g	2010	Internet	[165]
Cocaine analogues	Dimethocaine	€20–30	1 g	2011	Internet	[140]
FLYs	Bromo-dragonFLY	€10–30/ US$14–42	single dose blotter	2011	Unspecified	[106]
		€300/US$420	1 g	2011	Unspecified	[106]
		€14.50–19.50	1 g	2011	Internet	[1]
Party pills	E = XTC	€2–5.60	1 capsule	2008	Internet	[2]
	NXT Phase Blue	€1.40–12.50	1 capsule	2008	Internet	[2]
	London Underground products	€8–9	1 capsule	2008	Internet	[2]
	Happy Caps Groove E	€1.80–2.10	1 capsule	2008	Internet	[2]
Piperazines	Generic	£3.30–5.85	1 tablet/ capsule	2009	Internet	[99]
	BZP	€3–4	1 tablet	2011	Unspecified	[140]
Thiophenes	Methiopropramine	€11.50–13	1 g	2011	Internet	[1]
Tryptamines	AMT	€23–46	1 g	2011	Internet	[1]
	DMT	£25	0.125–0.5 g	2011	Unspecified	[150]
Other phenethylamine analogues	2C-B	US$10–US$30	1 tablet	2011	Unspecified	[151]
	MDAT	€11–13	1 g	2011	Internet	[1]

(Continued)

TABLE 3.1 (Continued)

Class	Substance	Price	Unit	Year reported	Source	Ref
Botanicals	Caffeine/ Ephedrine	£2.50–3.40	1 tablet/ capsule	2009	Internet	[99]
	Hallucinogenic mushrooms	€10–14	10g	2011	Internet	[1]
	Hawaiian baby woodrose	€3–4.50	5 seeds	2008	Internet	[2]
		€4–8	10 seeds	2011	Internet	[1]
	Khat	€5	1 bundle	2011	Unspecified	[140]
	Kratom	US$10–40	ounce	2008	Internet	[76]
		€2.1–10.3	1 g	2008	Internet	[2]
		€0.6–1.50	1 g	2011	Internet	[1]
	Morning glory	€0.30–1.00	1 g	2008	Internet	[2]
	Salvia, generic	€0.60–1.20	1 g	2011	Internet	[1]
	Salvia, 5 × extract	US$15	tin	2009	Head shops	[152]
		€11–12	0.5 g	2011	Internet	[140]
	Salvia, 10 × extract	€17.80–38	1 g	2008	Internet	[2]
	Salvia, 40 × extract	US$30	0.5 g	2011	Internet	[140]
	Salvia, 60 × extract	€32–63	1 g	2008	Internet	[2]
	Salvia, 80 × extract	US$80	tin	2009	Head shops	[152]

2C-B, 4-bromo-2,5-dimethoxyphenethylamine; 3,4-DMMC, 3,4-dimethylmethcathinone; 4-MEC, 4-methyl-N-ethylcathinone; 5- and 6- APB, 5- and 6-2-aminopropyl)benzofuran; 5-IAI, 5-iodo-2-aminoindane; AMT, alpha-methyl-tryptamine; BZP, 1-benzylpiperazine; DMT, dimethyltryptamine; MDAI, 5,6-methylenedioxy-2-aminoindane; MDAT, 6,7-methylene-dioxy-2-aminotetralin; MDPV, 3,4-methylenedioxypyrovalerone; MPA, methiopropramine.

powders, 'party' pills or tablets, smokable mixtures and plants/extracts is attractive to users seeking diverse, customisable, social enhancement products.

Misbranding

Vendor-supplied product content data varies greatly. In a 2008 study, only 63% of online retailers indicated formulation information [2]. In 2009, only 40.1% of available online products in the UK listed their constituents [30]. In 2010, 61% of Irish NPS products tested listed ingredients [62]. Even when listed, reported ingredients of many products are incomplete, confusing, or inconsistent, using alternative names to describe the same substance [30]. Recommended dosing, side-effects, warnings, user feedback and referrals to information sources are similarly lacking [2,30,102]. NPS misbranding takes several forms. First, as indicated previously, NPS employ a variety of descriptive names and labels to circumvent legal restraints, despite an intent for human exposure. Aside from this rather transparent misbranding strategy, unlisted pharmacologically active ingredients (e.g. cathinones, SCRAs and piperazines) may reside in products actively marketed with 'legal high' intent, risking legal infractions or health risks [30,99,133]. Products labelled as 'natural' or 'herbal highs' confiscated from Polish head shops contained undeclared SCRAs. Notably, the particular SCRA correlated with the preparation's 'flavour' and not brand name [134]. NPS active ingredients may display temporal inconsistencies and may also include products in which psychoactive substances are undetectable [99]. Identical 'brand name' NPS may demonstrate divergent content. This has been established for cathinones as well as SCRAs [122,156]. Spraying of SCRAs on vegetation may result in uneven distribution or 'hot spots' within individual packets [163]. NPS are also used to adulterate other illicit substances (amphetamine, cocaine, 'ecstasy,' LSD and methamphetamine) [60,148]. Lastly, NPS themselves, even when sold as such, are affected by mislabelling and impurities. The potential dangers and extent of contamination posed by adulterants of traditional 'illicit' drugs are well documented (e.g. antihistamines, caffeine, calcium channel antagonists, clenbuterol, levamisole, local anaesthetics, paracetamol, phenacetin, quinine, scopolamine, etc.) [164]. Decreased wholesale drug costs increase intermediation margins and potential trafficking profits. Consequently, to an extent that the market will tolerate, NPS are just as vulnerable to adulteration [145], with national proscriptions against 'deceptive advertising' carrying little weight. Multiple products advertised in August 2010 (MDAI, 5-IAI, 'Benzo Fury,' 'NRG-3,' and 'E2') did not contain the purported active ingredient: a single sample contained the labelled substance (MDAI), all others contained caffeine and a mixture of BZP and 3-TFMPP [165]. Similarly, 61% of tested Irish headshop products contained caffeine [62]. Following the UK ban on mephedrone, 'NRG-1,' advertised as naphyrone (naphthylpyrovalerone) and later generation 'NRGs' appeared. However, in many cases these products were cathinones or novel cathinone derivatives (4-FMC, pentylone, MDPBP,

MDPV, etc.) merely displaying a new label, carrying potential criminal or health effects [146,166]. All NPS in another sample from 2011 contained significant undeclared caffeine (up to 1 gram) or taurine [167]. Both seized and purchased synthetic cathinones have demonstrated contamination with multiple substances [58]. Mephedrone has been adulterated with caffeine, paracetamol, cocaine, amphetamine and ketamine [168]. Following the UK cathinone restrictions, most users noted decreased purity and increased adulteration [26]. Botanicals are not immune: O-desmethyltramadol was confirmed as an adulterant in 'Krypton,' an herbal kratom mixture [169].

Market Dynamics

NPS market dynamics are influenced by user effects, price, availability, quality, legality and competition [170]. These contribute to NPS life cycle stages: introduction, growth, maturity and/or decline. BZP displacement of MDMA in seizures analysed by the UK Forensic Science Service in 2007–2009 was followed by its own subsequent displacement by cathinone derivatives in 2010 [37]. This trend was similarly seen in the USA, where BZP NFLIS mentions peaked in 2009, with SCRA and cathinone seizures surging in 2010 [13,35,36]. Suppliers demonstrate flexibility in developing legal market alternatives as the landscape shifts [51,102]. This leads to second-, third-, fourth- and later 'generation' substitutions. Just four weeks after JWH-018's prohibition in Germany, second generation SCRAs were flooding the market [19]. Following mephedrone control in the UK, purported naphyrone and related compounds (often advertised as 'NRG'-series compounds) were rapidly introduced as '100 per cent legal' substitutes [171]. Upon its control, naphyrone was replaced by desoxypipradrol (2-DPMP) in some 'Ivory Wave' products, which was then followed by diphenylprolinol (D2PM) substitution for 2-DPMP [172]. MXE and ketamine use also increased following recent UK cathinone and piperazine controls [143]. Before the Irish ban of synthetic cannabinoids, benzylpiperazine derivatives, cathinones, several GHB precursors and ketamine in May 2010, then legal head shops were reportedly opening at a rate of one per week [170]. Subsequent to the ban, users reported a range of anticipated behaviours, including continued use of now-illegal NPS, a return to prior illicit drugs and intent to experiment with even newer NPS [162]. However, a November 2011 £1 million mephedrone seizure demonstrated ongoing UK cathinone demand [173]. 'BZP free' party pills, potentially containing other piperazine derivatives such as pFPP (parafluorophenylpiperazine), appeared following BZP control [2]. Legal control of NPS may depress utilisation or induce diversion to existing illicit supply networks [102]. Conversely, market pressures on other illicit drugs may propel NPS consumption. Low cocaine and MDMA purity in 2009 was thought to contribute to UK increases in mephedrone use [91]. Supply shocks in the European heroin market between November 2010 and March 2011 led some users to replace heroin with injectable cathinones as well as other 'legal highs' [174]. These various pressures, as well as additional cultural aspects, may ultimately shape national reported NPS and other drug preferences – e.g. more frequent use of SCRAs, mushrooms, LSD, methylphenidate and opiates by US respondents and more frequent ketamine, mephedrone, 'MDMA,' and MXE consumption by UK respondents [101].

CONCLUSIONS

NPS availability and supply is highly complex and volatile. New products are continually introduced and a global network of suppliers, distributors and vendors ensures a constantly adjusting market space. Technological advances have lowered, if not eliminated, barriers to

diffusion of NPS knowledge and properties. Both endogenous and exogenous forces exert ongoing pressures, which ensure the continuing evolution of the NPS market.

References

[1] EMCDDA. Online sales of new psychoactive substances/legal highs: Summary of results from the 2011 multilingual snapshots. European Monitoring Centre for Drugs and Drug Addiction, Lisbon, Portugal; 2011.

[2] Hillebrand J, Olszewski D, Sedefov R. Legal highs on the internet. Subst Use Misuse 2010;45:330–40.

[3] U.S. Drug Enforcement Administration, Department of justice.2006 General information regarding the combat methamphetamine epidemic act of 2005 [Title VII of Public Law 109–177].

[4] United Nations. United Nations convention against illicit traffic in NARCOTIC DRUGS AND PSYCHOACTIVE Substances, 1988. United Nations, Austria, Vienna; 2003.

[5] US Drug Enforcement Administration, Department of Justice Control of a chemical precursor used in the illicit manufacture of fentanyl as a list I chemical. Final rule. Fed Regist 2008;73:43355–43357.

[6] Centers for Disease Control and Prevention (CDC). Youth risk behavior surveillance – United States, 2011. MMWR Morb Mortal Wkly Rep 2012;61(SS–4):1–162.

[7] SAMHSA Center for behavioral health statistics and quality. The DAWN report: highlights of the 2010 drug abuse warning network (DAWN) findings on drug-related emergency department visits. Rockville, MD: Substance Abuse and Mental Health Services Administration; 2012.

[8] SAMHSA Results from the 2010 national survey on drug use and health: summary of national findings. NSDUH series H-41, HHS publication no. (SMA) 11-4658. Rockville, Maryland: Substance Abuse and Mental Health Services Administration; 2011.

[9] Johnston LD, O'Malley PM, Bachman JG, Schulenberg JE. Monitoring the future national results on adolescent drug use: Overview of key findings, 2011. Ann Arbor, Michigan: Institute for Social Research, The University of Michigan; 2012.

[10] U.S. Drug Enforcement Administration. Stats & facts. Available: <http://www.justice.gov/dea/statistics.html#seizures>; 2011 [accessed 06.07.12].

[11] European Monitoring Centre for Drugs and Drug Addiction (EMCDDA). Seizures data (SZR) Table SZR-21. Other substances seized, 2004 to 2009. Available: <http://www.emcdda.europa.eu/stats11/szrtab21>; 2011 [accessed 06.07.12].

[12] Bronstein AC, Spyker DA, Cantilena LR, Green JL, Rumack BH, Dart RC. Annual report of the American association of poison control centers' national poison data system (NPDS): 28th annual report. Clin Toxicol (Phila) 2010;49:910–41. 2011.

[13] US Drug Enforcement Administration, Office of Diversion Control. National forensic laboratory information system special report: Synthetic cannabinoids and synthetic cathinones reported in NFLIS, 2009–2010. US Drug Enforcement Administration, Springfield, Virginia; 2011.

[14] National Institute on Drug Abuse (NIDA). Drugs of abuse information. Available: <http://www.drugabuse.gov/drugs-abuse>; 2011 [accessed 06.07.12].

[15] Substance Abuse and Mental Health Services Administration. Substances. Available: <http://store.samhsa.gov/facet/Substances>; 2011 [accessed 06.07.12].

[16] Legleye S, Ben Lakhdar C, Spilka S. Two ways of estimating the euro value of the illicit market for cannabis in France. Drug Alcohol Rev 2008;27:466–72.

[17] United Nations Office on Drugs and Crime (UNODC). World drug report 2012. Sales No. E.12.XI.1. United Nations, New York, NY; 2012.

[18] Kerwin J. Doors of deception – The diaspora of designer drugs. Drug Test Anal 2011;3:527–31.

[19] Vardakou I, Pistos C, Spiliopoulou C. Spice drugs as a new trend: mode of action, identification and legislation. Toxicol Lett 2010;197:157–62.

[20] Hu X, Primack BA, Barnett TE, Cook RL. College students and use of K2: an emerging drug of abuse in young persons. Subst Abuse Treat Prev Policy 2011;6:16.

[21] Vandrey R, Dunn KE, Fry JA, Girling ER. A survey study to characterize use of Spice products (synthetic cannabinoids). Drug Alcohol Depend 2012;120:238–41.

[22] Lange JE, Reed MB, Croff JM, Clapp JD. College student use of Salvia divinorum. Drug Alcohol Depend 2008;94:263–6.

[23] Wu LT, Woody GE, Yang C, Li JH, Blazer DG. Recent national trends in Salvia divinorum use and substance-use disorders among recent and former Salvia divinorum users compared with nonusers. Subst Abuse Rehabil 2011;2011:53–68.

[24] Perron BE, Ahmedani BK, Vaughn MG, Glass JE, Abdon A, Wu LT. Use of Salvia divinorum in a nationally representative sample. Am J Drug Alcohol Abuse 2011;38:108–13.

[25] Dargan PI, Albert S, Wood DM. Mephedrone use and associated adverse effects in school and college/university students before the UK legislation change. QJM 2010;103:875–9.

[26] Winstock A. The 2011 Mixmag drugs survey. Mixmag 2011;238:49–59. Available: <http://issuu.com/mixmag-fashion/docs/drugsurvey> [accessed 06.07.12].

[27] Schep LJ, Slaughter RJ, Vale JA, Beasley DM, Gee P. The clinical toxicology of the designer party pills benzylpiperazine and trifluoromethylphenylpiperazine. Clin Toxicol (Phila) 2011;49:131–41.

[28] Wilkins C, Sweetsur P, Girling M. Patterns of benzylpiperazine/trifluoromethylphenylpiperazine party pill use and adverse effects in a population sample in New Zealand. Drug Alcohol Rev 2008;27:633–9.

[29] Cohen BM, Butler R. BZP-party pills: a review of research on benzylpiperazine as a recreational drug. Int J Drug Policy 2011;22:95–101.

[30] Schmidt MM, Sharma A, Schifano F, Feinmann C. Legal highs on the net-Evaluation of UK-based Websites, products and product information. Forensic Sci Int 2011;206:92–7.

[31] McCandless D. Exotic, legal highs become big business as 'headshops' boom. The Guardian 2006. Available: <http://www.guardian.co.uk/society/2006/jan/09/drugsandalcohol.drugs?INTCMP=SRCH> [accessed 06.07.12].

[32] North American Herbal Incense Trade Association. NAHITA statement of positions. Available: <http://keepitlegal.org/node/93>; 2011 [accessed 06.07.12].

[33] Paynter, B.2011 The big business of synthetic highs. Bloomberg Businessweek. Available: <http://www.businessweek.com/magazine/content/11_26/b4234058348635.htm> [accessed 06.07.12].

[34] Evaluation and Inspections Division. Office of the Inspector General. Follow-up review of the drug enforcement administration's efforts to control the diversion of controlled pharmaceuticals. Evaluations and inspections report I-2006-004. US Department of Justice, Available: <http://www.justice.gov/oig/reports/DEA/e0604/final.pdf>; 2006 [accessed 06.07.12].

[35] US Drug Enforcement Administration Office of diversion control. National forensic laboratory information system: year 2009 annual report. Springfield, Virginia: US Drug Enforcement Administration; 2010.

[36] US Drug Enforcement Administration Office of diversion control. National forensic laboratory information system: year 2010 annual report. Springfield, Virginia: US Drug Enforcement Administration; 2011.

[37] EMCDDA Report on the risk assessment of mephedrone in the framework of the council decision on new psychoactive substances. Lisbon, Portugal: European Monitoring Centre for Drugs and Drug Addiction; 2011.

[38] Office of Diversion Control, Drug and Chemical Evaluation Section. 2011 4-methylmethcathinone (mephedrone). US Department of Justice, Drug Enforcement Administration, Washington, D.C.

[39] Office of Diversion Control, Drug and Chemical Evaluation Section. 2011 3,4-methylenedioxypyrovalerone (MDPV). US Department of Justice, Drug Enforcement Administration, Washington, DC.

[40] Wiley JL, Compton DR, Dai D, et al. Structure-activity relationships of indole- and pyrrole-derived cannabinoids. J Pharmacol Exp Ther 1998;285:995–1004.

[41] Huffman JW, Dong D, Martin BR, Compton DR. Design, synthesis and pharmacology of cannabimimetic indoles. Bioorg Med Chem Lett 1994;4:563–6.

[42] Atwood BK, Huffman J, Straiker A, Mackie K. JWH018, a common constituent of 'Spice' herbal blends, is a potent and efficacious cannabinoid CB receptor agonist. Br J Pharmacol 2010;160:585–93.

[43] Monte AP, Marona-Lewicka D, Parker MA, Wainscott DB, Nelson DL, Nichols DE. Dihydrobenzofuran analogues of hallucinogens. 3. Models of 4-substituted (2,5-dimethoxyphenyl)alkylamine derivatives with rigidified methoxy groups. J Med Chem 1996;39:2953–61.

[44] Parker MA, Marona-Lewicka D, Lucaites VL, Nelson DL, Nichols DE. A novel (benzodifuranyl)aminoalkane with extremely potent activity at the 5-HT2A receptor. J Med Chem 1998;41:5148–9.

[45] Sainsbury PD, Kicman AT, Archer RP, King LA, Braithwaite RA. Aminoindanes – the next wave of 'legal highs'?. Drug Test Anal 2011;3:479–82.

[46] Whalen J. In quest of legal high, chemists outfox law. Wall St J 2010. Available: <http://online.wsj.com/article/SB10001424052748704763904575550200845267526.html> [accessed 06.07.12].

[47] Nichols D. Legal highs: the dark side of medicinal chemistry. Nature 2011;469:7.

[48] Elliott SP. Fatal poisoning with a new phenylethylamine: 4-methylthioamphetamine (4-MTA). J Anal Toxicol 2000;24:85–9.

[49] Andreasen MF, Telving R, Birkler RI, Schumacher B, Johannsen M. A fatal poisoning involving Bromo-Dragonfly. Forensic Sci Int 2009;183:91–6.

[50] Wood DM, Looker JJ, Shaikh L, et al. Delayed onset of seizures and toxicity associated with recreational use of Bromo-Dragonfly. J Med Toxicol 2009;5:226–9.

[51] Hill SL, Thomas SH. Clinical toxicology of newer recreational drugs. Clin Toxicol (Phila) 2011;49:705–19.

[52] Campbell H, Cline W, Evans M, Lloyd J, Peck AW. Comparison of the effects of dexamphetamine and 1-benzylpiperazine in former addicts. Eur J Clin Pharmacol 1973;6:170–6.

[53] Kahn RS, Wetzler S. m-Chlorophenylpiperazine as a probe of serotonin function. Biol Psychiatry 1991;30:1139–66.

[54] Bossong MG, Van Dijk JP, Niesink RJ. Methylone and mCPP, two new drugs of abuse?. Addict Biol 2005;10:321–3.

[55] Bossong MG, Brunt TM, Van Dijk JP, et al. mCPP: an undesired addition to the ecstasy market. J Psychopharmacol 2010;24:1395–401.

[56] Collins M. Some new psychoactive substances: precursor chemicals and synthesis-driven end-products. Drug Test Anal 2011;3:404–16.

[57] Saem de Burnaga Sanchez J. Sur un homologue de l'éphédrine [French]. Bulletin De La Societé Chimique De France 1929;45:284–6.

[58] US Drug Enforcement Administration, Department of Justice Schedules of controlled substances: temporary placement of three synthetic cathinones in Schedule I. Final order. Fed Regist 2011;76:65371–65375.

[59] Office of Diversion Control, Drug and Chemical Evaluation Section Background, data and analysis of synthetic cathinones: mephedrone (4-MMC), methylone (MDMC) and 3,4-methylenedioxypyrovalerone (MDPV). Washington, D.C.: U.S. Department of Justice, Drug Enforcement Administration; 2011.

[60] Brunt TM, Niesink RJ. The drug information and monitoring system (DIMS) in the Netherlands: implementation, results and international comparison. Drug Test Anal 2011;3:621–34.

[61] Begg WG, Reid AA. Meratran; a new stimulant drug. Br Med J 1956;1:946–9.

[62] Kelleher C, Christie R, Lalor K, Fox J, Bowden M, O'Donnell C. An overview of new psychoactive substances and the outlets supplying them. Dublin, Ireland: National Advisory Committee on Drugs (NACD); 2011.

[63] Dunne, P. Misuse of Drugs Act 1975. Temporary Class Drug Notice. New Zealand Gazette 29, 833. Government of New Zealand, Wellington, New Zealand; 2012.

[64] Shulgin A, Shulgin A. PiHKAL: a chemical love story. Berkeley, California: Transform Press; 1991.

[65] Shulgin A, Shulgin A. TiHKAL: the continuation. Berkeley, California: Transform Press; 1997.

[66] Shulgin AT, Manning T, Daley PF. The Shulgin Index. Volume 1: Psychedelic phenethylamines and related compounds. Berkeley, California: Transform Press; 2011.

[67] Morris, H. Interview with a ketamine chemist. Vice. Available: <http://www.vice.com/read/interview-with-ketamine-chemist-704-v18n2> ; 2011 [accessed 06.07.12].

[68] Archer RP. Fluoromethcathinone, a new substance of abuse. Forensic Sci Int 2009;185:10–20.

[69] EMCDDA–Europol. 2008 Annual report on the implementation of Council Decision 2005/387/JHA. European Monitoring Centre for Drugs and Drug Addiction-Europol, Lisbon, Portugal, 2009. Available: <http://www.emcdda.europa.eu/attachements.cfm/att_77263_EN_EMCDDA-Europol_Annual_Report_Art10_2008.pdf> [accessed 06.07.12].

[70] EMCDDA–Europol. 2012 2011 Annual report on the implementation of council decision 2005/387/JHA.

European Monitoring Centre for Drugs and Drug Addiction-Europol, Lisbon, Portugal. Available: <http://www.emcdda.europa.eu/attachements.cfm/att_155113_EN_EMCDDA-Europol%20Annual%20Report%202011_2012_final.pdf> [accessed 06.07.12].

[71] Kikura-Hanajiri R, Uchiyama N, Goda Y. Survey of current trends in the abuse of psychotropic substances and plants in Japan. Leg Med (Tokyo) 2011;13:109–15.

[72] US Drug Enforcement Administration, Office of Diversion Control, Drug and Chemical Evaluation Section. 2011 2,5-dimethoxy-4-(n)-propylthiophenethylamine. U. Department of Justice. Available: <http://www.deadiversion.usdoj.gov/drugs_concern/2ct7.pdf> [accessed 29.12.11].

[73] ACMD Consideration of the novel psychoactive substances ('legal highs'). London, UK: Advisory Council on the Misuse of Drugs; 2011.

[74] Vardakou I, Pistos C, Spiliopoulou C. Drugs for youth via Internet and the example of mephedrone. Toxicol Lett 2011;201:191–5.

[75] International Narcotics Control Board (INCB). Report of the international narcotics control board for 2010. E/INCB/2010/1. United Nations, New York, NY; 2011.

[76] Babu KM, McCurdy CR, Boyer EW. Opioid receptors and legal highs: *Salvia divinorum* and Kratom. Clin Toxicol (Phila) 2008;46:146–52.

[77] Boyer EW, Shannon M, Hibberd PL. The internet and psychoactive substance use among innovative drug users. Pediatrics 2005;115:302–5.

[78] Montagne M. Drugs on the internet. I: Introduction and web sites on psychedelic drugs. Subst use Misuse 2008;43:17–25.

[79] Wax PM. Just a click away: recreational drug Web sites on the Internet. Pediatrics 2002;109(e96).

[80] Deluca P, Schifano F, Psychonaut 2002 Research Group Searching the Internet for drug-related web sites: analysis of online available information on ecstasy (MDMA). Am J Addict 2007;16:479–83.

[81] Boyer EW, Lapen PT, Macalino G, Hibberd PL. Dissemination of psychoactive substance information by innovative drug users. Cyberpsychol Behav 2007;10:1–6.

[82] Halpern JH, Pope Jr HG. Hallucinogens on the Internet: a vast new source of underground drug information. Am J Psychiatry 2001;158:481–3.

[83] European Commission. Young people and drugs among 15–24 year-olds. Flash EB Series #233. European Commission, Dublin, Ireland; 2008.

[84] National Center for Addiction and Substance Abuse. National Survey of American Attitudes on Substance Abuse XVI: Teens and Parents. New York: Columbia University; 2011.

[85] Erowid. Documenting the complex relationship between humans and psychoactives. Available: <http://www.erowid.org>; 2012 [accessed 06.07.12].

[86] Dancesafe. Available: <http://dancesafe.org/>; 2012 [accessed 06.07.12].

[87] Multidisciplinary Association for Psychedelic Studies (MAPS). Available: <http://maps.org/>; 2012 [accessed 06.07.12].

[88] Bluelight. Available: <http://www.bluelight.ru/vb/>, 2012 [accessed 06.07.12].

[89] Lycaeum. Available: <http://www.lycaeum.org/>; 2011 [accessed 24.12.11].

[90] The Shroomery. Available: <http://www.shroomery.org>; 2012 [accessed 06.07.12].

[91] ACMD Consideration of the cathinones. London, UK: Advisory Council on the Misuse of Drugs; 2010.

[92] International Narcotics Control Board (INCB). Report of the international narcotics control board for 2001. E/INCB/2001/1. United Nations, New York, NY; 2001.

[93] International Narcotics Control Board (INCB). Report of the international narcotics control board for 1997. E/INCB/1997/1. United Nations, New York, NY; 1997.

[94] Klein CA. Psychotropics without borders: ethics and legal implications of internet-based access to psychiatric medications. J Am Acad Psychiatry Law 2011;39:104–11.

[95] Schepis TS, Marlowe DB, Forman RF. The availability and portrayal of stimulants over the Internet. J Adolesc Health 2008;42:458–65.

[96] Forman RF, Marlowe DB, McLellan AT. The Internet as a source of drugs of abuse. Curr Psychiatry Rep 2006;8:377–82.

[97] US Drug Enforcement Administration. FY 2012 performance budget congressional submission. US Department of Justice, Washington, DC; 2011.

[98] Dev CS, Schultz DE. In the mix: a customer-focused approach can bring the current marketing mix into the 21st century. Mark Manage 2005;14:16–22.

[99] Davies S, Wood DM, Smith G, et al. Purchasing 'legal highs' on the internet – is there consistency in what you get?. QJM 2010;103:489–93.

[100] Dick D, Torrance C. MixMag drugs survey. Mixmag 2010:44–53.

[101] Winstock A. Global drug survey. MixMag 2012;251:68–73. Available: <http://www.mixmag.net/drugssurvey> [accessed 05.07.12].

[102] EMCDDA 2011 annual report on the state of the drugs problem in Europe. Lisbon, Portugal: European Monitoring Centre for Drugs and Drug Addiction; 2011.

[103] Hoover V, Marlowe DB, Patapis NS, Festinger DS, Forman RF. Internet access to Salvia divinorum:

implications for policy, prevention, and treatment. J Subst Abuse Treat 2008;35:22–7.

[104] Topping A. Guardian/Mixmag drug survey reveals a generation happy to chance it. The Guardian 2012. Available: <http://www.guardian.co.uk/society/2012/mar/15/respondents-guardian-mixmag-drug-survey> [accessed 06.07.12].

[105] Maruyama T, Kawamura M, Kikura-Hanajiri R, Takayama H, Goda Y. The botanical origin of kratom (Mitragyna speciosa; Rubiaceae) available as abused drugs in the Japanese markets. J Nat Med 2009;63:340–4.

[106] Corazza O, Schifano F, Farre M, et al. Designer drugs on the internet: a phenomenon out-of-control? The emergence of hallucinogenic drug Bromo-Dragonfly. Curr Clin Pharmacol 2011;6:125–9.

[107] Plant Feed Shop. Available: <http://www.plantfeedshop.com/>; 2011 [accessed 30.12.11].

[108] Blackwell S. Selling drugs on Facebook and Twitter. Death and Taxes 2010. Available: <http://www.deathandtaxesmag.com/32309/selling-drugs-on-facebook-and-twitter/> [accessed 06.07.12].

[109] Wilber DQ. Following antisocial elements on social media. The Washington Post 2011. Available: <http://www.washingtonpost.com/local/antisocial-side-of-social-media-helps-police-track-gangs/2011/10/13/gIQAWzRgAM_story.html>; [accessed 06.07.12].

[110] Power M. How I caused an international drug panic. Mixmag 2011. Available: <http://mixmag.net/words/news/the-mixmag-drug-survey-launches> [accessed 06.07.12].

[111] US Department of Justice. National Drug Intelligence Center. Situation report. synthetic cathinones (bath salts): an emerging domestic threat. Product number 2011-S0787-004. National Drug Intelligence Center, Johnstown, Pennsylvania; 2011.

[112] Capone, J..2011 A poppy panic on eBay, Courtesy of the DEA. The Fix. Available: <http://www.thefix.com/content/poppy-panic?page=all> [accessed 06.07.12].

[113] reddit. Tell me about buying drugs from eBay [forum discussion]. Available: <http://www.reddit.com/r/Drugs/comments/fw1d6/tell_me_about_buying_drugs_from_ebay/>; 2011 [accessed 06.07.12].

[114] eBay. <Available: http://www.ebay.com/>; 2011 [accessed 30/12/11].

[115] Chen A. The underground website where you can buy any drug imaginable. Gawker com 2011. Available: <http://gawker.com/5805928/the-underground-website-where-you-can-buy-any-drug-imaginable> [accessed 06.07.12].

[116] Amsterdam Coffeeshop Directory.Smartshop, growshop, headshop. Available: <http://www.coffeeshop.freeuk.com/GenSS.html>; 2011 [accessed 06.07.12].

[117] Auwarter V, Dresen S, Weinmann W, Muller M, Putz M, Ferreiros N. 'Spice' and other herbal blends: harmless incense or cannabinoid designer drugs? J Mass Spectrom 2009;44:832–7.

[118] de Boer D, Bosman I. A new trend in drugs-of-abuse; the 2C-series of phenethylamine designer drugs. Pharm World Sci 2004;26:110–3.

[119] Schifano, F., Deluca, P., Agosti, L., Psychonaut 2002 Research Group, et al. and 2005 New trends in the cyber and street market of recreational drugs? The case of 2C-T-7 ('Blue Mystic'). J Psychopharmacol 19, 675-679.

[120] Gee P, Fountain J. Party on? BZP party pills in New Zealand. N Z Med J 2007;120:U2422.

[121] de Boer D, Bosman IJ, Hidvegi E, et al. Piperazine-like compounds: a new group of designer drugs-of-abuse on the European market. Forensic Sci Int 2001;121:47–56.

[122] Spiller HA, Ryan ML, Weston RG, Jansen J. Clinical experience with and analytical confirmation of bath salts and legal highs (synthetic cathinones) in the United States. Clin. Toxicol. (Phila) 2011;49:499–505.

[123] Headshops.com. Available: <http://www.headshops.com/>; 2011 [accessed 24.12.11].

[124] HeadShopFinder.com. Available: <http://www.headshopfinder.com/>; 2011 [accessed 24.12.11].

[125] Hughes B, Malczewski A. Poland passes new law to control head shops and legal highs. Drugnet Europe 2011;73:5. January–March.

[126] Health Research Board (HRB). New report reveals the latest drug trends across Europe. Available: <http://www.hrb.ie/about/media/media-archive/press-release/release/149/>; 2011 [accessed 06.07.12].

[127] McNamara S, Stokes S, Coleman N. Head shop compound abuse amongst attendees of the drug treatment centre board. Ir Med J 2010;103(134):136–7.

[128] Long J. Conference on psychoactive drugs sold in head shops and online. Drugnet Ireland Spring 2010;33:1–3.

[129] Nicholson PJ, Quinn MJ, Dodd JD. Headshop heartache: acute mephedrone 'meow' myocarditis. Heart 2010;96:2051–2.

[130] Lidder S, Dargan P, Sexton M, et al. Cardiovascular toxicity associated with recreational use of diphenylprolinol (diphenyl-2-pyrrolidinemethanol [D2PM]). J Med Toxicol 2008;4:167–9.

[131] Frohlich S, Lambe E, O'Dea J. Acute liver failure following recreational use of psychotropic head shop compounds. Ir J Med Sci 2011;180:263–4.

[132] Smith C, Cardile AP, Miller M. Bath salts as a legal high. Am J Med 2011;124:e7–e8.

[133] Ramsey J, Dargan PI, Smyllie M, et al. Buying 'legal' recreational drugs does not mean that you are not breaking the law. QJM 2010;103:777–83.

[134] Zuba D, Byrska B, Maciow M. Comparison of herbal highs composition. Anal Bioanal Chem 2011;400:119–26.

[135] Doughty C, Walker A, Brenchley J. Herbal mind altering substances: an unknown quantity? Emerg Med J 2004;21:253–5.

[136] Meatherall R, Sharma P. Foxy, a designer tryptamine hallucinogen. J Anal Toxicol 2003;27:313–7.

[137] Wood DM, Dargan PI, Button J, et al. Collapse, reported seizure – and an unexpected pill. Lancet 2007;369:1490.

[138] Winstock A, Mitcheson L, Marsden J. Mephedrone: still available and twice the price. Lancet 2010;376:1537.

[139] Baker RA. 22 indicted by Feds in Syracuse-based conspiracy to traffic drug Molly. The Post-Standard. Available: <http://www.syracuse.com/news/index.ssf/2011/08/post_466.html>; 2011 [accessed 06.07.12].

[140] European Monitoring Centre for Drugs and Drug Addiction. Drug profiles. Available: <http://www.emcdda.europa.eu/drug-profiles>; 2011 [accessed 29.12.11].

[141] Phillips J. Britain steps up its war on legal highs. Time. Available: <http://www.time.com/time/world/article/0,8599,2100285,00.html>; 2011 [accessed 06.07.12].

[142] Dargan PI, Sedefov R, Gallegos A, Wood DM. The pharmacology and toxicology of the synthetic cathinone mephedrone (4-methylmethcathinone). Drug Test Anal 2011;3:454–63.

[143] Daly M, Simonson P. Street drug trends survey 2011: the ketamine zone. Druglink 2011;November/December:6–9.

[144] Carhart-Harris RL, King LA, Nutt DJ. A web-based survey on mephedrone. Drug Alcohol Depend 2011;118:19–22.

[145] EMCDDA Thematic paper: Pilot study on wholesale drug prices in Europe. Lisbon, Portugal: European Monitoring Centre for Drugs and Drug Addiction; 2011.

[146] Brandt SD, Freeman S, Sumnall HR, Measham F, Cole J. Analysis of NRG 'legal highs' in the UK: identification and formation of novel cathinones. Drug Test Anal 2011;3:569–75.

[147] Centers for Disease Control and Prevention (CDC) Emergency department visits after use of a drug sold as bath salts – Michigan, November 13, 2010-March 31, 2011. MMWR Morb Mortal Wkly Rep 2011;60:624–7.

[148] Schifano F, Albanese A, Fergus S, et al. Mephedrone (4-methylmethcathinone; 'meow meow'): chemical, pharmacological and clinical issues. Psychopharmacology (Berl) 2011;214:593–602.

[149] ACMD Consideration of the naphthylpyrovalerone analogues and related compounds. London, UK: Advisory Council on the Misuse of Drugs; 2010.

[150] Drugscope. Drug information. DrugSearch home. Available: <http://www.drugscope.org.uk/resources/drugsearch>; 2011 [accessed 29.12.11].

[151] US Drug Enforcement Administration, Office of Diversion Control, Drug & Chemical Evaluation Section. 2011 4-bromo-2,5-dimethoxyphenethylamine. U.S. Department of Justice, Available: <http://www.deadiversion.usdoj.gov/drugs_concern/bromo_dmp/bromo_dmp.pdf> [accessed 06.07.12].

[152] Kelly BC. Legally tripping: a qualitative profile of Salvia divinorum use among young adults. J Psychoactive Drugs 2011;43:46–54.

[153] Wood DM, Davies S, Calapis A, Ramsey J, Dargan PI. Novel drugs –novel branding. QJM 2012;105:1125–6.

[154] Ivanitskaya L, Brookins-Fisher J, O Boyle I, Vibbert D, Erofeev D, Fulton L. Dirt cheap and without prescription: how susceptible are young US consumers to purchasing drugs from rogue internet pharmacies? J Med Internet Res 2010;12:e11.

[155] VIP-Legals.com. Available: <http://vip-legals.com/>; 2011 [accessed 31.12.11].

[156] United Nations Office on Drugs and Crime (UNODC). Synthetic cannabinoids in herbal products. UN document ID number: SCITEC/24. United Nations, New York, NY; 2011.

[157] National Drug Intelligence Center. Connecticut drug threat assessment. Document ID: 2002-S0377CT-001. National Drug Intelligence Center, Johnstown, Pennsylvania; 2002.

[158] Sifaneck SJ, Ream GL, Johnson BD, Dunlap E. Retail marijuana purchases in designer and commercial markets in New York City: sales units, weights, and prices per gram. Drug Alcohol Depend 2007;90(Suppl. 1):S40–51.

[159] Office of National Drug Control Policy. Pulse check: trends in drug abuse January 2004. Executive Office of the President, Washington, DC; 2004.

[160] Daveluy A, Miremont-Salame G, Rahis AC, et al. Medicine or ecstasy? The importance of the logo. Fundam Clin Pharmacol 2010;24:233–7.

[161] Schensul JJ, Diamond S, Disch W, Bermudez R, Eiserman J. The diffusion of ecstasy through urban youth networks. J Ethn Subst Abuse 2005;4:39–71.

[162] McElrath K, O'Neill C. Experiences with mephedrone pre- and post-legislative controls: perceptions of safety and sources of supply. Int J Drug Policy 2011;22:120–7.

[163] Wells DL, Ott CA. The new marijuana. Ann Pharmacother 2011;45:414–7.

[164] Cole C, Jones L, McVeigh J, Kicman A, Syed Q, Bellis MA. CUT. A guide to adulterants, bulking agents and other contaminants found in illicit drugs. Liverpool, UK: Centre for Public Health, Liverpool John Moores University; 2010.

[165] Baron M, Elie M, Elie L. An analysis of legal highs-do they contain what it says on the tin? Drug Test Anal 2011;3:576–81.

[166] Brandt SD, Sumnall HR, Measham F, Cole J. Analyses of second-generation 'legal highs' in the UK: initial findings. Drug Test Anal 2010;2:377–82.

[167] Davies S, Lee T, Ramsey J, Dargan PI, Wood DM. Risk of caffeine toxicity associated with the use of 'legal highs' (novel psychoactive substances). Eur J Clin Pharmacol 2011;68:435–9.

[168] Camilleri A, Johnston MR, Brennan M, Davis S, Caldicott DG. Chemical analysis of four capsules containing the controlled substance analogues 4-methylmethcathinone, 2-fluoromethamphetamine, alpha-phthalimidopropiophenone and N-ethylcathinone. Forensic Sci Int 2010;197:59–66.

[169] Arndt T, Claussen U, Gussregen B, Schrofel S, Sturzer B, Werle A, Wolf G. Kratom alkaloids and O-desmethyltramadol in urine of a Krypton herbal mixture consumer. Forensic Sci Int 2011;208:47–52.

[170] Van Hout MC, Brennan R. 'Heads held high': an exploratory study of legal highs in pre-legislation Ireland. J Ethn Subst Abuse 2011;10:256–72.

[171] Vardakou I, Pistos C, Dona A, Spiliopoulou C, Athanaselis S. Naphyrone: a legal high not legal any more. Drug Chem Toxicol 2012;35:467–71.

[172] ACMD Desoxypipradrol (2-DPMP) advice. London, UK: Advisory Council on the Misuse of Drugs; 2011.

[173] BBC News. Bristol arrest after mephedrone worth £1m found in car. Available: <http://www.bbc.co.uk/news/uk-england-bristol-15892016>; 2011 [accessed 6/7/12].

[174] EMCDDA Trendspotter summary report. Recent shocks in the European heroin market: explanations and ramifications. Lisbon, Portugal: European Monitoring Centre for Drugs and Drug Addiction; 2011.

[175] Fattore L, Fratta W. Beyond THC: the new generation of cannabinoid designer drugs. Front Behav Neurosci 2011;5:60.

This page intentionally left blank

Epidemiology of Use of Novel Psychoactive Substances

Harry Sumnall*, *James McVeigh** *and Michael J. Evans-Brown*†

*Centre for Public Health, Liverpool John Moores University, Liverpool, UK †European Monitoring Centre for Drugs and Drug Addiction, Cais do Sodré, Lisbon, Portugal

INTRODUCTION

This chapter aims to provide a non-technical overview of some of the key international epidemiological data on novel psychoactive substances (NPS). Major compounds and groups of compounds are considered and where possible, trends in use prevalence are presented. Although many well-established drugs (e.g. psilocybin mushrooms) are not subject to the United Nations (UN) drug conventions and are therefore considered NPS, this chapter generally only focuses on those that have received attention since 2000. Similarly, although some of the first reports of the psychoactive effects of *Salvia divinorum* (henceforth referred to as 'Salvia') were published in the mainstream scientific press in the 1960s (e.g. [1]), it is included here because of the re-emergence in its use in the previous decade [2] and cultural propagation through the Internet [3].

We present prevalence data from major general population surveys in adults and young people. These provide the most robust estimates of use of NPS. Many of the drugs discussed in this chapter only become the focus of

policy and, subsequently, epidemiological interest once they have become controlled under national drugs legislation. This results in two important considerations. Firstly, it means that comparable data obtained using high quality methodologies and sampling techniques, such as those collected as part of well-resourced general population prevalence surveys, are only available on a small number of compounds. Examples of such include drugs sold as mephedrone and 'Spice' (synthetic cannabinoid receptor agonists; but as discussed below there is great uncertainty about the exact prevalence the data on these drugs actually represents). Regardless, it is not yet possible to make international prevalence comparisons. Secondly, it means that little robust data is available on truly novel and emerging psychoactive substances. Whilst more established drugs will have greater potential to produce public health and social burdens (because of the larger total number of users), it might be argued that in territories that operate systems of reactive drug control (i.e. new drugs are controlled once evidence on use prevalence and potential societal harms have been established, compared

Novel Psychoactive Substances.
DOI: http://dx.doi.org/10.1016/B978-0-12-415816-0.00004-3

with systems of generic control), it would be better for health and drug services to be able to develop intervention responses before use becomes established. Whilst general population estimates are essential to illustrate the dispersion of drug trends, it is also important to identify and assess prevalence among key groups and target populations; criteria which few general population surveys are able to fulfill. We therefore also present the findings of a number of important smaller surveys, which tend to use non-probabilistic sampling. Although less useful for policy monitoring and understanding trends in use, the value of this type of research is that, done well, it can provide a rapid assessment of the emergence of NPS and associated user behaviours, which are the essential foundations in developing targeted legal, forensic, social and health responses.

The chapter then focuses on a number of important NPS, namely mephedrone, Salvia and 'Spice' (summarised in Table 4.1). This choice of compounds primarily reflects the limited published epidemiological data on NPS, and although small studies have been conducted with other (relatively obscure) compounds (e.g. 4-hydroxy-N-methyl-N-ethyl-tryptamine) [4], these tend to exclusively recruit self-identified users of that drug and not compare use profiles with the wider population of substance user. In order to preserve the focus of the chapter we have also chosen not to include ethnographic data or studies which present prevalence data as part of examinations of the acute and sub-acute harms of use (e.g. cognitive functioning in recent users compared with non-using controls), although some data is included on route of administration where relevant (e.g. injection of mephedrone).

Methodological development in this area is extremely important, and findings from forensic analysis of products sold as 'legal highs' indicate that caution is warranted when interpreting the prevalence estimates provided here. NPS products, particularly powders, have consistently been found to be mislabelled, or contain other unlisted products [5,6]. There is a lack of consistency in constituent product chemicals across time [7] and many are only known to users as generic slang names (e.g. bubble vs 2-aminoindane) [8]. Therefore many users, unless they have access to sophisticated testing equipment, will be unaware of the actual product(s) that they have consumed. Generic white powders cannot be differentiated by sight alone and thus endorsing use of a product in a survey is unlikely to provide accurate prevalence estimates. We present a number of novel methodologies which may assist in improving the robustness of NPS prevalence estimates and conclude the chapter with a data table summarising studies in this area.

GENERAL POPULATION SURVEYS

Surveys in Young People

The USA's annual Monitoring the Future is a large and robust annual school survey which assesses substance use prevalence in 50 000 secondary school children (school grade 8, 13–14 yrs of age; grade 10, 14–16 yrs; and grade 12, 17–18 yrs). Synthetic cannabinoid receptor agonists were included in the survey for 12th grade students for the first time in 2011 (reported as synthetic marijuana and under the generic trade names Spice and K2) [9]. Although a number of synthetic cannabinoid receptor agonists were legally controlled in the USA in 2011, the survey took place shortly after these measures were imposed, and so provide a good indication of pre-legislation prevalence. 11.4% of 12th grade students reported use of synthetic marijuana in the previous 12 months, and of these, 94% also reported cannabis use in the same period, 31% were current (past 30-day) daily users of cannabis, 60% reported some illicit drug other than cannabis in the past year, 54% smoked cigarettes in the past 30 days, and 79%

Summary of Main Findings Reported in the Text

ug	Country	Year	Period Prevalence (%)			Survey Type	Population	Sourc
			LTP	LYP	LMP			
neral								
›vel Psychoactive ›bstances (general)	Ireland/ Northern Ireland	2011	–	6.7; 1.0; 3.5	–	Representative household survey	15–34; 35–64; 15–64 yrs	[21]
›her hallucinogens ‹cluding DMT, ‹IT, 5-MeO-DIPT, ‹d Salvia divinorum)	USA	2006	–	0.2	–	Representative household survey (NSDUH)	12–65+	SAMF [12]
‹al high powders	Ireland	2010	57.0	–	–	Convenience sample of self-identified NPS users	24.8 ± 6.8 yrs, 67% male	[24]
‹al tablet or liquid	Ireland	2010	48.0	–	–	Convenience sample of self-identified NPS users	24.8 ± 6.8 yrs, 67% male	[24]
‹nobotanicals	Ireland	2010	60.0	–	–	Convenience sample of self-identified NPS users	24.8 ± 6.8 yrs, 67% male	[24]
›phedrone	UK	2011/12	–	3.3; 1.1	–	Representative household survey (CWES)	16–24 yrs; 16–59 yrs	[14]
	UK	2010/11	–	4.4; 1.4	–	Representative household survey (BCS)	16–24 yrs; 16–59 yrs	[13]
	Northern Ireland	2011	2.0	–	–	Representative household survey	15–64 yrs	[21]
	UK	2011/2012	–	30.0	13.0	Convenience sample of magazine/newspaper readers	Mean age 28.3 yrs; 69.7% male (UK respondents only)	[25]
	International	2011	61.0	51.0	–	Convenience sample of clubbers	Mean age 25 yrs; 69% male	[23]
	Australia	2010	21.0	–	–	Sentinel survey of ecstasy users	24.0 ± 6.0 yrs, 58% male	[26]
	UK	2011	63.8	–	53.2	Convenience sample of clubbers	Mean age 29.7 yrs; 82% male	[27]
	UK	2010	13.0	11.0	5.0	Convenience sample of nightlife patrons	Mean age 23.8 yrs	[8]

(Contir

TABLE 4.1 (Continued)

Drug	Country	Year	Period Prevalence (%)			Survey Type	Population	Sc
			LTP	LYP	LMP			
	UK	2010	54.0	52.0	41.0	Convenience sample of clubbers	29.8 ± 8.0; 82% male	[2
	Northern Ireland	2010	66.2	–	–	Convenience sample of self-identified NPS users	24.8 ± 6.8 yrs, 67% male	[2
	Scotland (UK)	2009/10	20.3	–	–	Convenience sample of students	Age range 13–24 yrs	[2

SYNTHETIC CANNABINOID RECEPTOR AGONISTS

Drug	Country	Year	LTP	LYP	LMP	Survey Type	Population	Sc
	USA	2012	–	11.4	–	Representative school survey (Monitoring the Future)	12th grade students	[9
	UK	2010/11	–	0.4; 0.2	–	Representative household survey (BCS)	16–24 yrs; 16–59 yrs	[1
	Germany	2009	7.0	–	1.0	Representative school survey (city region)	Age range 15–18	[3
	UK	2011/2012	–	5.0	2.0	Convenience sample of magazine/newspaper readers	Mean age 28.3 yrs; 69.7% male (UK respondents only)	[2
	UK	2011	9.0	–	0.6	Convenience sample of clubbers	Mean age 29.7 yrs; 82% male	[2
	International	2010	21.7	–	2.0	Convenience sample of clubbers	18–27 yrs; 65% male	[2
	International	2011	10.3	2.2	–	Convenience sample of clubbers	Mean age 25 yrs; 69% male	[2

SALVIA DIVINORUM

Drug	Country	Year	LTP	LYP	LMP	Survey Type	Population	Sc
	USA	2006	–	1.7	–	Representative household survey (NSDUH)	18–25 yrs	S 20
	USA	2012	–	1.6; 3.9; 5.9	–	Representative school survey (Monitoring the Future)	8th; 10th; 12th Grade students	[9
	International	2010	29.2	–	3.2	Convenience sample of clubbers	Mean age 25 yrs; 69% male	[2

	Country	Year				Sample	Demographics	Reference
	USA	2006/07	–	4.4	–	Random sample of University Students	19.2 ± 2.0 yrs	Lange et al., 2008
	USA	2006	6.7	3.0	0.5	Convenience sample of University Students	19.2 ± 2.0 yrs	[31]
	USA	2009	6.7	–	–	Convenience sample of University Students	20.0 ± 2.1 yrs	[32]
	Ireland	2010	61.1	–	–	Convenience sample of self-identified NPS users	24.8 ± 6.8 yrs, 67% male	[24]

HER DRUGS (SELECTED DATA)

	Country	Year				Sample	Demographics	Reference
J	Ireland	2010	1.6	–	–	Convenience sample of self-identified NPS users	24.8 ± 6.8 yrs, 67% male	[24]
B	Australia	2010	9.0	–	–	Sentinel survey of ecstasy users	24 ± 6 yrs, 58% male	[26]
	UK	2011/2012		12.0	4.0	Convenience sample of magazine/newspaper readers	Mean age 28.3 yrs; 69.7% male (UK respondents only)	[25]
	International	2011	18.0	8.3	–	Convenience sample of clubbers	Mean age 25 yrs; 69% male	[23]
E	Australia	2010	3.0	–	–	Sentinel survey of ecstasy users	24 ± 6 yrs, 58% male	[26]
I	Australia	2010	6.0	–	–	Sentinel survey of ecstasy users	24 ± 6 yrs, 58% male	[26]
	UK	2011/2012	–	6.0	1.5	Convenience sample of magazine/newspaper readers	Mean age 28.3 yrs; 69.7% male (UK respondents only)	[25]
	International	2010	11.4	–	1.3	Convenience sample of clubbers	18–27 yrs; 65% male	[22]
	International	2011	9.9	4.1	–	Convenience sample of clubbers	Mean age 25 yrs; 69% male	[23]
AI	Ireland	2010	1.0	–	–	Convenience sample of self-identified NPS users	24.8 ± 6.8 yrs, 67% male	[24]

(Contin

TABLE 4.1 (Continued)

Drug	Country	Year	Period Prevalence (%)			Survey Type	Population	S
			LTP	LYP	LMP			
5-MeO-DMT	Australia	2010	2.0	–	–	Sentinel survey of ecstasy users	24 ± 6 yrs, 58% male	[2
BZP	UK	2010/11	–	0.1; 0.2	–	Representative household survey (BCS)	16–24 yrs; 16–59 yrs	[1
	Australia	2010	8.0	–	–	Sentinel survey of ecstasy users	24 ± 6 yrs, 58% male	[2
	UK	2011/2012	–	1.5	0.5	Convenience sample of magazine/newspaper readers	Mean age 28.3 yrs; 69.7% male (UK respondents only)	[2
	UK	2011	9.3	–	0.6	Convenience sample of clubbers	Mean age 29.7 yrs; 82% male	[2
	International	2011	17.2	5.0	–	Convenience sample of clubbers	Mean age 25 yrs; 69% male	[2
	Ireland	2010	37.1	–	–	Convenience sample of self-identified NPS users	24.8 ± 6.8 yrs, 67% male	[2
DOI	Australia	2010	2.0	–	–	Sentinel survey of ecstasy users	24 ± 6 yrs, 58% male	[2
Ketamine	UK	2010/11	–	2.1; 0.6	–	Representative household survey (BCS)	16–24 yrs; 16–59 yrs	[1
	UK	2010	57.0	46.0	28.0	Convenience samples of clubbers	29.8 ± 8.0; 82% male	[2
Khat	UK	2010/11	–	0.2; 0.3	–	Representative household survey (BCS)	16–24 yrs; 16–59 yrs	[1
mCPP	Ireland	2010	2.5	–	–	Convenience sample of self-identified NPS users	24.8 ± 6.8 yrs, 67% male	[2
Methoxetamine	UK	2011	6.4	–	1.6	Convenience sample of clubbers	Mean age 29.7 yrs; 82% male	[2

	Region	Year	LTP	LYP	LMP	Sample	Demographics	Ref
...thylone	UK	2011/2012	–	2.0	0.5	Convenience sample of magazine/newspaper readers	Mean age 28.3 yrs; 69.7% male (UK respondents only)	[25]
	International	2011	13.7	9.0	–	Convenience sample of clubbers	Mean age 25 yrs; 69% male	[23]
	International	2010	10.8	–	7.5	Convenience sample of clubbers	18–27 yrs; 65% male	[22]
...DAI	UK	2011/2012	–	3.0	0.5	Convenience sample of magazine/newspaper readers	Mean age 28.3 yrs; 69.7% male (UK respondents only)	[25]
	UK	2011	7.7	–	0.0	Convenience sample of clubbers	Mean age 29.7 yrs; 82% male	[27]
	International	2011	6.7	4.7	–	Convenience sample of clubbers	Mean age 25 yrs; 69% male	[23]
	UK	2010	6.0	6.0	4.0	Convenience samples of clubbers	29.8 ± 8.0; 82% male	[28]
	Ireland	2010	2.7	–	–	Convenience sample of self-identified NPS users	24.8 ± 6.8 yrs, 67% male	[24]
...DPV	Australia	2010	1.0	–	–	Sentinel survey of ecstasy users	24 ± 6 yrs, 58% male	[26]
	UK	2011/2012	–	0.5	<0.1	Convenience sample of magazine/newspaper readers	Mean age 28.3 yrs; 69.7% male (UK respondents only)	[25]
	International	2011	4.4	3.0	–	Convenience sample of clubbers	Mean age 25 yrs; 69% male	[23]

..., lifetime prevalence; LYP, last year prevalence; LMP, last month prevalence; AMT, α-methyltryptamine; 2-AI, 2-Aminoindan; 2-CB, 2,5-dimethoxy-4-bromo-phenethylamine; 2-CE, ...-dimethoxy-4-ethyl-phenethylamine; 2-CI, 2,5-dimethoxy4-iodo-phenethylamine; 5-IAI, 5-Iodo-2-aminoindane; 5-MeO-DIPT, 5-methoxy-diisopropyltryptamine (foxy/foxy methoxy), ...1eO-DMT, 5-methoxy-N,N-dimethyltryptamine; BZP, 1-benzylpiperazine; DMT, Dimethyltryptamine; DOI, 2,5-dimethoxy-4-iodoamphetamine; mCPP, meta-Chlorophenylpiperazi...; ...phedrone, 4-methylmethcathinone; Methylone, 3,4-methylenedioxy-N-methylcathinone; MDAI, 5,6-Methylenedioxy-2-aminoindane; MDPV, 3,4-methylenedioxypyrovalerone.

used alcohol in the past 30 days (Johnson, personal communication). Although it is uncertain if some users of synthetic marijuana reported use of this substance alone, it is clear that typical users had experience with a wide range of substances and for this sample at least, being a synthetic marijuana user does not suggest emergence of a novel group of substance users with a unique preference for NPS. The plant hallucinogen Salvia was included in earlier surveys but 2011 data had not, at the time of writing, been released for secondary analysis. In 8th grade students, last year prevalence (LYP) of Salvia fell from 1.7% in 2010 to 1.6% in 2011; in 10th grade students the figures were 3.7% and 3.9%; whilst in 12th grade students last year prevalence in 2009 was 5.7%, 5.5% in 2010, and 5.9% in 2011 [9]. Of note, reporting of Salvia in 12th grade students over the previous three years was greater than that of ecstasy (4.3%; 4.5%; 5.3%).

The European School Survey Project on Alcohol and Other Drugs (ESPAD; www.espad.org) is the largest cross-national survey of adolescent substance use in the world [10]. In some respects, although there are differences in methodology, it may be regarded as a European equivalent to Monitoring the Future, and its publications include comparison with this US data. It reports every four years and the current wave, 2011, reporting in summer 2012, included the participation of 39 countries. In the 2011 questionnaire, there was the option of including items relating to the prevalence of drugs in addition to those typically assessed (e.g. tobacco, alcohol, ecstasy, cocaine, amphetamines, cannabis). Some countries included items relating to NPS, but at the time of writing this data has not been released. As other pan-EU surveys (e.g. Eurobarometer, see below) are opportunistic in nature, ESPAD potentially provides a good vehicle to assess time trended prevalence of some of the more well established NPS, although as shall be discussed later accurate user identification of NPS may mean that the utility of this data is limited.

Adult Population Surveys

Three nationally representative surveys (available in the English language), the USA National Survey on Drug Use and Health (NSDUH), the UK's British Crime Survey (BCS), and the Drug Use in Ireland and Northern Ireland Survey present data on NPS epidemiology in adult populations, but the number of drugs included are limited to those of most interest to policy makers (e.g. GHB, mephedrone, hallucinogens, piperazines, ketamine, khat, synthetic cannabinoid receptor agonists).

The NSDUH is a representative household survey including data from around 70 000 respondents aged 12 and older [11]. The survey is generalisable to the non-institutionalised population of the USA, and so underestimates use in populations as prisoners, the military and those in residential treatment. Although the survey focuses on classic illicit drugs, classified under 'other hallucinogens' are substances of relevance to this chapter such as dimethyltryptamine (DMT), alpha-methyltryptamine (AMT), N,N-diisopropyl-5-methoxy-tryptamine (Foxy), and Salvia. It was estimated in 2006 that almost 700 000 persons aged 12 or older had used DMT, AMT or 'Foxy' in their lifetime, and approximately 100 000 had done so in the past year [12]. Last year prevalence rates were less than 0.2% for all of these drugs. About 1.8 million persons aged 12 or older used Salvia in their lifetime and approximately 750 000 did so in the past year (1.7% last year use in 18–25 year olds, greater than LSD at 1.2%).

The BCS is a household survey of over 27 000 adults aged 16–59 in England and Wales [13]. It is the largest self-reported experiences of crime survey in the UK and through its dedicated drugs module is one of the most important tools for monitoring UK government drugs policy. In 2009, in parallel with their addition to the Misuse of Drugs Act, 1971 as Class B drugs), questions on benzylpiperazine (BZP), gamma-butyrolactone/gamma-hydroxybutyrate

(GBL/GHB), and Spice (i.e. generic synthetic cannabinoid receptor agonists) were added to the BCS (although not controlled in the UK, khat was also included). Ketamine had been added in 2006/07, again after control under the Misuse of Drugs Act. In 2011, and in response to it being made a Class B drug the year before, mephedrone was added to the survey. In 2010/2011 last year prevalence of Spice (0.2% in 16–59 year olds; 0.4% in 16–24 year olds), BZP (0.1%; 0.2%), GBL/GHB (0.0%; 0.1%) and khat (0.2%; 0.3%) were all extremely low [13]. Last year prevalence of ketamine in adults was 0.6% (doubling since 2006/07), and in 16–24 year olds it was 2.1%. Whilst last year use of mephedrone in the entire sample (16–59 year olds) was relatively low, 1.4% (comparable to ecstasy), use in 16–24 year olds was 4.4%, similar to the prevalence of powder cocaine (also 4.4%). Although the reporting period included dates before mephedrone was controlled in the UK (April 2010), it is clear that mephedrone has achieved substantial market penetration. Preliminary data released from the 2011/12 survey (renamed the Crime Survey for England and Wales; CSEW) indicated that last year prevalence of mephedrone in both 16–59 (1.1%) and 16–24 (3.3%) year olds had fallen (whereas powder cocaine had fallen only slightly to 4.2%) [14]. Last year use was highest among those who consumed alcohol on three or more days a week during the past month (1.9%) compared with those who drank less frequently; for example, those who drank less than a day a week in the past month (0.6%), and those who had not had a drink in the last month (0.2%). This data is important as it referred entirely to a reporting period of when mephedrone was controlled under UK law.

Further analysis of the 2011/2012 BCS survey indicated that in last year users of mephedrone, 91% had also taken another illicit drug; 72% reporting cannabis, 53% cocaine, and 48% ecstasy. It is uncertain whether the additional 9% had experienced a lifetime, but not last year, use of other illicit drugs as this data was not presented, or whether in fact mephedrone was their first experience with an (illicit) substance. Bird [15] extended this line of enquiry through a secondary analysis of 2010/2011 data. In the online blog *Straight Statistics*, she estimated that around 23% of 16–24 year olds had used mephedrone but not ecstasy or cocaine powder, and suggested that mephedrone had displaced the two drugs. Although plausible, there is little additional behavioural or forensic data to support this claim. It is worth noting that the prevalence of cocaine and ecstasy use in this age group had been falling for several years prior to the introduction of mephedrone [13] and the purity of street samples of both has been low for many years [16,17]. So although according to the BCS mephedrone may have at least temporarily displaced other club drugs, it is possible that this was due to a combination of lack of availability of other drugs of acceptable quality, rather than consumer choice or preference for the subjective effects of mephedrone [18,19]. Supporting this interpretation is data from UK street and nightclub surveys, conducted by Measham and colleagues [8,20], which indicates that nightlife patrons had added mephedrone to their existing repertoire of drug use.

'Drug use in Ireland and Northern Ireland' is a household survey conducted with adults aged 15–64 years of age, and has a sample size of 7669 across the two territories [21]. Last year use of new psychoactive substances was included for the first time in the 2010/11 survey. In Ireland this list included herbal smoking mixtures/incense, party pills or herbal highs, bath salts, plant feeders or other powders, Kratom (Krypton), Salvia, Magic Mint, Divine Mint or Sally D and any other new psychoactive substances mentioned by the respondent. In Northern Ireland, the category Legal Highs included party pills, herbal highs, party powders, Kratom and Salvia; it did not include mephedrone which was assessed separately. In the whole sample 3.5% reported use of NPS in this time period, predominately by 15–34 year olds (6.7% vs 1% in 35–64 years olds). In Northern Ireland lifetime use of mephedrone was estimated at 2.0%.

Non-probabilistic Convenience Sample Surveys

Convenience sample surveys are those studies which utilise a sample that is readily available and easy to access. They are non-probabilistic surveys, meaning that data obtained cannot be generalised to the wider population and are therefore only valid for the population studied. It is the most frequently employed sampling strategy in NPS epidemiology as it is a relatively easy way to quickly gather data of interest (often conducted online, mirroring the source of access to NPS for many, and sometimes run in conjunction with suppliers). Whilst providing some useful data, for example identifying the emergence of new drug trends and patterns of use within certain sub groups, they cannot be considered representative, are subject to many biases, and should be interpreted with caution. These types of survey are therefore of less value to national policy monitoring, but still of importance in formulating health and social responses to use.

One of the most well-known of these types of surveys is the annual Mixmag survey, commissioned by the (UK) dance music magazine of the same name in collaboration with UK researchers. The Mixmag survey has reported on mephedrone since 2010 [22] (data collected in 2009; Internet and paper delivery), when lifetime and last month prevalence of use were 42% and 34% respectively. In 2011 [23] (N = 2560; Internet delivery) prevalence of use in the two reporting periods was 61% and 25%, although as samples were independent no conclusions can be drawn about trends in use. Furthermore, the most recent reporting period was changed to the previous year, rather than previous month. Perhaps the most interesting observation from the 2011 Mixmag

data is the increase in the range of drugs used by respondents, and whilst these were not spontaneously reported and there are no controls to ensure veracity of reporting, a large number had lifetime prevalence rates of >1%. Rates of lifetime and last month/year were also similar, perhaps reflecting both the recent emergence of these drugs, recent initiation of use, and the continued popularity of many of these substances. Other data are summarised in Table 4.2. The Mixmag survey was repeated in 2012, but this time also in conjunction with the UK's *Guardian* newspaper and renamed the Global Drugs Survey [25]. As such, participants included individuals other than readers of *Mixmag* (the survey link was shared online and so cannot also be said to represent *Guardian* readers either). Although the number of respondents increased to 15 500 (primarily from the UK and USA), this is still an unrepresentative cross-sectional survey and, at the time of writing, data has not been scrutinised by peer review. Although period prevalence of mephedrone use appeared to have decreased since 2010, as participants were drawn from a number of countries it is not possible to determine whether there has been a decline in use since the UK ban in April 2010.

The Australian Ecstasy and Related Drugs Reporting System (EDRS) is a surveillance system for the ecstasy market, which aims to monitor emerging trends in drug use epidemiology and related issues in the capital cities of each Australian territory [33]. Data is obtained from interviews with users, key informants, and through the interrogation of indicator data sources such national household surveys and forensic systems [34]. This is a sentinel survey and the research population (n = 693 in 2010) was recruited through a purposive sampling strategy, and represents a group with particularly high use of ecstasy and related drugs (inclusion criteria is at least a monthly use in the previous 6 months). Although this means generalisability of the findings are

TABLE 4.2 Summary of the 2010/2011 Mixmag and the 2012 Global Drugs Survey NPS Survey Data

	Period prevalence %					
	Mixmag Survey				Global Drugs Survey	
	2010		2011		2012	
	LTP	LMP	LTP	LYP	LYP	LMP
Mephedrone	41.7	33.6	61.0	51.0	30.0	13.0
Salvia divinorum	29.2	3.2	NR	NR	NR	NR
Spice/synthetic cannabinoid receptor agonists	21.7	2.0	10.3	2.2	5.0	2.0
Methylone	10.8	7.5	13.7	9.0	2.0	0.5
2-CI	11.4	1.3	9.9	4.1	6.0	1.5
2-CB	NR	NR	18.0	8.3	12.0	4.0
BZP	NR	NR	17.2	5.0	1.5	0.5
MDAI[a]	NR	NR	6.7	4.7	3.0	0.5
MDPV	NR	NR	4.4	3.0	0.5	<0.1
Benzofury[b]	NR	NR	2.7	2.3	3.0	1.0

It is important to note that although there have been peer review publications derived from data extracts from the Mixmag surveys, the data here is extracted from popular media reporting (authored in collaboration with the survey PI). Furthermore, whilst the Mixmag surveys were primarily targeted at clubbers, participation in the 2012 Global Drugs Survey was apparently much broader, hence for comparison purposes only data from clubbers is included here (this was a classification variable used in the Global survey).
LTP, lifetime prevalence; LYP, last year prevalence; LMP, last month prevalence; NR, not reported in this year; Methylone, 3,4-methylenedioxy-N-methylcathinone; 2-CI, 4-iodo-2,5-dimethoxyphenethylamine; 2-CB, 4-bromo-2,5-dimethoxyphenethylamine; BZP, benzylpiperazine.
[a]MDAI, 5,6-Methylenedioxy-2-aminoindane (while analysis of some products have confirmed the presence of MDAI, others have been found to be misbranded and have contained mephedrone or BZP and 3-TFMPP)[24,14] MDPV, methylenedioxypyrovalerone.
[b]benzofury is a generic name for a range of synthetic psychostimulants. It is allegedly 6-(2-aminopropyl)benzofuran or 1-benzofuran-6-ylpropan-2-amine (6-APB), but samples have also been found to contain BZP, caffeine and 3-TFMPP[24].

limited, the methodology is robust and consistent, and therefore provides a rich and important insight into emerging patterns of NPS use and changes in use behaviours that are lacking in many small cross-sectional surveys. In 2010, 25% of the surveyed population (mean age 24 ± 6 years) reported a lifetime use of a stimulant-like NPS, whilst 23% reported use of a hallucinogen-like NPS [26,34]. Despite popular accounts of their ubiquity, it must be noted therefore that the majority of this sample, despite being frequent club drug users, did not report use of a NPS, and in those that did, frequency of use was low. Compared with

previous survey years, only mephedrone prevalence had appeared to increase substantially; perhaps related to the increased availability and publicity afforded to this drug in recent years. Mephedrone was used by 17% of the sample (compared with last year prevalence of use of 51% in the 2011 Mixmag survey [23] and 30% in the Global Drugs Survey [25]), with an equal proportion of the sample reporting self-administration by insufflation or oral routes (~65%); median days used in the last six months was three. Recent benzylpiperizine (BZP) prevalence of use was 4.5%, and it was reportedly used on a median of two days, and Ivory Wave

(MDVP) was used by 0.5%, on a median of 1.5 days in the last six months. Phenethylamines such 2,5-dimethoxy-4-iodophenethylamine (2C-I), 4-bromo-2,5-dimethoxyphenethylamine (2C-B), 2, 5-dimethoxy-4-ethylphenethylamine (2C-E), and 2, 5-dimethoxy-4-iodamphetamine (DOI) were used in the previous six months by 1–2% of the sample. Median days of use for all five drugs were one day in the last six months. Last six month prevalence of tryptamine hallucinogens was similar, but dimethyltryptamine (DMT; median days of 1.5 days in the last six months), had been used by 7% of the sample. Bruno and colleagues [26] examined differences in the characteristics of participants who had never used an NPS (71%) with recent stimulant (15%) and hallucinogen (13%) NPS users in the sample. Overall, there was little difference between the groups, although stimulant NPS users were more likely to be under the age of 21, report more frequent use of other stimulants, and report being more likely to engage in risky sex. Users of hallucinogen-like NPS were more likely to be male, initiate ecstasy use at a younger age, and reported greater use of a range of drugs than the non-NPS sample. This population was distinguished in other ways too; they were more likely to have accessed psychosocial assistance for drug related problems, and reported higher rates of criminal involvement such as property crime and drug dealing.

In their surveys of patrons of gay nightclub nights in London, UK, Measham and colleagues [27,28] (mephedrone data described below) recorded use of a wide variety of substances. This population is believed to be important because a greater proportion of gay men in particular report substance use than the general population [27], and they are also considered 'early adopters' of a wide range of new drugs (although this may reflect bias in funding and research priorities); hence use of emerging NPS in this population may reflect later trends in others [35]. In a convenience sample of 308 patrons conducted in 2010, lifetime use of mephedrone was reported by 54%; MDAI 6%; and NRG-1 16% (many samples of which have also been forensically identified as mephedrone) [6]. For comparison, 69% reported a lifetime use of ecstasy. Recent use (in the last month) of these drugs was 41%; 4%; and 11% respectively (vs 28% MDMA). Interestingly, on the survey day, more people reported use of mephedrone than ecstasy (27% vs 15%; perhaps reflecting the poor quality of ecstasy tablets available in Western Europe in recent years) [17]. In their 2011 follow-up survey, lifetime (63.8%) and last month (53.2%) use of mephedrone was higher (there was no indication whether those participating had also completed the previous years' study). Other drugs reported included BZP (9.3%; 1.6% lifetime, last month prevalence), MDAI (7.7%; 1.3%) and methoxetamine (6.4%; 1.9%). Methoxetamine, a ketamine derivative, was subject to the UK's first Temporary Class Drug Order in April 2012, but was legal to possess and sell at the time of survey.

Bodybuilders are often overlooked in investigations of NPS. This group contains many innovators and early adopters of a large range of different substances [36,37]. In the context of this chapter two psychoactive substances are worth noting. Gamma-hydroxybutyrate was first advertised to, and used by, bodybuilders [38] to try and increase growth hormone secretion [39–43], which is used to build muscle and strip fat. Shortly thereafter it was sold as an anti-ageing/life-extension 'treatment' to other groups [38–52] before it became more widespread on the party scene as a recreational drug [53]. More recently, in 2010, bodybuilding supplements containing the stimulant desoxy-diphenylprolinol were being sold in the UK [54,55].

The availability of NPS in Ireland is controlled by the Criminal Justice (Psychoactive Substances) Act 2010 (Criminal Justice [Psychoactive Substances] Act 2010. See Section 1 for definition). It came into effect on

23rd August 2010. SI No. 401/2010 – Criminal Justice (Psychoactive Substances) Act 2010 (Commencement) Order 2010 prohibits the importation, advertising and sale of 'psychoactive substances'(excluding medicinal products, alcohol, tobacco and certain foods, plant, fungus or natural organisms), including potentially all NPS (although exceptions exist to preserve *inter alia* medicines and legitimate research use; production, supply, and possession of named substances is still controlled under the Misuse of Drugs Act Ireland). This legislation was partly informed by a review of the availability and use of NPS in Ireland, including marketing, forensic data, and user surveys. The associated report provides a useful, if time limited insight into the availability of NPS in Ireland in early 2010, and the type of data available to policy makers who are constructing legislation [24]. A convenience sample of 333 respondents (66.6% male, mean age 25.0 ± 6.8 years; 63% students) was recruited through national newspaper advertisements, social networking sites, key informants, and online forums discussing drug related issues. Use of at least one legal high powder during lifetime was reported by 57%; 48% reported use of NPS in tablet or liquid form (e.g. ecstasy substitutes), 60% smoking blends (e.g. synthetic cannabinoid receptor agonists) and 38.3% ethnobotanicals (e.g. Salvia). Most substances reported were generic brand names available in Ireland at the time, and so it is difficult to assess prevalence of specific compounds. Of those mentioned by name, mephedrone was reported by 66.2%, *Salvia divinorum* 61.1%; BZP 37.1%; DMT 4.4%; GBL/GHB 4.4%; mCPP 2.5%; MDAI 2.7%; 2-AI 1.6%; and 5-IAI 1.0%. No one reported use of 6-APB, MDPV or butylone. Some data were collected on route of administration; the most frequently reported was insufflation (85.5% for powders, 15.1% for tablets); followed by rubbing on gums or the inside of the mouth (40.3%; 6.3%), and through bombs (drug wrapped in cigarette rolling paper and swallowed: 28%; 13.2%). No individuals reported administration by injection, which is in contrast to data reported by other researchers in Ireland [56], although that was a study of low threshold harm reduction service clients who were already injecting drugs.

Mephedrone

Mephedrone is perhaps the best characterised of the recent NPS, and studies have identified use in countries such as France, Slovenia, Australia, Ireland and the UK [54]. However, epidemiological data is still limited [19]. The most robust estimate come from the UK, and as described above, the CSEW estimated last year use in 16–24 year olds (2011/12 data sweep) to be 3.3%, less than powder cocaine, but the same as ecstasy [14]. In Northern Ireland lifetime use of mephedrone was estimated at 2.0% [21]. Dargan and colleagues [29] had provided an earlier estimate in school students surveyed in the Tayside area of Scotland (UK) before mephedrone became a controlled drug. Of the 1006 students that completed the survey (mean ages of the samples ranged from 13–24 years), 20.3% reported use of mephedrone on at least one occasion, while 4.4% reported daily use. Unsurprisingly, prevalence was age related and rates of occasional use increased from approximately 39% in 13–15 year olds (point prevalences were not reported) to 61% in 22–24 year olds. As this data was collected from a pre-ban sample (April 2010 in the UK), it would be interesting to repeat this survey in order to consider the effectiveness of legislation on mephedrone use, particularly by younger pupils, who may not have the same access to illegal drug sellers as older students.

One recent Internet survey examined mephedrone use behaviours in an older population in the UK and attempted to assess the potential effects of the 2010 ban [18]. The mean age of the 1506 respondents was 26.0 ± 9.0 years, 84% were male, and the median number of lifetime uses was 11–50. The authors also reported that 64% of self-identified UK mephedrone users (n = 1265) stated that they would use less mephedrone and 49% more MDMA after mephedrone was controlled. However, as we have argued elsewhere [19] this paper provided an incomplete assessment of the effects of legislation and so these data must be interpreted with caution. It is feasible though that the emergence of mephedrone may have significantly affected the use of ecstasy/MDMA, even if this many users were unaware of this. Brunt and colleagues [17] provided an indirect assessment of the penetration of mephedrone in the ecstasy/MDMA market in the Netherlands in 2009. Using submissions to the Dutch national Drug Information and Monitoring System (DIMS), a toxico-epidemiological monitor of illegal drug markets, it was found that the amount of ecstasy tablets submitted for analysis that contained MDMA began to fall substantially in mid 2008. By 2009, 11.5% of the total amount of ecstasy tablets submitted contained mephedrone at doses of between 96 and 155 mg; this was also the first and only substituted cathinone derivative identified. As the Netherlands are an important source of UK ecstasy, UK consumers may have unwittingly been consuming mephedrone whilst believing they had bought ecstasy/MDMA.

As expected [20,57], mephedrone use is highest in night club attendees, and the Mixmag survey (see above) estimated lifetime and last year prevalence of use at 61% and 51% respectively. In another convenience sample of 308 attendees of London gay clubs (82% male), 41% reported use in the previous month, and 27% reported an intention to use on the night of survey [28]. An Australian estimate of 572 'same sex attracted' participants, yielded a last month prevalence of use estimate of 1.4%, but these subjects were not specifically recruited on the basis of being nightclub attendees and this sample comprised a higher proportion of women (44.4%) than the London survey. The type of nightlife patron also seems important in determining mephedrone prevalence [20]; when Measham and colleagues [8] investigated use in more 'mainstream' UK nightlife (i.e. participants were surveyed on main thoroughfares in small towns/cities in the UK) they estimated last month mephedrone prevalence to be 5% (n − 207). Interestingly, when asked about use of the drug 'bubble' (local slang term for mephedrone), 9% reported last month use, which as discussed below has important implications for the development of NPS epidemiology.

A secondary analysis of the 2009 Mixmag survey (~950 mephedrone users), indicated that the majority (69.7%) used mephedrone monthly or less [58]. Some (15.1%) reported weekly or more frequent consumption and 15.2% reported use every two weeks. Carhart-Harris and colleagues [17], in keeping with findings of other surveys, reported that mephedrone was preferentially administered by the nasal (57%) and oral (28%) routes, and that only a minority (3%) claimed to inject it. This distribution of administration routes is also reflected in (UK) hospital presentations [59]. It is well established that existing injecting drug users (IDUs) add so-called recreational drugs to their injection repertoire, but some users also transition from non-injecting to injecting practices whether because they prefer this way of taking the substance, a desire for more rapid onset of effect, curiosity, peer influence, cost effectiveness, or avoidance of intranasal/gastric irritation (e.g. [60,61]). Van Hout and Bingham [56] examined mephedrone injection in established injectors in more detail. In an Irish treatment-seeking population they identified a number of reasons why mephedrone was injected, including mephedrone's legal status (uncontrolled at the time of interview), ease of availability, low price, perceived psychopharmacological similarities to MDMA and cocaine, decline in heroin availability in the region, the normalisation of injecting drug use, homelessness and other experiences of social exclusion.

Synthetic Cannabinoid Receptor Agonists

The Monitoring the Future survey estimated last year use of synthetic cannabinoid receptor agonists to be 11.4% in 12th grade students [9], and the BCS 0.4% in 16–24 year olds [13](for comparison the contemporaneous 2011 Mixmag survey [23] reported lifetime prevalence to be 10.3% and last year use 2.2%). A 2009 survey of 1157 students aged between 15 and 18 years in Frankfurt, Germany ('Spice' products such as JWH-018, CP 47,497and the C6, C8 and C9 homologues of CP 47,497 were made controlled drugs in Germany in 2009), found that 7% of respondents reported having used 'Spice' at least once in their lifetime, although there a was a fall in recent use from the year before (last 30 days, 1%) [30]. Of this subsample, 95% were existing cannabis users. The United Nations Office on Drugs and Crime [62] reported that synthetic cannabinoid receptor agonists were still popular in Germany after the 2009 ban, especially in users having to undergo regular drug screenings, as they are not detectable by most existing routine toxicological screens. For example, one toxicological analysis of German forensic psychiatric patients showed that 56.4% of 101 serum samples taken from 80 patients tested positive for synthetic cannabinoid receptor agonists (JWH-015, JWH-018, JWH-073, JWH-081, JWH-250) [63].

One small Internet-based survey of lifetime users of synthetic cannabinoid receptor agonists (specifically Spice, which is a generic brand name; n = 168) reported that lifetime use was associated with being male, white and having completed high levels of education (>12 years) [64], although this may also represent typical Internet user demographics [65]. Lifetime users also reported polysubstance use (e.g. 84% also reported use of cannabis), 21% reported Spice was their *drug of choice*, and it was generally obtained from retail vendors (87%), the Internet (38%), or from friends/relatives (29%). Importantly, although half lived in regions where access to Spice was restricted, only 2% reported purchasing the drug from a drug dealer. Whilst one quarter (25%) reported no future intentions to use, 37% met DSM IV criteria for abuse, and 12% dependence; using Spice in a hazardous situation was the most commonly endorsed abuse criteria (27%), and being unable to cut down or stop Spice use (38%), the most commonly reported dependence criteria.

Salvia divinorum

As use is more prevalent (partly because it remains legal in most USA states, and indeed internationally), Salvia is the most well characterised of the hallucinogens included in the NSDUH, and lifetime prevalence of use increased from 0.7% in 2006 to 1.3% in 2008 (an 83% increase) [66,67]. In the surveyed population (aged 12 and older), Salvia use was most commonly reported among recent users of LSD (53.7%), ecstasy (30.1%), heroin (24.2%), phencyclidine (22.4%), and cocaine (17.5%), and secondary regression analysis showed that polysubstance use was the strongest predictor of recent and lifetime use [30]. According to the survey, 43.0% of last year Salvia users and 28.9% of former users (i.e. reporting lifetime, but not last year use) had an illicit or nonmedical drug use disorder compared with 2.5% of nonusers. Ford and colleagues extended this analysis in younger populations to include 2009 data in which they also investigated correlates of use [68]. Data indicated 1.7% of adolescents (aged 12–17) and 5.1% of young adults (aged 18–34) reported a lifetime use of the drug. Among adolescent respondents 51% were male, 58% white, and 54% reported a total family income of $50 000 or greater. Age, sex, and family income were all significant predictor variables of adolescent lifetime Salvia use in a number of theoretically driven regression models. Adolescents who reported cannabis use were over four times more likely to also report Salvia use. In accordance with the social learning theory of substance use [69], lifetime use was predicted by reports of substance use among peers at school, less personal conservative attitudes toward substance use, and reporting parents and friends having less conservative attitudes toward substance use. Among adults 50% of lifetime Salvia users were male, 60% white, and 42% reported a total family income of $50 000 or greater. Predictors of adult use included reporting a major depressive episode, criminality, tobacco, cannabis, prescription and any other type of substance use; whilst being married or in full time employment led to reduced odds of reporting lifetime use. As noted at the start of this chapter, the 2011 Monitoring the Future survey indicated that 5.7% of 12th grade students reported Salvia use in the past year [9].

Other local and population-specific studies have been conducted. Lifetime prevalence of use was estimated to be 29.2% and last month prevalence of 3.2% in readers of the UK dance music magazine *Mixmag*; [22] modal frequency of use was monthly (Winstock, personal communication). In university students in Florida (where *Salvia divinorum* was legal at the time of the study), 11% of males and 4% of females reported a lifetime use [64]. A similar study in California estimated last-year prevalence at 4.4% [3]. Regression analysis in another US college sample showed users were most likely to be young white males with a high prevalence of cannabis use [66]. A study of Italian party goers (n = 2015; mean age 25.1 years), indicated that 11% had used Salvia in their lifetime, and had initiated use at around 20 years of age (vs 16 for cannabis, and 18 for LSD) [102]. Over half of respondents in an Internet-based survey reported reduction or cessation of use in the previous 12 months, most commonly citing dislike of the subjective effects or a loss of interest in the drug [103]. In an Internet-based retrospective survey of use patterns, participants (n =

500) reported a median number of six lifetime uses, and 80.6 reported that they would *probably* or *definitely* take the drug again [75]. Of those endorsing that they would *probably* or *definitely* not use Salvia again (6.6%), 36.4% reported that they didn't like the effects, 30.3% preferred other methods of altering consciousness, and 30.3% reported that the effects were too mild.

Behavioural Epidemiology

Behavioural epidemiology is the study of the distribution of behaviours, risk factors and social contexts that are associated with health related events [70,71]. In substance use research, such indicators may include use environments, attitudes and norms, and is often focused on community rather than clinical populations [72]. Regardless, there needs to be clear analytical and theoretical model linking potential determinants of behaviour with their outcomes [72]. Amongst other social variables, systematic review suggests that social network (e.g. friendship groups or those that share similar cultural interests) norms are one of the most robust (and most studied) social determinants of drug use [74], and there is no reason to suspect that this is not also true for NPS. A 2011 survey conducted on behalf of the European Commission questioned 12 000 young people (aged 15/16–24) in 27 European Member States about availability of, and attitudes towards (i.e. social norms), a range of substances, including NPS (defined as *substances that imitate the effects of illicit drugs*, data not reported by substance) [74]. Although this was a representative sample, the sample size in each country was small and ranged from 250 to 509, and so only the European results may be considered robust. Overall, 5% of the sample reported having used NPS, with most (54%) reporting being offered such substances by friends. Thirty-three per cent bought NPS in a specialist shop and 7% over the Internet, perhaps suggesting caution is warranted in assuming that NPS marketing is primarily driven by the Internet [75]. Respondents were also asked what they thought appropriate legislative responses to NPS should be; most (47%) believed that NPS should only be banned if they posed a risk to health, whilst 34% reported that they should be banned in all circumstances. Fifteen per cent thought that NPS should be subject to some form of 'regulation' (in a similar manner to alcohol and tobacco) although it is uncertain what younger respondents may have understood by this term, being as some professionals in the field also seem to struggle with the complexities of regulatory options available to policy makers [79]. Interestingly, respondents who reported a last year use of cannabis were more likely to endorse regulation as a policy option (24% vs 12%), and those reporting previous use of any NPS were more likely to believe that where bans were introduced, these should only apply to substances shown to be harmful to health.

The 2010 Mixmag survey (see above) included questions assessing the reasons why respondents took NPS [22]. Although point estimates were not reported, *lack of availability of other drugs; not illegal; being able to buy online; and consistent product* (i.e. perceived purity) were the most frequently endorsed; *thinking they are safe* was ranked 7th out of 11 reasons. The Irish survey of Kelleher and colleagues [24] summarised above presented a different set of justifications. For these users, *curiosity* was the most frequently cited reason for use, whilst legal concerns ("*I think they are less easily detected in drug screens than illegal drugs*"; "*I think they are less easily detected by sniffer dogs*") did not appear to influence the decision.

Data from Medical Monitoring Systems

Medical monitoring systems provide a useful indicator of the use of NPS relative to other substances in the general population, and may help identify whether NPS are associated with a change in the number of medical services presentations. A number of these sources are described below for reference. However, without accurate prevalence data, it is impossible to conclude whether the number of NPS presentations is disproportionate to that of other drugs. Where this data might be useful though, is as a data source in *capture–recapture* estimates of hidden population of substance user [77], which uses data from multiple sources (e.g. hospital admissions, drug service databases) to provide an estimate of the population of interest. However, one of the limitations of this technique is that data sources must be independent and data must be attributable to individuals (e.g. contains a pseudoanonymised identifying code) to allow matching of data across databases. Furthermore, non-problematic users of NPS are less likely to be identified on drug service databases, hence these techniques may be less suitable for identifying populations of non-problematic NPS users.

The Drug Abuse Warning Network (DAWN) is a nationally representative USA public health surveillance system that monitors drug related emergency department (ED) visits to hospitals [80]. Reuter [81] reported that the 2009 dataset contained no mention of the popular NPS BZP, Spice, mephedrone or naphyrone, when searched for using both slang and chemical names. The present authors confirm this observation but add that *Salvia divinorum* was mentioned a total of 52 times in the first five recorded drugs.

The American Association of Poison Control Centers (AAPCC) issues annual reports based upon its rapid monitoring system that includes all case data from all USA poison centres submitted to the National Poison Data System (NPDS) [82]. The number of recorded exposures to synthetic cannabinoid receptor agonists (coded as *THC homologs*) was 8264 in 2010/11, whilst for *Bath salts* (mephedrone, MDPV and methylone) the figure was 5624. Requests to the AAPCC for information regarding NPS also increased in this time period, with calls relating to synthetic cannabinoid receptor agonists rising from 3 in 2009 to around 3000 in 2010 [83].

The National Poisons Information Service (NPIS) (commissioned by the Health Protection Agency) in the UK provides advice on the diagnosis, treatment and care of patients who have been poisoned. Data from poisons centres should be interpreted with some caution as it can be difficult to differentiate between individual case presentations, repeat calls to the poisons centre for advice about a case and, particularly in the case of poisons services such as NPIS that have a component of Internet delivered poisons information, information requests related, for instance to general interest or educational reading regarding an NPS. Data may therefore reflect medical interest in the treatment of potential drug poisonings and are thus sensitive to the influence of wider societal discussions (e.g. media campaigns), rather than representing actual presentations and use. Between March 2009 and February 2010 the system had received 1821 mephedrone enquiries, 624 methedrone enquiries (notionally methoxymethcathinone), and 11 MDPV enquiries [59]. Data released in October 2011 suggested that there had been a fall in the number of enquires in how to treat users of mephedrone since April 2010 (when the drug was controlled in the UK), although this may of course have represented increased expertise in the field, and sharing of knowledge thus negating the need for new or repeat calls to the service [76]. Enquiries had peaked in March 2010 at 120, but this had fallen to around 10 a month by January 2011. Other drugs reported in enquiries to NPIS in 2010/2011 included naphyrone, 6-(2-aminopropyl)benzofuran (6-APB), *Ivory Wave*, methcathinone, GHB/GBL, MDPV, MDAI, methylone, methedrone, 2-CB, butylone, desoxypipradrol, BZP, PMA, bromo-dragonfly, 2CI, TFMPP, and synthetic cannabinoid receptor agonists. Of these, only BZP, methedrone, GHB, naphyrone and mephedrone resulted in more than 100 enquiries to NPIS in 2010/2011.

Finally, some data was identified from the Institute of National Toxicology and the Department of Forensic Medicine, University Szeged, Hungary [84]. Mirroring its emergence elsewhere, mephedrone was first identified in 2008 and of the 5386 biological samples submitted for analysis (purpose unknown), mephedrone was identified in 363 (7%).

Alternative Methodologies to Assess Prevalence of NPS

As it is not an aim of this chapter to provide guidance on general population prevalence study methodologies, the interested reader is referred to tools such as the European Monitoring Centre for Drugs and Drug Addiction (EMCDDA) General Population Surveys 'toolbox' (available from http://www.emcdda.europa.eu/themes/key-indicators/gps). However, some notable recent developments focusing upon targeted samples are described below. These may be particularly useful when applied to the study of NPS epidemiology, where use is often limited to specific (and hidden) sub populations and use environments, and which generally have low general population prevalence. Classic prevalence surveys often require very large sample sizes when the behaviour of interest is rare, and as is often the case with drug use surveys, the expected target group (e.g. students living away from home) may not form part of readily available sampling frames. Although the methodologies described below all require further refinement, they show potential in supporting more robust NPS prevalence estimates.

The *time-space sampling* (TSS) method is a probability based sampling method that allows for the systematic generation of a location based sample (e.g. nightclub attendees) [85]. The method is particularly useful as it allows for sampling of hard to reach populations (defined by behavioural or socioeconomic parameters) in specific locations and times where it would be costly or difficult to construct a sampling frame [85–87]. In time space sampling, a range of venue-day-time (VDT) units for the target population form the sampling frame; [88] for example a specific London night club, Friday evening, between 12 a.m. and 2 a.m. In the formative stage of sampling, qualitative research is undertaken to gain a better understanding of the ethnographic profile of the population of interest and indicators characterising the population are developed. Subsequently, VDTs for the population are identified and headcounts for the number of people congregating in the range of VDTs are taken (primary enumeration). Following primary enumeration, the number and proportion of people in the likely target population are estimated (secondary enumeration). Sampling then takes place in a two-stage process by randomly selecting enumerated VDTs in the sampling frame, and then randomly recruiting individuals in the VDT. This latter element is sometimes removed in order to increase response rates and to reduce cost [89]. This technique has been used successfully in a number of environments and populations, for example, the use of 'club drugs' in New York City nightclubs [90] and black and minority ethnic populations of men who have sex with men [87]. Some caution in interpretation of data obtained from such techniques is warranted as sampling is limited to attendees of sampled venues types (e.g. ecstasy is considered the prototypical 'club drug' but some studies indicate that use is in fact more prevalent in private homes; e.g. [91,92]), but its usefulness for increasing the robustness of NPS prevalence estimates is clear, and avoids the use of non-probabilistic methods (and therefore the biases associated with non-representativeness) such as convenience or snowball sampling where the individual approaches the researcher to take part in their work (e.g. [8,58]). Similarly, although (respondent driven sampling (RDS) is a form of convenience sampling, and proceeds through snowball sampling techniques, the inclusion of statistical weighting, which allows for calculation of sample selection probabilities,

means that it is considered a probability sampling method (e.g. [93]). The technique has also been refined to allow for online recruitment and assessment, which has the potential to reduce research costs considerably [94]). In RDS, referral chains are established, which if sufficiently long enough result in a sample (with characteristics and behaviours of interest) that are independent of the seed from which it began. The incorporation of snowball sampling potentially draws study participants from a wide variety of settings and cultures, thus avoiding one of the limitations of TSS. However, recent methodological work suggests that RDS produces prevalence estimates with much greater variance than simple random sampling, and so whilst preferable to convenience sampling, the technique does require further development [95].

Petróczi and colleagues [96] presented an innovative fuzzy response 'single sample count' model to provide an estimate of mephedrone use prevalence in a small community sample in Wales and Greater London, UK (n = 318), and this was compared with the well-established Forced Response model. Their technique was unique in that it did not directly ask respondents about the behaviour of interest (i.e. mephedrone use) thus potentially reducing response bias against a socially/legally proscribed behaviour. Although perhaps initially complex for the non-statistician to understand, the technique relies on including questions about the behaviour of interest amongst apparently innocuous questions where the population distribution of responses is expected to be binomial. The respondent is asked to indicate how many of the presented questions are true for them, without revealing which ones are true. As the population distribution is expected to be 50%, differences from this figure indicate the proportion endorsing the sensitive question (i.e. a period use of mephedrone). Using a Forced Response method a prevalence of use (previous three months) of between 2.6% and 15.0% was estimated but using the new technique the estimate range was 0–10%. Hair analysis on a sub-sample of participants suggested that the true prevalence was at least 4%; further confirmatory work using a complete forensic validation is therefore required.

The last example to be briefly referred to here is the use of epidemiological techniques to identify the presence of drugs in the environment, including untreated wastewater (e.g. sewage systems and rivers), ambient air (e.g. airborne particulates in nightclubs), and urine (e.g. from nightclub toilets [97–99]). These are not intended as public health monitoring strategies (i.e. to assess the exposure risk of the population to illicit drugs), but are presented as objective forensic methodologies that can improve the quality of local or regional prevalence estimates, increase population coverage, and with a reduced risk of sampling and reporting bias [100]. Again it must also be noted that to date these techniques have yet been refined with NPS, although some work has been published on the detection of steroids, β2-agonists, diuretics and phosphodiesterase type V inhibitors [101]. Waste water epidemiology is probably the most well developed technique, and several authors have reported that it produces reliable and useful time trend data in a number of different geographies [100,102]. Some preliminary work has also been conducted on the detection of illicit drugs in airborne particulates [103], which, combined with time-space-sampling, may provide a useful means of estimating drug use in specific environments. However, little is currently known about the behaviour of drugs in the air, and the range of compounds able to be measured is currently limited and does not at present include synthetic cannabinoid receptor agonists. Use of waste water and airborne techniques is also limited for NPS as generally little is known about their stability or metabolism. Therefore the use of different sampling methodology to detect NPS may be required. Recently, Archer and colleagues [97] have presented feasibility data on a methodology for detecting drugs, including NPS, in pooled urine collected from portable urinals at nightclubs. The advantage of collecting urine for analysis for NPS (rather than mixed sewage including faeces in waste water) is that it eliminates the unknown variable of bacterial metabolism of NPS. Over two nights 38 drugs and their metabolites were identified, including NPS such as mephedrone, trifluoromethylphenylpiperazine (TFMPP), and 2-aminoindane (2-AI). The compounds identified at the highest concentrations were mephedrone, ketamine, MDMA and nicotine.

Aside from technical considerations (e.g. the metabolic profile and pharmacokinetic parameters of analysed drugs needs to be known, a particular problem with regards to the limited data on NPS; variation in the purity of drugs available in a locality), one disadvantage of environmental techniques in particular is that they typically produce estimates for 'average' use patterns. Overall, in formulating responses to substance use it is advantageous to understand a range of use profiles (e.g. dose taken per episode; low ↔ regular frequency), as well as coinciding behaviours (e.g. polysubstance use). Environmental techniques are unable to provide this qualifying evidence and so these techniques must be supported by classic epidemiological techniques such as interview and self-report. Furthermore, the estimates assume a 'closed' environment: i.e. changes in the detection of drug metabolites are a result of changes in use patterns rather than, for example, movement of populations, which is likely to be the case in cities with dynamic entertainment centres [99].

SUMMARY AND CONCLUSIONS

This chapter has provided an overview of NPS epidemiology, including major sources of information on prevalence of use in general and specific populations, including those thought of as being 'hard to reach'. Of the NPS examined in this chapter, mephedrone, *Salvia divinorum* and synthetic cannabinoid receptor agonists have been the most frequently studied, and although data is limited, these drugs seem to have been added to existing drug use repertoires, rather than displacing use of other drugs or characterising a new type of drug user. However, an appreciable proportion of NPS users (e.g. approximately 6% of the 2011 Monitoring the Future *synthetic marijuana* respondents) apparently report use of no other illegal substance, and so it is important to conduct further research in order to better understand the role NPS play in their lives. Mephedrone and Salvia have become particularly popular in recent years (mephedrone was more frequently reported than ecstasy in the UK in 2010/11, and Salvia more prevalent than LSD in the USA in 2006), but further data is required in order to understand the impact of bringing these drugs under legal controls (particularly controlled drug/substance frameworks), which has happened in many countries and territories. Most other NPS appear not to have made a significant impact on the drug market with respect to user preference, and even in targeted populations where a more extensive drug use history is expected, remain infrequent choices, perhaps ingested because of their novelty, (quasi) legal status, curiosity into their psychopharmacological effects, and sudden availability on the market [19].

As there are so few nationally representative surveys that provide robust estimates of use in the general population, it is difficult to describe trends in use, to make international comparisons or informed assessments of the role of NPS in mainstream drug repertoires. This is largely because of the novel nature of these substances and their limited dispersion, but also because general population surveys tend to include substances that are already of policy interest, for example those listed under international drug control conventions, and by their very definition, this will exclude many NPS.

Most data on NPS epidemiology comes from (small) convenience samples, and although the results of such studies are frequently cited and often receive national media attention [104,105], it is important to prominently acknowledge their limitations – such data can only reveal drug use behaviours in the respondents surveyed, and cannot easily be generalised beyond the study. Rather than more data being collected on a wider range of NPS, it may be more advantageous to focus attention on improving survey methodologies in hard to reach populations (e.g. time space sampling of NPS users) or through the development of analytical techniques for new sources of indicator data (e.g. environmental monitoring). However, this should not be interpreted as a rejection of such surveys. In the absence of large national NPS research, which always underestimates use in hard to reach populations, such work presents a rapid (and here grey literature has a relative advantage over the protracted timescales in most academic publishing) and useful insight into emerging patterns of use, which may manifest in general population trends later. Drug policy and health service responses to use require rapid intelligence, and are often targeted at minority population segments (such as NPS users); hence this type of data, even with all of its acknowledged limitations is useful until more robust estimates can be established.

REFERENCES

[1] Wasso RG. Notes on the present status of ololiuhqui and the other hallucinogens of Mexico, 20. Botanical Museum Leaflets, Harvard University; 1962. p. 161–212.

[2] Sumnall HR, Measham F, Brandt SD, Cole JC. Salvia divinorum use and phenomenology: results from an online survey. J Psychopharmacol 2011;25:1496–550.

[3] Lange JE, Daniel J, Homer K, Reed MB, Clapp JD. Salvia divinorum: effects and use among YouTube users. Drug Alcohol Depend 2010;108:138–40.

[4] Kjellgren A, Soussana C. Heaven and hell – a phenomenological study of recreational use of 4-HO-MET in Sweden. J Psychoactive Drugs 2011;43:211–9.

[5] Baron M, Elie M, Elie L. An analysis of legal highs-do they contain what it says on the tin? Drug Test Anal 2011;3:576–81.

[6] Brandt SD, Sumnall HR, Measham F, Cole JC. The confusing case of NRG-1. BMJ 2010 Jul 6;341:c3564.

[7] Davies S, Wood DM, Smith G, Button J, Ramsey J, Archer R, et al. Purchasing 'legal highs' on the Internet is there consistency in what you get?. QJM 2010;103:489–93.

[8] Measham F, Moore K, Østergaard J. Mephedrone, 'Bubble" and unidentified white powders: the contested identities of synthetic 'legal highs'. Drugs Alcohol Today 2011;11:137–46.

[9] Johnston LD, O'Malley PM, Bachman JG, Schulenberg JE. Monitoring the Future national results on adolescent drug use: overview of key findings, 2011. Ann Arbor: Institute for Social Research, The University of Michigan; 2012.

[10] Hibell B, Guttormsson U, Ahlström S, Balakireva O, Bjarnason T. The ESPAD report 2007: alcohol and other drug use among students in 35 European countries. Stockholm: Swedish Council for Information on Alcohol and Other Drugs (CAN); 2009.

[11] Substance Abuse and Mental Health Services Administration Results from the 2010 National survey on drug use and health: summary of national findings. Rockville, MD: Substance Abuse and Mental Health Services Administration; 2011.

[12] Substance Abuse and Mental Health Services Administration, Office of Applied Studies The NSDUH report – Use of specific hallucinogens: 2006. Rockville, MD: Substance Abuse and Mental Health Services Administration; 2008.

[13] Smith K, Flatley J. Drug misuse declared; findings from the 2010/11 British Crime Survey. London: Home Office; 2011.

[14] Home Office Statistics 2012 Drug misuse declared; findings from the 2011/12 crime survey for England and wales. Available: <http://www.homeoffice.gov.uk/publications/science-research-statistics/research-statistics/crime-research/drugs-misuse-dec-1112/> [accessed 26.07.12].

[15] Bird S. More insights on mephedrone from british crime survey. Straight Statistics Blog 2011 Available: <http://www.straightstatistics.org/article/more-insights-mephedrone-british-crime-survey> [accessed 09.02.12].

[16] Brunt T, Rigter S, Hoek J, Vogels N, VanDijk P, Niensk R. An analysis of cocaine powder in the Netherlands: content and health hazards due to adulterants. Addiction 2009;104:798–805.

[17] Brunt T, Poortman A, Niesink R, Van den Brink W. Instability of the ecstasy market and a new kid on the block: mephedrone. J Psychopharmacol 2011;25:1543–7.

[18] Carhart-Harris R, King L, Nutt D. A web-based survey on mephedrone. Drug Alcohol Depend 2011;118:19–22.

[19] Sumnall HR, McVeigh J, Evans-Brown MJ. Social, policy, and public health perspectives on new psychoactive substances. Drug Test Anal 2011;3:515–23.

[20] Measham F, Moore K. Repertoires of distinction: exploring patterns of weekend polydrug use within local leisure scenes across the English night time economy. Criminol Crim Justice 2009;9:437–64.

[21] National Advisory Committee on Drugs (NACD) and Public Health Information and Research Branch (PHIRB). Drug use in Ireland and Northern Ireland: first results from the 2010/11 drug prevalence survey. Dublin: NACD & PHIRB; 2011.

[22] Mixmag. Mixmag survey 2010. Available:<http://www.mixmag.net/words/news/the-mixmag-drug-survey-launches>; 2010 [accessed 08.02.12].

[23] Mixmag. Mixmag survey 2011. Available: <http://www.mixmag.net/words/news/the-mixmag-drug-survey-launches>; 2011 [accessed 08.02.12].

[24] Kelleher C, Christie R, Lalor K, Fox J, Bowden M, O'Donnell C. An overview of new psychoactive substances and the outlets supplying them. Dublin: NACD; 2011.

[25] Global drugs survey 2012. Available: <http://globaldrugsurvey.com/> [accessed 20.07.12].

[26] Bruno R, Matthews AJ, Dunn M, Alati R, McIlwraith F, Hickey S, et al. Emerging psychoactive substance use among regular ecstasy users in Australia. Drug Alcohol Depend 2011. doi:10.1016/j.drugalcdep.2011.11.020.

[27] Wood DM, Hunter L, Measham F, Dargan PI. Limited use of novel psychoactive substances in South London nightclubs. Q J Med 2012. doi:10.1093/qjmed/hcs107. Advance Access published 19/6/12.

[28] Measham F, Wood D, Dargan P, Moore K. The rise in legal highs: prevalence and patterns in the use of illegal drugs and first and second generation 'legal highs' in south London gay dance clubs. J Subst Use 2011;16:263–72.

[29] Dargan PI, Albert S, Wood DM. Mephedrone use and associated adverse effects in school and college/university students before the UK legislation change. QJM 2010;103:875–9.

[30] Werse B, Müller O. Spice, smoke, sence & co. – smoking mixtures containing cannabinoids: consumption and motivation for consumption against the backdrop of changing laws. Frankfurt: Centre for Drug Research; 2010. English summary available: <http://www.uni-frankfurt.de/fb/fb04/forschung/cdr/download/Spice_Werse_2010_english.pdf> [accessed 08.02.12].

[31] Khey DN, Miller BL, Griffin OH. Salvia divinorum use among a college student sample. J Drug Educ 2008;38:297–306.

[32] Miller BL, Griffin OH, Gibson GL, Khey DN. Trippin' on Sally D: exploring predictors of *Salvia divinorum* experimentation. J Crim Justice 2009;37:396–403.

[33] Topp L, Breen C, Kaye S, Darke S. Adapting the illicit drug reporting system (IDRS) to examine the feasibility of monitoring trends in the markets for 'party drugs'. Drug Alcohol Depend 2004;73:189–97.

[34] Sindicich N, Burns L. Australian trends in ecstasy and related drug markets 2010: findings from the ecstasy and related drugs reporting system (EDRS). Sydney: National Drug and Alcohol Research Centre, University of New South Wales; 2011.

[35] Beddoes D, Sheikh S, Pralat R, Sloman J. The impact of drugs on different minority groups; a review of UK literature. Part 2: lesbian, gay, bisexual and transgender (LGBT) groups. London: UKDPC; 2010.

[36] Evans-Brown MJ, McVeigh J. Anabolic steroid use in the general population of the United Kingdom Møller V, Dimeo P, McNamee M, editors. Elite sport, doping and public health. Odense: University of Southern Denmark Press; 2009. p. 75–97.

[37] Evans-Brown M, Kimergård A, McVeigh J. Elephant in the room? The methodological implications for public health research of performance-enhancing drugs derived from the illicit market. Drug Test Anal 2009;1:323–6.

[38] Assael S. Steroid nation. Venice beach, CA: ESPN Books; 2007.

[39] Takahashi Y, Kipnis DM, Daughaday WH. Growth hormone secretion during sleep. J Clin Invest 1968;47:2079–90.

[40] Luby S, Jones J, Zalewski A. GHB use in South Carolina. Am J Public Health 1992;82:128.

[41] Oyama T, Takiguchi M. Effects of gamma-hydroxybutyrate and surgery on plasma human growth hormone and insulin levels. Agressologie 1970;11:289–98.

[42] Chin MY, Kreutzer RA, Dyer JE. Acute poisoning from gamma-hydroxybutyrate in California. West J Med 1992;156:380–4.

[43] Philen RM, Ortiz DI, Auerbach SB, Falk H. Survey of advertising for nutritional supplements in health and bodybuilding magazines. J Am Med Assoc 1992;268:1008–11.

[44] Dyer JE. γ-Hydroxybutyrate: a health-food product producing coma and seizure like activity. Am J Emerg Med 1991;9:321–4.

[45] Krawczeniuk A. The occurrence of gamma-hydroxybutyric acid (GHB) in a steroid seizure. Microgram 1993;26:160–6.

[46] Steele MT, Watson WA. Acute poisoning from gamma hydroxybutyrate (GHB). Mo Med 1995;92:354–7.

[47] Myrenfors P. Ten cases of poisoning with gamma hydroxybutyrate. An endogenous substance used by body builders. Lakartidningen 1996;93:1973–4.

[48] Burton C. Anabolic steroid use among the gym population in Clwyd. Pharm J 1996;256:557–9.

[49] Klatz R, Kahn C. Grow young with HGH. New York: HarperCollins; 1998.

[50] Medicines Control Agency Mail. The MCA updating service. 2000;118 (March/April): 1–31.

[51] Medicines Control Agency Mail. The MCA updating service. 2001;128 (Nov/Dec): 1–26.

[52] Medicines Control Agency Mail. The MCA updating service. 2003;135 (Jan/Feb): 1–25.

[53] EMCDDA Report on the risk assessment of GHB in the framework of the joint action on new synthetic drugs. Luxembourg: Office for Official Publications of the European Communities; 2002.

[54] EMCDDA, and Europol 2011 EMCDDA–Europol 2010 annual report on the implementation of Council Decision 2005/387/JHA. Lisbon: European Monitoring Centre for Drugs and Drug Addiction.

[55] Bailey DJ, O'Hagan D, Tavasli M. A short synthesis of (S)-2-(diphenylmethyl)pyrrolidine, a chiral solvating agent for NMR analysis. Tetrahedron: Asymmetry 2003;8:149–53.

[56] Van Hout MC, Bingham T. A costly turn on: patterns of use and perceived consequences of mephedrone based head shop products amongst Irish injectors. Int J Drug Policy 2012;23:188–97.

[57] Deehan A, Saville E. Calculating the risk: recreational drug use among clubbers in the South East of England. London: Home Office; 2003.

[58] Winstock AR, Mitcheson LR, Deluca P, Davey Z, Corazza O, Schifano F. Mephedrone, new kid for the chop?. Addiction 2011;106:154–61.

[59] James D, Adams RD, Spears R, Cooper G, Lupton DJ, Thompson JP, et al. Clinical characteristics of mephedrone toxicity reported to the UK National Poisons Information Service. Emerg Med J 2011;28:686–9.

[60] Dunn M, Degenhardt L, Bruno R. Transition to and from injecting drug use among regular ecstasy users. Addict Behav 2010;35:909–12.

[61] Topp L, Hando J, Dillon P, Roche A, Solowij N. Ecstasy use in Australia: patterns of use and associated harm. Drug Alcohol Depend 1999;55:105–15.

[62] The United Nations Office on Drugs and Crime Synthetic cannabinoid receptor agonists in herbal products. Geneva: UNODC; 2011.

[63] Dresen S, Kneisel S, Weinmann W, Zimmermann R, Auwärter V. Development and validation of a liquid chromatography–tandem mass spectrometry method for the quantitation of synthetic cannabinoid receptor agonists of the aminoalkylindole type and methanandamide in serum and its application to forensic samples. J Mass Spectrom 2011;46:163–71.

[64] Vandrey R, Dunn KE, Fry JA, Girling ER. A survey study to characterise use of Spice products (synthetic

cannabinoid receptor agonists). Drug Alcohol Depend 2012;120:238–41.

[65] Office for National Statistics Internet access quarterly update 2011 Q2. London: ONS; 2011.

[66] Perron BE, Ahmedani BK, Vaughn MG, Glass JE. Use of salvia divinorum in a nationally representative sample. Am J Drug Alcohol Abuse 2011:1–6. Early Online.

[67] Wu L-T, Woody GE, Chongming Y, Li J-H, Blazer DG. Recent national trends in *Salvia divinorum* use and substance-use disorders among recent and former Salvia divinorum users compared with nonusers. Subst Abuse Rehabil 2011;2:53–68.

[68] Ford JA, Watkins WC, Blumenstein L. Correlates of *Salvia divinorum* use in a national sample: findings from the 2009 national survey on drug use and health. Addict Behav 2011;36:1032–7.

[69] Akers R. Deviant behavior: a social learning approach. Belmont, CA: Wadsworth; 1985.

[70] Raymond JS. Behavioral epidemiology: the science of health promotion. Health Promot 1989;4:281–6.

[71] Sallis JF, Owen N, Fotheringham MJ. Behavioral epidemiology: a systematic framework to classify phases of research on health promotion and disease prevention. Ann Behav Med 2000;22:294–8.

[72] Perry CL, Murray DM. The prevention of adolescent drug abuse: implications from etiological, developmental, behavioral, and environmental models. J Prim Prev 1985;6:31–52.

[73] Galea S, Nandi A, Vlahov D. The social epidemiology of substance use. Epidemiol Rev 2004;26:36–52.

[74] Gallup 2011 Youth Attitudes on Drugs. Analytical report. Flash barometer 330. Brussels: European Commission.

[75] Baggot MJ, Erowid E, Erowid F, Galloway GP, Mendelson J. Use patterns and self-reported effects of Salvia divinorum: an internet-based survey. Drug Alcohol Depend 2010;111:250–6.

[76] Health Protection Agency (HPA) National poisons information service – annual report 2010/2011. London: HPA; 2011.

[77] EMCDDA Guidelines for the prevalence of problem drug use (PDU) key indicator at national level. Luxembourg: Office for Official Publications of the European Communities; 2004.

[78] Schmidt MM, Sharma A, Schifano F, Feinmann C. Legal highs on the net-evaluation of UK-based websites, products and product information. Forensic Sci Int 2011;206:92–7.

[79] Birdwell J, Chapman J, Singleton N. Taking drugs seriously: a demos and UK drug policy commission report on legal highs. London, UK: UKDPC; 2010.

[80] United States Department of Health and Human Services, Substance Abuse and Mental Health Services Administration, Center for Behavioral Health Statistics and Quality 2011 Drug Abuse Warning Network (DAWN), 2009 [Computer file ICPSR31921-v1] Ann Arbor, MI: Inter-university Consortium for Political and Social Research.

[81] Reuter P. Options for regulating new psychoactive drugs: a review of recent experiences and an analytic framework. London: UK Drug Policy Commission; 2011.

[82] Bronstein AC, Spyker DA, Cantilena LR, Green JL. Annual report of the American Association of Poison Control Centers' National Poison Data System (NPDS): 28th annual report. Clin Toxicol 2011;49:910–41.

[83] American Association of Poison Control Centers Fake marijuana spurs more than 2500 calls to US poison centers this year alone. Alexandria, VA: American Association of Poison Control Centers; 2010.

[84] Toth AR, Hideg Z, Institoris L. Mephedrone – an old-new drug of abuse. Orv Hetil 2011;152:1192–6.

[85] MacKellar D, Valleroy L, Karon J, Lemp G, Janssen R. The young men's survey: methods for estimating HIV seroprevalence and risk factors among young men who have sex with men. Public Health Rep 2006;11:138–44.

[86] Muhib FB, Lin LS, Stueve A, Ford WL, Miller RL, Johnson WD. Avenue-based method for sampling hard-to-reach populations. Public Health Rep 2001;116:216–22.

[87] Stueve A, O'Donnell L, Duran R, Sandoval A, Blome J. Time-space sampling in minority communities: results with young Latino men who have sex with men. Am J Public Health 2001;91:922–6.

[88] Semaan S. Time-space sampling and respondent-driven sampling with hard-to-reach populations. Methodol Innov Online 2010;5:60–75.

[89] Parsons JT, Grov C, Kelly BC. Comparing the effectiveness of two forms of time space sampling to identify club drug-using young adults. J Drug 2008;38:1061–801.

[90] Ramo DE, Grov C, Delucchi K, Kelly BC, Parsons JT. Typology of club drug use among young adults recruited using time-space sampling. Drug Alcohol Depend 2010;107:119–27.

[91] Boeri M, Sterk C, Elifson K. Rolling beyond raves: ecstasy use outside the rave setting. J Drug 2004;34:831–60.

[92] Fendrich M, Wislar JS, Johnson TP, Hubbell A. A contextual profile of club drug use among adults in Chicago. Addiction 2003;98:1693–703.

[93] Heckathorn DD, Jeffri J. Finding the beat: using respondent driven sampling to study jazz musicians. Poetics 2001;28:307–29.

[94] Wejnert C, Heckathorn D. Web-based network sampling: efficiency and efficacy of respondent-driven sampling for online research. Sociol Methods Res 2008;37:105–34.

[95] Goela S, Salganik MJ. Assessing respondent-driven sampling. PNAS 2010;107:6743–7.

[96] Petróczi A, Nepusz T, Cross P, Taft H, Shah S, Deshmukh N, et al. New non-randomised model to assess the prevalence of discriminating behaviour: a pilot study on mephedrone. Subst Abuse Treat Prev Policy 2011;6:20.

[97] Archer JRH, Dargan PI, Rintoul-Hoad S, Hudson S, Wood DM. Using urinals in the night-time economy to determine what recreational drugs people are actually using. Br J Clin Pharmacol 2012;73:985.

[98] Daughton CG. Illicit drugs: contaminants in the environment and utility in forensic epidemiology. Rev Environ Contam Toxicol 2011;210:59–110.

[99] EMCDDA Assessing illicit drugs in wastewater. Luxembourg: Office for Official Publications of the European Communities; 2008.

[100] van Nuijs ALN, Castiglioni S, Tarcomnicu I, et al. Illicit drug consumption estimations derived from wastewater analysis: a critical review. Sci Total Environ 2011;409:3564–77.

[101] Schroeder HF, Gebhardt W, Thevis M. Anabolic doping, and lifestyle drugs, and selected metabolites in wastewater – detection, quantification, and behaviour monitored by high resolution MS and MS before and after sewage treatment. Anal Bioanal Chem 2010;398:1207–29.

[102] Zuccato E, Chiabrando C, Castiglioni S, Bagnati R, Fanelli R. Estimating community drug abuse by wastewater analysis. Environ Health Perspect 2011;116:1027–32.

[103] Cecinato A, Balducci C, Guerriero E, Sprovieri F, Cofone F. Possible social relevance of illicit psychotropic substances present in the atmosphere. Sci Total Environ 2011 2011;412-413:87–92.

[104] Bates C. Cash-strapped young adults turning to cheaper 'bubble' drugs following mephedrone ban. Dly Mail 2011;Nov 8. Available from: <http://www.dailymail.co.uk/health/article-2058877/Mephedrone-ban-causes-young-adults-turn-cheap-bubble-drug.html>.

[105] Butler P. Drug use survey: tell us what you take. Guardian 2011;Nov 23. Available from: <http://www.guardian.co.uk/society/2011/nov/23/mixmag-drugs-survey-2012>.

This page intentionally left blank

Social Issues in the Use of Novel Psychoactive Substances
Differentiated Demand and Ideological Supply

Fiona Measham

Professor of Criminology, School of Applied Social Sciences, Durham University, Durham, UK

INTRODUCTION

This chapter explores some of the social issues in the novel psychoactive substances (NPS) debate including availability, prevalence and contexts of use, motivations for use, social harms, policy options and the impact of legislative control. In its consideration of the prevalence of use of three key NPS in three different UK survey samples, this chapter presents previously unpublished data from the author's annual surveys at English music festivals in the summers of 2010 and 2011, suggesting a picture of differentiated demand with pockets of popularity. The characteristics of the evolving NPS market combine legitimate Internet trading with features of the international trade in counterfeit prescription medications as well as aspects of the ideological motivation of some suppliers of ecstasy in early 1990s dance culture. A key theme of this chapter is speed: the speed of emergence of NPS echoes their predominantly stimulant effects and contrasts with the sluggishness of the academic and policy debate surrounding them. The caveat here is that the extent of our knowledge is limited by the recency of the phenomenon combined with the time lag between emergent trends in drug use and the development of a scientific evidence base. I return to this point in the final section of this chapter where I cast a critical eye on the development of the NPS debate itself and draw analogies with other high profile debates on substance use such as the alcohol and 'binge drinking' debate, arguing that the pace of academic research and publishing combine with historical stagnation and vested interests to hamper scientific developments in this field. Firstly, I consider the availability and use of NPS, focusing on Europe.

AVAILABILITY

Both the range of NPS available for purchase and the number of retail outlets which sell them have expanded rapidly since 2009. Although there were some early indications of NPS appearing in Israel, Sweden and some other countries before this date, it was in 2008 that drugs such as mephedrone were first mentioned on the Internet, predominantly from Scandinavian sources [1], and by early 2009 their availability rapidly spread to markets, festivals, high street 'head shops' and through Internet websites dedicated to the sale of these drugs. A key feature, and indeed a key political concern, regarding NPS has been how readily available they have become through the development of an online 'industry' in substances variously described as retail chemicals, 'herbal highs', 'bath salts', 'plant food' and other products purportedly 'not for human consumption' [2–5]. The European Monitoring Centre for Drugs and Drug Addiction (EMCDDA) noted in its annual report that "'legal highs'" have become a global phenomenon which is developing at an unprecedented pace' [6]. Given the reach, speed and limited regulation of the Internet, there are limitations to any assessment of availability of NPS. However, two organisations which have attempted to chart the emergence and availability of NPS on the Internet are the EMCDDA and the UK-based Psychonaut Web Mapping Research Group.

The EMCDDA monitors the NPS market through both the emergence of new unregulated psychoactive substances and the online availability of products through Internet retailers. In its most recent annual report it noted the identification of 73 new products in 2012, compared with 49 new products in 2011, 41 in 2010 and 24 in 2009 through its pan European Early warning monitoring system, a trebling in the 36-month period from 2009 to 2012 [6]. In 2012, these new registered NPS were notably all synthetic and were made up of predominantly two groups of drugs, the synthetic cannabinoids and synthetic cathinones. The number of online shops selling NPS also similarly increased, from 68 in 2008, to 115 in 2009, 170 in January

2010, 314 in January 2011 and 693 by January 2012, more or less doubling in each 12-month monitoring period, 2010–11 and 2011–12, suggesting that Internet retailers are mushrooming at a much greater rate than the drug manufacturers' product development.

Regarding the location of online retailers, the EMCDDA found that 52% of the online retailers of these products were UK-based in 2008. By 2011 the EMCDDA found that the USA had leapfrogged to be the most likely country of origin for the majority of online retailers, increasing from 65 to 197 US-based online retailers in the six-month period from January to July 2011 [7]. In second place, UK-based retailers increased from 74 to 121 online retailers in the same six months period. Larger increases but from a smaller base line were evident in Canada and New Zealand, both of which saw their online retailers more than treble in size in less than six months. However, whilst the raw numbers indicate that there are more websites based in the USA than European countries, the number of retailers does not necessarily denote the size of the operation or the country to which sales are delivered. Manufacturers and retailers of NPS, as with counterfeit prescription medicines, may transport their products from producer nations to consumer nations via circuitous routes through intermediary countries that have fewer restrictions, to reduce the suspicions of law enforcement agencies, as more and more countries are legislating to control NPS.

The European Commission-funded Psychonaut Web Mapping Research Group used online resources to detect emerging trends and develop a profile of NPS from January 2008 to December 2009. In the 24-month monitoring period, 412 novel compounds and combinations appeared, consisting of 151 chemical compounds (synthetic, semi-synthetic or pharmaceutical), 121 herbal compounds and 140 combination products [8]. A related study by the same research group identified 39 unique

UK-based websites selling 'legal highs' in the period April–June 2009 [4]. The UK-based websites advertised 346 unique products between them at an average price of £9.69 pounds sterling. Concern was expressed that 40% of these products failed to list their ingredients and 92% failed to list possible side effects. The researchers concluded that not only were 'legal highs' easily available and affordable in this period but consumer product information was poor resulting in increased risks for significant numbers of inexperienced users.

Therefore we can chart a well-documented rapid increase in Internet retailers and NPS products in recent years, illustrating both growing global availability and a shift away from UK-based websites to continental Europe and North America. The next consideration is whether NPS consumers use these online retailers or show a preference for high street 'head shop' retailers instead, regarding *legal* NPS. And what do we mean by 'head shops'? In the UK the Association of Chief Police Officers defines head shops as:

> "a commercial retail outlet (including online businesses) specializing in the sale or supply of equipment, paraphernalia or literature relating to the growing, production or consumption of cannabis, or other drugs, and includes the sale or supply of 'New Age' herbs, exotic plant materials or other 'New Psychoactive Substances' (aka 'Legal or Herbal Highs') that are intended to be consumed by the user to mimic the effect of an illicit drug, e.g. Cocaine, Ecstasy or Amphetamine" [9].

What appears to be happening is that there are temporal swings between use of Internet retailers and head shops as well as national, regional and user-related variations in retail patterns. For example, regarding age and choice of retail outlet, there are more Internet sales amongst over 18-year-olds who have greater access to credit/debit cards combined with lesser surveillance of postal deliveries to their home. This was supported by a study of 1006 Scottish school and college pupils by Dargan and colleagues conducted before mephedrone was controlled in the UK, which found significant sourcing from street dealers and 'friends' amongst under 18s even though it was legal to buy, whereas sourcing from the Internet was more common amongst respondents over 18 years of age [10].

In terms of the impact of legislative control on choice of retail outlets, an annual survey of the readership of a dance music magazine (*Mixmag*) and associated website (www.dontstayin.com) found that 33% had purchased mephedrone from the Internet and 5% from high street head shops before legislative control in the UK and a further 38% bought mephedrone from friends and 24% from 'dealers' [11]. A year later, after legislative control, only 1% reported buying NPS from either the Internet or head shops, whilst mephedrone sales from friends had increased slightly to 41% and from 'dealers' had more than doubled to 58% [12]. This suggests that legislative control restricted availability through legitimate channels and that Internet retailers and head shops appeared to have switched from selling mephedrone and other substituted cathinones to selling legal NPS. In terms of timing, mephedrone and other substituted cathinones became Class B controlled substances under the Misuse of Drugs Act, 1971 in the UK in April 2010; and naphyrone and other substituted pyrovalerones became Class B controlled substances in July 2010.

In some countries, sales of NPS have been concentrated in head shops rather than on the Internet. For example, in Northern Ireland most respondents in McElrath and O'Neill's study had never bought NPS off the Internet even before legislative control, instead preferring the comparative anonymity of head shops [13]. In the USA, Spiller and colleagues, in a poisons centre study, reported that the majority of patients with acute toxicity after exposure to 'bath salts' had purchased the products from small local independent stores, head shops and gas stations rather than from a 'dealer' or over the Internet [14]. However head shops appear to have been less significant in the rise of NPS in countries such as the UK. The annual South London Surveys of gay dance clubbers asked respondents whether they had ever purchased drugs from head shops [15]; whilst 28% of respondents in July 2010 had bought drugs off the Internet (over one-third of whom had bought mephedrone), only 19% of the sample had ever bought drugs from a head shop (nearly one-quarter of whom had bought mephedrone). Furthermore a sizeable minority of the South London Survey respondents who had not used head shops did not know what a head shop was when they were asked. In those countries where head shops have featured as key retailers of NPS, they have faced increased monitoring and threat of closure, with head shops closed across Poland in 2010 and across both Northern Ireland and the Republic of Ireland in 2012.

Whilst there was a brief honeymoon period in some countries of door-to-door delivery of pure, legal NPS stimulants it ended as abruptly as it started due to a combination of legislative control, increased wariness of consumers regarding anonymity of Internet sales and increased disillusionment regarding the content of such products. However, statistics on NPS sales are not publicly available and information on the NPS trade is elusive. The author conducted interviews with UK-based NPS retailers in 2012-13, one of who's pseudonyms for the purposes of this chapter will be the 'Rave Florist'. The Rave Florist suggested that in terms of currently legal NPS, in his experience, the UK has seen a swing away from Internet retailers and towards head shops in recent months. This swing towards head shops in the UK may be partly due to the consolidation of NPS into a smaller number of larger online retailers as some of the small, independent retailers that sprung up with the initial 'mephedrone madness' stopped trading, unable to make a significant profit due to economic pressures in the recession, increased competition

between retailers and an increasingly complex legal situation. Also for NPS users, growing concerns about privacy and customer identification when shopping online using a credit/debit card combined with high profile police operations against NPS retailers using card transaction details have led to the growing appeal of purchasing NPS in head shops, some of which trade under aliases. Of course for those without access to credit/debit cards and for under-18s, the evident advantages of buying NPS (or indeed any drug-related paraphernalia) from head shops rather than the Internet have existed since NPS first appeared [10].

In terms of the availability of controlled NPS such as mephedrone, legislative change has resulted in a predictable shift from online legal retailers to street 'dealers'. For example in Carhart-Harris et al.'s online survey of over 1000 mephedrone users, 47% said mephedrone was noticeably less available after the ban [16]. The *Mixmag* dance music magazine survey also showed a reduction in availability in the UK in 2010–11, with 75% reporting that it was easy or very easy to obtain mephedrone before the ban compared with 38% after the ban [12]. Whilst this is a cross-sectional rather than longitudinal survey with a self-selecting online sample, it does nevertheless suggest that legislative control may have had an impact on the availability of mephedrone, with an associated switch from online to street supply routes.

Overall, the trade in NPS – produced in China and South East Asia to be sold online via Internet retailers based in Europe and increasingly North America to customers across the world – has mushroomed in the last five years [2]. This has resulted in a wide range of legal NPS being available to purchase without restriction, both on the Internet and in the high street head shops, as well as the development of a street trade in NPS once they are controlled (discussed further below). What is clear from the limited research on availability of NPS is that recent years have seen a very rapid increase in the range of psychoactive drugs that are available to the consumer and rapid innovation in response to legislative change. Considerable time, effort and funding are being invested in monitoring NPS products and their retailers. The key question is, what is the uptake in use of these NPS?

PREVALENCE IN CONTEXT: DIFFERENTIATED DEMAND IN DIFFERENT LEISURE VENUES

General population surveys capture data on adult consumption patterns at population level and in the UK the key annual national household survey was the British Crime Survey (BCS) by the Home Office, replaced in 2012 by the Crime Survey for England and Wales (CSEW). Whilst such surveys are invaluable for identifying national and regional trends in the use of illicit drugs using large and representative samples each year, they can be slow at adapting to changes in consumption patterns, for example of the kind we have seen since the arrival of NPS. National household surveys such as the BCS and CSEW are also likely to underestimate drug use due to their sampling methods and non-random non-response rates [17]. For example, household surveys exclude those who do not live in private households including those living in halls of residence, hostels, prisons and other institutions, students, the homeless and other transitory populations. They may also miss adults with more active social lives who by definition are more likely to be out socialising when household surveys are conducted and who are also more likely to have higher levels of drinking and drug use than others, indicated by their more frequent attendance at bars and clubs [18]. The annual dance music magazine surveys also found higher levels of drug use amongst regular clubbers, for example with over three quarters of UK respondents who had been clubbing in the past month having taken ecstasy (powder or pills) in the past

year compared to 54% of the overall sample of UK respondents [19]. These two broad groups – the geographically mobile and socially active – may also overlap. Given that these groups have been found to have higher rates of drug use than national figures, we should approach general population surveys with caution.

Nevertheless, bearing in mind this caveat, in terms of current knowledge on use of NPS in the UK, the only national data to date on mephedrone use comes from the BCS and CSEW, to which was added a question on mephedrone use in 2010. Table 5.1 shows that past year mephedrone use in the months after prohibition had surpassed ecstasy and was at a similar level to cocaine amongst the young adult population, with about one in 25 having taken it in the past year.

Self-report surveys, as distinct from general population surveys such as the BCS and CSEW, are better able to adapt to rapid changes in availability and use of emergent drugs such as NPS in the 'Internet age' [4,20]. If we look at self-report surveys of specific groups we can see considerably higher levels of reported use of NPS than in the general population, depending on who was sampled and where.

School and College Students

One of the very few surveys of school and college students' use of NPS before legislative control in the UK was conducted by Dargan et al. in Scotland in February 2010 [10]. Of 1006 respondents who completed the survey, 43% reported having tried mephedrone at least once in their lifetime and 4% reported daily use. In terms of access, nearly half of users sourced mephedrone from street 'dealers' and 11% from the Internet.

Although there are several European surveys of prevalence of NPS they tend to be mostly after the introduction of legislative control. The Eurobarometer survey is one of the largest, surveying 12000 young people aged 15–24 in 2011 [21]. It found that about 1 in 20 respondents had used 'legal highs' in most countries, although closer to double that in the UK, Poland and Latvia, and with the highest prevalence of all in Ireland at 16%.

A Polish study of 1250 18–19 year old students found that 4% reported having used 'legal highs' at least once in their life in 2008, increasing to 11% in 2010. Self-reported past year use of 'legal highs' was 3% of the Polish students in 2008 which trebled to 7% in 2010. However, past month use fell from 2% in 2008 to 1% in 2010. This suggests that there was a period of experimentation and occasional use in Poland in 2008–10 which may have peaked and which may have been affected by a ban resulting in the closure of Polish head shops in 2010 [21]. A Spanish study of 25 000 14–18 year olds in 2010 found that 4% of students had consumed one or more of a list of nine NPS that included mephedrone, magic mushrooms and 'legal highs' [21].

Bar Customers

Use of NPS such as mephedrone appears to be higher amongst bar customers than amongst the general population (see Table 5.1). For example a study of UK bar customers stopped at random in the streets of towns and cities of Lancashire (a county in north-west England) on a Friday night in the autumn of 2010 found that about one in 10 had taken mephedrone within the past year and about one in 20 had taken it within the previous month [22]. By comparison, uptake of subsequent NPS that were unregulated at the time of the survey (e.g. MDAI (methylene-dioxyaminoindane, advertised by retailers as a euphoric stimulant with effects similar to MDMA) were much lower (see Table 5.2). Therefore surveys of NPS use by bar customers support earlier surveys of drug use by bar customers [23] in suggesting that bar customers are more likely to report taking illicit drugs than the general population in the UK, including NPS, as well as being more likely to drink alcohol regularly and smoke cigarettes.

TABLE 5.1 Self-reported Past Year Prevalence of Three Most Popular Stimulants, UK Young Adults (%) (Date of Data Collection in Brackets)

	Cocaine	Ecstasy (pills if specified)	Mephedrone
General population survey (2010/11)[a]	4.4	3.8	4.4
(2011/12)[b]	4.2	3.3	3.3
Lancashire bars (2010)[c]	25	18	11
Lancashire nightclubs (2012)[d]	24	16	7
NW music festival (2010)[e]	44	40	32
(2011)[e]	39	34	21
London gay dance clubs (2010),[f]	59	49	52
(2011)[g]	66	45	58
Mixmag magazine (2010),[h]	63	75	51
(2011)[i]	42	39	20

[a]Smith K., Flatley J, eds. Drug Misuse Declared: Findings from the 2010/11 British Crime Survey, England and Wales, Home Office Statistical Bulletin 12/11. London: Home Office; 2011 – data for 16–24-year-olds.
[b]Home Office. Drug Misuse Declared: Findings from the 2011/12 Crime Survey for England and Wales, London: Home Office; 2012. Data for 16–24-year-olds
[c]Measham F, Moore, K, Østergaard, J. Mephedrone, 'Bubble' and unidentified white powders: the contested identities of synthetic 'legal highs'. Drugs and Alcohol Today 2011; 11 (3): 137–147. Average age 24.
[d]Measham F, Moore K, Welch Z. Emerging drug trends in Lancashire: club surveys. Phase Three Report, Lancaster: Lancaster University and Lancashire Drug and Alcohol Action Team; 2012. Average age 23.
[e]Unpublished data from Measham F. Alcohol and drug use at an English festival in 2011: Year Two Report, Lancaster University; 2012. Unpublished report of research conducted at the same music festival in 2010 and 2011 with an average age of 27 in the 2010 survey and 29 in the 2011 survey.
[f]Measham F, Wood D, Dargan P, Moore K. The rise in legal highs: prevalence and patterns in the use of illegal drugs and first and second generation 'legal highs' in south London gay dance clubs. Journal of Substance Use 2011;16 (4): 263–272. Average age 30.
[g]Wood D, Measham F, Dargan P. 'Our favourite drug': prevalence of use and preference for mephedrone in the London night time economy one year after control. Journal of Substance Use 2012; 17 (2): 91–97. Average age 30.
[h]Mixmag The 2011 Drugs Survey. 2011; March 238: 49–59.
[i]Mixmag. Mixmag's Drug Survey: the results. 2012. UK respondents, average age 28. Available: http://www.mixmag.net/drugssurvey [accessed 22/9/12].

TABLE 5.2   Past Year Prevalence of Three Popular Novel Psychoactive Substances, UK Young Adults (%; Date of Data Collection in Brackets)

	Mephedrone	NRG-1	MDAI
General population survey (2010/11) [a]	4.4	n/a	n/a
(2011/12) [b]	3.3	n/a	n/a
Lancashire bars (2010)[c]	11	1	1
Lancashire nightclubs (2012)[d]	7	n/a	n/a
NW music festival (2010) [e]	32	3	1
(2011) [e]	21	n/a	1
London gay dance clubs (2010) [f]	52	13	6
(2011) [g]	58	n/a	7
Mixmag magazine (2010) [h]	51	n/a	5
(2011) [i]	20	n/a	2

[a]Smith K., Flatley J, eds. Drug Misuse Declared: Findings from the 2010/11 British Crime Survey, England and Wales, Home Office Statistical Bulletin 12/11. London: Home Office; 2011. Data for 16–24-year-olds.
[b]Home Office. Drug Misuse Declared: Findings from the 2011/12 Crime Survey for England and Wales, London: Home Office; 2012 – data for 16–24-year-olds
[c]Measham F, Moore, K, Østergaard, J. Mephedrone, 'Bubble' and unidentified white powders: the contested identities of synthetic 'legal highs'. Drugs and Alcohol Today 2011; 11 (3): 137–147. Average age 24.
[d]Measham F, Moore K, Welch Z. Emerging drug trends in Lancashire: club surveys. Phase Three Report, Lancaster: Lancaster University and Lancashire Drug and Alcohol Action Team; 2012. Average age 23.
[e]Unpublished data from Measham F. Alcohol and drug use at an English festival in 2011: Year Two Report, Lancaster University; 2012. Unpublished report of research conducted at the same music festival in 2010 and 2011 with an average age of 27 in the 2010 survey and 29 in the 2011 survey.
[f]Measham F, Wood D, Dargan P, Moore K. The rise in legal highs: prevalence and patterns in the use of illegal drugs and first and second generation 'legal highs' in south London gay dance clubs. Journal of Substance Use 2011; 16 (4): 263–272. Average age 30.
[g]Wood D, Measham F, Dargan P. 'Our favourite drug': prevalence of use and preference for mephedrone in the London night time economy one year after control. Journal of Substance Use 2012; 17 (2): 91–97. Average age 30.
[h]Mixmag The 2011 Drugs Survey. 2011; March 238: 49–59.
[i]Mixmag. Mixmag's Drug Survey: the results. 2012. UK respondents, average age 28. Available: http://www.mixmag.net/drugssurvey [accessed 22/9/12].

Festival-goers

Annual surveys of drinking and drug use at a summer music festival in the north-west of England undertaken by the author provide rare data on the prevalence of use of mephedrone at festivals just after legislative control (April 2010) in the UK, as well as providing a comparison with a NPS that was, and to date still is, legal in the UK (MDAI). Tables 5.1 and 5.2 show data on use of the Class A stimulants (cocaine and ecstasy), mephedrone and other NPS by festival-goers at the same festival in 2010 and 2011 using the same survey instrument and research design both years. Festival-goers were stopped at random and interviewed across the course of three days at the festival by a mixed sex research team using a survey instrument similar to that used in the author's surveys of bar customers [22,24] and dance club customers [15,25]. Amongst festival-goers, as with bar customers, mephedrone was by far the most popular of the NPS that emerged in the UK in 2009–11. In 2010, one-third of festival respondents reported having had mephedrone in the past year compared with 44% having had cocaine and 40% ecstasy pills (Table 5.1). In 2011 self-reported drug use by festival-goers had fallen: only one in five respondents reported having had mephedrone in the past year, compared to four in 10 having cocaine and a third having ecstasy pills. This suggests that although mephedrone use appeared to have fallen in the 12-month period since it was banned there was still a small minority of festival-goers who were both willing and able to access the drug. By comparison use of other NPS was negligible. In 2010 3% of festival goers had taken NRG-1 in the previous year (when its supposed active ingredient naphyrone was still legal in the UK) and 1% had used MDAI. A similar number reported past year use of MDAI in 2011 (Table 5.2). This suggests that use of the unregulated NPS MDAI did not increase amongst festival-goers in the 12-month period after both mephedrone and naphyrone were banned in the UK and therefore the legal status of a NPS is not the only or indeed primary motivating force in its use.

Dance Club Customers

An annual survey of its readership by the dance music magazine *Mixmag* provides times-series data on self-reported drug use by an international sample of several thousand young people. Data presented for the UK subsample of 2295 respondents who responded to the survey in the autumn of 2009 found that whilst 90% had tried ecstasy pills at least once and 85% had tried cocaine, 41% had tried mephedrone with 39% having taken it within the past year and one-third within the past month [26]. For the sample as a whole, the survey conducted just after legislative control in autumn 2010 found that three-quarters of respondents had taken ecstasy pills within the past year, nearly two-thirds had taken cocaine and over half had taken mephedrone [12] (see Table 5.1). A year later, figures for past year drug use were much lower with only about four in 10 respondents having taken ecstasy pills or cocaine in the past year and two in 10 having taken mephedrone [19]. Caution is needed in interpreting differences between years in the annual dance music magazine survey, however, as the survey conducted in the autumn of 2011 used a different sampling method which included a national broadsheet newspaper and its associated website (www.guardian.co.uk), resulting in a diversified sample that was not directly comparable with earlier years [19]. Also whilst ecstasy remains very popular overall, there has been a shift in the UK from ecstasy pills to MDMA powder in recent years, with MDMA in powder form perceived to be a higher priced, higher purity premium product in the wake of the declining MDMA content in pills across the 2000s [27].

Annual surveys undertaken by the author and colleagues in South London gay dance clubs suggest that use of NPS, particularly mephedrone, remains much higher amongst clubbers than in the general population, amongst school pupils, bar customers or festival-goers. In July 2010 and 2011 self-reported past year prevalence of use of cocaine, ecstasy pills and mephedrone were at a similar level, with five in six respondents having had cocaine and mephedrone in the past year and four in five respondents having had ecstasy pills (see Table 5.1). In terms of drugs being taken and/or planned on the fieldwork night, mephedrone was the most popular drug taken by South London gay dance clubbers, ahead of cocaine and ecstasy, taken or planned by 27% of respondents on the fieldwork night in 2010 and by 41% of respondents in 2011, whereas ecstasy pills taken or planned on the fieldwork night was 15% in 2010 and had fallen to 6% in 2011, mirroring the downward trend that was seen in the same time period amongst festival-goers.

In terms of the impact of legislative control on prevalence of use, the annual dance music magazine surveys undertaken suggested that whilst lifetime and past year mephedrone use increased between the pre-ban survey conducted in autumn 2009 [11] and the post-ban survey conducted in autumn 2010 [12], more recent, past month use fell from 34% to 25% in the same period [11,12]. In the post-ban period from the autumn of 2010 [12] to the autumn of 2011 [19] self-reported past year mephedrone use by UK respondents fell from 51% to 20% (Table 5.1). A follow-up survey of mephedrone users from the dance music magazine sample by Winstock et al. found that whilst 63% reported continued use and 55% intended to continue using the same amount, 40% intended on using lesser amounts of mephedrone after control [28]. The online survey by Carhart-Harris et al. in 2011 of over 1000 UK mephedrone users also found reduced use of mephedrone after control with nearly two-thirds (64%) of those surveyed reporting that they used mephedrone less after it was controlled [16].

By contrast, in the South London gay nightclub surveys, mephedrone use in the past month and on the fieldwork night *increased* in the year after control suggesting that mephedrone continued to be available and used after its control, with pockets of popularity such as the gay clubbing scene [29]. However, there were very much lower levels of use of subsequent NPS that had not (yet) been controlled, such as naphyrone and MDAI (Table 5.2) [30]. The low uptake of MDAI again suggests that legal status alone is not the most significant factor in deciding whether or not to take psychoactive drugs, a point discussed further below. The different trends between the annual dance music magazine and annual gay clubbing surveys may indicate methodological differences between online/self-selecting and *in situ*/convenience sampling methods, as well as differences in NPS use within different dance club 'scenes' [23], suggesting highly differentiated demand.

Aside from general population surveys and Internet surveys, the surveys discussed above are the only surveys to date assessing prevalence of NPS use amongst non-treatment based users. Conducted *in situ* with students, bar customers, festival-goers and dance club customers, they suggest that overall then, whilst NPS use is low in the general population with only about 1 in 10 to 1 in 20 having tried them across Europe [21], there have been rapid increases in use amongst certain sub-populations. For example clubbers – who are often experienced users of established street drugs and particularly Class A stimulants – appear to be amongst the most prolific users of NPS. However, even amongst clubbers there seem to be large variations in uptake of NPS. Whilst mephedrone use appears to be falling across the dance music magazine sample, prevalence remains robust amongst gay clubbers. Indeed the South

London Surveys found that self-reported mephedrone use was higher than that of established street drugs such as ecstasy and cocaine, whereas for all other survey samples mephedrone use was lower than that of established Class A stimulants.

In summary, what is clear from a consideration of even a handful of self-report surveys of mephedrone amongst different survey samples in the UK and Europe is that mephedrone – often considered the 'original legal high' – very quickly became popular amongst a range of sociodemographic groups but particularly amongst young adult recreational drug users frequenting a range of social settings who already used established stimulant drugs such as ecstasy and cocaine. By contrast, subsequent NPS such as naphyrone methoxetamine and MDAI, some of which have been advertised as substitutes both for mephedrone and for the more established street drugs such as cocaine, ketamine and ecstasy, have failed to achieve the level of uptake that occurred with mephedrone. This suggests a picture of differentiated demand for NPS, both by product and by user group, with legal status not a key motivating force. Motivations for use are considered further below.

UNDERSTANDING MOTIVATIONS: THE DISPLACEMENT QUESTION

Key to understanding the social dimensions of the NPS debate is an understanding of the motivations for use, a critical but still under researched area. In conjunction with understanding the motivations for using NPS, it is also important to explore the wider context to use and particularly changes in both the legal and illegal drug markets given the considerable overlap that has been identified between users of NPS and established street drugs.

Early studies of NPS that explored motivations for use [13,20,31] suggested that the consumer's quest for legal psychoactive drugs reflected their perception that these products were more likely to be of higher purity than street drugs, be more easily available, carry a lower risk of physical harm and avoid the risk of detection or criminal sanction associated with the consumption of controlled drugs. Analyses of police seizures, test purchases and user monitoring/'early warning' systems confirm that the purity levels of street drugs such as cocaine and ecstasy pills reached a low point in the UK and elsewhere in Europe in 2009–10, suggesting that less cocaine and MDMA were available and less was being consumed [32]. Whilst some commentators suggested that this was due to suppliers deciding and acting in unison to add increasing amounts of bulking agents and adulterants to their products to maximise profits, a more convincing explanation for the cross-national trends relates to interdiction successes along the length of the supply chain reducing availability of both raw materials and precursor ingredients [20,33,34], alongside increased border security as a by-product of anti-terrorist and anti-trafficking measures. Consequently the purity of heroin [35], cocaine [36] and ecstasy [37] all fell significantly in the late 2000s.

Despite an upturn in the purity levels of seizures of Class A drugs in the last couple of years in the UK and continental Europe [38], users continue to feel that the quality of cocaine and ecstasy are declining. In the 2011 dance music magazine survey [19], 52% of cocaine users believed that the quality had gone down in the previous 12 months (autumn 2010–11) compared with only 5% feeling that the quality had increased. User impressions of the quality of ecstasy pills were more mixed, with 45% of ecstasy users believing that the quality had gone down in the previous 12 months compared with 28% feeling that it had increased. This fits with the much predicted 'return of the £10 pill'; [11] with nearly three quarters of ecstasy users reporting in the 2011 dance music magazine survey [19] that they had been

offered higher priced ecstasy pills with the promise of better quality. Over three quarters of those offered had then gone on to buy them and over eight in 10 reported that they *did* think the pills were of better quality.

This perceived inability to source street drugs, or to source street drugs of satisfactory purity levels, also emerges as a motivation for use of more recent NPS. For example, the reasons given by users for taking methoxetamine (the ketamine analogue that became the first drug to receive a Temporary Class Drug Order in the UK in 2012) in the 2011 dance music magazine survey include 73% saying it was easier to get hold of than ketamine, 20% saying it was better value for money than ketamine, 18% believing it caused less damage to the liver or kidneys than ketamine and 20% either being curious to try methoxetamine or being mis-sold what they thought was ketamine [19].

Given that NPS can offer similar opportunities for users to experience stimulant, depressant or hallucinogenic effects as can be produced by established street drugs, we would expect the reasons for taking NPS to be similar to those for taking established street drugs and depending on their effects, taken to enhance various user activities such as dancing, socialising, partying or relaxing. At an individual level NPS users may utilise a cost-benefit analysis of specific NPS as has been noted previously in relation to adolescent decision making concerning their use or non-use of illegal drugs [39] although as with illegal drugs, any cost-benefit analyses are constrained by the unpredictable content and consequences of NPS. There are also broader socio-economic and cultural reasons for consumption of NPS as with established illicit drugs that have yet to be fully explored, which may include a complex mix of associations and influences including family, childhood, peers, friendship networks, educational aspirations and attainment, religious and other influences, as well as the potential impact of social exclusion, deprivation and poverty in local communities, and alternative valued activities [40–43].

Recent research suggests that the emergence and rapid growth in NPS may be more complex than simply a wholesale displacement from established street drugs to NPS due to their reduced purity/availability as was first supposed. For example, analyses of the South London Survey data suggest that there is a statistically significant relationship between recent mephedrone use and recent use of cocaine, ecstasy pills and MDMA powder [44]. This is supported by a survey of bar customers conducted about four months later that also found that recent mephedrone use was associated with recent use of cocaine, ecstasy pills and MDMA powder [45]. These studies suggest that NPS may act to supplement rather than wholesale displace established street drugs and thus provide a means to bolster and diversify polydrug repertoires.

Therefore, whilst the initial popularity of NPS such as mephedrone was clear: it was relatively cheap, easily available, a novelty, and for experienced users, an eager supplement to the reduced purity/availability and resultant disillusionment with street drugs [20], the growth in the trade creates a much more complex picture. What we are seeing is widespread polydrug use, with different drugs taken for different effects in different social contexts, and also influenced by regional and temporal variations in price, purity and availability of both NPS and street drugs.

Furthermore, as well as questioning the occurrence of a wholesale displacement from established illegal drugs to NPS, the widely predicted displacement from first generation NPS such as mephedrone to currently legal NPS, such as MDAI, also did not occur in the UK, according to annual surveys conducted with dance music magazine readers, with festival-goers and with South London gay clubbers. This picture is further complicated by the reported use of generic 'pills' and unidentified white powders with stimulant effects. The annual dance music magazine survey reported for its 2011

cohort that 15% of respondents had consumed a mystery white powder in the past year without knowing what it was or what it was originally sold as, with younger respondents more likely than older respondents to report taking these mystery powders [19]. Furthermore some users were not just unclear but also unconcerned about which NPS they consumed, illustrated by the growth in the north-west of England in self-reported use of 'Bubble', a white powder with disputed and varied contents according to users [45]. This was confirmed by analyses of police seizures which suggest that a wide range of stimulant drugs have been found in 'Bubble', from NPS such as mephedrone through to Class A drugs such as *para*-methoxymethamphetamine (PMMA).

Finally, although there was some evidence of mephedrone users planning to switch to Class A drugs when mephedrone was banned, this does not seem to have materialised. Amongst the online drug survey by Carhart-Harris et al. of over 1000 UK mephedrone users, nearly half (49%) said that they would use more MDMA once mephedrone was banned [16]. However, trends in ecstasy use in the UK, the BCS and CSEW household surveys do not provide evidence of this, even though the MDMA content of ecstasy pills is supposedly increasing. Self-reported lifetime use of ecstasy has been consistently but gradually falling from a peak in 2002/3 of 13% down to 9% in 2011/12 amongst 16–24 year olds, a statistically significant fall [46].

There are therefore grounds to question all three predicted types of displacement: from established illegal drugs to NPS, from first to second generation NPS and from NPS back to established illegal drugs. From the limited survey evidence that exists (most of which is from the UK), it seems instead that users adopt more fluid and flexible polydrug repertoires involving supplementing and adding various drugs to their consumption profiles rather than wholesale displacement.

EXPLORING SOCIAL HARMS: LEGISLATIVE CONTROL AND THE SUPPLY OF NPS

As the use of NPS increased, so more physical and social harms have come to light. Some of these, particularly the physical harms, are addressed in Section 3 of this book. This chapter highlights three key concerns relating to the social harms of NPS: the effects of legislative control on legal NPS; the development of an illegal trade in controlled NPS; and dependency.

Regarding the impact of legislation on NPS, there is evidence of increased variability in contents with the introduction of increased controls. For example, whereas mephedrone sold before legislative control was assessed to have high purity levels (approximately 95%) [47], the content and purity of NPS have become more variable as increasing numbers of drugs have been controlled by increasing numbers of countries. This is evident in a number of test purchasing studies by academics, chemists and law enforcement agencies, as well as the perceptions of users and NPS retailers. For example, a study by Brandt et al. in the UK in 2010 involved test purchase of 20 samples of products sold as NRG-1 and NRG-2 to see whether they contained the contents claimed in their packaging and marketing. Brandt and colleagues found that 70% of the NRG-1 and NRG-2 products they had purchased online contained a mixture of cathinones banned in the UK in April 2010 which had been rebranded as naphyrone, a legal NPS at the time of the test purchases. In fact, only one of the 17 products that had been sold as naphyrone was accurately labelled and contained naphyrone. Most samples purchased by the team in the immediate aftermath of April 2010 contained substituted cathinones that had been banned in the UK, with some samples also containing legal stimulants such as caffeine and inorganic bulking agents [48–50]. This may be because retailers were using up their old stocks of the (now)

controlled drugs or because misbranding by producers higher up the supply chain was leading to confusion and uncertainty amongst retailers about the contents of their own products. Similar concerns have been raised in other test purchase studies in the UK such as those of Davies and colleagues [51] and Ramsey and colleagues [52]. In a test purchase scheme by the Serious Organized Crime Agency, it was found that 19% of supposedly 'legal highs' contained controlled substances [2]. Furthermore, studies in Europe and the USA have shown similar results with variable content of NPS test purchased over the Internet, in head shops and in patient samples obtained in hospital admissions. For example, Spiller and colleagues analyzed the clinical effects and samples of 'bath salts' products for patients admitted to two poison centres with neurological and cardiovascular problems after exposure to 'bath salts' [14]. Whilst MDPV was evident in 13 out of 17 blood and/or urine samples, no mephedrone or methylone was found in any sample. Some NPS retailers have reported that since the banning of mephedrone across Europe and in China in 2010, illicit mephedrone is now as likely to contain 4-methylethcathinone (4-MEC), still legal in much of Europe and China, as mephedrone (4-methylmethcathinone, 4-MMC). For example, the Rave Florist described how: 'After mephedrone was banned... it was immediately replaced in the black market with methedrone. ... The market that was created by the legitimate sale of mephedrone by law abiding business was immediately taken over by criminals as 4-MEC, a dirtier high with more negative side effects and greater dangers, remained legal through most of the EU so could be imported to the border of the UK legally. The vast majority of vendors who legally sold mephedrone have nothing to do with the illegal 4-MEC trade. It is now a trade in the UK run exclusively by criminals.' This is supported by UK police seizure and drug-related deaths statistics. For example, analyses of NPS seizures by Lancashire

Constabulary in the 12 month period ending February 2013 contained 40% methedrone, 29% mephedrone and 12% pentedrone.

Amongst users, the annual dance music magazine survey reported growing suspicions that mephedrone was cut with other products or bulking agents, rising from 30% suspecting mephedrone was adulterated in 2010 to 80% in 2011, combined with the average price rising from about £12 per gram when mephedrone was legal, to approximately £19 per gram about six months after it was banned [12], and to approximately £20 per gram 18 months after it was banned [19]. By the autumn of 2011, 18 months after the ban, nearly half of dance music magazine respondents thought that the purity of mephedrone had fallen in the previous 12 months, i.e. from the autumn of 2010 to 2011 [19]. In the annual South London Surveys in gay clubs, respondents reported to the researchers that they were able to buy mephedrone from illicit suppliers within weeks of the ban and the researchers observed suppliers operating both inside and nearby to the dance clubs in the study, selling mephedrone at £20–25 per gram in July 2010 [15]. Interestingly the South London Survey respondents reported in the follow-up surveys paying a similar amount for mephedrone in July 2011 and 2012 as in 2010, suggesting that whilst the initial ban had an immediate impact upon availability, any changes in the availability of mephedrone (or its chemical cousins) in the subsequent 24 months had not been significant enough to influence the price of street mephedrone. The South London Survey respondents also did not think that there had been a notable deterioration in purity levels, citing the existence of crystals as visual evidence that street mephedrone had not been excessively cut with bulking agents, presumed by respondents usually to be powders.

It has been noted by users and online retailers interviewed by the author and also by enforcement agencies that to date organised crime has not taken over the illegal trade in NPS after legislative control in a replica of the criminal networks selling heroin and cocaine [53]. The current trade in controlled NPS such as mephedrone appears to be a hybrid trade, still largely separate from the heroin and cocaine trades, following circuitous routes from producer to user countries more akin to the trade in counterfeit prescription medicines (also manufactured in China and south east Asia) than the criminal organisations that dominate the heroin and cocaine trades. However, in relation to the trade in synthetic drugs more generally, continental European police have noted partial displacement prompted by the relative scarcity along the supply chain of some precursors and other essential ingredients [54]. This has combined both spatial displacement from Europe to China and also methodological displacement in terms of extending manufacturing to precursor ingredients as well, if certain chemicals become increasingly scarce due to stricter control measures.

Furthermore, in user countries, the competitiveness of the drug trade combines with the vast profits involved and the pressure of law enforcement agencies to produce a complex, fluid and entrepreneurial structure where ingenuity, loyalty and friendship reap greater rewards than aggression and fear [55–58]. Within these diverse and fluid drug enterprises there also exist independent retailers that have been characterised as evangelical 'trading charities' because they possess an ideological commitment to the use of drugs or an altruistic desire to procure drugs as a service to the community rather than being motivated simply by profit [55]. This is similar to the sorts of independent retailers noted in the 1990s ecstasy market of the UK acid house and rave scene, with the author interviewing ecstasy 'dealers' who exhibited an evangelical zeal in providing high quality ecstasy to clubbers because of what they saw as its positive and empathic influence on dance club culture [25]. Such trading charities were also associated with the supply of LSD in the late 1960s hippy counterculture and the supply of cannabis through 1970s Rastafarianism [55]. Also illustrative of this approach was the Rave Florist NPS retailer interviewed by the author in the spring of 2012 and again a year later. With a history of participating in the Manchester Hacienda rave scene himself, he had no desire to sell *illegal* drugs but instead expressed a wish that one day he would be able to sell a NPS that had been identified, tested and confirmed as a safe, legal alternative to ecstasy, combining stimulant and euphoric effects to enhance weekend partying, saying that "for myself, this business began as I enjoyed taking recreational substances but did not want to break the law." This is not dissimilar to the attempts by Nutt 'to make better recreational drugs' such as a safer or modified version of alcohol without the hangover [59].

Online NPS retailers have emphasised that in order to maintain their competitive edge over rival retailers they ensure quality through the regular laboratory testing of their products as well as maintaining high levels of customer service [60]. In an interview with the author, the Rave Florist retailer emphasied that he wanted to sell NPS that were accurately labelled and safety tested and accurately labelled but he also expressed frustration in relation to both of these issues. In terms of safety testing, pressure from enforcement agencies made some laboratories reluctant to test his products and that this consequently reduced his ability to sell accurately labelled, (or even knowingly legal) NPS. Furthermore, in relation to labelling, he highlighted how currently the law discourages the provision of explicit user advice because recognition that products are for human consumption increases the risks of prosecution: "I certainly want to offer products that are accurately labelled and tested and would like to provide relevant safety information however current legislation prevents this. It seems that certain government agencies have decided to take a moral crusade against us. ... We have

experienced problems with trading standards when we have tried to display safety information. When no safety information is displayed they seem to have no problem however when safety information is displayed they are fast to attempt to prevent sales by focusing on mistakes and refusing to offer advice as to how products should be correctly labelled. ... We are very resourceful and have invested in advice and now have all our products labelled in full accordance with current legislation." The recent response by the New Zealand government to the challenges of controlling NPS moves some way towards incorporating this perspective into a national regulatory framework, discussed further below.

In terms of dependency, even the earliest qualitative studies with users raised concerns about the negative stimulant effects of mephedrone [20,31]. The first UK survey of school and college students noted that 18% reported symptoms associated with addiction or dependence [10]. Strong cravings for the drug (redosing or 'fiending'), dependency and suicidal tendencies have all been noted by mephedrone users and in animal laboratory studies with suggestions that this may be due to its strong dopaminergic effects [3,61]. Aside from newspaper headlines linking NPS with, for example, violent crime, there is a small but growing body of evidence suggesting that small numbers of NPS users are developing problem use, dependency and associated problems, as well as growing concerns about the route of administration with evidence of binge intravenous injecting in Eastern Europe, Ireland [62] and the UK, both amongst established heroin injectors [63] and clubbers [64]. Indeed the first confirmed mephedrone-related death in the UK was the result of multiple injecting of illicit drugs [65].

It is perhaps salutary to consider the historical precedents to these debates. Warner noted in relation to the 'gin epidemic' of the 18th century that government attempts to legislate to reduce consumption failed, but that consumption fell once a critical mass of the general public had experienced gin and had come to observe in themselves or others significant negative effects [66]. After several years, it could be that there develops a significant body of experiential evidence of this in relation to NPS. Growing numbers of users are reporting NPS-related social problems in terms of excessive use resulting in a negative impact on work, studies, finances and relationships. One festival-goer interviewed by the author in 2012 noted with pride how well he controlled his cocaine use but that he became 'addicted' to mephedrone when it was legal in the UK. An optimist might deduce that the low uptake in use of contemporary NPS could be in part a reflection on some of the problems users have already experienced with mephedrone and other NPS when they first emerged. However, it could be due to wider processes of cultural change going beyond first hand experience of NPS. It has been suggested that the declining uptake in NPS mirrors a broader fall in illegal drug use in recent years and that this can be explained by changing fashions which are themselves influenced by the increased availability and cultural accommodation of recreational drug use in the UK in the 1990s and 2000s [39,67]. For example, Shapiro suggests that:

"drug use having become more normalized in society, might then be just as prey to fashion as any other cultural artefact. Drugs don't appear to be 'cool' these days as they once were" [68].

POLICY OPTIONS IN A CHANGING WORLD

Novel psychoactive substances represent a unique set of challenges for policy makers because of the combination of the role of the Internet in facilitating the retail trade and user fora; a robust demand to experiment with psychoactive substances amongst some sections of youth and young adults; resulting in an 'irrepressible market' [55] with multi-million pound profits for retailers; and tensions between the speed of emerging NPS and the development of an evidence base to inform policy. Therefore despite the very low levels of use of many of the NPS on offer (discussed above), policy makers, commentators and campaigners have been exercised in the consequent debate about policy options or 'solutions' to this perceived new 'threat'.

For many legislatures, prohibition remains the basis for drug control and includes a system of risk assessment of each individual new drug which is seen to pose a threat with consequent legislative control of those deemed to cause significant physical or social harm. Before 2009, this assessment process happened on occasion – for example in the UK GHB, ketamine [69], GBL [70], and BZP were reviewed and banned in the 2000s – at a quickening but manageable rate. The sudden explosion of NPS from 2009 onwards has challenged what was arguably a creaky and decades-old system of legislative control across Europe, North America and Australasia, limited by the minimal knowledge base for so many of these new drugs that have become readily available to curious potential consumers. Concerns about lack of scientific rigour in the risk assessment process [71] combined with the pooling of drug harms with drug policy harms [72] have further overshadowed the assessment process. The need for an independent assessment of drug-related harm which is entirely 'decoupled' from determining punishments has been recommended by amongst others, the UK Parliamentary Science and Technology Select Committee [73].

Further disadvantages of the prohibition model include the high costs of enforcement, the adverse consequences for individuals of criminalisation and the potential for the law to produce greater problems than it solves in the form of incentives for criminal organisations to profit from the international supply of illegal drugs to meet continued demand. Social costs include infringements of human rights and individual liberties, the adverse consequences of criminalising potentially otherwise law-abiding citizens and divisive effects on families and communities [74]. Organisations advocating for reform of the drug laws such as the Transform Drug Policy Foundation have argued that not only has the prohibition paradigm failed to achieve its goal of eradicating drug supply or demand, but it has created new harms associated with the illegal trade and in many cases is actively counterproductive [75]. Other non-governmental organisation groupings recently included in the Count the Costs initiative have attempted to detail and describe the multiple harms either created or exacerbated by prohibition (both domestically and internationally) [76]. The costs that they categorise include the fuelling of crime at all levels from street crime to international organised crime; an array of human rights abuses; stigma and discrimination of marginalised and vulnerable populations; health costs in terms of increased risk to users from consumption of unregulated products in unsupervised venues; obstacles to effective harm reduction interventions; and the costs of enforcement with the resultant loss of resources for public health interventions. The Rave Florist and a consortium of NPS online and High Street retailers recently estimated that the NPS trade potentially could raise £200 million in revenue for the Exchequer in the tax year 2012/3 through various contributions such as Value Added Tax, Pay As You Earn and so forth, although given the disincentives to open trading, it is doubtful that the full revenue is currently being raised.

Furthermore there is little evidence that one of the key purposes of legislative control – deterrence – is effective. In relation to cannabis, policy analyses suggest that the various policy options relating to control of cannabis appear to have little if any effect on rates of use or levels of associated social or physical harm, as Single and colleagues concluded in their comparison of cannabis decriminalisation in Australia and the

USA [74]. The Portuguese decriminalisation of the use and possession of all illicit drugs in 2001 also did not result in an increase in drug use, when compared to trends in neighbouring countries such as Spain and Italy [77], with evidence that problem drug use and drug-related harms actually lessened in subsequent years. However, Hughes and Stevens warn against attributing any and all positive changes in Portugal as due to decriminalisation alone, however, and cite expanding drug service provision and an ageing population of opiate users contributing to some of the positive outcomes that have been identified in Portugal, as elsewhere in Europe [78]. They also note that the benefits of policy change can often be cumulative, as was the case with Portugal.

The deterrent value of legislative control produces scant evidence in its support [72,73,79]. Furthermore, studies with non-drug users indicate that abstainers rarely cite illegality as a key reason for their non-use of drugs, with health, disinterest and alternative leisure pursuits such as sports being cited as motivating factors instead [39]. This also has an important corollary in relation to 'legal highs': as noted above, NPS users do not tend to consider the legal status of NPS as the key motivating factor in their use. This can be illustrated in the South London gay clubbers' study in that self-reported use of mephedrone and preference for mephedrone continued many months after the drug was controlled under the Misuse of Drugs Act, 1971 in 2010 in the UK, with mephedrone being identified as not just the most widely consumed illegal drug on the fieldwork night but also users' favourite drug [29]. An EMCDDA study of changes in cannabis policy across Europe, across a 10-year period showed no association between policy change and drug prevalence and concluded that for the 2000s, for the countries in question, 'no simple association can be observed between legal changes and cannabis use prevalence' [80]. However if there is little evidence of a deterrent effect, contemporary studies of polydrug use suggest that changes in availability/purity of one drug can have unintended consequences in terms of supplementary use (if not wholesale displacement) to other drugs, both legal and illegal.

In terms of NPS the time and resources involved in the risk assessment and review of individual emergent drugs have led to increased pressures on government and scientific advisory bodies at a time of increased economic pressures on the public sector more generally. In the UK the trafficking and sale of NPS of unknown harm can be temporarily banned whilst they are reviewed by the Advisory Council on the Misuse of Drugs, using Temporary Class Drug Orders introduced in the Police Reform and Social Responsibility Act 2011 amending the Misuse of Drugs Act, 1971, with the aim of reducing availability and thereby protecting the general public whilst the review process is undertaken. For the first drug which received a Temporary Class Drug Order in the UK – methoxetamine – there is evidence that the temporary ban on trafficking and sales may have achieved its goal of reducing availability without criminalising users [81]. However, frustrations with the 'cat and mouse' process of reviewing and potentially banning each individual NPS have led governments and policy bodies to contemplate other, more comprehensive and potentially more effective methods of controlling NPS. Two contrasting models are illustrated from the USA and New Zealand.

The USA passed the Controlled Substance Analogue Enforcement Act in 1986 which outlaws the supply of drugs which are 'substantially similar' in chemical structure or effect to stimulant, depressant or hallucinogenic drugs that are already controlled. The appeal of the Analogue Act is that it potentially avoids a full risk assessment for each individual emergent drug for which there may be little scientific evidence available. However, American and British critics have suggested that it has failed to act as a deterrent; it is too unwieldy and reliant on expensive expert witnesses and the vagaries of the adversarial court system;

it is a blunt tool which can stifle future medical and psychotherapeutic developments; and now nearly 30 years after the Analogue Act was passed not only is most of the case law surrounding it dated but the US Drug Enforcement Administration itself sees the act as an 'imperfect law' that is ill equipped to address the contemporary challenges of NPS hence new legislation was required [82–84]. Also as Moore and Measham have noted [85], by criminalising all drugs 'similar' to those already controlled, the Analogue Act in effect results in the criminalisation of recreational drug use as an activity, regardless of the possible harms (or possible lack of harms) associated with any individual emergent drug. Thus it precludes the possibility of identification of an analogue that does not cause undue harm to the user and therefore outlaws drug use in and of itself, whether for the purposes of spiritual, creative, recreational or 'psychonautic' pursuits [86].

An indication of the perceived limitations of the Analogue Act can be seen in the rapidity of the introduction of state and federal legislation in recent years: within eight months of the appearance of 'bath salts' 16 states added synthetic cathinones to their list of controlled substances as a Schedule 1 drug [14] and furthermore in 2011 US Congress proposed that the two main groups of NPS – cathinone derivatives and cannabinoid agonists – should be placed in Schedule 1 of the Controlled Substances Act without relying on the Analogue Act [83]. This flurry of legislation to ban NPS is all the more surprising given the approach towards cannabis control in the USA, with 18 states now allowing medical cannabis use, 14 of these states also having taken steps towards decriminalisation of personal possession of cannabis and 2 states allowing the legal sale and possession of cannabis for recreational as well as medical purposes, and the distinction between medically sanctioned and recreational cannabis use becoming increasingly blurred. The bigger danger in terms of conceptualising drug policy, as Moore and Measham note, is that in highlighting the laborious process of individual risk assessment, alongside the logistical problems of operating temporary banning orders and the populist appeal of the precautionary principle, authorities could be pushed towards analogue legislation like that currently on the US statute books, or generic legislation as increasingly favoured in the UK, which

> "extends the long arm of criminalization to potentially *all* stimulant, depressant and hallucinogenic drugs *regardless* of their individual potential for harm and *before* any problems emerge (emphasis in original)" [85].

In contrast to analogue legislation, New Zealand has adopted a novel alternative approach to blanket prohibition. In this new regulatory regime due to be introduced in 2013, all NPS will be banned unless and until there is satisfactory clinical evidence of their safety [87]. This places the commercial burden of proof for safety on NPS manufacturers and retailers, it provides strong consumer protection but it also recognises the possibility that one day a recreational drug *could* be identified that is deemed to be 'safe enough' for its sale to be approved by the New Zealand health ministry. Potentially, then, a regulated market could develop in New Zealand whereby manufacturers would assess the risks of individual NPS through human trials; the state would approve those rare products that pass such safety tests as are established; and manufacturers would pay the government a substantial application or licensing fee. This would result in the onus being on manufacturers rather than retailers, users or the state to assess the risks; and on the state rather than retailers, 'dealers' or enforcement agencies to control the availability of NPS.

As outlined by Rolles and by Demos and the UK Drug Policy Commission (UKDPC) [75,88], there is a broad spectrum of policy options between prohibition and legalisation, with regulation achieved through a range of possible controls over products, vendors and availability which could utilise existing mechanisms such as prescriptions, licensing or consumer trading standards. Demos and UKDPC [88] have called for a review of the Misuse of Drugs Act 1971 in the UK and suggested instead the introduction of a new Harmful Substances Control Act covering all legal and illegal drugs. Beyond a review of existing drug laws, drugs researchers have also called for a more robust framework for the regulation of trading standards, consumer protection and Internet sales, including appropriate user information on contents, dosage, risks and harm reduction. As Rolles has argued [75], a responsible government may be better able to control NPS than a free market and a state imposed regulatory framework may be better able than individual manufacturers and retailers to prioritise minimising the physical and social harms from drugs rather than maximising profits.

A cornerstone of education, prevention and harm reduction is to inform the user or potential user about the effects and consequences of use. However, whilst some of these drugs have a pre-existing body of research in relation to their clinical development as potential medicines (e.g. methoxetamine), many more are chemicals with little or no research base upon which to draw (e.g. mephedrone). As with risk assessment, this makes the task of education, prevention and harm reduction more challenging for NPS than for established street drugs for which there are scores, if not hundreds of years, of writings on these drugs and their effects [89].

Whilst support for evidence-based policy in general and risk assessments in particular is widespread, limited research on many NPS hampers the formation of drug policy in this field. There is a resulting temptation to fill this void with personal memorandums, anecdotes, individual case studies and 'killer charts' in support of a particular policy option [90,91], whereby certain drug discourses are considered to be tactically and structurally useful to policy makers in positions of power and which thereby further enhance pre-existing asymmetrical power relations [92]. Added to this the electorate (whether or not influenced by the media) may demand a government response to a perceived problem, as was evident with emerging mephedrone-related health and social problems in early 2010 in the UK. Doing nothing is rarely a viable option for politicians in the heat of a public health scare and an impending general election, even if 'wait and see' and 'do more research' is often the preferred academic response.

THE NPS DEBATE: HISTORICAL STAGNATION AND VESTED INTERESTS (A UK PERSPECTIVE)

As evident from the data presented in this chapter, the prevalence of use of individual NPS varies enormously. Whilst there is evidence that mephedrone use rapidly increased and embedded itself in the repertoires of some weekend recreational drug users in the UK, for more recent NPS, their uptake and popularity has not matched that of mephedrone. Therefore whilst it is important to recognise the policy significance of the emergence of these drugs, their challenges to existing methods of control and the development of global manufacture and Internet trading, it is also notable that prevalence and indeed interest in more recent NPS might be very low amongst even established drug users. The control of such drugs may be justified on the grounds of the precautionary principle, to mediate against future uptake and harms, but for many of these NPS, we do not have resounding evidence of either significant prevalence or existing harms.

The concern here is with how the NPS debate is unfolding and why. Lessons can be learned from public health scares surrounding alcohol. As Reinarman noted in relation to the anti-drunk driving movement of 1980s America, 'social problems have careers that

ebb and flow independent of the "objective" incidence of the behaviours thought to constitute them' [93]. In the UK the 'binge drinking epidemic' of the late 1990s and early 2000s was emblematic of the millennial zeitgeist with pervasive media images of intoxicated young people stumbling through city streets at night in an orgy of hedonistic consumption [94]. A few years earlier, the warning signs of this change in consumption patterns were starting to develop. Young people's weekly alcohol consumption doubled across the 1990s, for both over and under 16 s, with increased consumption concentrated into weekend binges facilitated by an expanding market of high strength alcoholic drinks marketed to young people, and which revealed itself to the author when working on a longitudinal study of young people's alcohol and drug use in the early 1990s [95]. What was interesting was the response of social scientists: an eagerness to reassert the historical continuities to binge drinking, both that Britain had a longstanding 'binge and brawl' culture dating back to Victorian times and earlier, and that Britain had a longstanding press culture which vilified young people and their leisure patterns [96], resulting in a cyclical reproduction of the 'problem of youth' beautifully expounded in the classic history of the hooligan by Pearson [97]. Thus, for alcohol and drugs researchers, to identify a change in consumption practices (specifically an increase) was to risk falling prey to accusations from liberal progressives of fuelling a media 'moral panic' [45]. Added to this are legitimate concerns about drawing attention to the practices of a specific user group and risking (further) problematising or stigmatising that group, a point noted elsewhere by the author in relation to GBL [98]. The danger, of course, is that the time lag in the academic process from proposal writing to funding, data collection and eventual publication in a peer reviewed journal can take several years, added to which those most successful in obtaining funding for such research may be least likely to have their finger on the pulse of youth at play. This can result in academic stagnation particularly in the field of drug and alcohol research. Then as academics and service providers play catch-up with a new trend another dynamic comes into play: resources, research profiles and careers increasingly become tied up with the new trend, resulting in a reluctance to acknowledge the (almost inevitable) decline in consumption that follows. This has been the case with binge drinking. Despite clear evidence across the 2000s that alcohol consumption has been falling in the UK and particularly binge drinking by young people, alongside falling levels of violent crime, confirmed in the General Lifestyles Surveys, BCS/CSEW, offender statistics and hospital admissions, for some academics and health service providers, their heels are lodged firmly in the ground [99,100].

Therefore, despite much of the criticism of evidence-based policy being of policy makers not taking enough notice of academic research, we also need to shine a light on the 'producers' (of evidence) and 'providers' (of services) in order to better understand how the processes of knowledge production and health and criminal justice service provision (with their inbuilt aversion to change) can impact upon NPS policy.

CONCLUSION

A growing patchwork of studies of NPS use by different sociodemographic groups in different leisure venues paints a picture of differentiated demand, variable uptake, with pockets of popularity contrasted with considerable indifference, influenced by the different drugs on offer, their relative price, purity, availability and the broader social context to use. Whether or not many people are interested in taking NPS – and there is evidence that in Europe it remains a minority activity occurring predominantly amongst experienced young adult recreational drug users – the large number of drugs on offer and the ease of access to uncontrolled NPS makes this a new challenge for policy makers as well as a pharmacological experiment with the youthful minds and bodies of the next generation. The scientific evidence base – on the social and physical harms of NPS and evaluation of policy responses – remains slim and therefore it is the general public who currently act as guinea pigs until an effective system of control can be established which successfully regulates a sphere of enormous international political, commercial and scientific interest. The challenge is to balance the continuities and change in drug use, veering neither towards oversimplified vilification of NPS users and retailers, nor towards historical stagnation in asserting an essential and unvarying pattern of intoxication and excess.

REFERENCES

[1] Gustavsson D, Escher C. Mephedrone – internet drug which seems to have come and stay. Fatal cases in Sweden have drawn attention to previously unknown substance. Lakartidningen 2009;106(43):2769–71. [Article in Swedish].

[2] Advisory Council on the Misuse of Drugs (ACMD). Consideration of the novel psychoactive substances ('Legal Highs'). London: Home Office; 2011.

[3] Schifano F, Albanese A, Fergus S, Stair JL, Deluca P, Corazza O, et al. Mephedrone (4-methylmethcathinone; 'meow meow'): chemical, pharmacological and clinical issues. Psychopharmacology 2011;214:593–602.

[4] Schmidt M, Sharma A, Schifano F, Feinmann C. 'Legal highs' on the net – evaluation of UK-based websites, products and product information. Forensic Sci Int 2011;206(1):92–7.

[5] Vardakou I, Pistos C, Spiliopoulou Ch. Drugs for youth via internet and the example of mephedrone. Toxicol Lett 2011;201(3):191–5.

[6] European Monitoring Centre for Drugs and Drug Addiction (EMCDDA). EU Drug Markets Report: A Strategic Analysis. Lisbon: EMCDDA; 2013.

[7] European Monitoring Centre for Drugs and Drug Addiction (EMCDDA). Online sales of new psychoactive substances/ 'legal highs': summary of

results from the 2011 multilingual snapshots. Lisbon: EMCDDA; 2011.

[8] Davey Z, Corazza O, Schifano F, Deluca P, on behalf of the Psychonaut Web Mapping Group. Mass-information: mephedrone, myths, and the new generation of legal highs. Drugs Alcohol Today 2010;10(3):24–8.

[9] Association of Chief Police Officers of England, Wales and Northern Ireland. ACPO guidance on policing new psychoactive substances including temporary class drugs. ACPO, 2011.

[10] Dargan P, Albert S, Wood D. Mephedrone use and associated adverse effects in school and college/university students before the UK legislation change. QJM 2010;103:875–9.

[11] Mixmag, Mixmag drugs survey. February 2010;225: 44–53.

[12] Mixmag, The 2011 drugs survey. March 2011;238:49–59.

[13] McElrath K, O'Neill C. Experiences with mephedrone pre and post-legislative controls: Perceptions of safety and sources of supply. Int J Drug Policy 2011;22(2):120–7.

[14] Spiller H, Ryan M, Weston R, Jansen J. Clinical experience with and analytical confirmation of 'bath salts' and 'legal highs' (synthetic cathinones) in the United States. Clin Toxicol (Phila.) 2011;49(6):499–505.

[15] Measham F, Wood D, Dargan P, Moore K. The rise in legal highs: prevalence and patterns in the use of illegal drugs and first and second generation 'legal highs' in south london gay dance clubs. J Subst Use 2011;16(4):263–72.

[16] Carhart-Harris R, King L, Nutt D. A web-based survey on mephedrone. Drug Alcohol Depend 2011;118(1):19–22.

[17] Newcombe R. Trends in the prevalence of illicit drug use in Britain Simpson M, Shildrick T, MacDonald R, editors. Drugs in Britain: supply, consumption and control. Basingstoke: Palgrave Macmillan; 2007.

[18] Hoare J. Drug misuse declared: findings from the 2008/09 British crime survey. London: Home Office; 2009. Available: <http://rds.homeoffice.gov.uk/rds/pdfs09/hosb1209.pdf>. [accessed 23.09.12].

[19] Mixmag. Mixmag's drug survey: the results. Available: <http://www.mixmag.net/drugssurvey>; 2012 [accessed 22.09.12].

[20] Measham F, Moore K, Newcombe R, Welch Z. Tweaking, bombing, dabbing and stockpiling: the emergence of mephedrone and the perversity of prohibition. Drugs Alcohol Today 2010;10(1):14–21.

[21] European Monitoring Centre for Drugs and Drug Addiction (EMCDDA). EMCDDA-Europol 2011 Annual Report on the implementation of Council Decision 2005/387/JHA. Lisbon: EMCDDA; 2012.

[22] Measham F, Moore K, Østergaard J. Emerging drug trends in Lancashire: night time economy surveys. phase one report. Lancaster: Lancaster University and Lancashire Drug and Alcohol Action Team; 2011. pp. 1–62. Available: <http://www.ldaat.org/files/emerging_trends_report.pdf> [accessed 04.03.13].

[23] Measham F, Moore K. Repertoires of distinction: exploring patterns of weekend polydrug use within local leisure scenes across the English night time economy. Criminol Crim Justice 2009;9(4):437–64.

[24] Measham F, Brain K. 'Binge' drinking, British alcohol policy and the new culture of intoxication. Crime, Media, Culture: An international journal 2005;1(3):263–84.

[25] Measham F, Aldridge J, Parker H. Dancing on drugs: risk, health and hedonism in the British club scene. London: Free Association Books; 2001.

[26] Winstock A, Mitcheson L, Deluca P, Davey Z, Corazza O, Schifano F. Mephedrone, new kid for the chop? Addiction 2011;106:154–61.

[27] Smith Z, Moore K, Measham F. MDMA powder, pills and crystal: the persistence of ecstasy and the poverty of policy. Drugs Alcohol Today 2009;9(1):13–19.

[28] Winstock A, Mitcheson L, Marsden J. Mephedrone: still available and twice the price. Lancet 2010;376: 1537.

[29] Wood D, Measham F, Dargan P. 'Our favourite drug': prevalence of use and preference for mephedrone in the London night time economy one year after control. J Subst Use 2012;17(2):91–7.

[30] Wood D, Hunter L, Measham F, Dargan P. Limited use of novel psychoactive substances in South London nightclubs. QJM: An international journal of medicine 2012;105(10):959–64.

[31] Newcombe R. Mephedrone: the use of mephedrone in middlesbrough. Manchester: Lifeline Publications and Research; 2009.

[32] Brunt T, Poortman A, Niesink R, van den Brink W. Instability of the ecstasy market and a new kid on the block: mephedrone. J Psychopharmacol 2011;25(11):1543–7.

[33] Power M. How I caused an international drug panic. In: Mixmag, The 2011 drugs survey. 2011; March: 238.

[34] Power M. Drugs 2.0: The web revolution that's changing how the world gets high. London: Portobello; 2013.

[35] Hallam C. The heroin shortage in the UK and Europe. International Drug Policy Consortium Briefing Paper. London: IDPC; 2011.

[36] Hand T, Rishiraj A. Seizures of drugs in England and Wales, 2008/9, Home Office Statistical Bulletin 16/09. London: Home Office; 2009.

[37] SOCA (Serious and Organised Crime Agency) The United Kingdom threat assessment of serious organised crime, 2008/9. London: The Stationary Office; 2008.

[38] Brunt T, Niesink R. The drug information and monitoring system (DIMS) in the Netherlands:

implementation, results, and international comparison. Drug Test Anal 2011;3(9):621–34.

[39] Parker H, Aldridge J, Measham F. Illegal leisure: the normalisation of adolescent recreational drug use. London: Routledge; 1998.

[40] Graham J, Bowling B. Young people and crime. Home Office Research Study No.145. London: Home Office; 1995.

[41] Kosterman R, Hawkins D, Guo J, Catalano R, Abbott R. The dynamics of alcohol and marijuana initiation: patterns and predictors of first use in adolescence. Am J Public Health 2000;90(3):360–6.

[42] Storr C, Chen C-Y, Anthony J. 'Unequal opportunity': neighbourhood disadvantage and the chance to buy illegal drugs. J Epidemiol Community Health 2004;58(3):231–7.

[43] McIntosh J, MacDonald F, McKeganey N. The reasons why children in their pre and early teenage years do or do not use illegal drugs. Int J Drug Policy 2005;16(4):254–61.

[44] Moore K, Dargan P, Wood D, Measham F. Do novel psychoactive substances (NPS) displace established street drugs, supplement them or act as drugs of initiation? The relationship between mephedrone, ecstasy and cocaine. Eur Addict Res 2013;19:276–282.

[45] Measham F, Moore K, Østergaard J. Mephedrone, 'Bubble' and unidentified white powders: the contested identities of synthetic 'legal highs'. Drugs Alcohol Today 2011;11(3):137–47.

[46] Home Office Drug misuse declared: findings from the 2011/12 crime survey for England and Wales. London: Home Office; 2012.

[47] European Monitoring Centre for Drugs and Drug Addiction (EMCDDA) Europol–EMCDDA joint report on a new psychoactive substance: 4-methylmethcathinone (mephedrone). Lisbon: EMCDDA; 2010.

[48] Brandt S, Sumnall H, Measham F, Cole J. Letter: second generation mephedrone: the confusing case of NRG-1. BMJ 2010;341:c3564.

[49] Brandt S, Sumnall H, Measham F, Cole J. Analyses of second-generation 'legal highs' in the UK: initial findings. Drug Test Anal 2010;2(8):377–82.

[50] Brandt S, Freeman S, Sumnall H, Measham F, Cole J. Analysis of NRG 'legal highs' in the UK: identification and formulation of novel cathinones. Drug Test Anal 2011;3(9):569–75.

[51] Davies S, Wood D, Smith G, Button J, Ramsey J, Archer R, Holt D, Dargan P. Purchasing 'legal highs' on the internet – is there consistency in what you get?. QJM 2010;103:489–93.

[52] Ramsey J, Dargan P, Smyllie S, Davies S, Button J, Holt D, Wood D. Buying 'legal' recreational drugs does not mean that you are not breaking the law. QJM 2010;103:777–83.

[53] SOCA (Serious and Organised Crime Agency) Drugs: Mephedrone – Update on the Threat, Harms and Involvement of Organised Criminals, SOCA Report, KCAD-6427(01/13), pp.1–9. 2013.

[54] Vijlbrief M. Looking for displacement effects: exploring the case of ecstasy and amphetamine in the Netherlands. Trends Organized Crime 2012;15:198–214.

[55] Dorn N, South N. Drug markets and law enforcement. Br J Criminol 1990;30(2):171–88.

[56] Coomber R. Pusher myths: re-situating the drug dealer. London: Free Association Books; 2006.

[57] Pearson G, Hobbs D. King pin? A case study of a middle market drug broker. Howard J Crim Justice 2003;42(4):335–47.

[58] Pearson G, Hobbs D. 'E' is for enterprise: middle level drug markets in ecstasy and stimulants. Addict Res Theory 2004;12(6):565–76.

[59] Nutt D. The future of drugs: safe and non-addictive? The Guardian, 2012 June 11:G2 [Extract from Nutt D. Drugs without the hot air: minimising the harms of legal and illegal drugs. Cambridge: UIT; 2012].

[60] Hobson S. Drugs and dotcoms: how the legal highs industry exploded online. LondonlovesBusiness, 23 August 2012. Available: <http://www.londonlovesbusiness.com/business-news/business/drugs-and-dotcoms-how-the-legal-highs-industry-exploded-online/3267.article> [accessed 23.09.12].

[61] Kehr J, Ichinose F, Yoshitake S, et al. Mephedrone, compared with MDMA (ecstasy) and amphetamine, rapidly increases both dopamine and 5-HT levels in nucleus accumbens of awake rats. Br J Pharmacol 2011;164(8):1949–58.

[62] Van Hout M, Bingham T. 'A costly turn on': patterns of use and perceived consequences of mephedrone based head shop products amongst Irish injectors. Int J Drug Policy 2012;23(3):188–97.

[63] Daly M. Mephedrone: the rise of heroin's cheap rival, The Guardian, 2nd December; 2012. Available at: <http://www.guardian.co.uk/society/shortcuts/2012/dec/02/mephedrone-heroin-cheap-rival> [Last accessed 14.04.13.].

[64] Kirby T, Thornber-Dunwell M. High-risk drug practices tighten grip on London gay scene. Lancet 2013;381(9861):101–2.

[65] BBC News online (2010), Man from Hove died after injecting mephedrone, 27 May. Available: <http://www.bbc.co.uk/news/10176982> [accessed 23.09.12].

[66] Warner J, Her M, Gmel G, Rehm J. Can legislation prevent debauchery? mother gin and public health in 18th-century England. Am J Public Health 2001;91(3):375–84.

[67] Aldridge J, Measham F, Williams L. Illegal leisure revisited. London: Routledge; 2011.

[68] Shapiro H. Druglink, 2012; September

[69] Advisory Council on the Misuse of Drugs ACMD technical committee: report on ketamine. London: Home Office; 2004.

[70] ACMD GBL & 1,4-BD: assessment of risk to the individual and communities in the UK. London: Home Office; 2007.

[71] Nutt D, King L, Phillips L. Drug harms in the UK: a multicriteria decision analysis. Lancet 2010;376(9752):1558–65.

[72] Rolles S, Measham F. Questioning the method and utility of ranking drug harms in drug policy. Int J Drug Policy 2011;22(4):243–6.

[73] House of Commons Science and Technology Committee Drug classification: making a hash of it? fifth report of session 2005–06, HC 1031. London: The Stationery Office; 2006. Available: <http://www.publications.parliament.uk/pa/cm200506/cmselect/cmsctech/1031/103102.htm> [accessed 04.03.13].

[74] Single E, Christie P, Ali R. The impact of cannabis decriminalisation in Australia and the United States. J Pub Health Pol 2000;21(2):157–86.

[75] Rolles S. After the war on drugs: blueprint for regulation. Bristol: Transform Drug Policy Foundation; 2009.

[76] Count The Costs. The alternative world drug report, count the costs initiative. Available: <www.countthecosts.org>; 2012.

[77] Greenwald G. Drug decriminalisation in Portugal: lessons for creating fair and successful drug policies, Washington DC: CATO Institute 2009.

[78] Hughes C, Stevens A. What can we learn from the Portuguese decriminalisation of illicit drugs? Br J Criminol 2010;50(6):999–1022.

[79] House of Commons Public Accounts Committee, 2010, Tackling problem drug use, Thirtieth Report of Session 2009–10, HC 456, 7 April 2010, London: The Stationery Office.

[80] European Monitoring Centre for Drugs and Drug Addiction (EMCDDA) Annual report: the state of the drugs problem in Europe. Lisbon: EMCDDA; 2011.

[81] Advisory Council on the Misuse of Drugs. Methoxetamine. London: Home Office, 2012.

[82] Kau G. Flashback to the Federal Analog Act of 1986: Mixing rules and standards in the cauldron. Univ PA Law Rev 2008;156:1077–115.

[83] King L, Nutt D, Singleton N. Analogue controls: an imperfect law. UK drug policy commission and independent scientific committee on drugs. Available: <http://www.ukdpc.org.uk/wp-content/uploads/Analogue-control-19.06.12.pdf>; 2012 [accessed 26.02.13].

[84] Wong L, Dormont D, Matz, H. United States controlled substance analogue act: legal and scientific overview of an imperfect law. Presentation to ACMD, July 2010.

[85] Moore K, Measham F. Impermissible pleasures in UK leisure: exploring policy developments in alcohol and illicit drugs Jones C, Barclay E, Mawby R, editors. The problem of pleasure: leisure, tourism and crime. Cullompton: Willan; 2011. p. 62–76.

[86] Newcombe R, Johnson M. Psychonautics: a model and method for exploring the subjective effects of psychoactive drugs. Liverpool: 3D Research Bureau; 1999.

[87] TVNZ. Party pills must be proved safe for sale – Dunne, 16 July. Available: <http://tvnz.co.nz/politics-news/party-pills-must-proved-safe-sale-dunne-4970945>; 2012.

[88] Birdwell J, Chapman J, Singleton N. Taking drugs seriously: a Demos and UK drug policy commission report on legal highs. London: Demos; 2011.

[89] Plant S. Writing on drugs. London: Faber and Faber; 1999.

[90] Stevens A. Survival of the ideas that fit: an evolutionary analogy for the use of evidence in policy. Soc Policy Soci 2007;6(1):25–35.

[91] Stevens A. Telling policy stories: an ethnographic study of the use of evidence in policy-making in the UK. J Soc Policy 2011;40(2):237–55.

[92] Stevens A. When two dark figures collide: evidence and discourse on drug-related crime. Crit Soc Pol 2007;27(1):77–99.

[93] Reinarman C. The social construction of an alcohol problem: the case of mothers against drunk drivers and social control in the 1980s. Theory Soc 1988;17(1):91–120. Quote p.91.

[94] Hayward K, Hobbs D. Beyond the binge in 'Booze Britain': market-led climinalization and the spectacle of binge drinking. Br J Sociol 2007;58(3):437–56.

[95] Measham F. The 'Big Bang' approach to sessional drinking: changing patterns of alcohol consumption amongst young people in North West England. Addict Res 1996;4(3):283–99.

[96] Herring R, Berridge V, Thom B. Binge drinking: an exploration of a confused concept. J Epidemiol Community Health 2008;62:476–9.

[97] Pearson G. Hooligan: a history of respectable fears, 1991. Basingstoke: Macmillan; 1983.

[98] Moore K, Measham F. The silent 'G': the absence of GHB/GBL from research and policy. Contemp Drug Probl 2012;39:565–90.

[99] Measham F. The turning tides of intoxication: young people's drinking in Britain in the 2000s. Health Educ 2008;108(3):207–22.

[100] Measham F, Østergaard J. The public face of binge drinking: British and Danish young women, recent trends in alcohol consumption and the European binge drinking debate. Probat J 2009;56(4):415–34.

This page intentionally left blank

ANALYTICAL TECHNIQUES

This page intentionally left blank

Analytical Techniques for the Detection of Novel Psychoactive Substances and Their Metabolites

Frank T. Peters* and Markus R. Meyer**

Institute of Forensic Medicine, University Hospital Jena, Jena, Germany,
**Department of Experimental and Clinical Toxicology, Institute of Experimental and Clinical Pharmacology and Toxicology, Saarland University, Homburg (Saar), Germany

INTRODUCTION

In the last decade, numerous novel psychoactive substances have appeared on the semi-legal drugs market being sold as, for example, bath salts, research chemicals, plant food, incense, or simply as 'legal highs'. In Table 6.1, the chemical structures, names and acronym names that will henceforth be used in this chapter are given for drugs from the most important classes of novel psychoactive substances: piperazines, 2,5-dimethoxyphenethylamiens (2Cs), 2,5-dimethoxyamphetamines, 4-substituted amphetamines, β-ketoamphetamines (cathinones), pyrrolidinophenones, tryptamines, and synthetic cannabinoid receptor agonists.

This emergence of ever-new designer drugs is an ongoing challenge for analytical toxicologists in forensic as well as clinical toxicology, because most of the new drugs may cause serious toxicity or impairment but are not covered by established analytical methods for several reasons. Firstly, immunoassay (IA)-based techniques targeting the classical drugs do not reliably pick up most of the novel psychoactive substances due to low cross-reactivity [1–11]. Secondly, even with more sophisticated techniques such as mass spectrometry (MS)-based screening techniques, the new drugs may be missed, either because reference mass spectra are not (yet) included in the respective reference libraries, or because established methods employing selected-ion monitoring (SIM) or multiple-reaction monitoring (MRM) focusing on fragments of the classic drugs generally do not include typical fragments of novel psychoactive substances. In order to keep up with recent developments on the recreational drug market, it is therefore important to continuously adapt existing analytical methods or to develop new ones that allow determination of these new compounds. This, however, is associated with a

Novel Psychoactive Substances.
DOI: http://dx.doi.org/10.1016/B978-0-12-415816-0.00006-7

TABLE 6.1   Chemical Structures, Names, and Acronyms of Novel Psychoactive Substances

Drug Class	Chemical Structure		Compound Names, Acronyms
Piperazines		$R1=R2=R3=H$	Benzylpiperazine, BZP, A2
		$R1=R2=CH_2-O-CH_2$, $R3=H$	3,4-methylenedioxybenzylpiperazine, MDBP
		$R1=O-CH_3, R2=H$	4-methoxyphenylpiperazine, MeOPP
		$R1=H, R2=Cl$	3-chlorophenylpiperazine, mCPP
		$R1=H, R2=CF_3$	3-trifluoromethylphenylpiperazine, TFMPP
Phenethylamines (2Cs)		$R=H$	2,5-dimethoxyphenethylamine, 2C-H
		$R=CH_3$	4-methyl-2,5-dimethoxyphenethylamine, 2C-D
		$R=C_2H_5$	4-ethyl-2,5-dimethoxyphenethylamine, 2C-E
		$R=C_3H_7$	4-propyl-2,5-dimethoxyphenethylamine, 2C-P
		$R=Br$	4-bromo-2,5-dimethoxyphenethylamine, 2C-B
		$R=I$	4-iodo-2,5-dimethoxyphenethylamine, 2C-I
		$R=S-CH_3$	4-methylthio-2,5-dimethoxy-phenethylamine, 2C-T
		$R=S-C_2H_5$	4-ethylthio-2,5-dimethoxyphenethylamine, 2C-T-2
		$R=S-C_3H_7$	4-propylthio-2,5-dimethoxy-phenethylamine, 2C-T-7
		$R=S-cyclohexyl$	4-cyclohexylthio-2,5-dimethoxy-phenethylamine, 2C-T-5
		$R=S-CH(CH_3)_2$	4-isopropylthio-2,5-dimethoxy-phenethylamine, 2C-T-4
		$R=S-methyl-cyclopropyl$	4-cylopropylmethylthio-2,5-dimethoxy-phenethylamine, 2C-T-8
		$R=S-C_2H_4-O-CH_3$	4-(2-methoxyethyl)thio-2,5-dimethoxy-phenethylamine, 2C-T-13
		$R=S-CH_2-CH(CH_3)_2$	4-isobutylthio-2,5-dimethoxy-phenethylamine, 2C-T-17

(Continued)

TABLE 6.1 (Continued)

Drug Class	Chemical Structure	Compound Names, Acronyms
FLYs	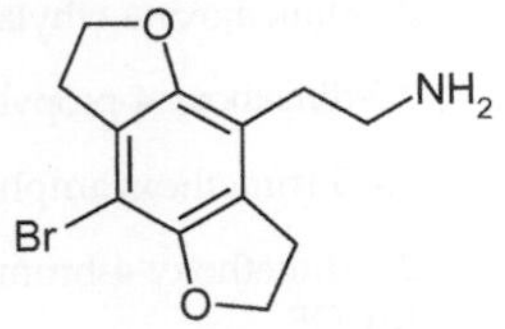	Bromo-FLY, 2C-B-FLY
		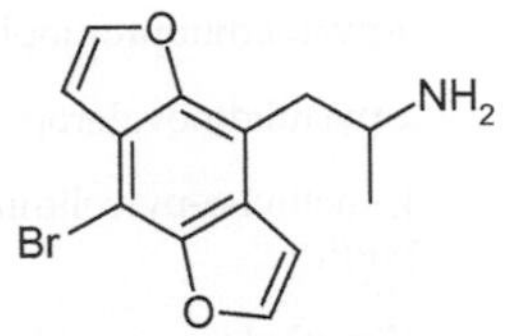 Bromo-dragonFLY
4-substituted amphetamines	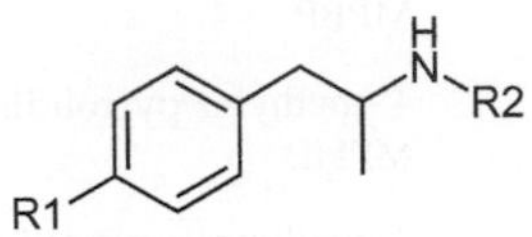	R1=O–CH$_3$, R2=H $\quad$ *Para*-methoxyamphetamine, PMA
		R1=O–CH$_3$, R2=CH$_3$ $\quad$ *Para*-methoxymethamphetamine, PMMA
		R1=O–CH$_3$, R2=C$_2$H$_5$ $\quad$ *Para*-methoxyethylamphetamine, PMEA
		R1=S–CH$_3$, R2=H $\quad$ 4-methylthioamphetamine, 4-MTA
		R1=CH$_3$ $\quad$ 4-methylamphetamine, 4-MA
		R1=F $\quad$ 4-fluoroamphetamine, 4-FA
		R1=Cl $\quad$ *Para*-chloroamphetamine, *p*CA
β-keto-amphetamines	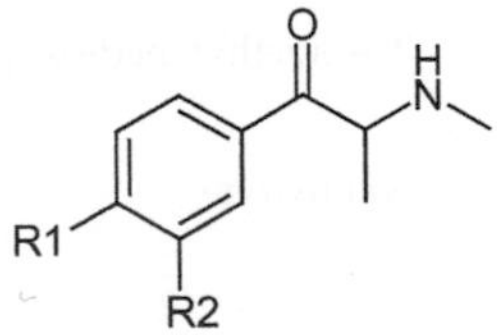	R1=H, R2=H $\quad$ Ephedrone, methcathinone
		R1=CH$_3$, R2=H $\quad$ Mephedrone, 4-MMC
		R1=OCH$_3$, R2=H $\quad$ Methedrone
		R1=F, R2=H $\quad$ Flephedrone, 4-FMC
		R1=H, R2=F $\quad$ 3-fluoromethcathinone, 3-FMC
		R1=H, R2=Br $\quad$ 3-bromomethcathinone
	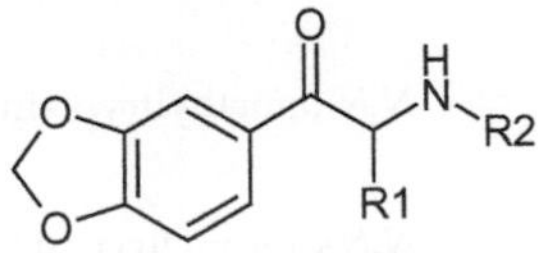	R1=R2=CH$_3$ $\quad$ Methylone, bk-MDMA
		R1=CH$_3$, R2=C$_2$H$_5$ $\quad$ Ethylone, bk-MDEA
		R1=C$_2$H$_5$, R2=CH$_3$ $\quad$ Butylone, bk-MBDB
2,5-dimethoxy-amphetamines	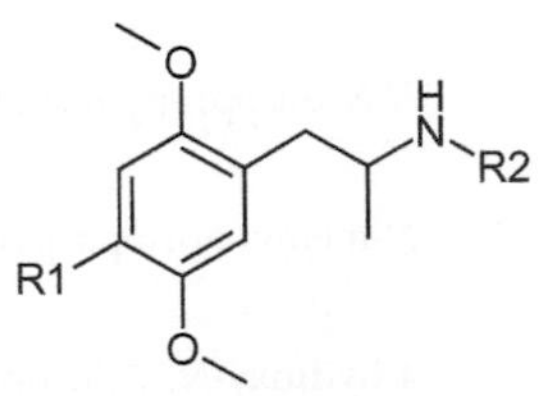	R1=R2=H $\quad$ 2,5-dimethoxyamphetamine (DMA)
		R1=Cl, R2=H $\quad$ 2,5-dimethoxy-4-chloroamphetamine, DOC
		R1=Br, R2=H $\quad$ 2,5-dimethoxy-4-bromoamphetamine, DOB
		R1=I, R2=H $\quad$ 2,5-dimethoxy-4-iodoamphetamine, DOI
		R1=CH$_3$, R2=H $\quad$ 2,5-dimethoxy-4-methylamphetamine, DOM

(*Continued*)

TABLE 6.1 (Continued)

Drug Class	Chemical Structure		Compound Names, Acronyms
		R1=C$_2$H$_5$, R2=H	2,5-dimethoxy-4-ethylamphetamine, DOET
		R1=C$_3$H$_7$, R2=H	2,5-dimethoxy-4-propylamphetamine, DOPR
		R1=OCH$_3$, R2=H	2,4,5-trimethoxyamphetamine, TMA-2
		R1=Br, R2=CH$_3$	2,5-dimethoxy-4-bromomethamphetamine, MDOB
Pyrrolidinophenones	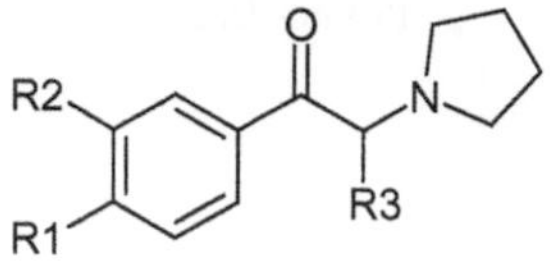	R1=R2=H, R3=CH$_3$	α-pyrrolidinopropiophenone, PPP
		R1=R2=H, R3=C$_3$H$_5$	α-pyrrolidinovalerophenone, PVP
		R1=R3=CH$_3$, R2=H	4'-methyl-α-pyrrolidinopropiophenone, MPPP
		R1=CH$_3$, R2=H, R3=C$_2$H$_5$	4'-methyl-α-pyrrolidinobutyrophenone, MPBP
		R1=CH$_3$, R2=H, R3=C$_4$H$_9$	4'-methyl-α-pyrrolidinohexanophenone, MPHP
		R1=OCH$_3$, R2=H, R3=CH$_3$	4'-methoxy-α-pyrrolidinopropiophenone, MOPPP
		R1=R2=CH$_2$–O–CH$_2$, R3=CH$_3$	3',4'-methylenedioxy-α-pyrrolidinopropiophenone, MDPPP
		R1=R2=CH$_2$–O–CH$_2$, R3=C$_2$H$_5$	3',4'-methylenedioxy-α-pyrrolidinobutyrophenone, MDBP
		R1=R2=CH$_2$–O–CH$_2$, R3=C$_3$H$_7$	3',4'-methylenedioxypyrovalerone, MDPV
	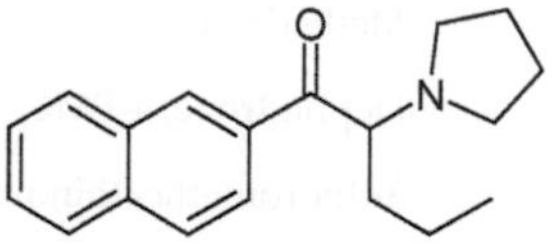		Naphyrone
Tryptamines	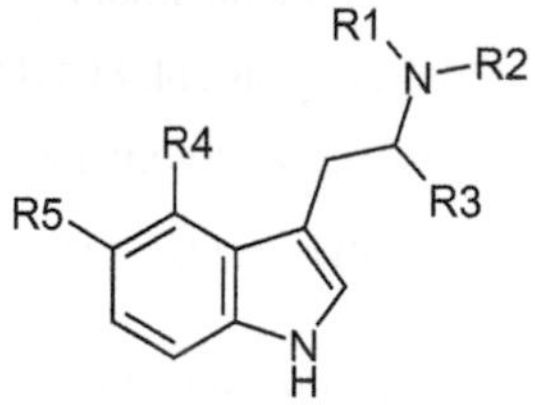	R1=R2=R4=R5=H, R3=CH$_3$	α-methyl-tryptamine, AMT
		R1=R2=CH$_3$, R3=R4=R5=H	N,N-dimethyltryptamine, DMT
		R1=R2=C$_3$H$_7$, R3=R4=R5=H	N,N-dipropyltryptamine, DPT
		R1=R2=CH(CH$_3$)$_2$ R3=R4=R5=H	N,N-diisopropyltryptamine, DIPT
		R1=CH$_3$, R2=CH(CH$_3$)$_2$ R3=R4=R5=H	N-methyl-N-isopropyltryptamine, MIPT
		R1=R2=CH(CH$_3$)$_2$, R3=R5=H, R4=OH	4-hydroxy-N,N-diisopropyltryptamine, 4-HO-DIPT

(*Continued*)

TABLE 6.1 (Continued)

Drug Class	Chemical Structure		Compound Names, Acronyms
		R1=R2=CH(CH$_3$)$_2$, R3=R5=H, R4=OOCCH$_3$,	4-acetoxy-*N,N*-diisopropyltryptamine, 4-acetoxy-DIPT
		R1=R2=CH(CH$_3$)$_2$, R3=R4=H, R5=OCH$_3$	5-methoxy-*N,N*-dimethyltryptamine, 5-MeO-DMT
		R1=R2=CH(CH$_3$)$_2$, R3=R4=H, R5=OCH$_3$	5-methoxy-*N,N*-Diisopropyltryptamine, 5-MeO-DIPT
Synthetic cannabinoids	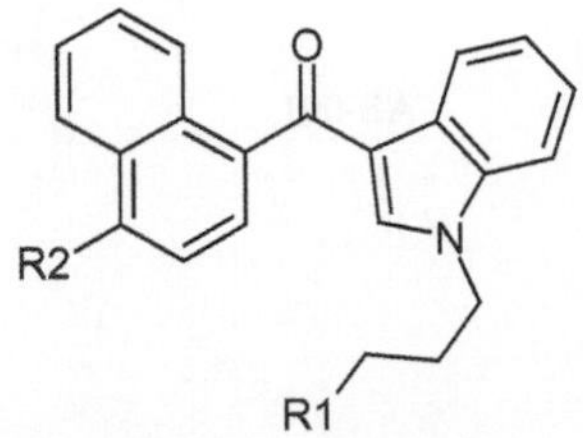	R1=CH$_3$, R2=H	JWH-073
		R1=C$_2$H$_5$, R2=H	JWH-018
		R1=C$_3$H$_7$, R2=H	JWH-019
		R1=C$_4$H$_9$, R2=H	JWH-020
		R1=C$_3$H$_7$, R2=OCH$_3$	JWH-081
	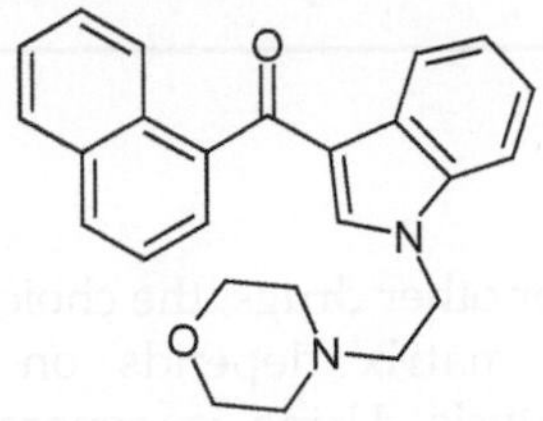		JWH-200
	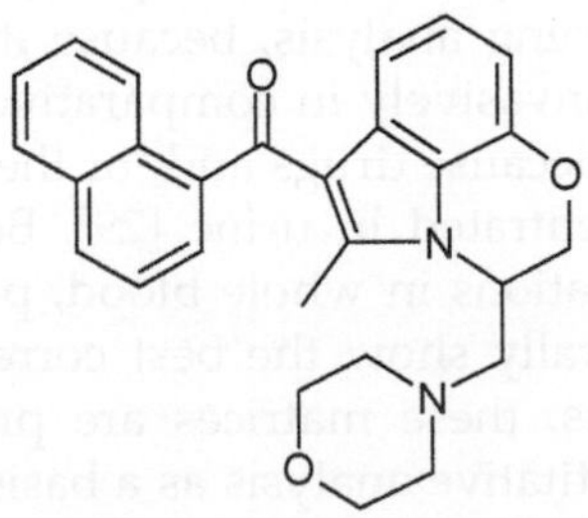		WIN 55,212-2
	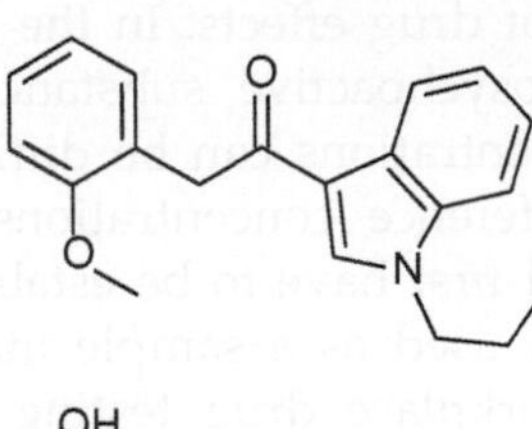		JWH-250
	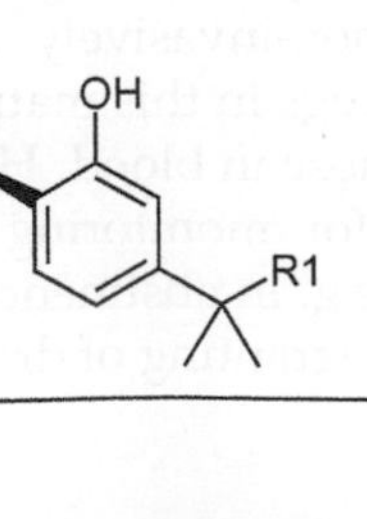	R1=C$_6$H$_{14}$	CP 47,497
		R1=C$_7$H$_{16}$	CP 47,497 C8-homologue

(Continued)

TABLE 6.1   (Continued)

Drug Class	Chemical Structure	Compound Names, Acronyms
		HU-210
		AB-001

number of problems. When such drugs are first detected in suspicious powders, pills or herbal preparations, their chemical structures are often unknown and must first be elucidated by combining information of various spectroscopic, mass spectrometric, and chemical methods [12–28]. Once the structures are known, it generally takes quite some time before reference standards of the respective drugs become commercially available and can be obtained by toxicology laboratories to set up methods for analysis of the drugs in biological matrices. However, even if such standards are available, method development can still be difficult. One reason is that, at this stage, it may be entirely unknown at which concentrations the novel psychoactive substances may be expected in the different biological matrices. Moreover, most often there is no information about their pharmaco-/toxicokinetic properties such as the metabolic pathways or ways of excretion of these drugs in humans. Such information may, however, be essential, especially for the development of toxicological screening procedures for urine.

As for other drugs, the choice of the biological sample matrix depends on the purpose of the analysis. Urine is generally preferred for screening analysis, because it can be obtained non-invasively in comparatively large volumes and because drugs and/or their metabolites are concentrated in urine [29]. Because drug concentrations in whole blood, plasma and serum generally show the best correlation with drug effects, these matrices are primarily used for quantitative analysis as a basis for the interpretation of drug effects. In the particular case of novel psychoactive substances, interpretation of concentrations can be difficult because reliable reference concentrations are often missing and first have to be established. Oral fluid is often used as a sample matrix in road-side and workplace drug testing. It can be easily taken non-invasively and detection times of most drugs in this matrix are more or less similar to those in blood. Hair is a biomatrix ideally suited for monitoring long-term exposure to drugs, e.g. in abstinence monitoring in the context of regranting of driver's licences.

In the following, published methods for analysis of novel psychoactive substances will be summarised and discussed, starting with a section on analytical methods allowing simultaneous screening, identification and/or quantification of new drugs from various drugs classes and/or their metabolites. Further sections are dedicated to specific drugs classes, focusing on matrices, sample preparation and analytical methods used for analysis of the respective drugs in biosamples. Important pitfalls and interpretation issues will also be addressed. Key information of selected analytical methods published in the literature is summarised in Tables 6.2 and 6.3. The chapter will close with some recommendations on how set up a procedure for analysis of novel psychoactive substances in biological samples.

ANALYTICAL METHODS COVERING SEVERAL CLASSES OF NOVEL PSYCHOACTIVE SUBSTANCES

A series of papers on the analysis of different classes of novel psychoactive substances in urine has been published by Maurer et al. [1,30–50] The general approach was the same in all of these studies. Novel psychoactive substances having appeared on the drugs market were administered to rats and urine was collected over a 24h period. The urine samples were analysed by gas chromatography-mass spectrometry (GC-MS) after different workup and derivatisation procedures. Structure elucidation of urinary metabolites of the drugs was achieved by mass spectral interpretation. The identified metabolites were subsequently added to the reference library [94]. In order to check whether the drugs and/or their metabolites were detectable with the group's routine method for Systematic Toxicological Analysis (STA) in urine, the urine samples were worked up according to this method based on acid

hydrolysis followed by liquid–liquid extraction (LLE) and acetylation [95]. Analysis was performed by GC-MS in the electron ionisation (EI) fullscan mode and library searching. These studies have so far shown that this method is applicable for simultaneous screening of piperazines [1,30–33], 2,5-dimethoxyphenethylamines [34–39], 2,5-dimethoxy-amphetamines [45–49], 4-substituted amphetamines [40–42], cathinone-derivatives [43,50], and the pyrrolidinophenone-type drug 3′,4′-methylenedioxypyrovalerone (MDPV) [44], at least in rat urine. However, in all cases where authentic urine samples from subjects having ingested a new psychoactive drug were available, the applicability could also be confirmed for human urine [1,32,40,42–44]. Similar findings were also reported in studies with other drugs [56,96,97]. These findings strongly suggest that the drugs and/or metabolites detectable in rat urine can generally be expected to be also detectable in human urine, although the excreted quantities may differ between the two species.

Kerrigan et al. [51] recently described a GC-MS method for detection of five 2,5-dimethoxy-phenethylamine (2C) drugs, four 2,5-dimethoxy-amphetamines, and 4-methylthio-amphetamine (4-MTA) in human urine. Metabolites were not included due to the lack of commercial reference standards. Consequently, the method was designed as a targeted screening employing a selective solid-phase extraction (SPE) and operating the MS in the SIM mode. Adequate quantitative performance was demonstrated in an extensive validation study. Nevertheless, metabolites being not detectable may limit the general applicability of the method, because at least in rats, most of the targeted drugs are extensively metabolised before being excreted in urine [98].

A procedure for targeted screening of four tryptamine-derived hallucinogens and two 2C drugs in human urine was described by Vorce et al. [52]. It involved selective mixed-mode SPE followed by pentafluoropropionylation and

TABLE 6.2 Key Parameters of Quadrupole GC-MS-based Analytical Methods for Analysis of Novel Psychoactive Substances in Biological M

Analyte(s)	Matrix	Sample Preparation	Stationary Face	Detection	Validation	Re
BZP, mCPP, TFMPP, MeOPP, MDBP, 2C-D, 2C-E, 2C-B, 2C-I, 2C-T-2, 2C-T-7, PMA, PMMA, PMEA, 4-MTA, mephedrone, bk-MDMA, bk-MBDB, 3-FMC, 3-BMC, DOM, DOC, DOB, MDOB, DOI, TMA-2, MDPV	U	acHy, LLE, AC	HP-1	EI, fullscan	LOD, low dose rat experiments	[1,]
2C-B, 2C-H, 2C-I, 2C-T-2, 2C-T-7; 4-MTA, DOB, DOET, DOI, DOM,	U	SPE (polymer)	DB-5MS	EI, SIM	SEL, RE, LIN, ACC, PRC, LOQ, LOD	[51
AMT, DMT, DPT, 5-MeO-DIPT, 2C-B, 2C-T-7	U, B	SPE (HCX), PFPA	ZB-1	EI, SIM	LIN, LOD	[52
PMA, PMMA, 4-MTA, BZP, mCPP, TFMPP, MeOPP, MDBP and 5 classic amphetamines 2C-D, 2C-E, 2C-P, 2C-B, 2C-I, 2C-T-2, 2C-T-7, mescaline	P	SPE (HCX), HFBA	HP-5MS	EI, SIM	SEL, LIN, ACC, PRC, LOQ, LOD, STB	[53
BZP, TFMPP	U	SPE (HCX)	Synergi Polar RP18	ESI+, SIM	SEL, RE, ME, LIN, ACC, PRC, LOQ, LOD	[3]
TFMPP, mCPP, MeOPP	H	SPE (HCX), TMS	HP-5MS	EI, SIM	SEL, RE, LIN, ACC, PRC, LOQ	[55
DOB	U, S, G	acHy for U SPE (SCX) for U and G, LLE for S and hydrolysed U, AC	HP-5MS	EI, fullscan, SIM (for S)	LIN, PRC, LOD	[56
PMA, PMMA, 4-MTA and 10 other drugs	B	PP/SPE (polymer), AC	HP-5MS	EI, fullscan	RE, LIN, ACC, PRC, LOD	[57
Mephedrone	H	LLE	Nonpolar	EI, SIM	LIN, ACC, PRC, SEL, LOD, LOQ	[58
Mephedrone, methedrone	B	LLE, TFA	HP-5MS	EI, SIM	LIN, ACC, PRC	[59
PPP, MPPP, MOPPP, MDPPP, MPBP, PVP, MDPV, MPHP	U	enHy, SPE (HCX), TMS	HP-1	EI, fullscan	LOD, low dose rat experiments	[44
MDPV	U	LLE, HFB	HP-5MS	EI, SIM	LIN, ACC, PRC, LOQ, LOD	[66
AMT, 5-MeO-DIPT	B, U	LLE, AC	HP-1ms	EI, SIM	SEL, RE, LIN, PRC, LOD	[67
AB-001 metabolites	U	HY + LLE	EVDX-5MS	EI, fullscan		[68

Abbreviations: for compound acronyms see Table 6.1; U, urine; B, blood; P, plasma; H, hair; VH, vitreous humour; G, gastric content; S, serum; OF, oral fluid; acHy, acidic hydrolys for conjugate cleavage; enHy, enzymatic hydrolysis for conjugate cleavage; LLE, liquid–liquid extraction; SPE, solid-phase extraction; RP, reversed phase; AC, acetylation; TFA, trifluoroacetylation; PFP, pentafluoropropionylation; HFB, heptafluorobutyrylation; TMS, trimethylsilylation; EI, electron ionisation; SIM, selected ion monitoring; SEL, selectivity matrix effects; LOD, limit of detection; RE, recovery; LOQ, limit of quantification; PRC, precision; ACC, accuracy; LIN, linearity; STB, stability.

TABLE 6.3   Key Parameters of LC-MS(/MS)-based Analytical Methods Covering Several Classes of Novel Psychoactive Substances

Analyte(s)	Matrix	Sample Preparation	Stationary Face	Instrument Type	Detection	Validation	Reference
2C-D, 2C-E, 2C-B, 2C-B-Fly, 2C-I, 2C-T-2, MDMA, 4-OH-DIPT, 4-acetoxy-DIPT	U	enHy, SPE (HCX)	Hypersil Gold ultra	Q	ESI+, SIM	SEL, ME, RE, LIN, ACC, PRC, LOQ, LOD, STB	[69]
BZP, 2C-B, TMA-2, DOB, DOI, and 2,5-DMA and 17 other drugs/metabolites	U	Screening: dilution Confirmation: SPE (C18)	HyPURITY Advance	QQQ	APCI, MRM	RE, PRC, STB, cut-offs	[70,71]
DMA, 2C-H, DOM, DOC, 2C-B, DOB, 2C-I, DOI	U	SPE (C18)	Fused silica caprillary (CE)	Q	ESI+; fullscan	SEL, RE, LIN, ACC, PRC, LOQ, LOD,	[72]
2,5-DMA, 2C-B, 2C-D, 2C-H, 2C-I, 2C-P, 2C-T-2, 2C-T-4, 2C-T-7, 3,4,5-TMA, 3,4-DMA, 4-MTA, 5MeO-DMT, AMT, BZP, DiPT, DMT, DOB, DOET, DOM, DPT, mCPP, MDBP, MDDMA, MeOPP, MiPT, PMA, PMMA, TFMPP, TMA-6	P	SPE (HCX), PFPA	Synergi Polar RP	QQQ	ESI+, MRM	SEL, RE, ME, LOD	[73]
BZP, mCPP, oCPP, pCPP, TFMPP, 4-TFMPP	B, U	LLE	Phenomenex Gemini	QQQ	ESI+, EPI	LIN, ACC, PRC, LOQ, LOD	[74]
BZP, TFMPP	U	SPE (HCX)	Synergi Polar RP18	Q	ESI+, SIM	SEL, RE, ME, LIN, ACC, PRC, LOQ, LOD	[3]
BZP, TFMPP and hydroxyl metabolites	U	enHy, SPE (HLB)	semi-micro SCX	Q	ESI+, SIM	RE, LIN, ACC, PRC, LOD	[75]
mCPP	P, U	PP	Zorbax SB-C18	QQQ	APCI+, MRM	LIN, ACC, PRC, LOQ, LOD	[76]
BZP, TFMPP, 3-HO-BZP, 4-HO-BZP, 4-HO-TFMPP	P U	PP (enHy), centrifugation	Zorbax C18	Q	ESI+, SIM	SEL, RE, ME, LIN, ACC, PRC, LOQ, STB	[77]
BZP	H	SPE (HCX)	Luna SCX ACN/AF/FA	Q	ESI+, MRM	RE, LIN, ACC, PRC, LOQ	[78]
2C-T-4, 2C-T-8, 2C-T-13, 2C-T-17	U	SPE (C18)	fused silica CE capillary	Q	ESI+, SIM	SEL, RE, ME, LIN, ACC, PRC, LOQ, LOD, STB	[79]
2C-T, 2C-T-2, 2CT-5, 2C-T-7	P	PP/LLE	fused silica CE capillary	Q	ESI+, SIM	SEL, RE, ME, LIN, ACC, PRC, LOQ, LOD	[80]

(Continued)

TABLE 6.3 (Continued)

Analyte(s)	Matrix	Sample Preparation	Stationary Face	Instrument Type	Detection	Validation	Reference
Bromo-dragonFLY	B, U, VH	enHy (for U), LLE	Acquity UPLC BEH C18	TOF	ESI+, fullscan	SEL, RE, ME, LIN, ACC, PRC, LOQ, LOD	[81]
DOM, DOET, DOPR	U	LLE	fused silica capillary	Q	ESI+, SIM	SEL, RE, LIN, ACC, PRC, LOQ, LOD	[82]
TMA, TMA-2, TMA-6	U	SPE (C18)	Polar Plus column	Q	ESI+, SIM	SEL, RE, ME, LIN, ACC, PRC, LOQ, LOD	[83]
PMEA, PMA and metabolites	U, B	(enHy), online SPE (polymer)	ODS semi-micro	Q	ESI+, SIM	SEL, RE, LIN, ACC, PRC, LOQ, LOD	[84]
4-MTA	B, U, VH, tissues	LLE	Hypersil BDS phenyl	QTOF	ESI+, PIS	LIN, ACC, PRC, LOQ, LOD	[85]
Mephedrone, methylone, butylone, methedrone, BZP, TFMPP, mCPP, MDPV	U	Dilution	BEH C18	QQQ	ESI+, SRM	LIN, ACC, PRC, LOD, LOQ, ME,	[5]
cathinone, methcathinone, ethcathinone, amfepramone, mephedrone, flephedrone, methylone, methedrone, butylone, cathine, norephedrine, ephedrine, pseudoephedrine, methylephedrine, methylpseudoephedrine	B	LLE	Prodigy Phenyl 3	QQQ	ESI+, SRM	LIN, ACC, PRC, SEL, REC, LOD, LOQ, STA, ME	[86]
Mephedrone	H	LLE	SB-C18	QQQ	ESI+, MRM	LIN, ACC, PRC, SEL, LOD, LOQ	[87]
Mephedrone, methedrone	H	LLE	BEH C18	QQQ	ESI+, MRM	LIN, ACC, PRC	[59]
MDPV	S	SPE (HLB)	Phenyl hexyl column	QQQ	ESI+, MRM	SEL, RE, ME, LIN, ACC, PRC, LOQ, LOD	[88]
DMT, NMT, 5-MeO-DMT, bufotenine	U	SPE (HLB)	Speri-5 RP18	QQQ	ESI+, MRM	SEL, RE, LIN, ACC, PRC	[89]
5-MeO-DIPT and metabolites	U	PP	BDS Hypersil C18	QQQ	ESI+, MRM	RE, LIN, ACC, PRC, LOQ, STB	[90]

(*Continued*)

TABLE 6.3 (Continued)

Analyte(s)	Matrix	Sample Preparation	Stationary Face	Instrument Type	Detection	Validation	Reference
JWH-018, -073 and their metabolites	U	enHy, SPE (polymer HCX)	ZorbaxEclipse XDB-C18	QQQ	ESI+, MRM	LIN, ACC, PRC, SEL, LOD, LOQ	[91]
JWH-018, JWH-250, JWH-073, CP47497 C8, CP47497, HU-210	OF	SPE Trace-N	Extend C18	QQQ	ESI+ and ESI−, MRM	LIN, ACC, PRC, SEL, LOD, LOQ, STB, ME	[92]
JWH-015, -018, -019, -020, -073, -081, -200, -250, Methanandamide, WIN55.212-2	S	LLE	Luna Phenyl Hexyl	QQQ	ESI+, MRM	LIN, ACC, PRC, SEL, LOD, LOQ, STB, ME	[93]

Abbreviations: for compound acronyms see Table 6.1; U, urine; B, blood; P, plasma; H, hair; VH, vitreous humour; G, gastric content, S, serum; OF, oral fluid; acHy, acidic hydrolysis for conjugate cleavage; enHy, enzymatic hydrolysis for conjugate cleavage; LLE, liquid–liquid extraction; SPE, solid-phase extraction; RP, reversed phase; PP, protein precipitation; CE, capillary electrophoresis; Q, quadrupole; QQQ, triple quadrupole; TOF, time of flight mass spectrometer; ESI, electrospray ionisation; APCI, atmospheric pressure chemical ionisation; SIM, selected ion monitoring; MRM, multiple reaction monitoring; PIS, product ion scanning; SEL, selectivity; ME, matrix effects; LOD, limit of detection; RE, recovery; LOQ, limit of quantification; PRC, precision; ACC, accuracy; LIN, linearity; STB, stability.

analysis by GC-MS in the EI SIM mode. This method allows sensitive detection of the parent drugs down to concentrations of 10 ng/ml in urine, but also has the limitation of not covering metabolites expected to be the primary analytes in urine. A positive finding for 5-methoxy-*N,N*-diisopropyltryptamine (5-MeO-DIPT) was confirmed at a concentration of about 200 ng/ml by a quantitative LC-MS-based confirmation method.

Pichini et al. [69] used liquid chromatography-mass spectrometry (LC-MS) with electrospray ionisation (ESI) for a multi-target screening and quantification of five 2Cs, bromo-FLY (2C-B-FLY), three tryptamine-derived drugs, and the piperazine *meta*-chlorophenylpiperazine (mCPP). Urine samples were either worked up without further pre-treatment or after an enzymatic deconjugation step. The analytes were again isolated from urine using selective mixed-mode solid-phase extraction (SPE) and LC-MS analysis was performed in the SIM mode. After an extensive validation study to demonstrate the qualitative and quantitative performance of the method it was applied to authentic human urine samples obtained from subjects who had reported to have taken one of the targeted compounds. In these samples, the concentrations of all analytes except 4-acetoxy-*N,N*-diisopropyltryptamine (4-acetoxy-DIPT) were found at in part much higher concentrations after enzymatic hydrolysis. This suggests that they are all excreted to certain extent in conjugated form and that a deconjugation step is not only useful when phase I metabolites are included in the method, but also when the respective parent drugs offer sites for direct *N*-glucuronidation such as the 2Cs or *O*-conjugation such as 4-HO-DIPT.

Nordgren et al. [70,71] described a screening procedure for multiple drugs in urine using a liquid chromatography-tandem mass spectrometry (LC-MS/MS) system with atmospheric pressure chemical ionisation (APCI). Apart from several other drugs, this method included three 2,5-dimethoxyampehtamines as well as 1-benzylpiperazine (BZP) and 4-bromo-2,5-dimethoxy-phenethylamine (2C-B). In a first screening step, urine samples were simply diluted with water by a factor of 10 and directly injected into the LC-MS/MS operated in the multiple reaction monitoring (MRM) mode with one transition being monitored for all analytes except 4-bromo-2,5-dimethoxy-amphetamine (DOB) for which three transitions were monitored. If positive peaks above a certain cut-off level were present at the correct retention time, confirmation analysis was performed. For this purpose, a more selective SPE-based sample workup and different chromatographic conditions were employed for better separation, while the MS/MS settings were left unchanged. The latter is a major drawback of the method at least in a forensic setting, where monitoring of a single transition per compound is not considered sufficient for unambiguous compound identification [99].

Another procedure using LC-MS/MS after simple dilution (1:4, v/v) of urine samples was reported by Bell et al. [5]. This method covers three piperazine-derived drugs, three cathinone-derived drugs, and the pyrrolidinophenone MDPV. The MS/MS system is operated in the SRM mode monitoring two transitions per compound as considered sufficient for identification even in a forensic setting. The method is purely qualitative in nature and has been validated with respect to limit of detection (LOD), limit of quantification (LOQ), recovery and matrix effects. Matrix effects were found to be present, but only to a minor to moderate extent. Like most other methods mentioned above, this methods allows a sensitive targeted screening for the respective parent drugs, but not of metabolites or other non-targeted compounds.

Boatto et al. [72] published an analytical method for simultaneous determination of six 2,5-dimethoxyamphetamines and four 2C drugs in human urine. After SPE with a reversed-phase sorbent, instrumental analysis was performed by capillary electrophoresis interfaced to a quadrupole mass spectrometer by an ESI source. The analytes were detected in the fullscan mode and the method's quantitative performance

was evaluated in an extensive validation study. Considering the high sensitivity of the method with LODs in the low ng/ml range, the method should be applicable for screening of the targeted drugs in authentic urine samples. However, once again metabolites were not included in the method, although the targeted drugs are primarily excreted in metabolised form.

Peters et al. [53] described a multi-target procedure for screening and quantification of amphetamines and piperazines in plasma samples. The monitored amphetamines included the 4-substitutes amphetamines para-methoxy-amphetamine (PMA) and para-methoxy-methamphetamine (PMMA), their O-demethyl metabolites 4-HO-amphetamine and 4-HO-methampetamine, as well as 4-MTA. The analytes were isolated from plasma using selective mixed-mode SPE followed by heptafluorobutyrylation to improve the chromatographic properties and thus the detectability of the analytes by GC-MS. Detection was performed in the SIM mode monitoring three ions per analyte. The method was extensively validated and found applicable for quantification of all analytes with exception of the piperazine drug MDBP, which could only be determined semi-quantitatively due to unacceptably high variability. The same approach was later used by Habrdova et al. [54] for screening and quantification of seven 2C drugs and mescaline in plasma. The samples preparation was the same as in the previous method but GC-MS settings were adapted for analysis of the phenethylamines. Validation of the final method showed that all studied analytes but the thioalkyl phenethylamines 4-ethylthio-2,5-dimethoxy-phenethylamine (2C-T-2) and 4-propylthio-2,5-dimethoxy-phenethylamine (2C-T-7) can be reliably quantitated with the adapted method. More recently, Derungs et al. [100] used the same workup for determination of naphyrone in a case of sympathomimetic toxicity after recreational use of this drug. This shows that the general workup procedure should be applicable for other novel psychoactive substances with basic properties and that hence the same extract should be applicable for analysis of a wide range new drugs.

To date (December 2011), the most comprehensive screening procedure covering 32 new designer drugs including 2,5-dimethoxy-amphetamines, 4-substituted amphetamines, 2C drugs, piperazine-derived drugs and tryptamines was described by Wohlfahrt et al. [73]. The method is also designed as targeted screening for the respective parent drugs. As in most other methods discussed above, mixed-mode SPE was used for selective extraction of the basic drugs from plasma and detection was performed in the MRM mode. The method is qualitative in nature, but very sensitive with LODs in the low ng/ml range. It should hence allow detection even of low-dosed drugs such as some of the 2,5-dimethoxy-amphetamines, at least in case of acute use. As expected for a qualitative method, the validation study for this procedure was limited to evaluation of selectivity, recovery, matrix effects, and LODs.

METHODS FOR SCREENING AND/OR QUANTIFICATION OF NOVEL PSYCHOACTIVE SUBSTANCES FROM SPECIFIC DRUG CLASSES

Piperazine-Derived Designer Drugs

Although IA-based techniques are widely used in clinical and forensic toxicology, few studies have been dedicated to IA-based screenings for piperazine-derived drugs. The cross-reactivity of BZP [1–5], its metabolite N-benzylethylenediamine [1], and 3-trifluoromethyl-phenylpiperazine (TFMPP) [3,5] with various IAs targeting (meth)amphetamines/ecstasy was reported to be low. Nevertheless, using the Roche Abuscreen® Amphetamines assay, Vorce et al. obtained positive or nearly positive screening results at a 500 ng/ml cutoff in seven urine samples from routine casework that contained high concentrations of BZP (13–429 mg/l) and TFMPP (0.83–26 mg/l), but confirmed negative for amphetamine and methamphetamine [3]. Moreover, mCPP [4,101] and TFMPP [4] showed sufficient cross-reactivity with some IAs to potentially cause false positive results.

Piperazine-derived drugs and/or their metabolites have been analysed in post-mortem blood [74], plasma [53,73,76,77], urine [1,3,5,30–33,69,70,74–76,102–104], and hair [55,78]. Authors describing analysis of piperazines in urine, partly performed acidic [1,30–33] or enzymatic [1,30–33,69,75,102,103] cleavage of conjugates prior to extraction. Sample preparation on the one hand included very simple workup methods like dilution of urine [5,70] or protein precipitation of plasma [76,77] and on the other hand LLE [1,3,30–33,74,102–104] or SPE with reversed-phase [70], hydrophilic/lipophilic polymer (HLB) [75,102,103] or mixed-mode reversed-phase/cation exchange (HCX) sorbents [53,55,69,73,78,103] Derivatisation of the extracts for GC-MS analysis included acetylation [1,30–33], trifluoroacetylation [75,102,103], pentafluoropropionylation [104], heptafluorobutyrylation [53], and silylation [55].

Chromatographic techniques employed for analysis of piperazine-derived drugs in biosamples include liquid chromatography with diode-array detection (LC-DAD) [2,74], and gas chromatography with nitrogen-phosphorous detection (GC-NPD) [2], GC-MS [1–3,30–33,53,55,74,75,102–104], and/or LC-MS (/MS) [3,69,70,74–78,102,103]. Reversed-phase stationary phases were used by most authors in LC-DAD and LC-MS(/MS) analysis, while Tsutsumi et al. [75,102,103] and Bassindale et al. [78] used a strong-cation exchange (SCX) stationary phase in order to more effectively retain the studied piperazines and their metabolites, which are poorly retained by reversed-phase chromatography. In the studies where less selective DAD and NPD were used, the results were confirmed by MS-based techniques [2,74]. With the exception of the method described by Nordgren et al. [70,71] and Antia et al. [77], in which only a single transition or ion, respectively, was monitored, all methods used a detection mode generally considered sufficient for MS-based compound identification, i.e., fullscan [1,3,30–33,75,102,103], SIM with three ions per analyte [53,55,69,104], product ion scanning [74], or MRM with at least two transitions per analyte [73,76,78].

Studies by Staack et al. [1,30–33] and Tsutsumi et al. [75,102,103] using rat urine showed that some of the piperazines are extensively metabolised and that the target analytes in urine are metabolites rather than the parent drugs. Nevertheless, rather high concentrations of BZP [3,74], TFMPP [3,74,77], and mCPP [76] have been found in human urine samples from users. This suggests that the piperazines may be less extensively metabolised in humans and that the parent compound should at least be detectable in individuals presenting with piperazine toxicity.

Finally, when analysing for piperazine-derived designer drugs, it is important to consider that many of these drugs or their positional isomers are also metabolites of therapeutic drugs and might thus result from ingestion of the latter [2,32,33,105,106]. Most importantly, mCPP is a metabolite of the antidepressants trazodone and nefazodone. At least in cases of doubt it is therefore important to check for the presence of the respective precursor compounds or more specific metabolites of the latter to avoid false interpretation. Moreover, UV spectra may help to differentiate between positional isomers of mCPP and TFMPP [74].

Phenethylamine-Derived Drugs

The 2,5-dimethoxyphenethylamine-derived drugs (2Cs) cannot be detected with commercially available IA techniques [6–8]. Analysis of these drugs must therefore be performed by more sophisticated techniques.

Methods for analysis of phenethylamines have been described for various biological matrices such as (post-mortem) blood [52,81,107], plasma [54,73,80], urine [34–39,51,52,69,72,107–111],

and tissue samples [81,107]. All methods employed an extraction step using LLE [34–39,80,81,107,108] or SPE with classical reversed-phase [70,72,79], polymer-based [51], or mixed-mode HCX sorbents [52,54,69,73], while the screening part of the method described by Nordgren et al. [70] only involved simple dilution of urine samples. Some authors performed acidic [34–39] or enzymatic [69,81] conjugate cleavage prior to extraction and/or acetylation [34–39], pentafluoropropionylation [52], or heptafluorobutyrylation [54] after extraction.

The instrumental techniques used for separation and detection of the 2C drugs ranged from less selective capillary electrophoresis (CE) with UV detection [108] or GC-NPD [107] to mass spectrometry-based techniques such as CE-MS [72,80], GC-MS [34–39,52,54,107,109,110], and LC-MS(/MS) [69,70,73,81,112,113]. Used detection modes were fullscan [34–39,80,109,112], SIM [52,54,69,107], and MRM [70,73,81,113].

Very little is known about the disposition of phenethylamines in humans, but the results of metabolism studies in the rat suggest that these drugs are extensively metabolised and hence metabolites rather than the parent drugs should be the target analytes in urine analysis [34–39,109–112]. Indeed, Pichini et al. [69] who targeted the parent drugs in urine found only comparatively low concentration of 4-methyl-2,5-dimethoxyphenethylamine (2C-D), 4-ethyl-2,5-dimethoxyphenethylamine (2C-E), 4-iodo-2,5-dimethoxyphenethylamine (2C-I), 2C-T-2, and 2C-B-FLY in authentic human urine samples. Another potential pitfall in the analyses of 2C drugs is illustrated by the findings of Curtis et al. [107]. These authors analysed a methylene artifact of 2C-T-7 resulting from injection of underivatised extracts reconstituted in methanol and not the parent compound as intended. Such an artifact formation might explain why only rather low 2C-T-7 concentrations of 57 and 100ng/mL were found in heart and femoral blood, respectively, although the case was considered a fatal poisoning with 2C-T-7. The authors attributed the low concentrations to potential partial degradation of the analyte between autopsy and analysis. Andreasen et al. [81] reported postmortem bromo-dragonFLY concentrations of 4.7 and 22 μg/kg in femoral blood and urine, respectively. Nevertheless, the case was considered a fatal poisoning with bromo-dragonFLY based on the case history, the absence of an obvious alternative cause of death, and reports that bromo-dragonFLY is effective even after very low dosages. This clearly shows that concentrations of some phenethylamines after recreational use may be very low and hence a negative analytical finding may not always exclude the presence of 'recreational' concentrations of these drugs unless a method with LODs in the low ng/ml or even in the pg/ml range has been used.

2,5-Dimethoxyamphetamines

The 2,5-dimethoxyamphetamines are closely related to the 2Cs but carry an amphetamine rather than a '2C' phenethylamine side-chain. Nevertheless, their cross-reactivity with commercial enzyme-linked immunosorbent assays designed for analysis of amphetamine, methamphetamine, and/or methylenedioxyamphetamine was found to be extremely low so that these drugs cannot be screened for by these techniques [8].

Analysis of 2,5-dimethoxyamphetamines and/or their metabolites has been described using serum [56], plasma [73], gastric content [56], and urine [45–49,51,56,72,82,83,114]. In the methods explicitly including detection of metabolites, an acid hydrolysis step for cleavage of conjugates was performed prior to extraction whereas acetylation was performed after extraction [45–49,56,114]. Extraction was based on LLE [45–49,56,114] or SPE with classic reversed-phase [72,82,83], polymer-based [51] or mixed-mode HCX [73] sorbents.

Fullscan GC-MS was primarily used for separation and detection [45–49,56,114]. Nieddu et al. also used fullscan MS as detection mode, but separation was based on CE [82] and LC [83].

Kerrigan et al. [51] employed GC for separation but operated the MS system in the SIM mode with three ions being monitored per analyte, whereas Boatto et al. [72] combined CE and fullscan MS.

Information on the metabolism of 2,5-dimethoxyamphetamines in humans are scarce, but the findings reported for DOB by Balikova et al. [56] and Berankova et al. [114] show that a considerable percentage of this drug is excreted in form of metabolites. Moreover, the reported serum DOB concentrations of 13 and 19 ng/ml in two cases of non-fatal and fatal DOB poisoning indicate, that the concentrations of 2,5-dimethoxyamphetamines may be very low after recreational use, so that very sensitive methods are required for analysis of these drugs, at least in serum or plasma [56].

4-Substituted Amphetamines

The 4-substituted amphetamines have a close structural relation to amphetamine, methamphetamine and the MDMA-type designer drugs and hence exhibit sufficient cross-reactivity to be detectable with IAs targeting these drug classes [7,41,115–118].

Chromatography-based methods for analysis of 4-substituted amphetamines have been described for blood [57,84,85,115,119–127], plasma [53,73], serum [117], urine [40–42,84,85,118–121, 123–126], and/or post-mortem tissue samples [85,121,123,125]. Sample preparation included simple dilution [70], protein precipitation and dilution [127], LLE [40–42,57,84,85,115,118–123, 125,126] or (online) SPE using reversed-phase [70,117,124], polymer-based [84], or mixed-mode HCX [53] sorbents. Some authors describing analysis of urine samples further performed acidic [40–42] or enzymatic [84,119] conjugate cleavage. The majority of those employing GC-MS improved the gas chromatographic properties and fragmentation by acetylation [40–42,57,122], trifluoroacetylation [84,118], pentafluoropropionylation [117,124], or heptafluorobutyrylation [53,115,126].

After (fatal) overdoses of 4-substituted amphetamines, their concentrations in blood were in the middle ng/ml to low µg/l range [115,119–122,124,126,128–130], so the sensitivity required to test for these compounds in overdose cases can be achieved with less sophisticated analytical instruments such as HPLC-DAD [119,120] or GC-NPD [119,121]. However, most analytical methods for analysis of these compounds in biological matrices, also after recreational use, are based on GC-MS [40–42,53,57,84,115,117–119,121,122,124,126], and/or LC-MS(/MS) [73,84,85,118,123,125,127]. However, the detection modes of the MS instruments differ considerably with fullscan [40–42,57,84,117–119,121] or product ion scanning [84,85] being preferred for compound identification and SIM [84,115,117,121,124,126] or MRM [118,123,125,127] being used when the primary focus is quantification.

For the correct interpretation of findings of 4-substituted amphetamines it is important to note that *para*-methoxy-ethylamphetamine (PMEA) and its metabolites *p*-hydroxyethyl-amphetamine as well as *p*-hydroxyamphet-amine are also metabolites of the antispasmodic drugs mebeverine [131]. Therefore, it is important to check the case history for any possible mebeverine ingestion and to check for the parent compound and/or mebeverine-specific metabolites. Vice versa, it must be considered that the *O*-demethyl metabolite of PMMA is identical with the antihypotensive drug pholedrine (4-hydroxy-methamphetamine) and hence detection of the latter may result from PMMA.

β-Keto-Amphetamines (Cathinones)

Bell et al. [5] studied the cross-reactivity of several cathinones (methylone, butylone, mephedrone, and methedrone) with a commercial IA. With exception of butylone, none of the tested drugs was able to cause a positive result with this IA. However, at a high concentration of 10000 ng/ml butylone produced a reading of 1600 ng/ml (amphetamine), which was above the applied cut-off concentration of 1000 ng/ml. These findings are in line with those of Marais et al. [132] who described only poor detectability of ephedrine with an on-site drug-screening device. Low cross-reactivity of approximately 1–3% was also reported for mephedrone with different commercial ELISA methods for amphetamine and methamphetamine testing [133]. Nevertheless, samples from four cases of fatal mephedrone poisonings triggered a positive result in the methamphetamine assay.

More sophisticated methods for analysis of cathinone-derived drugs have been described for (post-mortem) blood [59,86,133–135], serum [136], plasma [73], urine [5,43,50,132,135,137,138], and hair [58,59,87,139]. Sample preparation consisted of simple dilution of urine samples [5], protein precipitation [86], LLE [43,50,58, 59,134,135,137,138], mixed-mode SPE [73,139], extractive pentafluropropionylation [132], acidic [43,50,137,138] cleavage of conjugates and derivatisation: acetylation [43,50], trifluoroacetylation [59,137,138], pentafluoropropionylation [132–134, 139], or heptafluorobutyrylation [58].

Instrumental analysis was performed by HPLC-DAD [135], GC-MS [43,59,132–134, 137–139], or LC-MS(/MS) [59,73,86,87,137,138]. The MS(/MS) instruments were operated in fullscan [43,50,59,132,137], SIM with three ions per compound [59,132–134], MRM with at least two transitions per compound [59,73,86,87], or product ion scanning mode [138]. Hence, the minimum criteria for MS-based compound identification were in principle fulfilled. Kamata et al. [137] also used the SIM mode with one ion per compound, but only for LC-MS quantification of methylone and its metabolites after these had been identified by GC-MS.

Altogether, identification and/or quantification of cathinones in biological samples is not a particular challenge for toxicological laboratories. Reported concentrations are generally in the high ng/ml to low µg/ml range, even in blood or blood-derived matrices [59,133–135]. Hence, the sensitivity of basic instrumentation of a modern toxicology laboratory should be sufficient to analyse these drugs in biomatrices.

However, it should be considered that there is evidence of instability of cathinones in biological samples. Sorensen investigated the stability of cathinones in fortified live whole blood samples [86]. The latter had been preserved either with sodium fluoride/potassium oxalate or sodium fluoride/citrate buffer in sample pH values of 7.4 and 5.9, respectively. Stability was tested at storage temperatures of 20°C and 5°C. It could be shown that stability of cathinones in blood samples was pH-dependent. In blood samples preserved with sodium fluoride/potassium oxalate, the measured concentrations of cathinone, methcathinone, ethcathinone, mephedrone, and flephedrone declined by approximately 30% after 2 days of storage at 20°C. When the blood samples were preserved with sodium fluoride/citrate buffer, the loss was reduced to approximately 10%. At a storage temperature of 5°C, the decomposition proceeded with a markedly lower rate, but a trend was still observed after 3–6 days of storage for samples preserved with sodium fluoride/potassium oxalate.

Pyrrolidinophenones

Information on the cross-reactivity of α-pyrrolidiniophenone-type designer drugs with commercial IAs is not available in the literature. The published methods for analysis of these drugs and/or their metabolites in serum [88], plasma [100,140], or urine are all either based on GC-MS [44,60–66,100,140–142] or LC-MS(/MS) [44,88,142] operated in the fullscan [44,60–65,100,140–142], SIM [66], or MRM [88] mode. With exception of Ojanpera et al. [66] and Strano-Rossi et al. [142] all authors describing analysis of pyrrolidinophenones in urine had included acidic [44] or enzymatic [44,60–65,141]

conjugate cleavage in their methods prior to extraction. The latter was achieved by LLE [44,66,142] or SPE with polymer-based HLB [88] or mixed-mode HCX [44,60–65,100,140,141] sorbents. In almost all GC-MS-based methods, the sample extracts were further derivatised by acetylation [44,60–65,141], methylation [44,64,65], ethylation [60–63,141], trifluoroacetylation [44,65], heptafluorobutyrylation [65,66], or trimethyl-silylation [44,60–65]. However, it should be noted that the pyrrolidinophenones themselves can neither be conjugated, nor do they react with the common derivatisation reagents. Thus, conjugate cleavage and derivatisation steps are not required when targeting only the parent compounds of this drug class.

Little information is available about expected concentrations of pyrrolidinophenones in serum or plasma. However, MDPV concentrations in serum samples from drivers apprehended for driving under the influence of drugs in Finland [88] and plasma concentrations of 4′-methyl-α-pyrrolidinohexanophenone (MPHP) [140] and naphyrone [100] in published overdose cases suggest that these should most likely be in the medium to high ng/ml range.

When analysing for pyrrolidinophenones in urine it is important to note that all drugs of this class and especially those with a 4′-methyl moiety such as 4′-methyl-α-pyrrolidinopropiophenone (MPPP), 4′-methyl-α-pyrrolidinopropiophenone (MPBP), or MPHP are extensively metabolised and hence are detectable in urine primarily or even exclusively in form of metabolites, at least in the rat [44,60–65,141]. In a urine sample from an overdose case, albeit one taken only on day 3 after ingestion, only the 4′-carboxy metabolite and no parent drug of MPHP could be detected [140]. In contrast, the amount of MDPV being excreted unchanged in urine seems to be high enough for using MDPV as a target analyte when screening for this drug in urine, at least when a sensitive method is used [44,66].

Tryptamines

Systematic studies on the cross-reactivity of tryptamine-derived drugs with commercial IAs for drugs of abuse testing are not available in the literature. In case reports, Boland et al. [143] reported positive IA results for amphetamines in gastric content and urine from a fatal poisoning with AMT, while in published fatal poisonings with 5-MeO-DIPT results of IA-based amphetamine screenings were negative [9,10].

Bioanalytical methods for analysis of tryptamines in serum [143], plasma [73], (post-mortem) blood [67,143], urine [9,52,67,69, 89,90,144,145], gastric content [143], and post-mortem tissues [143] have been described. Enzymatic hydrolysis of urine samples was only described by two groups [69,144]. Their findings show, that at least 4-HO-DIPT [69] and 6-HO-5-MeO-DIPT [144] are primarily excreted in conjugated form. Isolation of tryptamines from biological samples has been achieved by protein precipitation [90,144], LLE [9,67,143–145], and SPE with mixed-mode HCX [52,69,73,143], and polymer-based HLB [89] sorbents. Acetylation [67], pentafluoropropionylation [52,143], and trimethylsilylation [144] were used for derivatisation.

Sample analysis was performed with less sophisticated techniques such as GC-NPD [143] and micellar electrokinetic chromatography (MEKC)-UV [145] as well as with GC-MS [9,52,67,143,144] and LC-MS(/MS) [52,69,73,89, 90,144,145] in the fullscan [9,143,144], SIM [52,67, 69,143,144], and MRM [73,89,90] mode. Reported concentrations of tryptamines in blood or serum from overdose cases ranged from the medium ng/ml to the low µg/ml range [10,143,145].

Synthetic Cannabinoid Receptor Agonists

By the end of 2011, six major groups of synthetic cannabinoid receptor agonists had been reported: naphthoylindoles (n = 74), naphthylmethylindoles (n = 9), naphthoylpyrroles (n = 32), naphthylmethylindenes (n = 3), phenylacetylindoles (n = 28) and cyclohexylphenols (n = 16) [93,146]. The numbers of compounds known to be bioactive as CB_1-receptor agonists are given in brackets. In addition, methanandamide a derivative from anandamide (an endogenous cannabinoid) was described to be active and have increased metabolic stability [147].

The cross-reactivity of synthetic cannabinoid receptor agonists and their metabolites in urine samples was investigated by Grigoryev et al. [11] who checked the immunochromatographic test strips ICA-4-MULTI-FACTOR, ICA-MARIJUANA-FACTOR, and ICA-TAD-FACTOR. As expected, none of these tests were able to indicate the consumption of synthetic cannabinoid receptor agonists.

Bioanalytical methods for synthetic cannabinoid receptor agonists have been described for the following matrices: blood [148], serum [11,17,93,149], urine [11,68,91,150–155], and oral fluid [92]. Sample preparation consisted of simple dilution and centrifugation [152], LLE [11,68,93,148–151,154,155], or SPE with reversed-phase [11,17,150], mixed-mode reversed-phase/anion exchange (HAX) [92], or SCX [91] sorbents.

Methods for analysis of synthetic cannabinoid receptor agonists in blood or oral fluid must be very sensitive, because the concentrations of these compounds in these matrices must be expected to be in the lower ng/ml or even sub ng/ml range [92,93,148]. For analysis of synthetic cannabinoid receptor agonists in urine, it is very important to note that the target compounds must be metabolites, at least for the JWH compounds. In fact, the parent compounds of JWH-018 and JWH-073 were not detectable in the three authentic urine samples analysed by Moran et al. [153], although the detection limit of the employed analytical method was below 2ng/ml [153]. This is in line with the findings of other authors who were able to identify numerous metabolites of JWH-018, JWH-073, JWH-250 or AB-001 in authentic human urine, but not the respective parent drugs [11,68,150,154,155]. Extensive metabolism of these and other synthetic cannabinoid receptor agonists was also confirmed and reported by several *in vitro* metabolism studies [156–159].

Owing to the fact that the hydroxylated metabolites of JWH-018 [91,150,153], JWH-073 [91,150,153], JWH-250 [11] were found to be excreted in conjugated form to a very high percentage, it further seems essential to include a conjugate cleavage step when analysing urine samples for synthetic cannabinoid receptor agonists. This can either be achieved by acidic [11,68,150] or enzymatic [151,153–155,160] hydrolysis of the samples.

All methods employing GC-MS analysis also included derivatisation of the sample extracts prior to analysis: acetylation [11,150], trifluoroacetylation [11,150], trimethylsilylation [11,17,68,150], or methylation [68]. According to Grigoryev et al., acetylation should be preferred despite the slightly lower sensitivity to be achieved when using this derivatisation, because trifluoroacetyl derivatives of JWH-018, JWH-073, JWH-250 and their metabolites were found to be thermally unstable [11,150] as well as the trimethylsily derivatives of N-dealkyl metabolites of JWH-250 [11].

Some authors have used GC-MS for analysis of synthetic cannabinoid receptor agonists in biosamples [11,17,68,150], but LC-MS(/MS)-based methods have been used by most authors [11,91–93,148–155]. MS detection was performed in the fullscan [11,68,150,154,155], SIM [11,150], (enhanced) product ion scanning [11,91,150,153,155], or MRM [11,92,93,148–152, 155] mode. In contrast to almost all other new psychoactive active compounds, the negative ionisation mode was used for LC-MS/MS analysis of CP 47,497, its homologue, and HU-210 [92,152], which can be easily explained by the absence of nitrogen in the chemical structures of these compounds.

Dresen et al. [93] observed a degradation of JWH-081 and JWH-018 in processed serum samples kept in the autosampler at 4°C which exceeded 15%. A similar degradation was observed for the respective internal standard JWH-018-d_{11}. The same authors reported that the stability of the analytes in serum stored at room temperature in glass and polypropylene tubes was independent from the material of the container. In both types of vessels degradation >15% was observed for JWH-073, JWH-081 and methanandamide after 72 h of storage. However, all tested analytes were reportedly stable when stored at −20°C, so that these authors recommended to keep samples frozen whenever possible. In oral fluid samples stored at room temperature for 7 days, Coulter et al. [92] observed a loss of 25% of the initial concentration of JWH-250 and also some degradation for JWH-018 and JWH-073, while in refrigerated samples the first showed a loss of only 10% and the latter two were stable. In contrast CP 47,497, its C8 homologue and HU-210 showed small to moderate degradation of 9 to 14% that was similar at both storage conditions. In contrast to these findings, Kacinko et al. [148] observed no relevant degradation of the synthetic cannabinoid receptor agonists JWH-018, JWH-073, JWH-019, and JWH-250 in spiked whole blood over a period of 30 days, no matter if kept at ambient temperature, refrigerated, or frozen at −10°C.

RECOMMENDATIONS FOR SETTING UP ANALYTICAL METHODS FOR ANALYSIS OF NOVEL PSYCHOACTIVE SUBSTANCES

As shown by the structures given in Table 6.1, with exception of the synthetic cannabinoid receptor agonists, all novel psychoactive substances are basic, lipophilic compounds with a comparatively low molecular weight. For sample preparation, it therefore seems reasonable to start method development with an in-house extraction method for basic drugs, if available. If not, LLE with a classic extraction solvent such as chlorobutane at basic pH or, alternatively, SPE with a mixed-mode SPE sorbent should be applicable for this purpose. If urine is to be used as sample matrix, acidic or enzymatic conjugate cleavage should be performed to enhance the sensitivity of detection for highly conjugated phase I metabolites or parent drugs. For GC methods it is further recommended to include a derivatisation step to improve the chromatographic properties, especially of primary and secondary amines or hydroxy metabolites. The perfluroroacyl derivatisation is generally well-suited for such purposes with heptafluorobutyrylation yielding derivatives and fragment ions with comparatively high molecular masses which may enhance the methods selectivity. For the majority of compounds, standard single-stage quadrupole GC-MS instruments should be sufficiently sensitive to cover the expected analyte concentrations, even when using blood or the blood-derived matrices plasma or serum as sample matrix and operating the MS in the full-scan mode. However, analysis of some of the 2,5-dimethoxyamphetamines or compounds such as bromo-dragonFLY require higher sensitivity and hence analysis in the SIM mode or switching to tandem MS. Of course, up-to-date LC-MS(/MS) instrumentation is also applicable for sensitive analysis of these basic drugs in extracts of biological samples. In fact, it may even be sufficient to perform a simple dilution step prior to analysis of urine.

In contrast to the mentioned basic drugs, the synthetic cannabinoid receptor agonists are rather neutral to weakly acidic compounds of a comparatively high molecular weight. Especially because of the latter, they are not ideal compounds for GC-MS analysis, but should rather be targeted by LC-MS(/MS) if possible. Development of the extraction method can be started by LLE at neutral or acidic pH or, alternatively, by SPE with a reversed-phase, anion exchange or mixed mode HAX sorbent. If urine is to be used as sample matrix, it seems essential to include a step of conjugate cleavage in the method and to focus on metabolites rather than the parent compounds, which may not be detectable in urine even after recent uptake of synthetic cannabinoid receptor agonists.

REFERENCES

[1] Staack RF, Fritschi G, Maurer HH. Studies on the metabolism and toxicological detection of the new designer drug N-benzylpiperazine in urine using gas chromatography-mass spectrometry. J Chromatogr B Analyt Technol Biomed Life Sci 2002;773:35–46.

[2] de Boer D, Bosman IJ, Hidvegi E, et al. Piperazine-like compounds: a new group of designer drugs-of-abuse on the European market. Forensic Sci Int 2001;121:47–56.

[3] Vorce SP, Holler JM, Levine B, Past MR. Detection of 1-benzylpiperazine and 1-(3-trifluoromethylphenyl)-piperazine in urine analysis specimens using GC-MS and LC-ESI-MS. J Anal Toxicol 2008;32:444–50.

[4] Logan BK, Costantino AG, Rieders EF, Sanders D. Trazodone, meta-chlorophenylpiperazine (an hallucinogenic drug and trazodone metabolite), and the hallucinogen trifluoromethylphenylpiperazine cross-react with the EMIT((R))II ecstasy immunoassay in urine. J Anal Toxicol 2010;34:587–9.

[5] Bell C, George C, Kicman AT, Traynor A. Development of a rapid LC-MS/MS method for direct urinalysis of designer drugs. Drug Test Anal 2011;3:496–504.

[6] de Boer D, Gijzels MJ, Bosman IJ, Maes RA. More data about the new psychoactive drug 2C-B. J Anal Toxicol 1999;23:227–8.

[7] Apollonio LG, Whittall IR, Pianca DJ, Kyd JM, Maher WA. Matrix effect and cross-reactivity of select amphetamine-type substances, designer analogues, and putrefactive amines using the Bio-Quant direct ELISA presumptive assays for amphetamine and methamphetamine. J Anal Toxicol 2007;31:208–13.

[8] Kerrigan S, Mellon MB, Banuelos S, Arndt C. Evaluation of commercial enzyme-linked immuno-sorbent assays to identify psychedelic phenethyl-amines. J Anal Toxicol 2011;35:444–51.

[9] Meatherall R, Sharma P. Foxy, a designer tryptamine hallucinogen. J Anal Toxicol 2003;27:313–7.

[10] Wilson JM, McGeorge F, Smolinske S, Meatherall R. A foxy intoxication. Forensic Sci Int 2005;148:31–6.

[11] Grigoryev A, Melnik A, Savchuk S, Simonov A, Rozhanets V. Gas and liquid chromatography-mass spectrometry studies on the metabolism of the synthetic phenylacetylindole cannabimimetic JWH-250, the psychoactive component of smoking mixtures.

J Chromatogr B Analyt Technol Biomed Life Sci 2011;879:2519–26.

[12] Abdel-Hay KM, Awad T, DeRuiter J, Clark CR. Differentiation of methylenedioxybenzylpiperazines (MDBP) by GC-IRD and GC-MS. Forensic Sci Int 2010;195:78–85.

[13] Abdel-Hay KM, Awad T, Deruiter J, Clark CR. Differentiation of methylenedioxybenzylpiperazines (MDBPs) and methoxymethylbenzylpiperazines (MMBPs) By GC-IRD and GC-MS. Forensic Sci Int 2011;210:122–8.

[14] Abdel-Hay KM, Deruiter J, Randall Clark C. Differentiation of methoxybenzoylpiperazines (OMeBzPs) and methylenedioxybenzylpiperazines (MDBPs) By GC-IRD and GC-MS. Drug Test Anal 2011.

[15] Al-Hossaini AM, Awad T, DeRuiter J, Clark CR. GC-MS and GC-IRD analysis of ring and side chain regio-isomers of ethoxyphenethylamines related to the controlled substances MDEA, MDMMA and MBDB. Forensic Sci Int 2010;200:73–86.

[16] Archer RP. Fluoromethcathinone, a new substance of abuse. Forensic Sci Int 2009;185:10–20.

[17] Auwarter V, Dresen S, Weinmann W, Muller M, Putz M, Ferreiros N. 'Spice' and other herbal blends: harmless incense or cannabinoid designer drugs? J Mass Spectrom 2009;44:832–7.

[18] Brandt SD, Freeman S, Sumnall HR, Measham F, Cole J. Analysis of NRG 'legal highs' in the UK: identification and formation of novel cathinones. Drug Test Anal 2011;3:569–75.

[19] Brandt SD, Sumnall HR, Measham F, Cole J. Analyses of second-generation 'legal highs' in the UK: initial findings. Drug Test Anal 2010;2:377–82.

[20] Jankovics P, Varadi A, Tolgyesi L, Lohner S, Nemeth-Palotas J, Balla J. Detection and identification of the new potential synthetic cannabinoids 1-pentyl-3-(2-iodobenzoyl)indole and 1-pentyl-3-(1-adamantoyl) indole in seized bulk powders in Hungary. Forensic Sci Int 2012;214:27–32.

[21] Maher HM, Awad T, Clark CR. Differentiation of the regioisomeric 2-, 3-, and 4-trifluoromethylphenylpiperazines (TFMPP) by GC-IRD and GC-MS. Forensic Sci Int 2009;188:31–9.

[22] Maher HM, Awad T, DeRuiter J, Clark CR. GC-MS and GC-IRD studies on dimethoxyamphetamines (DMA): regioisomers related to 2,5-DMA. Forensic Sci Int 2009;192:115–25.

[23] McDermott SD, Power JD, Kavanagh P, O'Brien J. The analysis of substituted cathinones. Part 2: An investigation into the phenylacetone based isomers of 4-methylmethcathinone and N-ethylcathinone. Forensic Sci Int 2011;212:13–21.

[24] Power JD, McGlynn P, Clarke K, McDermott SD, Kavanagh P, O'Brien J. The analysis of substituted cathinones. Part 1: Chemical analysis of 2-, 3- and 4-methylmethcathinone. Forensic Sci Int 2011;212:6–12.

[25] Rosner P, Quednow B, Girreser U, Junge T. Isomeric fluoro-methoxy-phenylalkylamines: a new series of controlled-substance analogues (designer drugs). Forensic Sci Int 2005;148:143–56.

[26] Westphal F, Junge T, Girreser U, Stobbe S, Perez SB. Structure elucidation of a new designer benzylpiperazine: 4-bromo-2,5-dimethoxybenzylpiperazine. Forensic Sci Int 2009;187:87–96.

[27] Westphal F, Junge T, Klein B, Fritschi G, Girreser U. Spectroscopic characterization of 3,4-methylenedioxypyrrolidinobutyrophenone: a new designer drug with alpha-pyrrolidinophenone structure. Forensic Sci Int 2011;209:126–32.

[28] Westphal F, Junge T, Rosner P, Sonnichsen F, Schuster F. Mass and NMR spectroscopic characterization of 3,4-methylenedioxypyrovalerone: a designer drug with alpha-pyrrolidinophenone structure. Forensic Sci Int 2009;190:1–8.

[29] Maurer HH. Position of chromatographic techniques in screening for detection of drugs or poisons in clinical and forensic toxicology and/or doping control [review]. Clin Chem Lab Med 2004;42:1310–24.

[30] Staack RF, Fritschi G, Maurer HH. New designer drug 1-(3-trifluoromethylphenyl) piperazine (TFMPP): gas chromatography/mass spectrometry and liquid chromatography/mass spectrometry studies on its phase I and II metabolism and on its toxicological detection in rat urine. J Mass Spectrom 2003;38:971–81.

[31] Staack RF, Maurer HH. New designer drug 1-(3,4-methylenedioxybenzyl) piperazine (MDBP): studies on its metabolism and toxicological detection in rat urine using gas chromatography/mass spectrometry. J Mass Spectrom 2004;39:255–61.

[32] Staack RF, Maurer HH. Toxicological detection of the new designer drug 1-(4-methoxyphenyl) piperazine and its metabolites in urine and differentiation from an intake of structurally related medicaments using gas chromatography-mass spectrometry. J Chromatogr B Analyt Technol Biomed Life Sci 2003;798:333–42.

[33] Staack RF, Maurer HH. Piperazine-derived designer drug 1-(3-chlorophenyl)piperazine (mCPP): GC-MS studies on its metabolism and its toxicological detection in rat urine including analytical differentiation from its precursor drugs trazodone and nefazodone. J Anal Toxicol 2003;27:560–8.

[34] Theobald DS, Fehn S, Maurer HH. New designer drug, 2,5-dimethoxy-4-propylthio-beta-phenethylamine (2C-T-7): studies on its metabolism and toxicological detection in rat urine using gas chromatography/mass spectrometry. J Mass Spectrom 2005;40:105–16.

[35] Theobald DS, Fritschi G, Maurer HH. Studies on the toxicological detection of the designer drug 4-bromo-2,5-dimethoxy-beta-phenethylamine (2C-B) in rat urine using gas chromatography-mass spectrometry. J Chromatogr B Analyt Technol Biomed Life Sci 2007;846:374–7.

[36] Theobald DS, Maurer HH. Studies on the metabolism and toxicological detection of the designer drug 2,5-dimethoxy-4-methyl-beta- phenethylamine (2C-D) in rat urine using gas chromatographic/ mass spectrometric techniques. J Mass Spectrom 2006;41:1509–19.

[37] Theobald DS, Maurer HH. Studies on the metabolism and toxicological detection of the designer drug 4-ethyl-2,5-dimethoxy-beta-phenethylamine (2C-E) in rat urine using gas chromatographic-mass spectrometric techniques. J Chromatogr B Analyt Technol Biomed Life Sci 2006;842:76–90.

[38] Theobald DS, Putz M, Schneider E, Maurer HH. New designer drug 4-iodo-2,5-dimethoxy-beta-phenethylamine (2C-I): studies on its metabolism and toxicological detection in rat urine using gas chromatographic/mass spectrometric and capillary electrophoretic/mass spectrometric techniques. J Mass Spectrom 2006;41:872–86.

[39] Theobald DS, Staack RF, Puetz M, Maurer HH. New designer drug 2,5-dimethoxy-4-ethylthio-beta-phenethylamine (2C-T-2): studies on its metabolism and toxicological detection in rat urine using gas chromatography/mass spectrometry. J Mass Spectrom 2005;40:1157–72.

[40] Ewald AH, Peters FT, Weise M, Maurer HH. Studies on the metabolism and toxicological detection of the designer drug 4-methylthioamphetamine (4-MTA) in human urine using gas chromatography-mass spectrometry. J Chromatogr B Analyt Technol Biomed Life Sci 2005;824:123–31.

[41] Staack RF, Fehn J, Maurer HH. New designer drug p-methoxymethamphetamine: studies on its metabolism and toxicological detection in urine using gas chromatography-mass spectrometry. J Chromatogr B Analyt Technol Biomed Life Sci 2003;789:27–41.

[42] Kraemer T, Bickeboeller-Friedrich J, Maurer HH. On the metabolism of the amphetamine-derived antispasmodic drug mebeverine: gas chromatography-mass spectrometry studies on rat liver microsomes and on human urine. Drug Metab Dispos 2000;28:339–47.

[43] Meyer MR, Wilhelm J, Peters FT, Maurer HH. Beta-keto amphetamines: studies on the metabolism of the designer drug mephedrone and toxicological detection of mephedrone, butylone, and methylone in urine using gas chromatography-mass spectrometry. Anal Bioanal Chem 2010;397:1225–33.

[44] Meyer MR, Du P, Schuster F, Maurer HH. Studies on the metabolism of the alpha-pyrrolidinophenone designer drug methylenedioxy-pyrovalerone (MDPV) in rat and human urine and human liver microsomes using GC-MS and LC-high-resolution-MS and its detectability in urine by GC-MS. J Mass Spectrom 2010;45:1426.

[45] Ewald AH, Fritschi G, Maurer HH. Designer drug 2,4,5-trimethoxyamphetamine (TMA-2): studies on its metabolism and toxicological detection in rat urine using gas chromatographic/mass spectrometric techniques. J Mass Spectrom 2006;41:1140–8.

[46] Ewald AH, Fritschi G, Bork WR, Maurer HH. Designer drugs 2,5-dimethoxy-4-bromo-amphetamine (DOB) and 2,5-dimethoxy-4-bromo-methamphetamine (MDOB): studies on their metabolism and toxicological detection in rat urine using gas chromatographic/mass spectrometric techniques. J Mass Spectrom 2006;41:487–98.

[47] Ewald AH, Fritschi G, Maurer HH. Metabolism and toxicological detection of the designer drug 4-iodo-2,5-dimethoxy-amphetamine (DOI) in rat urine using gas chromatography-mass spectrometry. J Chromatogr B Analyt Technol Biomed Life Sci 2007;857:170–4.

[48] Ewald AH, Ehlers D, Maurer HH. Metabolism and toxicological detection of the designer drug 4-chloro-2,5-dimethoxyamphetamine in rat urine using gas chromatography-mass spectrometry. Anal Bioanal Chem 2008;390:1837–42.

[49] Ewald AH, Puetz M, Maurer HH. Designer drug 2,5-dimethoxy-4-methyl-amphetamine (DOM, STP): involvement of the cytochrome P450 isoenzymes in formation of its main metabolite and detection of the latter in rat urine as proof of a drug intake using gas chromatography-mass spectrometry. J Chromatogr B Analyt Technol Biomed Life Sci 2008;862:252–6.

[50] Meyer MR, Vollmar C, Schwaninger AE. Wolf E. Maurer HH. New cathinone-derived designer drugs 3-bromomethcathinone and 3-fluoromethcathinone: studies on their metabolism in rat urine and human liver microsomes using GC-MS and LC-high-resolution MS and their detectability in urine. J Mass Spectrom 2012;47:253–62.

[51] Kerrigan S, Banuelos S, Perrella L, Hardy B. Simultaneous detection of ten psychedelic phenethylamines in urine by gas chromatography-mass spectrometry. J Anal Toxicol 2011;35:459–69.

[52] Vorce SP, Sklerov JH. A general screening and confirmation approach to the analysis of designer tryptamines and phenethylamines in blood and urine using GC-EI-MS and HPLC-electrospray-MS. J Anal Toxicol 2004;28:407–10.

[53] Peters FT, Schaefer S, Staack RF, Kraemer T, Maurer HH. Screening for and validated quantification of amphetamines and of amphetamine- and piperazine-derived designer drugs in human blood plasma by gas chromatography/mass spectrometry. J Mass Spectrom 2003;38:659–76.

[54] Habrdova V, Peters FT, Theobald DS, Maurer HH. Screening for and validated quantification of phenethylamine-type designer drugs and mescaline in human blood plasma by gas chromatography/mass spectrometry. J Mass Spectrom 2005;40:785–95.

[55] Barroso M, Costa S, Dias M, Vieira DN, Queiroz JA, Lopez-Rivadulla M. Analysis of phenylpiperazine-like stimulants in human hair as trimethylsilyl derivatives by gas chromatography-mass spectrometry. J Chromatogr A 2010;1217:6274–80.

[56] Balikova M. Nonfatal and fatal DOB (2,5-dimethoxy-4-bromoamphetamine) overdose. Forensic Sci Int 2005;153:85–91.

[57] Kudo K, Ishida T, Hara K, Kashimura S, Tsuji A, Ikeda N. Simultaneous determination of 13 amphetamine related drugs in human whole blood using an enhanced polymer column and gas chromatography-mass spectrometry. J Chromatogr B Analyt Technol Biomed Life Sci 2007;855:115–20.

[58] Martin M, Muller JF, Turner K, Duez M, Cirimele V. Evidence of mephedrone chronic abuse through hair analysis using GC/MS. Forensic Sci Int 2011

[59] Wikstrom M, Thelander G, Nystrom I, Kronstrand R. Two fatal intoxications with the new designer drug methedrone (4-methoxymethcathinone). J Anal Toxicol 2010;34:594–8.

[60] Springer D, Peters FT, Fritschi G, Maurer HH. New designer drug 4'-methyl-alpha-pyrrolidinohexanophenone: studies on its metabolism and toxicological detection in urine using gas chromatography-mass spectrometry. J Chromatogr B Analyt Technol Biomed Life Sci 2003;789:79–91.

[61] Springer D, Fritschi G, Maurer HH. Metabolism of the new designer drug alpha-pyrrolidinopropiophenone (PPP) and the toxicological detection of PPP and 4'-methyl-alpha-pyrrolidinopropiophenone (MPPP) studied in rat urine using gas chromatography-mass spectrometry. J Chromatogr B Analyt Technol Biomed Life Sci 2003;796:253–66.

[62] Springer D, Fritschi G, Maurer HH. Metabolism and toxicological detection of the new designer drug 3',4'-methylenedioxy-alpha-pyrrolidinopropiophenone studied in urine using gas chromatography-mass spectrometry. J Chromatogr B Analyt Technol Biomed Life Sci 2003;793:377–88.

[63] Springer D, Fritschi G, Maurer HH. Metabolism and toxicological detection of the new designer drug 4'-methoxy-alpha-pyrrolidinopropiophenone studied in rat urine using gas chromatography-mass spectrometry. J Chromatogr B Analyt Technol Biomed Life Sci 2003;793:331–42.

[64] Peters FT, Meyer MR, Fritschi G, Maurer HH. Studies on the metabolism and toxicological detection of the new designer drug 4'-methyl-alpha-pyrrolidinobutyrophenone (MPBP) in rat urine using gas chromatography-mass spectrometry. J Chromatogr B Analyt Technol Biomed Life Sci 2005;824:81–91.

[65] Sauer C, Peters FT, Haas C, Meyer MR, Fritschi G, Maurer HH. New designer drug alpha-pyrrolidinovalerophenone (PVP): studies on its metabolism and toxicological detection in rat urine using gas chromatographic/mass spectrometric techniques. J Mass Spectrom 2009;44:952–64.

[66] Ojanpera IA, Heikman PK, Rasanen IJ. Urine analysis of 3,4-methylenedioxypyrovalerone in opioid-dependent patients by gas chromatography-mass spectrometry. Ther Drug Monit 2011;33:257–63.

[67] Ishida T, Kudo K, Kiyoshima A, Inoue H, Tsuji A, Ikeda N. Sensitive determination of alpha-methyltryptamine (AMT) and 5-methoxy-N,N-diisopropyltryptamine (5MeO-DIPT) in whole blood and urine using gas chromatography-mass spectrometry. J Chromatogr B Analyt Technol Biomed Life Sci 2005;823:47–52.

[68] Grigoryev A, Kavanagh P, Melnik A. The detection of the urinary metabolites of 3-[(adamantan-1-yl)carbonyl]-1-pentylindole (AB-001), a novel cannabimimetic, by gas chromatography-mass spectrometry. Drug Test Anal 2011.

[69] Pichini S, Pujadas M, Marchei E, et al. Liquid chromatography-atmospheric pressure ionization electrospray mass spectrometry determination of 'hallucinogenic designer drugs' in urine of consumers. J Pharm Biomed Anal 2008;47:335–42.

[70] Nordgren HK, Holmgren P, Liljeberg P, Eriksson N, Beck O. Application of direct urine LC-MS-MS analysis for screening of novel substances in drug abusers. J Anal Toxicol 2005;29:234–9.

[71] Nordgren HK, Beck O. Multicomponent screening for drugs of abuse: direct analysis of urine by LC-MS-MS. Ther Drug Monit 2004;26:90–7.

[72] Boatto G, Nieddu M, Carta A, et al. Determination of amphetamine-derived designer drugs in human urine by SPE extraction and capillary electrophoresis with mass spectrometry detection. J Chromatogr B Analyt Technol Biomed Life Sci 2005;814:93–8.

[73] Wohlfarth A, Weinmann W, Dresen S. LC-MS/MS screening method for designer amphetamines, tryptamines, and piperazines in serum. Anal Bioanal Chem 2010;396:2403–14.

[74] Elliott S, Smith C. Investigation of the first deaths in the United Kingdom involving the detection and

quantitation of the piperazines BZP and 3-TFMPP. J Anal Toxicol 2008;32:172–7.

[75] Tsutsumi H, Katagi M, Miki A, et al. Isolation, identification and excretion profile of the principal urinary metabolite of the recently banned designer drug 1-(3-trifluoromethylphenyl)piperazine (TFMPP) in rats. Xenobiotica 2005;35:107–16.

[76] Kovaleva J, Devuyst E, De Paepe P, Verstraete A. Acute chlorophenylpiperazine overdose: a case report and review of the literature. Ther Drug Monit 2008;30:394–8.

[77] Antia U, Tingle MD, Russell BR. Validation of an LC-MS method for the detection and quantification of BZP and TFMPP and their hydroxylated metabolites in human plasma and its application to the pharmacokinetic study of TFMPP in humans. J Forensic Sci 2010;55:1311–8.

[78] Bassindale TA, Berezowski R. Quantitative analysis of hair samples for 1-benzylpiperazine (BZP) using high-performance liquid chromatography triple quadrupole mass spectrometry (LC-MS/MS) detection. Anal Bioanal Chem 2011;401:2013–7.

[79] Nieddu M, Boatto G, Pirisi MA, Dessi G. Determination of four thiophenethylamine designer drugs (2C-T-4, 2C-T-8, 2C-T-13, 2C-T-17) in human urine by capillary electrophoresis/mass spectrometry. Rapid Commun Mass Spectrom 2010;24:2357–62.

[80] Boatto G, Nieddu M, Dessi G, Manconi P, Cerri R. Determination of four thiophenethylamine designer drugs (2C-T-series) in human plasma by capillary electrophoresis with mass spectrometry detection. J Chromatogr A 2007;1159:198–202.

[81] Andreasen MF, Telving R, Birkler RI, Schumacher B, Johannsen M. A fatal poisoning involving Bromo-Dragonfly. Forensic Sci Int 2009;183:91–6.

[82] Nieddu M, Boatto G, Dessi G. Determination of 4-alkyl 2,5 dimethoxy-amphetamine derivatives by capillary electrophoresis with mass spectrometry detection from urine samples. J Chromatogr B Analyt Technol Biomed Life Sci 2007;852:578–81.

[83] Nieddu M, Boatto G, Pirisi MA, Azara E, Marchetti M. LC-MS analysis of trimethoxyamphetamine designer drugs (TMA series) from urine samples. J Chromatogr B Analyt Technol Biomed Life Sci 2008;867:126–30.

[84] Zaitsu K, Katagi M, Kamata T, et al. Determination of a newly encountered designer drug 'p-methoxyethylamphetamine' and its metabolites in human urine and blood. Forensic Sci Int 2008;177:77–84.

[85] Decaestecker T, De Letter E, Clauwaert K, et al. Fatal 4-MTA intoxication: development of a liquid chromatographic-tandem mass spectrometric assay for multiple matrices. J Anal Toxicol 2001;25:705–10.

[86] Sorensen LK. Determination of cathinones and related ephedrines in forensic whole-blood samples

by liquid-chromatography-electrospray tandem mass spectrometry. J Chromatogr B Analyt Technol Biomed Life Sci 2011;879:727–36.

[87] Shah SA, Deshmukh NI, Barker J, et al. Quantitative analysis of mephedrone using liquid chromatography tandem mass spectroscopy: application to human hair. J Pharm Biomed Anal 2012;61:64–9.

[88] Kriikku P, Wilhelm L, Schwarz O, Rintatalo J. New designer drug of abuse: 3,4-Methylenedioxypyrovalerone (MDPV). Findings from apprehended drivers in Finland. Forensic Sci Int 2011;210:195–200.

[89] Forsstrom T, Tuominen J, Karkkainen J. Determination of potentially hallucinogenic N-dimethylated indoleamines in human urine by HPLC/ESI-MS-MS. Scand J Clin Lab Invest 2001;61:547–56.

[90] Jin MJ, Jin C, Kim JY, In MK, Kwon OS, Yoo HH. A quantitative method for simultaneous determination of 5-methoxy-N,N-diisopropyltryptamine and its metabolites in urine using liquid chromatography-electrospray ionization-tandem mass spectrometry. J Forensic Sci 2011;56:1044–8.

[91] Chimalakonda KC, Moran CL, Kennedy PD, et al. Solid-Phase extraction and quantitative measurement of omega and omega-1 metabolites of JWH-018 and JWH-073 in human urine. Anal Chem 2011;83:6381–8.

[92] Coulter C, Garnier M, Moore C. Synthetic cannabinoids in oral fluid. J Anal Toxicol 2011;35:424–30.

[93] Dresen S, Kneisel S, Weinmann W, Zimmermann R, Auwarter V. Development and validation of a liquid chromatography-tandem mass spectrometry method for the quantitation of synthetic cannabinoids of the aminoalkylindole type and methanandamide in serum and its application to forensic samples. J Mass Spectrom 2011;46:163–71.

[94] Maurer HH, Pfleger K, Weber AA. Mass spectral library of drugs, poisons, pesticides, pollutants and their metabolites 2011. Weinheim: Wiley-VCH; 2011.

[95] Maurer HH, Pfleger K, Weber AA. Mass spectral and GC data of drugs, poisons, pesticides, pollutants and their metabolites, 4th revised and enlarged ed. Weinheim: Wiley-VCH; 2011.

[96] Bickeboeller-Friedrich J, Maurer HH. Screening for detection of new antidepressants, neuroleptics, hypnotics, and their metabolites in urine by GC-MS developed using rat liver microsomes. Ther Drug Monit 2001;23:61–70.

[97] Beyer J, Ehlers D, Maurer HH. Abuse of nutmeg (Myristica fragrans Houtt.): studies on the metabolism and the toxicologic detection of its ingredients elemicin, myristicin, and safrole in rat and human urine using gas chromatography/mass spectrometry. Ther Drug Monit 2006;28:568–75.

[98] Meyer MR, Maurer HH. Metabolism of designer drugs of abuse: an updated review. Curr Drug Metab 2010;11:468–82.

[99] Paul LD, Musshoff F. Richtilinie der GTFCh zur Qualitätssicherung bei forensisch-toxikologischen Untersuchungen. Toxichem Krimtech 2009;76:142–76.

[100] Derungs A, Schietzel S, Meyer MR, Maurer HH, Krahenbuhl S, Liechti ME. Sympathomimetic toxicity in a case of analytically confirmed recreational use of naphyrone (naphthylpyrovalerone). Clin Toxicol (Phila) 2011;49:691–3.

[101] Baron JM, Griggs DA, Nixon AL, Long WH, Flood JG. The trazodone metabolite meta-chlorophenylpiperazine can cause false-positive urine amphetamine immunoassay results. J Anal Toxicol 2011;35:364–8.

[102] Tsutsumi H, Katagi M, Miki A, et al. Metabolism and the urinary excretion profile of the recently scheduled designer drug N-Benzylpiperazine (BZP) in the rat. J Anal Toxicol 2006;30:38–43.

[103] Tsutsumi H, Katagi M, Miki A, et al. Development of simultaneous gas chromatography-mass spectrometric and liquid chromatography-electrospray ionization mass spectrometric determination method for the new designer drugs, N-benzylpiperazine (BZP), 1-(3-trifluoromethylphenyl)piperazine (TFMPP) and their main metabolites in urine. J Chromatogr B Analyt Technol Biomed Life Sci 2005;819:315–22.

[104] Dickson AJ, Vorce SP, Holler JM, Lyons TP. Detection of 1-benzylpiperazine, 1-(3-trifluoromethylphenyl)-piperazine, and 1-(3-chlorophenyl)-piperazine in 3,4-methylenedioxymethamphetamine-positive urine samples. J Anal Toxicol 2010;34:464–9.

[105] Staack RF, Maurer HH. Studies on the metabolism and the toxicological analysis of the nootropic drug fipexide in rat urine using gas chromatography-mass spectrometry. J Chromatogr B Analyt Technol Biomed Life Sci 2004;804:337–43.

[106] Staack RF, Theobald DS, Maurer HH. Studies on the human metabolism and the toxicologic detection of the cough suppressant dropropizine in urine using gas chromatography-mass spectrometry. Ther Drug Monit 2004;26:441–9.

[107] Curtis B, Kemp P, Harty L, Choi C, Christensen D. Postmortem identification and quantitation of 2,5-dimethoxy-4-n-propylthiophenethylamine using GC-MSD and GC-NPD. J Anal Toxicol 2003;27:493–8.

[108] Chiu YC, Chou SH, Liu JT, Lin CH. The bioactivity of 2,5-dimethoxy-4-ethylthiophenethylamine (2C-T-2) and its detection in rat urine by capillary electrophoresis combined with an on-line sample concentration technique. J Chromatogr B Analyt Technol Biomed Life Sci 2004;811:127–33.

[109] Kanamori T, Inoue H, Iwata Y, Ohmae Y, Kishi T. In vivo metabolism of 4-bromo-2,5-dimethoxyphenethylamine (2C-B) in the rat: identification of urinary metabolites. J Anal Toxicol 2002;26:61–6.

[110] Kanamori T, Tsujikawa K, Ohmae Y, et al. Excretory profile of 4-bromo-2,5-dimethoxy-phenethylamine (2C-B) in rat. J Health Sci 2003;49:166–9.

[111] Kanamori T, Kuwayama K, Tsujikawa K, Miyaguchi H, Iwata YT, Inoue H. Synthesis and identification of urinary metabolites of 4-iodo-2,5-dimethoxyphenethylamine. J Forensic Sci 2011;56:1319–23.

[112] Kanamori T, Kuwayama K, Tsujikawa K, et al. In vivo metabolism of 2,5-dimethoxy-4-propylthiophenethylamine in rat. Xenobiotica 2007;37:679–92.

[113] Nieddu M, Boatto G, Pirisi MA, Baralla E. Multi-residue analysis of eight thioamphetamine designer drugs in human urine by liquid chromatography/tandem mass spectrometry. Rapid Commun Mass Spectrom 2009;23:3051–6.

[114] Berankova K, Balikova M. Study on metabolites of 2,5-dimethoxy-4-bromamphetamine (DOB) in human urine using gas chromatography-mass spectrometry. Biomed Pap Med Fac Univ Palacky Olomouc Czech Repub 2005;149:465–8.

[115] Kraner JC, McCoy DJ, Evans MA, Evans LE, Sweeney BJ. Fatalities caused by the MDMA-related drug paramethoxyamphetamine (PMA). J Anal Toxicol 2001;25:645–8.

[116] Loor R, Lingenfelter C, Wason PP, Tang K, Davoudzadeh D. Multiplex assay of amphetamine, methamphetamine, and ecstasy drug using CEDIA technology. J Anal Toxicol 2002;26:267–73.

[117] Rohrich J, Becker J, Kaufmann T, Zorntlein S, Urban R. Detection of the synthetic drug 4-fluoroamphetamine (4-FA) in serum and urine. Forensic Sci Int 2011

[118] Lin TC, Lin DL, Lua AC. Detection of p-Chloroamphetamine in Urine Samples with Mass Spectrometry. J Anal Toxicol 2011;35:205–10.

[119] Elliott SP. Fatal poisoning with a new phenylethylamine: 4-methylthioamphetamine (4-MTA). J Anal Toxicol 2000;24:85–9.

[120] Elliott SP. Analysis of 4-methylthioamphetamine in clinical specimens. Ann Clin Biochem 2001;38: 339–47.

[121] Johansen SS, Hansen AC, Muller IB, Lundemose JB, Franzmann MB. Three fatal cases of PMA and PMMA poisoning in Denmark. J Anal Toxicol 2003;27:253–6.

[122] Martin TL. Three cases of fatal paramethoxyamphetamine overdose. J Anal Toxicol 2001;25:649–51.

[123] Mortier KA, Dams R, Lambert WE, De Letter EA, Van Calenbergh S, De Leenheer AP. Determination of paramethoxyamphetamine and other amphetamine-related designer drugs by liquid chromatography/sonic spray ionization mass spectrometry. Rapid Commun Mass Spectrom 2002;16:865–70.

[124] Becker J, Neis P, Rohrich J, Zorntlein S. A fatal para-methoxymethamphetamine intoxication. Leg Med (Tokyo) 2003;5(Suppl. 1):S138–41.

[125] Dams R, De Letter EA, Mortier KA, et al. Fatality due to combined use of the designer drugs MDMA and PMA: a distribution study. J Anal Toxicol 2003;27:318–22.

[126] Lin DL, Liu HC, Yin HL. Recent paramethoxymeth-amphetamine (PMMA) deaths in Taiwan. J Anal Toxicol 2007;31:109–13.

[127] Vevelstad M, Oiestad EL, Middelkoop G, et al. The PMMA epidemic in Norway: comparison of fatal and non-fatal intoxications. Forensic Sci Int 2012.

[128] De Letter EA, Coopman VA, Cordonnier JA, Piette MH. One fatal and seven non-fatal cases of 4-methyl-thioamphetamine (4-MTA) intoxication: clinico-path-ological findings. Int J Legal Med 2001;114:352–6.

[129] Ling LH, Marchant C, Buckley NA, Prior M, Irvine RJ. Poisoning with the recreational drug paramethoxyam-phetamine ('death'). Med J Aust 2001;174:453–5.

[130] Refstad S. Paramethoxyamphetamine (PMA) poi-soning; a 'party drug' with lethal effects. Acta Anaesthesiol Scand 2003;47:1298–9.

[131] Kraemer T, Wennig R, Maurer HH. The anti-spasmodic drug mebeverine leads to positive amphetamine results by fluorescence polarization immunoassay (FPIA) – studies on the toxicological analysis of urine by FPIA and GC-MS. J Anal Toxicol 2001;25:333–8.

[132] Marais AA, Laurens JB. Rapid GC-MS confirmation of amphetamines in urine by extractive acylation. Forensic Sci Int 2009;183:78–86.

[133] Torrance H, Cooper G. The detection of mephedrone (4-methylmethcathinone) in 4 fatalities in Scotland. Forensic Sci Int 2010;202:e62–3.

[134] Dickson AJ, Vorce SP, Levine B, Past MR. Multiple-drug toxicity caused by the coadministration of 4-methylmethcathinone (mephedrone) and heroin. J Anal Toxicol 2010;34:162–8.

[135] Maskell PD, De Paoli G, Seneviratne C, Pounder DJ. Mephedrone (4-methylmethcathinone)-related deaths. J Anal Toxicol 2011;35:188–91.

[136] Wood DM, Davies S, Greene SL, et al. Case series of individuals with analytically confirmed acute mephe-drone toxicity. Clin Toxicol (Phila) 2010;48:924–7.

[137] Kamata HT, Shima N, Zaitsu K, et al. Metabolism of the recently encountered designer drug, methylone, in humans and rats. Xenobiotica 2006;36:709–23.

[138] Zaitsu K, Katagi M, Kamata HT, et al. Determination of the metabolites of the new designer drugs bk-MBDB and bk-MDEA in human urine. Forensic Sci Int 2009;188:131–9.

[139] Kikura-Hanajiri R, Kawamura M, Saisho K, Kodama Y, Goda Y. The disposition into hair of new designer drugs;

[140] Sauer C, Hoffmann K, Schimmel U, Peters FT. Acute poisoning involving the pyrrolidinophenone-type designer drug 4′-methyl-alpha-pyrrolidinohex-anophenone (MPHP). Forensic Sci Int 2011;208:e20–5.

[141] Springer D, Peters FT, Fritschi G, Maurer HH. Studies on the metabolism and toxicological detection of the new designer drug 4′-methyl-alpha-pyrrolidinopro-piophenone in urine using gas chromatography-mass spectrometry. J Chromatogr B Analyt Technol Biomed Life Sci 2002;773:25–33.

[142] Strano-Rossi S, Cadwallader AB, de la Torre X, Botre F. Toxicological determination and in vitro metabolism of the designer drug methylenedi-oxypyrovalerone (MPDV) by gas chromatography/mass spectrometry and liquid chromatography/quadrupole time-of-flight mass spectrometry. Rapid Commun Mass Spectrom 2010;24:2706–14.

[143] Boland DM, Andollo W, Hime GW, Hearn WL. Fatality due to acute alpha-methyltryptamine intoxi-cation. J Anal Toxicol 2005;29:394–7.

[144] Kamata T, Katagi M, Kamata HT, et al. Metabolism of the psychotomimetic tryptamine derivative 5-meth-oxy-N,N-diisopropyltryptamine in humans: identifi-cation and quantification of its urinary metabolites. Drug Metab Dispos 2006;34:281–7.

[145] Tanaka E, Kamata T, Katagi M, Tsuchihashi H, Honda K. A fatal poisoning with 5-methoxy-N,N-diisopro-pyltryptamine, Foxy. Forensic Sci Int 2006;163:152–4.

[146] Consideration of the major cannabinoid agonists. In: Drugs ACotMo, editor. 2009.

[147] Palmer SL, Khanolkar AD, Makriyannis A. Natural and synthetic endocannabinoids and their struc-ture-activity relationships. Curr Pharm Des 2000;6:1381–97.

[148] Kacinko SL, Xu A, Homan JW, McMullin MM, Warrington DM, Logan BK. Development and vali-dation of a liquid chromatography-tandem mass spectrometry method for the identification and quantification of JWH-018, JWH-073, JWH-019, and JWH-250 in human whole blood. J Anal Toxicol 2011;35:386–93.

[149] Teske J, Weller JP, Fieguth A, Rothamel T, Schulz Y, Troger HD. Sensitive and rapid quantification of the cannabinoid receptor agonist naphthalen-1-yl-(1-pen-tylindol-3-yl)methanone (JWH-018) in human serum by liquid chromatography-tandem mass spectrom-etry. J Chromatogr B Analyt Technol Biomed Life Sci 2010;878:2659–63.

[150] Grigoryev A, Savchuk S, Melnik A, et al. Chromatography-mass spectrometry studies on the metabolism of synthetic cannabinoids JWH-018 and JWH-073, psychoactive components of smoking

methylone, MBDB and methcathinone. J Chromatogr B Analyt Technol Biomed Life Sci 2007;855:121–6.

mixtures. J Chromatogr B Analyt Technol Biomed Life Sci 2011;879:1126–36.

[151] Elsohly MA, Gul W, Elsohly KM, Murphy TP, Madgula VL, Khan SI. Liquid chromatography-tandem mass spectrometry analysis of urine specimens for K2 (JWH-018) Metabolites. J Anal Toxicol 2011;35:487–95.

[152] Dowling G, Regan L. A method for CP 47, 497 a synthetic non-traditional cannabinoid in human urine using liquid chromatography tandem mass spectrometry. J Chromatogr B Analyt Technol Biomed Life Sci 2011;879:253–9.

[153] Moran CL, Le VH, Chimalakonda KC, et al. Quantitative measurement of JWH-018 and JWH-073 metabolites excreted in human urine. Anal Chem 2011;83:4228–36.

[154] Moller I, Wintermeyer A, Bender K, et al. Screening for the synthetic cannabinoid JWH-018 and its major metabolites in human doping controls. Drug Test Anal 2011;3:609–20.

[155] Sobolevsky T, Prasolov I, Rodchenkov G. Detection of JWH-018 metabolites in smoking mixture post-administration urine. Forensic Sci Int 2010;200:141–7.

[156] Wintermeyer A, Moller I, Thevis M, et al. In vitro phase I metabolism of the synthetic cannabimimetic JWH-018. Anal Bioanal Chem 2010;398:2141–53.

[157] Zhang Q, Ma P, Cole RB, Wang G. Identification of in vitro metabolites of JWH-015, an aminoalkylindole agonist for the peripheral cannabinoid receptor (CB2) by HPLC-MS/MS. Anal Bioanal Chem 2006;386:1345–55.

[158] Zhang Q, Ma P, Iszard M, Cole RB, Wang W, Wang G. In vitro metabolism of R(+)-[2,3-dihydro-5-methyl-3-[(morpholinyl)methyl]pyrrolo [1,2,3-de]1,4-benzoxazinyl]-(1-naphthalenyl) methanone mesylate, a cannabinoid receptor agonist. Drug Metab Dispos 2002;30:1077–86.

[159] Zhang Q, Ma P, Wang W, Cole RB, Wang G. Characterization of rat liver microsomal metabolites of AM-630, a potent cannabinoid receptor antagonist, by high-performance liquid chromatography/electrospray ionization tandem mass spectrometry. J Mass Spectrom 2004;39:672–81.

[160] Brents LK, Gallus-Zawada A, Radominska-Pandya A, et al. Monohydroxylated metabolites of the K2 synthetic cannabinoid JWH-073 retain intermediate to high cannabinoid 1 receptor (CB1R) affinity and exhibit neutral antagonist to partial agonist activity. Biochem Pharmacol 2012.

This page intentionally left blank

INDIVIDUAL NOVEL PSYCHOACTIVE SUBSTANCES

This page intentionally left blank

7

Synthetic Amphetamine Derivatives

Jeff Lapoint, * *Paul I. Dargan*[†] *and Robert S. Hoffman*[**]

*Emergency Medicine, Medical Toxicology, Southern California Permanente
Medical Group, San Diego, CA
[†]Guy's and St Thomas' NHS Foundation Trust; Reader in Toxicology,
King's College London, London, UK
[**]Department of Emergency Medicine, New York University School of Medicine, New York City
Poison Control Center, Bellevue Hospital Center, New York, NY

PHARMACOLOGY

Physical and Chemical Description

When considering the myriad structures of synthetic amphetamines and MDMA derivatives it is essential to return to the structural backbone, or pharmocaphore, that is common amongst them – this is the phenylethylamine molecule. To that end, it is extraordinarily appropriate that the word 'phenylethylamine' comes from the Greek root Phainein (meaning to show or illuminate) – as a brief review of its structure and general principles of substitution to that structure can greatly aid both the classification of newer amphetamine based drugs and the anticipation of their likely clinical effects.

Amphetamine derivatives cause their pharmacodynamic effects through various mechanisms that are in part predictable based on their chemical structure. Drugs in this chapter are organised by structural similarities to reinforce the clinical similarities within each class and to facilitate the prediction of clinical and toxicological effects that may be expected as newer drugs continue to emerge.

Phenylethylamine is the term used to describe any structure derived from an aromatic group adjoined to a terminal amine by an ethyl group (Fig. 7.1). This apparent structural simplicity belies the vast number of novel psychoactive substances, and their corresponding and varied psychoactive effects that can be produced from minor modifications to the phenylethylamine backbone. For the purposes of this section we have only included phenylethylamine derivatives that have an alpha methyl group that is to say are true amphetamine (alpha-methyl-phenyl-ethyl-amine) derivatives.

Certain key substitutions discovered mainly through animal discrimination studies have biological significance worth mentioning. Substitution of a methyl group at the alpha carbon increases the duration of sympathomimetic action as it confers protection against metabolism by monoamine oxidase (MAO) [1–3].

Novel Psychoactive Substances.
DOI: http://dx.doi.org/10.1016/B978-0-12-415816-0.00007-9

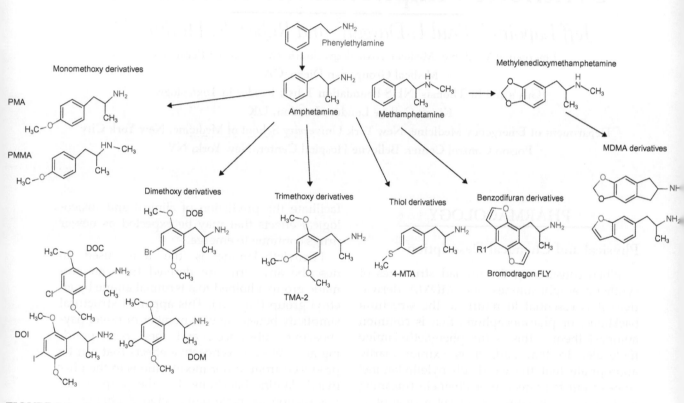

FIGURE 7.1 Synthetic Amphetamine Derivatives by structure and relationship to Amphetamine.

Additional methylation of the terminal amine greatly increases central nervous system activity, evidenced by the comparison of amphetamine to methamphetamine [4].

In general, additional substitutions to the alpha carbon, beta carbon, or terminal amine decrease alpha and beta adrenergic activity [5]. Substitution to the 2 and/or 5 positions of the aromatic ring confer serotonergic agonism that is further potentiated by addition of a halogen in combination with the above mentioned alpha methyl group [6]. Hydroxy substitution to the 3 and/or 4 position of the aromatic ring increases alpha adrenergic agonism, but defines the structure of a catecholamine and thereby increases susceptibility to metabolism by catechol-O-methyltransferase (COMT) [5].

Classification of Synthetic Amphetamines

Figure 7.1 shows a graphical representation of the synthetic amphetamines separated by structural classification. We will now consider each of the groups represented in the Figure separately.

MONOMETHOXY DERIVATIVES

Physical and Chemical Description

Paramethoxyamphetamine (PMA) and para-methoxymethamphetamine (PMMA) are monomethoxy amphetamine derivatives that have less sympathomimetic potency compared to their unsubstituted parent amphetamine. Methoxy ring substitution of amphetamine or methamphetamine at the 3 or 4 position yields the final structure with para substitution being the most commonly encountered and more potent compound. The most common physical form is a pressed pill or powder and that is often sold as 'ecstasy' or 'MDMA'.

Pharmacokinetics

Animal models demonstrate good oral bioavailability with rapid absorption from the gastrointestinal tract, although some reports indicate slower onset of action compared to MDMA. As is the case with many of synthetic amphetamine analogues, metabolism is through o-demethylation to active metabolites, namely 4-hydroxy amphetamine [7], occurring via CYP2D6 which potentially makes genetic polymorphism an important clinical consideration in cases of toxicity although there is no clinical data available to be able to determine whether this is in fact the case [8,9]. One human metabolic study contained a participant with a documented deficit in drug oxidation who excreted unchanged drug rather than the 4-hydroxy metabolite [8]. Subsequently, renal elimination predominates; PMA is a minor metabolite of PMMA [7,10].

Pharmacodynamics

As stated, both PMA and PMMA are less potent sympathomimetics than amphetamine (AM) [11]. Rank order of adrenergic potency as compared to amphetamine is AM > PMA > PMMA with PMMA exhibiting the least effect [12]. 2-methoxy substituted amphetamines have even less sympathomimetic activity. Serotonergic activity occurs through specific reuptake inhibition and rank potency is essentially the opposite to sympathomimetic effects with PMMA's increased effect resulting from its N-methylated structure in addition to para-methoxy ring substitution [12]. *In vitro*, these drugs demonstrate no significant effect on dopamine reuptake [13]. Both PMA and PMMA also act to reversibly inhibit monoamine oxidase type A (MAO-A) mediated 5-HT metabolism in addition to stimulating its release, which likely account for the described clinical effects of hyperthermia and convulsions in both animals and man [14]. Monomethoxy amphetamine derivatives are not strong p-glycoprotein inhibitors [15].

Prevalence of Use

The prevalence of use of these drugs is uncertain. PMA and PMMA users typically mistake these drugs for MDMA/'ecstasy' and take them inadvertently when purchasing what they think is MDMA/'ecstasy; no discrete market for them has been established [16].

Acute Toxicity

Animal Data

Dose dependent sympathomimetic effects are reported, as well as hyperthermia and death at high doses. Increases in heart rate and contractility were observed after PMA administration in a dog model and were reversed with propranolol [9]. In some studies, as discussed further below, hypothermia followed hyperthermic events in a dose dependent fashion. Behavioural markers used in animal models to approximate hallucinogenic effects are reported. Dogs given PMA at IV doses ranging from 4–15 mg/kg developed ataxia, mydriasis, disturbances in visual tracking, vocalisation, and appeared to react to stimuli that were not present and behaviour were less pronounced than those observed in subjects given mescaline or MDMA but were greater than in subjects given amphetamine alone [17]. The intravenous 24 hour LD_{50} for these dogs was 7 mg/kg (range 6.5–7.6 mg/kg) [17]. Primate data from the same study reported an intravenous 24 hour LD_{50} for PMA of 10 mg/kg [17]. The peak time to effect in monkeys was 1–2 hours after IV administration with a duration of effect up to 6 hours [17]. Lethal doses resulted in hyperthermia and convulsions preceding death for both dogs and monkeys [17].

Human Data

USER REPORTS

User reports available on Internet discussion forums cannot be verified factually or analytically as users are generally not aware whether they are in fact using PMA or PMMA.

Nevertheless, patterns amongst them may provide certain patterns worth mentioning [18]. These reports describe slower onset of effects of PMA and PMMA compared with MDMA, a fact that led several users to ingest additional pills. In these accounts, users expected onset of effect within one hour according to stated past experiences with MDMA and decided to ingest additional pills when no effect was experienced by that time. Nausea, vomiting and dysphoria are frequently reported with the most severe adverse effects of psychosis, hyperthermia and seizures.

CLINICAL REPORTS

A case series of 22 patients with analytically confirmed PMA exposure described multiple consequential effects ranging from tachycardia (64% of patients) to hyperthermia (temperature greater than 40°C) (36% of patients) and seizures (34% of patients) [19]. In addition, arrhythmias and QRS prolongation were also reported but the exact degree of prolongation or the criteria for determining the presence of QRS prolongation are not mentioned in the article. Two patients in the series presented with hypoglycaemia and hypokalaemia. Unfortunately, patients with coingestants were included in this series, but these coingestants or the number of individuals with coingestants were not specified.

Most case reports report recovery from initial toxicity within 24 hours. In a case series of 19 patients, 8 of the 10 patients who were intubated were extubated on the same day or the day after presentation to the Emergency Department and were discharged either same day or the following day [20]. Criteria for intubation were not described. Presenting features described in the majority of cases were agitation, tachycardia, hypertension and diaphoresis.

DEATH

Appropriate to PMA's street name 'death', a seemingly high number of fatalities are reported from the monomethoxy amphetamines PMA and PMMA (Table 7.1). One almost uniform

TABLE 7.1 Summary of published fatalities

Drug	Reference	n = number of cases	Concentration	Sample type	Other drug involvement
PMA					
	Lamberth 2008 [21]	1	2.3 mcg/ml	Post-mortem blood	Methylecgonine
	Refstad 2003 [22]	1	0.825 mcg/ml	Post-mortem blood	
	Dams 2003 [23]	1	2.01 mcg/ml	Post-mortem blood	MDMA
	Martin 2001 [24]	2	0.6, 1.3 mcg/ml	Pre-mortem serum	
	Kraner 2001 [25]	3	0.6, 1.07, 1.9 mcg/ml	Post-mortem blood	
	Felgate 1998 [26]	6	0.24–4.9 mcg/ml (mean, 2.3 mcg/ml)	Post-mortem blood	6/6 cases. Cocaine, MDMA, amphetamine
	Cimbura 1974 [27]	9	0.03–0.19 mg/dl	Post-mortem blood	
PMMA					
	Vevelstad 2012 [28]	12	1.92 mcg/ml (0.17–3.30 mcg/ml)	Post-mortem blood	9/12 cases
	Lurie 2012 [29]	24	2.72 ± 1.67 mcg/ml	Post-mortem blood	17/24 cases: 3,4-methylenedioxy-methamphetamine (MDMA), 3,4-methylenedioxy-amphetamine (MDA), cocaine
	Johansen 2003 [30]	3	3.3 mg/kg	Post-mortem liver sample	
	Lin 2007 [31]	8	4.3 ± 4.8 mcg/ml	Post-mortem blood	7/8 cases. MDA, MDMA, ketamine, amphetamine
	Becker 2003 [32]	1	0.85 mcg/ml	Post-mortem blood	No
DOB					
	Balíková 2005 [33]	1	19 ng/ml	Pre-mortem serum	Tetrahydrocannabinol (THC), alcohol
4-MTA					
	Decaestecker 2001 [34]	1	5.23 mcg/ml	Post-mortem blood	MDMA
	De Letter 2001 [66]	1	8.38 mcg/ml	Post-mortem blood	MDMA
	Elliott 2000 [36]	1	4.6 mcg/ml	Post-mortem blood	No
BDF					
	Andreasen 2009 [37]	1	0.0047 mg/kg	Post-mortem blood	No
4-MA	EMCDDA Risk Assessment [38]	16	0.5–5.8 mg/L	Post-mortem blood	Amphetamine, caffeine, THC, MDMA, olanzapine, cocaine, morphine, methadone, meta-chlorophenylpiperazine (mCPP)

similarity between these cases is the presence of multiple drug ingestions suggesting the possibility of synergy with other sympathomimetics or serotonergics, most commonly MDMA. This suggests there is likely little intentional market for PMA and users generally ingest it mistaking it for MDMA. The additive toxicity of MAO inhibition and increased serotonin release make coingestion with other sympathomimetics or serotenergic drugs especially dangerous. Unfortunately, details of CYP2D6 polymorphism are not reported in fatalities to be able to determine whether or not this may have been a factor.

Chronic Toxicity

Animal Data

Administration of 30 mg/kg of PMA to rats over four weeks resulted in a significant decrease in 5HT transporter binding but not in the overall amount of serotonin [39]. This is suggestive of chronic neurotoxicity through a mechanism that is discretely different from the 5HT neuronal degradation that is described with chronic MDMA use [39].

Human Data

To date, there are no reports of chronic PMA or PMMA use in the literature.

Dependence and Abuse Potential

Animal data

PMA and PMMA produced neither reinforcement behaviour nor preferential selection in a rat model when compared with cocaine and amphetamine [40].

Human data

No human data exists on the dependence and abuse potential of PMA and PMMA.

DIMETHOXY DERIVATIVES

Pharmacology

Physical and Chemical Description

Dimethoxy amphetamine derivatives, or 'D' series compounds, are structurally characterised by methyoxy ring substitution at the 2 and 5 positions on the aromatic ring with varying additional substitution of hydrophobic moieties at the 4 position (Fig. 7.1). Drugs included in this group include DOM (2,5-Dimethoxy-4-methylamphetamine), DOI (2,5-Dimethoxy-4-iodoamphetamine), DOB (2,5-Dimethoxy-4-bromoamphetamine) and DOC (2,5-Dimethoxy-4-chloroamphetamine). Typical forms are as pressed pills, powders, or as liquid embedded onto paper.

Pharmacokinetics

Kinetic data in humans are limited but animal models suggest good oral bioavailability and absorption from the GI tract with delayed onset on effects, sometimes in excess of one hour [41,42]. Significant first pass effect from hepatic metabolism was observed in one study comparing peak plasma concentrations of DOB measured after oral and subcutaneous dosing in rats [42]. Metabolism is at least partially due to demethylation of the methoxy group residing on the 3 or 5 position of the ring by CYP2D6 followed by oxidative deamination with the formation of active metabolites the O-demethyl DOB isomer [43].

Pharmacodynamics

Drugs of the D series are potent 5-HT_2 agonists and act as full agonists at the 5-HT_{2A} and 5-HT_{2C} receptor subtypes [44]. Studies comparing DOM and DOI to non N-methylated phenylethylamines ('C' series) reveal that potency of serotonin agonism is more dependent on N-methylation rather than the particular hydrophobic substitution at the 4 position of the aromatic ring [45]. Agonism at these receptors produces potent hallucinogenic effects in addition to prolonged vasoconstriction [3,46]. In contrast to the monomethoxy amphetamine derivatives, 'D' series compounds exhibit dopaminergic agonism in these animal models [46].

Acute Toxicity

Animal Data

Both DOM and DOB caused vasoconstriction in a dog metatarsal vein study that compared vascular responses to PMA, DMA, norpeinephrine and DOB [44]. DOB's effect was second only to norepinephrine and was not reversed with phentolamine administration but was considerably reduced with application of the serotonin inhibitor cineserin suggesting little direct alpha adrenergic activity.

Human Data

USER REPORTS

Online user accounts of DOB and DOI describe a delay to onset of effects in excess of one hour with potent hallucinogenic effects and dysphoria [47,48]. Examination of these accounts evokes uncanny comparisons to medieval reports of ergotism and St Anthony's fire. Indeed, several of the accounts sampled relate events that resulted in emergency department presentations for extreme dysphoria with sensations of limb and generalised body pain that prompted users to seek medical attention [47]. Another user experienced hallucinations, dysphoria, agitation and vomiting that were refractory to self-administered repeated doses of lorazepam [48].

CLINICAL REPORTS

Case reports detail potent hallucinogenic effects lasting from 12–24 hours followed in one instance by coma [33]. In another case of DOB use, a patient developed progressive vasospasm in both upper and lower extremities over 24 hours after ingestion, which was confirmed by angiography and required hospital admission. His vasospasm resolved with phentolamine and nitroprusside infusions [49]. In one case of analytically confirmed DOC use, a patient who had ingested what he believed to be DOI and MDMA presented with convulsions, hypertension, tachycardia and mydriasis [50]. The patient was managed with supportive care and discharged 22 hours after admission.

Fatalities associated with the dimethoxy amphetamine derivatives are summarised in Table 7.1.

Chronic Toxicity

Animal Data

Rats subjected to chronic subcutaneous injections of DOM (4 doses over 1 day) showed a rapid decrease of drug effect and down regulation of $5\text{-}HT_2$ receptors as demonstrated by behavioural observation and decreased radiolabeled ketanserin binding in a manner that was dose and chronicity dependent when compared to controls [51]. Over a 7-day period, both $5\text{-}HT_2$ receptors and corticotropin-releasing factor CRF receptors underwent significant down regulation in a study with continuous DOB administration to rats [35].

Human Data

There is no human data which can determine the chronic toxicity of the dimethoxy amphetamine derivates.

Dependence and Abuse Potential

Animal Data

Cross tolerance between and DOB and LSD was demonstrated in a canine model, suggesting that these two drugs produce their effects through modulation of the same receptors [52].

Human Data

There is no human data which can determine the dependence and abuse liability of the dimethoxy amphetamine derivates.

TRIMETHOXY DERIVATIVES

Pharmacology

Physical and Chemical Description

The structure of trimethoxy amphetamine derivatives is closely related to that of mescaline. Rearrangement of the methoxy groups to various positions on the aromatic ring yields six distinct compounds: TMA (3,4,5-trimethoxyamphetamine), TMA-2 (2,4,5-trimethoxyamphetamine), TMA-3 (2,3,4-trimethoxyamphetamine), TMA-4 (2,3,5-trimethoxyamphetamine), TMA-5 (2,3,6-trimethoxyamphetamine) and TMA-6 (2,4,6-trimethoxyamphetamine). Relative to the other trimethoxy derivatives, TMA-2 is the best studied, but even for this compound, only sparse human and animal data are available.

Accounts of TMA-2 interdictions describe white powder and various coloured pills. TMA can be synthesised using an extract from the rhizome of *Acorus calamus* as well as by fully synthetic means [53]. Elemicin, a fraction of nutmeg oil, can be used as a precursor to TMA-2 [53].

Pharmacokinetics

Animal studies reveal hepatic metabolism that occurs through both single and double O-demethylation with subsequent oxidative deamination and renal excretion [54]. Data concerning which specific P450 isozymes are involved are not available.

Pharmacodynamics

In the murine model TMA-2 binds 5HT receptors and has mild inhibitory effects on MAO activity [55]. In one study, rats injected with high doses of TMA (50 mg/kg and 100 mg/kg) experienced adrenocortical stimulation evidenced by elevations in measured epinephrine and cortisone concentrations [55].

Prevalence of Use

While comprehensive epidemiological data are unavailable, use of trimethoxy amphetamine derivatives among the general population appears uncommon; there does not appear to be a specific market for these compounds and most users are likely to encounter the trimethoxy amphetamine derivates when attempting to purchase and use 'ecstasy'/MDMA. No case reports of human toxicity or fatalities are available in the peer reviewed medical literature. While this lack of information alone cannot be definitively used as a surrogate marker for prevalence, it does agree with the low number of interdiction reports and user accounts available.

Acute Toxicity

Animal Data

Increasing doses of TMA-2 administered to rats resulted in increased locomotor activity and stimulation at doses of 2.5 mg/kg with blood pressure elevation and convulsions at doses of 80 mg/kg [56]. In the same study, fatal doses were reported to be 120 mg/kg.

Human Data

USER REPORTS

After taking 13 mg of TMA-2, Shulgin reported mild nausea and less hallucinogenic effect when compared with mescaline and DMT (N,N-Dimethyltryptamine) [57]. Online user accounts detail mydriasis, muscle tremor, shivering and intense hallucinogenic effects with a dose of 80 mg in a 75 kg man [58]. The onset of action is generally reported to be within 1–1.5 h of ingestion with effects lasting 8–9 h in the majority of online user experiences sampled [58].

CASE REPORTS AND FATALITIES

No case reports of toxicity or fatality exist in the medical literature.

CHRONIC EFFECTS AND DEPENDENCE AND ABUSE LIABILITY

No data on chronic effects or dependence and abuse liability of the trimethoxy amphetamine derivatives in animals or humans exist.

THIOL-DERIVATIVES

Pharmacology

Physical and Chemical Description

The thiol derivatives are characterised by a sulphur containing thiol group in the para position of the aromatic ring as the only ring substitution with varying amino carbon substitutions (Fig. 7.1). They were initially developed as research chemicals and antidepressants because of their specific serotonergic agonism and Nichols et al. synthesised 4-methylthioamphetamine (4-MTA) for research use in 1992 [59]. A bulky thiol group was chosen because the researchers theorised that, when given chronically, drugs with a less electronegative group might produce more selective serotonergic effects without neurotoxic effects when compared with halogenated amphetamines [59]. 4-MTA is described as both powder and pill form with the latter being sold as 'ecstasy' or 'flatliners'.

Pharmacokinetics

Data in humans are incomplete but case and user reports describe ingestion and insufflation as common routes of administration. A delay to onset of effect is in excess of one hour is reported, increasing the risk of additional drug dosing by users expecting rapid symptoms [60]. In vitro studies show increased cellular toxicity among cell lines expressing high activity CYP2D6 alleles compared to cells with impaired isoenzyme activity [61]. The same study showed no increase in toxicity with cells expressing the CYP3A4 isoenzyme. 4-methiobenzoic acid is the major human metabolite identified [62].

Pharmacodynamics

4-MTA is a potent and selective inhibitor of serotonin reuptake. In a rat study comparing 4-MTA to the halogenated amphetamine parachloroamphetamine (PCA), MTA was twice as effective at blocking synaptic serotonin uptake [63]. In contrast, MTA was far less potent at blocking dopamine and norepinephrine reuptake and about equipotent at causing serotonin release compared to PCA. MTA also acts as a MAO-A inhibitor and affects peripheral norepinephrine regulation through both stimulation of release and inhibition of reuptake [63,64]. When administered to rats continuously over one week, no decreases in cortical, striatal, or hippocampal serotonin was observed, suggesting less neurotoxicity in chronic MTAb dosing compared with MDMA and PCA [64].

Acute Toxicity

Animal Data

Rats given 4-MTA at doses of 40 mg/kg and 80 mg/kg showed significant hyperthermia that was greatly potentiated by pretreatment with the MAO inhibitor pargyline and prevented by pretreatment with the alpha-1 adrenergic blocker prazosin [64]. Pretreatment with yohimbine had no effect on hyperthermia and treatment with the nonspecific serotonergic blocker methysergide potentiated hyperthermia. In studies on rat aorta, MTA produced a dose dependent increase in contraction that was inhibited by alpha adrenergic blockade and unaffected by administration of the $5-HT_2$ selective serotonin antagonist ketanserin. MTA blocked norepinephrine reuptake at the level of the neurotransmitter transporter [63]. Higher doses administered to rats resulted in convulsions and death.

Whether these results can be extrapolated to humans is uncertain, but in the rat model MTA mediated hyperthermia and vasoconstriction appear to result largely from alpha-1 adrenergic stimulation and to a lesser extent serotonin release.

Drug discrimination studies demonstrate that 4-MTA completely substituted for MDMA and PCA but not for LSD or amphetamine [59], indicating similarities in receptor systems between ring substituted amphetamines. In the same study, high doses of MTA (21 mg/kg) produced salivation, changes in body posture, and hindlimb abduction, which are all markers for serotonin mediated behavioural changes [59].

Human Data
USER REPORTS

Online user reports are scarce. One online user account details ingestion of one tablet purchased as 'flatliner', with an unspecified dosage [65]. Onset of effects were reported to occur after one hour. Intense visual hallucinations with dysphoria, diaphoresis, and 'speed like amphetamine' effects were described.

CASE REPORTS

The medical case literature contains several cases of analytically confirmed MTA toxicity characterised by delayed onset of symptoms, hyperthermia, and diaphoresis but little objective clinical data is provided in these reports [66]. One report documents a user experiencing palpations and a feeling of 'slowness' one hour after MTA ingestion with amnesia and insomnia that persisted for one week [67]. In one case reported as a fatality, witnesses describe the time course and appearance of a 22 year-old man after he may have ingested pills that were referred to as MDMA-like [68]. He was noted to be asymptomatic at 22:00 hours but reported having gastrointestinal upset and feeling unwell at 22:30. By 01:30 the next day he was shaking. At 06:15 he was described as sweaty and shaking with difficulty speaking and within one hour was very hot

and unable to stand. He experienced convulsions and breathing problems at 10:00 and was taken to a hospital where he later died at 14:00 hours. MTA was confirmed in his blood by HPLC.

See Table 7.1 for a summary of reported fatalities associated with these compounds.

CHRONIC EFFECTS AND DEPENDENCE AND ABUSE LIABILITY

No data are available regarding chronic effects or dependence and abuse liability of these compounds in either animals or humans.

4-FLUOROAMPHETAMINE

Pharmacology

Physical and Chemical Description

4-Fluoroamphetamine (4-FA, 1-(4-Fluorophenyl)propan-2-amine) is a para positional ring substitute of amphetamine. Pharmaceutical research in the last century developed several para halogen substituted amphetamines with the thought that such positional substitution may decrease or delay para oxidative attack during metabolism [3]. There are no commercially available pharmaceutical formulations for approved medicinal use in this class. Multiple preparations appear to be available as illustrated by law enforcement agency seizure data. Reports detail seizure of white powder, crystalline material, pressed tablets, liquid, capsule and paste-like preparations; it is not clear whether seized liquid preparations had been sold with the 4-fluoroamphetamine already dissolved or whether this is undertaken at user level [69].

Pharmacokinetics

There is no formal published human pharmacokinetic data available, but online user reports offer a possible picture of onset of action, dosage, and duration of action. Online reports on Erowid contain five accounts where time was documented and a single substance was involved [70]. Of these, time to onset for oral administration was between 'a few minutes' to one hour with duration of effect ranging from seven hours 45 minutes to 20 hours. Onset of effect was noted to be five minutes after insufflation and one attempt at pyrolisation yielded no effect although successful inhalation by vaporisation is described. Rectal administration is discussed in one Internet forum post where one user details 80 mg rectal insertion of 4-FA with an onset of effect in 10–20 minutes with peak effect in 20–30 minutes [71].

Pharmacodynamics

4-FA increases the release of norepinephrine, dopamine, and serotonin [72]. Halogenated amphetamines decrease serotonin concentrations in rat brain after drug administration, but unlike *p*-chloroamphetamine and fenfluoroamine, 4-FA administration does not appear to result in permanent serotonergic neuronal destruction [73]. In a study comparing decreases of rat brain serotonin at various timed interval after drug administration, doses of up to 100 microgram/kg resulted in only a 1% decrease of brain serotonin compared to controls and no histological evidence of neuronal destruction was observed as was the case for chloroamphetamine, fenfluramine, and *p*-bromoamphetamine [73].

Prevalence of Use

No data regarding the prevalence of 4-FA use is available but surrogate information regarding law enforcement seizures suggest increasing availability. Whether this data represents a discrete market for 4-FA or increased unintentional use by contamination of amphetamine is unclear; it does not appear that there is a discrete market for 4-FA amongst users. Fifteen cases of exposure were reported over a 2-year period in Eastern Denmark among drivers being investigated for driving under the influence of drugs or alcohol [74]. 4-FA concentration ranged from 0.006 mg/kg to 0.43 mg/kg and there were polypharmacy involved with every case.

Acute Toxicity

Human Data

USER REPORTS

User reports detail experiences with 4-FA doses ranging from 50 mg to 250 mg orally and most accounts involved polydrug use [70]. Effects are described as 'mild stimulant-like' in nature with users describing less euphoria and enactogenic symptoms compared to MDMA. Adverse effects reported include bruxism, tachycardia, nausea, vomiting, diaphoresis, mydriasis, and muscle pain. One account involving 180 mg of 4-FA taken orally plus the hallucinogenic phenylethylamine 2C-D (4-methyl-2,5-dimethoxyphenylethylamine) resulting in intense nausea, vomiting, dysphoria, and visual hallucinations lasting several hours. Sharp nasal pain lasting 20–30 minutes is mentioned in most account involving nasal insufflation as the route of administration; following oral use of 4-FA powder, users also report an unpleasant taste/sourness after oral use of 4-FA powder [70].

CLINICAL REPORTS

There are currently no reported cases of acute self-reported or analytically confirmed 4-FA toxicity in the published medical literature.

DEATHS

4-FA has been isolated during post-mortem analysis in one fatality but the presence of significant polydrug use make evaluating its role in the case difficult [74]. The whole blood post-mortem concentration of 4-FA obtained was 0.53 mg/kg but amphetamines, benzodiazepines and methadone were also detected. No clinical information from the case is available to be able to further determine the potential role of stimulants such as 4-FA in the fatality.

MONOMETHYL DERIVATIVES

Pharmacology

Physical and Chemical Description

4-methylamphetamine (4-MA) and 4-methylmethamphetamine (4-MMA) are monomethyl para ring-substituted derivatives of amphetamine and methamphetamine respectively. 4-MA underwent initial investigation as an appetite suppressant under the trade name 'Aptrol' during the early 1950s [75].

4-MA has predominantly been seized in powder and paste form, but liquids and tablets containing 4-MA have also been occasionally encountered in law enforcement seizures [38]. Generally 4-MA is found together with amphetamine and caffeine in varying ratios.

Pharmacokinetics

Animal Data

There are no published animal pharmacokinetic data for either 4-MA or 4-MMA.

Human Data

There is no published human data or user report data to be able to determine the pharmacokinetics of 4-MMA. Limited information is available for 4-MA from one user report and one human volunteer study. In the human volunteer study, anorectic effects lasted 6–10 hours after 1.5 mg/kg of 4-MA and severe hypertension lasted 20–30 minutes after 2.0 mg/kg 4-MA [76]. In a personal communication to Shulgin et al., a user reported psychedelic effects after 4-MA use with a plateau at two hours and return to baseline four hours after self-reported use of oral (160 mg) and intramuscular (80–120 mg) 4-MA [77].

Pharmacodynamics

4-MA induces dopamine (DA) and norepinephrine (NE) release similarly to amphetamine but in a less potent fashion. The EC_{50} for dopamine and norepinephrine were compared for amphetamine and 4-MA in rat nucleus accumbens after administration of amphetamine or 4-MA dopamine: dopamine EC_{50}: 4-methylamphetamine 44.1 ± 2.6nM, amphetamine 8.0 ± 0.43nM; noradrenaline EC_{50}: 4-methylamphetamine 22.2 ± 1.3nM, amphetamine 7.2 ± 0.44nM [78]. Conversely, 4-MA is a much more potent at inducing serotonin release as seen in the same study with a EC_{50} of 53.4 ± 4.1nM compared to the EC_{50} for amphetamine of 1756 ± 94nM [78]. The consequences of higher serotonergic release appear to result in a decrease in reinforcement behaviours observed in animal studies. In this study, self-administration of 4-MA was compared with other amphetamine analogues including amphetamine, 3-methylamphetamine, 4-fluoroamphetamine and 3-fluoroamphetamine [78]. 4-MA was the least likely of all the compounds to cause self-administration.

The relative effect of 4-MA on sympathetic neurotransmitters was confirmed in another study investigating the effects of 4-MA compared to m-fluoroamphetamine, p-fluoroamphetamine and m-methylamphetamine on extracellular DA and 5-HT concentrations in microdialysis of rat nucleus accumbens [78]. 4-MA was the most potent at increasing extracellular 5-HT concentrations and the least potent at increasing extracellular DA concentrations.

In another study, rhesus monkeys self-administered 4-MA but at a less frequent rate when compared to ortho and meta methyl substituted amphetamines as well as 4-FA [79]. 4-MA given by the intra-peritoneal route to rats at doses of 2, 4, 6, 8, and 16 µmol/kg exhibited minimal reinforcing behaviour, in this case forward locomotion, when compared to similar doses of 4-FA but both drugs resulted in equal measure of anorexia [80]. The effects of 4-MA were compared with 4-chloroamphetamine and

4-fluoroamphetamine in a whole rat model at doses of 5–10 mg/kg [81]. 4-MA was the least potent, with 'low stimulant' effects at 5 mg/kg and 'high dose' stimulant effects at 10 mg/kg. All of the 4-fluoroamphetamine and 4-chloroamphetamine treated rats died but there were no deaths in 4-MA treated rats.

In a further study measuring motor activity, 4-MA was compared to amphetamine and other ring substituted amphetamines [82]. The dose of amphetamine analogue required to increase motor activity by 200% was 38 µmol/kg for 4-MA, 16 µmol/kg for amphetamine and 24 µmol/kg for both 2- and 3-chloroamphetamine.

Prevalence of Use

There is limited data available on the prevalence of use of 4-MA. Seizures of 4-MA have been reported from 17 European countries suggesting that it is widely available in Europe [38]. However it is generally found together with amphetamine and caffeine and it is likely that most 4-MA users are seeking amphetamine/'speed' rather than specifically 4-MA [38].

Only two sub-population studies have investigated the prevalence of use of 4-MA. The first of these was an online survey in Hungary in 2012 [83]. 4 (2.1%) of 194 individuals had used 4-MA. However, there is the potential that this is an over-estimate as the street names provided in two of these cases suggest other drugs: 'formek' (generally associated with 4-methylethcathinone) and 'piko' (generally associated with methamphetamine). In another study conducted in nightclubs in London in 2012, 16.2% of 327 surveyed had heard of 4-MA, 5.8% reported having ever used it and 4.0% reported using it in the last year [38].

Acute Toxicity

Animal Data

A number of mouse models have investigated the LD_{50} of 4-MA and demonstrated that it has a similar median lethal dose to amphetamine. Studies with animals kept in isolation have shown an LD_{50} for intraperitoneal (IP), intravenous (IV) and subcutaneous (SC) administration of 4-MA and amphetamine respectively of IP: 136 mg/kg and 101 mg/L; IV 31.0 mg/kg, 12.5 mg/kg; SC 160 mg/kg, 205 mg/kg [76,84,85]. Studies have shown that, as previously noted for a number of other stimulants, that crowding increases the lethality of 4-MA [85]. The LD_{50} of 4-MA for mice kept in isolation administered subcutaneous 4-MA was 160 mg/kg and for mice kept in groups of ten was 35.0 mg/kg. 4-MA administration in mice at high doses (LD_{50} 136 mg/kg) resulted in piloerection and pulmonary haemorrhage [76]. In the same study, 4-MA demonstrated analeptic effects in mice pretreated with pentobarbital at half the potency of amphetamine, similar waking times were seen with 10 mg of amphetamine compared to 20 mg of 4-MA [76].

Human Data
USER REPORTS

There are limited user reports discussing 4-MA and/or 4-MMA; this is likely to be because these compounds are generally sold as amphetamine ('speed') and so users are unaware that they are taking 4-MA/4-MMA. In one report on *Drugs Forum*, a user described taking 10–50 mg of a product that they thought contained 4-MA or 4-MMA [86]. They reported the following symptoms: headache, 'din' heart, nausea, nervousness and stimulation that they felt resembled ephedra. In a separate report from users in France on *Bluelight*, users reported negative symptoms including anxiety, sweating, nausea, abdominal pain, headache, paranoia, hallucinations and depression followed by post-use effects that included insomnia, cognitive and mood disorders [87].

VOLUNTEER STUDY AND CLINICAL TRIALS

In a study in 14 human volunteers, 4-MA was compared to amphetamine [76]. 4-MA at a dose of 1 mg/kg increased systolic blood pressure 14 mmHg and diastolic blood pressure 4 mmHg, but did not change heart rate. The change in blood pressure with 1 mg/kg 4-MA was similar to that report with a dose of 0.25 mg/kg amphetamine. At 1.5 mg/kg doses of 4-MA there were reports of nausea and sweating and an increase in systolic blood pressure of 18 mmHg and diastolic blood pressure of 16 mmHg. At 4-MA doses of 2.0 mg/kg there were reports of 'severe and prolonged anorexia' and volunteers 'complained bitterly of gastric distress with much salivation, expectoration and coughing, terminating in copious vomiting of mucus secretions'. At this dose, there was a greater increase in blood pressure (systolic increase of 50 mmHg and diastolic of 34 mmHg).

In a clinical trial of 4-MA for weight reduction [80], individuals were recruited: 48 were administered 4-MA and then placebo, 14 only 4-MA and 11 only placebo [75]. 4-MA or placebo was given at 25 mg three times a day, and increased if tolerated to 50 mg three times a day. There was no reported difference in blood pressure and heart rate between placebo and 4-MA at either dose. Adverse effects reported only with 4-MA were headache, pruritus and palpitations.

HUMAN ACUTE TOXICITY CLINICAL REPORTS

Oral administration of 1.0 mg/kg of 4-MA in healthy human volunteers produced no CNS stimulation and only mild anorexia but at a dose of 2.0 mg/kg there was hypertension with observed average increase of both systolic blood pressure (50 mmHg over baseline) and diastolic blood pressure (34 mmHg over baseline) [76]. Psychomotor agitation, diaphoresis and 'gastric distress' with vomiting were also observed after the 2.0 mg/kg dose. In one subject, anorexia persisted almost three days.

A single confirmed clinical report of 4-MA toxicity exists in the medical literature. A 40 year-old man and habitual amphetamine user presented to the emergency department by ambulance one hour after insufflating an undisclosed amount of what he believed to be amphetamine [88]. He also drank several glasses of lager and unknown amounts of sherry and barley wine. Shortly after insufflating the powder, he began to experience palpitations and xerostomia and an impending feeling of doom. On arrival to the emergency department he was noted to appear distressed and to smell of ethanol. His heart rate was 150 b.p.m., blood pressure 200/120 mmHg, and he had an oral temperature of 37.2°C and dilated pupils. Chest X-ray was described as normal as were initial laboratory values with the exception of hyperglycaemia (glucose 13.0 mmol/l) and leucocytosis (white blood cell count 17.5×10^9 per litre). 5 hours after presentation the Poison Center was consulted and IV beta blocker was recommended (10 mg practolol IV) and administered. In response to this relatively cardioselective beta blocker the patient experienced a decrease in heart rate to 115 b.p.m. and an increase in blood pressure to 240/160 mmHg. The patient was discharged in good health 48 hours after admission, but reported difficulty in sleeping and persistent feelings of anxiety for several weeks after use. He later produced the powder which upon confirmatory testing with gas chromatography and mass spectroscopy revealed the powder to contain 4-MA and 4-MMA; there was no analysis of biological samples from the patient to confirm use of these and exclude concomitant use of other substances. It is therefore not possible to determine whether the effects were due to 4-MA, 4-MMA, alcohol or potentially other substances that the patient may have used.

DEATHS

At the time of the EMCDDA risk assessment of 4-MA there had been 16 deaths in Europe (Belgium: 6; Denmark: 1; the Netherlands: 6; and UK: 3) where 4-MA had been detected in post-mortem samples; the first of these was from the UK in October 2010 [38,89]. Limited information is available for most of these cases and in many cases other drugs were also detected and so it is not possible to determine the cause of death or the role of 4-MA in these deaths. The 4-MA concentrations in the post-mortem samples from these deaths varied from 0.5–5.8 mg/L.

CHRONIC TOXICITY

No human or animal data are available on the potential patters of chronic toxicity associated with 4-MA or 4-MMA.

Dependence and Abuse Potential

Animal Data

One animal study has compared self-administration of 4-MA with other amphetamine analogues including amphetamine, 3-methylamphetamine, 4-fluoroamphetamine and 3-fluoroamphetamine. 4-MA was the least likely of these drugs to be associated with a self-administration for both a fixed-ratio dosing schedule and a progressive ratio dosing schedule [79].

Human Data

There are no user reports or published cases in the scientific or grey literature describing the potential for dependence or abuse potential for 4-MA or 4-MMA.

SUMMARY

Synthetic amphetamine derivatives are comprised of a structurally diverse groups of drugs that all interact with serotonin, norepinephrine and dopamine neurotransporter systems to produce effects that can be, in part, predicted from their chemical structure and similarity to amphetamine/phenylethylamine. Many were originally synthesised as research chemicals. Data on effects in humans of most of these drugs is very limited and both the dynamic nature of novel psychoactive substances and the lack of widely available confirmatory testing in individuals presenting to hospital with acute toxicity present challenges to effective case reporting. Continued vigilance through monitoring systems and clinical reporting are key to understanding and anticipating public health risks from emerging synthetic amphetamine use.

REFERENCES

[1] Freeman F, Alder J. Psychotropic recreational drugs: a chemical perspective. Eur J Med Chem 2002;37:527–39.

[2] Nichols DE. Medicinal chemistry and structure activity relationships. In: Cho AK, Segal DS, editors. Amphetamine and its analogs. Psychopharmacology, toxicology and abuse. San Diego, CA: Academic Press Inc.; 1994. p. 3–33.

[3] Shulgin AT. Psychotomimetic drugs: structure activity relationships. In: Iverson LL, Iverson SD, Snyder SH, editors. Handbook of psychopharmacology. New York: Plenum Press; 1978. p. 243–86.

[4] van der shoot JB, Ariens EJ, van Possum JM, Hurkmans JA. Phenylisopropylamine derivatives, structure and action. Arzneim Foresting 1961;9:902–7.

[5] Gunn JA, Gourmand MR, Sachs J. The action of some amines related to adrenaline: methoxy-phenylisopropylamines. J Physiol 1939;95:485–500.

[6] Monte AP, Marona-Lewicka D, Parker MA. Dihydrobenzofuran analogues of hallucinogens. 3. Models of 4-substituted (2,5-dimethoxy-phenyl) alkylamine derivatives with rigidified methoxy groups. J Med Chem 1996;39:2953–61.

[7] Kaminskas LM, Irvine RJ, Callaghan PD, White JM, Kirkbride P. The contribution of the metabolite p-hydroxyamphetamine to the central actions of p-methoxyamphetamine. Psychopharmacology 2002;160:155–6.

[8] Kitchen I, Tremblay J, André J, Dring LG, Idle JR, Smith RL, et al. Interindividual and interspecies variation in the metabolism of the hallucinogen 4-methoxyamphetamine. Xenobiotica 1979;9(7):397–404.

[9] Cheng HC, Long JP, Nichols DE, Barfknecht CF. Effects of para-methoxyamphetamine (PMA) on the cardiovascular system of the dog. Arch Int Pharmacodyn Ther 1974;212(1):83–8.

[10] Rohanova M, Balikova M. Studies on distribution and metabolism of para-methoxymethamphetamine (PMMA) in rats after subcutaneous administration. Toxicology 2009;259(1–2):61–8.

[11] Tseng LF, Menon MK, Loh HH. Comparative actions of monomethoxyamphetamines on the release and

uptake of biogenic amines in brain tissue. J Pharmacol Exp Ther 1976;197(2):263–71.

[12] Stack RF, Maurer HH. Metabolism of designer drugs of abuse. Drug Metabolism 2005;6:259–274 259.

[13] Tseng LF, Menon MK, Loh HH. Comparative actions of monomethoxyamphetamines on the release and uptake of biogenic amines in brain tissue. JPET 1976;197(2):263–71.

[14] Green AL, El Hait MA. P-Methoxyamphetamine, a potent reversible inhibitor of Type-A monoamine oxidase in vitro and in vivio. J Parm Pharmacol 1980;32(4):262–6.

[15] Ketabi-Kiyanvash N, Weiss J, Haefeli WE, Mikus G. P-glycoprotein modulation by the designer drugs methylenedioxymethamphetamine, methylenedioxy-ethylamphetamine and paramethoxyamphetamine. Addict Biol 2003;8(4):413–8.

[16] European Monitoring Centre for Drugs and Drug Addiction. Risk assessments, report on the risk assessment of PMMA in the framework of the joint action on new synthetic drugs, 2003.

[17] Davis WM, Bedford JA, Buelke JL, Guinn MM, Hatoum HT, Waters IW, et al. Acute toxicity and gross behavioral effects of amphetamine, four methoxyamphetamines, and mescaline in rodents, dogs, and monkeys'. Toxicol Appl Pharmacol 1978;45(1):49–62.

[18] Erowid. Available: <http://www.erowid.org/experiences/subs/exp_PMA.shtml>.

[19] Ling LH, Marchant C, Buckley NA, Prior M, Irvine RJ. Poisoning with the recreational drug paramethoxyamphetamine ('death'). Med J Aust 2001;174(9):453–5.

[20] Caldicott DG, Edwards NA, Kruys A, Kirkbride KP, Sims DN, Byard RW, et al. Dancing with 'death': p-methoxyamphetamine overdose and its acute management. J Toxicol Clin Toxicol 2003;41(2):143–54.

[21] Lamberth PG, Ding GK, Nurmi LA. Fatal paramethoxy-amphetamine (PMA) poisoning in the Australian Capital Territory. Med J Aust 2008;188(7):426.

[22] Refstad S. Paramethoxyamphetamine (PMA) poisoning; a 'party drug' with lethal effects. Acta Anaesthesiol Scand 2003;47(10):1298–9.

[23] Dams R, De Letter EA, Mortier KA, Cordonnier JA, Lambert WE, Piette MH, et al. Fatality due to combined use of the designer drugs MDMA and PMA: a distribution study. J Anal Toxicol 2003;27(5):318–22.

[24] Martin TL. Three cases of fatal paramethoxyamphetamine overdose. J Anal Toxicol 2001;25(7):649–51.

[25] Kraner JC, McCoy DJ, Evans MA, Evans LE, Sweeney BJ. Fatalities caused by the MDMA-related drug paramethoxyamphetamine (PMA). J Anal Toxicol 2001;25(7):645–8.

[26] Felgate HE, Felgate PD, James RA, Sims DN, Vozzo DC. Recent paramethoxyamphetamine deaths. J Anal Toxicol 1998;22(2):169–72.

[27] Cimbura G. PMA deaths in Ontario. Can Med Assoc J 1974;110(11):1263–7.

[28] Vevelstad M, Øiestad EL, Middelkoop G, Hasvold I, Lilleng P, Delaveris GJ, et al. The PMMA epidemic in Norway: comparison of fatal and non-fatal intoxications. Forensic Sci Int 2012

[29] Lurie Y, Gopher A, Lavon O, Almog S, Sulimani L, Bentur Y. Severe paramethoxymethamphetamine (PMMA) and paramethoxyamphetamine (PMA) outbreak in Israel. Clin Toxicol (Phila) 2012;50(1):39–43.

[30] Johansen SS, Hansen AC, Müller IB, Lundemose JB, Franzmann MB. Three fatal cases of PMA and PMMA poisoning in Denmark. J Anal Toxicol 2003;27(4):253–6.

[31] Lin DL, Liu HC, Yin HL. Recent paramethoxymethamphetamine (PMMA) deaths in Taiwan. J Anal Toxicol 2007;31(2):109–13.

[32] Becker J, Neis P, Rührich J, Zörntlein S. A fatal paramethoxymethamphetamine intoxication. Leg Med (Tokyo) 2003;5(Suppl. 1):S138–41.

[33] Balikova M. Nonfatal and fatal DOB (2,5-dimethoxy-4-bromam phetamine) overdose. Forensic Sci Int 2005;153:85–91.

[34] Decaestecker T, De Letter E, Clauwaert K, Bouche MP, Lambert W, Van Bocxlaer J, et al. Fatal 4-MTA intoxication: development of a liquid chromatographic-tandem mass spectrometric assay for multiple matrices. J Anal Toxicol 2001;25(8):705–10.

[35] Owens MJ, Knight DL, Ritchie JC, Nemeroff CB. The 5-hydroxytryptamine 2 agonist, 1-(2,5-dimethoxy-4-bromo-phenyl)-2-aminopropane stimulates the hypothalamic-pituitary-adrenal (HPA) axis. II. J Pharmacol Exp Ther 1991;256:795–800.

[36] Elliott SP. Fatal poisoning with a new phenylethylamine: 4-methylthioamphetamine (4-MTA). J Anal Toxicol 2000;24(2):85–9.

[37] Andreasen MF, Telving R, Birkler RI, Schumacher B, Johannsen M. A fatal poisoning involving Bromo-Dragonfly. Forensic Sci Int 2009;183(1–3):91–6.

[38] EMCDDA-Europol Joint Report on a new psychoactive substance: 4-methylamphetamine. EMCDDA, Lisbon, November 2012 [accessed 21.09.12].

[39] Steele TD, Katz JL, Ricaurte GA. Evaluation of the neurotoxicity of N-methyl-1-(4-methoxyphenyl)-2-aminopropane (para-methoxymethamphetamine, PMMA). Brain Res 1992;589(2):349–52.

[40] Corrigall WA, Robertson JM, Coen KM, Lodge BA. The reinforcing and discriminative stimulus properties of para-ethoxy- and para-methoxyamphetamin. Pharmacol Biochem Behav 1992;41(1):165–9.

[41] Ewald AH, Maurer HH. 2,5-Dimethoxyamphetamine-derived designer drugs: studies on the identification of cytochrome P450 (CYP) isoenzymes involved in formation of their main metabolites and on their capability to inhibit CYP2D6. Toxicol Lett 2008;183:52–7.

[42] Beránková K, Szkutová M, Balíková M. Distribution profile of 2,5-dimethoxy-4-bromoamphetamine (DOB) in rats after oral and subcutaneous doses. Forensic Sci Int 2007;170(2–3):94–9.

[43] Ewald AH, Fritschi G, Bork WR, Maurer HH. Designer drugs 2,5-dimethoxy-4-bromo-amphetamine DOB and 2,5-dimethoxy-4-bromo-methamphetamine MDOB: studies on their metabolism and toxicological detection in rat urine using gas chromatographic/mass spectrometric techniques. J Mass Spectrom 2006;41(4):487–98.

[44] Cheng HC, Long JP, Nichols DE, Barfknecht CF. Effects of psychotomimetics on vascular strips: Studies of methoxylated amphetamines andoptical isomers of 2,5-dimethoxy-4-methylam-phetamine and 2,5-dimethoxy-4-bromo-amphet-amine. J Pharmacol Exp Ther 1974;188:114.

[45] Acuna-Castillo C, Villalobos C, Moya PR, Saez P, Cassels BK, Huidobro-Toro JP. Differences in potency and efficacy of a series of phenylisopropylamine/phenylethylamine pairs at 5-HT(2A) and 5-HT(2C) receptors. Br J Pharmacol 2002;136(4):510–9.

[46] Rusterholz DB, Spratt JL, Long JP, Kelly TF. Serotonergic and dop-aminergic involvement in the mechanism of action of R-(−)-2, 5-dime- thoxy-4-bromoamphetamine (DOB) in cats. Life Sci 1978:1499–506.

[47] Erowid. Available: <http://www.erowid.org/experiences/subs/exp_DOB.shtml> [accessed 13.9.12].

[48] Erowid. Available: <http://www.erowid.org/experiences/exp.php?ID=64401> [accessed 06.03.13].

[49] Bowen JS, Davis GB, Kearney TE, Bardin J. Diffuse vascular spasm associated with 4-bromo-2,5-dimethoxyamphetamine ingestion. JAMA 1983;249:1477–9.

[50] Ovaska H, Viljoen A, Puchnarewicz M, et al. First case report of recreational use of 2,5-dimethoxy-4-chloroamphetamine (DOC) confirmed by toxicological screening. Eur J Emerg Med 2008;15:354–6.

[51] Leysen JE, Janssen PFM. Niemegeers CJE. Rapid desensitisation and down-regulation of 5-HT2 receptors by DOM treatment. Eur J Pharmacol 1989;163(1): 145–149.

[52] Martin WR, Vaupel DB, Nozaki M, Bright LD. The identification of LSD-like hallucinogens using the chronic spinal dog. Drug Alcohol Depend 1978;3(2):113–23.

[53] Shulgin AT, Sargent T, Naranjo C. The chemistry and psychopharmacology of nutmeg and of several related phenylisopropylamines. Psychopharmacol Bull 1967;4(3):13.

[54] Mitoma C. Metabolic studies on trimethoxyamphetamines. Proc Soc Exp Biol Med 1970;134:1162.

[55] Weltman AS, Sackler AM, Pandhi V, Johnson L. Behavior and endocrine effects of 3,4,5-trimethoxyamphetamine in male mice. Experientia 1976;32(3):357–9.

[56] Otis LS, Pryor GT, William J. Preclinical identification of hallucinogenic compounds Stillman RC, Willette RE, editors. The psychopharmacology of hallucinogens. New York: Pergamon Press; 1978. p. 126–49.

[57] Shulgin AT, Shulgin A. Number 158: TMA-2. In: Joy D, editor. PIHKAL: a chemical love story. Berkely, CA: Transform Press; 1991. p. 864–8.

[58] Erowid. Available: <http://www.erowid.org/experiences/subs/exp_TMA2.shtml> [accessed 06.03.13].

[59] Huang D, Marona-Lewicka DE, Nichols X. p-methylthioamphetamine is a potent new non-neurotoxic serotonin-releasing agent. Eur J Pharmacol 1992;229(1):31–8.

[60] Carmo H, Brulport M, Hermes M, et al. CYP2D6 increases toxicity of the designer drug 4-methylthioamphetamine (4-MTA). Toxicology 2007;229(3): 236–44.

[61] Carmo H, Hengstler JG, de Boer D, Ringel M, Carvalho F, Fernandes E, et al. Comparative metabolism of the designer drug 4-methylthioamphetamine by hepatocytes from man, monkey, dog, rabbit, rat and mouse. Naunyn Schmiedebergs Arch Pharmacol 2004;369(2):198–205.

[62] Scorza C, Silveira R, Nichols DE, Reyes-Parada M. Effects of 5-HT-releasing agents on the extracellullar hippocampal 5-HT of rats. Implications for the development of novel antidepressants with a short onset of action. Neuropharmacology 1999;38(7):1055–61.

[63] Quinn ST, Guiry PJ, Schwab T, Keenan AK, McBean GJ. Blockade of noradrenaline transport abolishes 4-methylthioamphetamine-induced contraction of the rat aorta in vitro. Auton Autacoid Pharmacol 2006;26(4):335–44.

[64] Carmo H, Remião F, Carvalho F, Fernandes E, de Boer D, dos Reys LA, et al. 4-Methylthioamphetamine-induced hyperthermia in mice: influence of serotonergic and catecholaminergic pathways. Toxicol Appl Pharmacol 2003;190(3):262–71.

[65] Erowid. Available: <http://www.erowid.org/experiences/exp.php?ID=83154> [accessed 12.9.12].

[66] De Letter EA, Coopman VA, Cordonnier JA, Piette MH. One fatal and seven non-fatal cases of 4-methylthioamphetamine (4-MTA) intoxication: clinico-pathological findings. Int J Legal Med 2001;114(6):352–6.

[67] de Boer D, Egberts T, Maes RA. Para-methylthioamphetamine, a new amphetamine designer drug of abuse. Pharm World Sci 1999;21(1):47–8.

[68] Elliott SP. Fatal poisoning with a new phenylethylamine: 4-Methylthioamphetamine (4-MTA). J Anal Toxicol 2000;24(2):85–9.

[69] Europol drugs newsletter. Alert 2009-001, Project synergy, 4-fluoroamphetamine 1. 2009; July 1–9.

[70] Erowid.Available: <http://www.erowid.org/experiences/subs/exp_4Fluoroamphetamine.shtml> [accessed 06.03.13].

[71] Bluelight. Available: <http://www.bluelight.ru/vb/archive/index.php/t-271982.html>.

[72] Marona-Lewicka D, Rhee GS, Sprague JE, Nichols DE. Psychostimulant-like effects of p-fluoroamphetamine in the rat. Eur J Pharmacol 1995;287(2):105–13.

[73] Harvey JA. Neurotoxic action of halogenated amphetamines. Ann N Y Acad Sci 1978;305(1):289–304.

[74] Johansen S, Hansen T. Isomers of fluoroamphetamines detected in forensic cases in Denmark. Int J Legal Med 2012;126:541–7.

[75] Gelvin EP, McGavack TH. 2-Amino-1-(p-methylphenyl)-propane (aptrol) as an anorexigenic agent in weight reduction. N Y State J Med 1952;52:223–6.

[76] Marsh DF, Herring DA. The pharmacological activity of the ring methyl substituted phenisopropylamines. J Pharmacol Exp Ther 1950;100:298–308.

[77] Shulgin AT, Manning T, Daley PE. The Shulgin Index, 1. Berkeley: Transform Press; 2011.

[78] Baumann MH, Clark RD, Woolverton WL, Wee S, Blough BE, Rothman RB. Vivo effects of amphetamine analogs reveal evidence for serotonergic inhibition of mesolimbic dopamine transmission in the rat. J Pharmacol Exp Ther 2011;337(1):218–25.

[79] Wee S, Anderson KG, Baumann MH, Rothman RB, Blough BE, Woolverton WL. Relationship between the serotonergic activity and reinforcing effects of a series of amphetamine analogs. J Pharmacol Exp Ther 2005;313(2):848–54.

[80] Wellman PJ, Davis KW, Clifford PS, Rothman RB, Blough BE. Changes in feeding and locomotion induced by amphetamine analogs in rats. Drug Alcohol Depend 2009;100(3):234–9.

[81] Beaton JM, Smythies JR, Benington F, Morin RD, Clarke Jr. LC. Behavioural effects of some 4-substituted amphetamines. Nature 1968;220(5169):800–1.

[82] Ögren SO, Ross SB. Substituted amphetamine derivatives. II Behavioural effects in mice related to monoaminergic neurones. Acta Pharmacol Toxicol (Copenh) 1977;41(4):353–68.

[83] EMCDDA–Europol Joint report on a new psychoactive substance: 4-methylamphetamine. Lisbon: European Monitoring Centre for Drugs and Drug Addiction; 2012.

[84] Haas H, Forth W. Ein Beitrag zur Analyse der zentral erregenden Wirkungskomponenten einiger sympathicomimetischer Amine. Arzneimittelforschung 1956;6:436–42.

[85] Riva M, Kabir Naimzada M, Pirola C, Mantegazza P. Rapporti tra attività anoressigena, ipertermizzante ed eccitomotoria di composti strutturalmente correlati all'ampfetamina. Il Farmaco Edizione Scientifica 1969;24(2):238–48.

[86] Drugs Forum. 4-methylamphetamine and 4-methylmethamphetamine. Available at: <http://www.drugs-forum.com/forum/showthread.php?t=71790>; 2008 [accessed 15.10 12].

[87] Bluelight. p-alkyl-amphetamines. Available at: <http://www.bluelight.ru/vb/threads/419580-4-methylamphetamine>; 2008 [accessed 15.10.12].

[88] Bal TS, Gutteridge DR, Johnson B. Adverse effects of the use of unusual phenylethylamine compounds sold as illicit amphetamine. Med Sci Law 1989;29:3.

[89] Belgian Early Warning System on Drugs (BEWSD). Available: <http://ewsd.wiv-isp.be/Main/4-methylamphetamine%20alert%20by%20BEWSD.aspx> [accessed 06.03.13].

1-Benzylpiperazine and other Piperazine-based Derivatives

Paul Gee* and Leo Schep†

*Christchurch Hospital, University of Otago, Christchurch, New Zealand, †National Poisons Centre, Department of Preventive and Social Medicine, University of Otago, Dunedin, New Zealand

INTRODUCTION

Over the last decade, a number of piperazine derivatives (PZDs) have emerged as a new group of recreational drugs. They are sought both for their amphetamine- and 3,4-methylenedioxymethamphetamine (MDMA)-like effects. Those PZDs reported as drugs of abuse include 1-benzylpiperazine (BZP), 1,3-trifluoromethylphenylpiperazine (TFMPP), 1-(3-chlorophenyl)piperazine (mCPP) and 1-(4-methoxyphenyl)piperazine (MeOPP). When introduced as recreational drugs, they were not classified under existing drug or analogue laws in most jurisdictions. Use of these derivatives was first reported in the USA and Scandinavia in the 1990s and quickly spread across Europe and Australasia [1–4].

The backbone of the PZDs is the piperazine moiety attached to an aromatic group. Piperazine itself was widely used as a potent anthelminthic agent used to treat animals for intestinal roundworms [5]. It acts by altering cell membrane permeability and causes hyperpolarisation of cells. It is one of the base compounds used in the production of fluoroquinolones such as ciprofloxacin and norfloxacin [6]. Piperazine itself does not appear to have psychoactive properties [7]. The piperazine ring and derivatives are also used as raw materials in industry for manufacturing epoxy resins, insecticides, rubber compounds and antioxidants [8].

The most widely used PZDs are BZP, and to a lesser extent TFMPP. BZP was originally synthesised by researchers at Burroughs Wellcome & Co. in 1944 [9], although not as an anthelminthic agent [10]. A patent was finally submitted by Wellcome researchers in 1968 for use as a potential antidepressant agent [11]. Later human studies suggested the subjective and psychomotor effects of BZP were similar to dexamphetamine, though doses necessary to achieve this were 10 times greater than that of dexamphetamine [12,13].

The first mention of BZP as a potential recreational drug was by Alexander Shulgin, who listed it amongst other synthetic psychoactive substances and noted that it had 'an acceptability close to amphetamine' [14].

Novel Psychoactive Substances.
DOI: http://dx.doi.org/10.1016/B978-0-12-415816-0.00008-0

Research into a potential antidepressant N-benzyl-piperazine-piconyl fumarate (EGYT-475) was conducted in Hungary in the early 1970s. This agent was found to act as a prodrug, with BZP forming the active metabolite [15–19]. It was developed and marketed by EGIS (EGIS Gyogyszergyar Reszvenytarsasag Keresztúri ut 30-38, H-1106 Budapest, Hungary) briefly under the trade name Trelibet® but was subsequently withdrawn.

A number of other pharmaceutical piperazine derivatives and their precursors form metabolites that are recognised as piperazine-based recreational drugs. mCPP is an important intermediate in the production of the antidepressant trazdone as well as nefazadone, etoperidone and mepiprazole [20,21]. A primary metabolite of Fipexide, a cognitive enhancer considered for use treating asthenia and memory disorders in the elderly, is 1-(3,4-methylenedioxybenzyl)piperazine (MDBZP) [22]. Fipexide was marketed as a nootropic drug by Lab Bouchara [23], but is no longer available due to adverse reactions such as fever [24] and hepatitis [25,22].

Niaprazine is a hypnosedative drug of the phenyl-piperazine class and metabolises to form 1-(4-fluorophenyl)piperazine (pFPP) – another piperazine derivative reportedly used as a drug of abuse [26]. Niaprazine (Nopron; Sanofi-Aventis SpA) is available in Europe as a sedative-hypnotic drug [27].

PHARMACOLOGY

Physical Description

Piperazine (1,4-hexahydropyrazine) is a cyclic organic molecule possessing two nitrogen atoms in opposite positions within a 6-member heterocyclic ring. Piperazine forms the backbone for the various piperazine derivatives used as recreational drugs; it is typically bound to an aromatic ring (Fig. 8.1). The piperazine

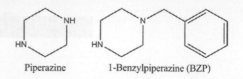

Piperazine　　　1-Benzylpiperazine (BZP)

FIGURE 8.1 The chemical structure of piperazine and 1-benzylpiperazine.

derivatives can be divided into two groups. The benzyl-piperazines are those distinguished by a methyl group that bridges the piperazine with an aromatic ring. The phenyl-piperazines are classified as a phenyl group bound to the piperazine moiety. BZP and TFMPP are representatives of the benzyl- and phenyl-derivatives respectively (Figs 8.2 and 8.3).

1-Benzylpiperazine (synonyms: BZP, N-benzylpiperazine, CAS 2759-28-6), is a synthetic benzyl-analogue of piperazine (Fig. 8.2) and has a molecular formula of $C_{11}H_{16}N_2$ and a molecular weight of 176.26 g/mol. In its freebase form it is a pale yellow viscous liquid that reacts with air and light. It has a pH of nine and is therefore an irritant to living tissue in this form. It is usually sold as BZP dihydrochloride and presented in the powdered form (in capsules or bagged). BZP may also be mixed with TFMPP, caffeine and binders and made into tablets, and has in recent years been substituted for other stimulants sold to users as 'ecstasy' [28,29].

1,3-trifluoromethylphenylpiperazine (synonyms: TFMPP, CAS 15532-75-9) is a fluorinated phenyl-piperazine derivative (Fig. 8.3), that has a molecular formula of $C_{11}H_{13}F_3N_2$ and a molecular weight of 230.23 g/mol. TFMPP is typically taken in combination with BZP; evidence with animal models suggests the combination may evoke the cumulative release of dopamine and serotonin, comparable to the effects of ecstasy [30,31].

1-(3-chlorophenyl) piperazine (synonyms: mCPP, meta-Chlorophenylpiperazine, CAS 51639-49-7) is a chlorinated phenyl-analogue of piperazine (Fig. 8.3) and has a molecular

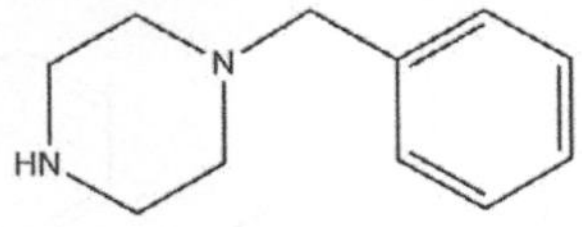

1-Benzylpiperazine

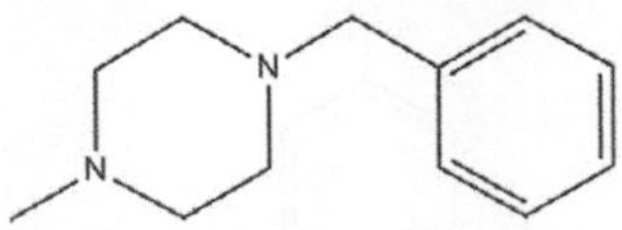

4-Methyl-1-benzylpiperazine

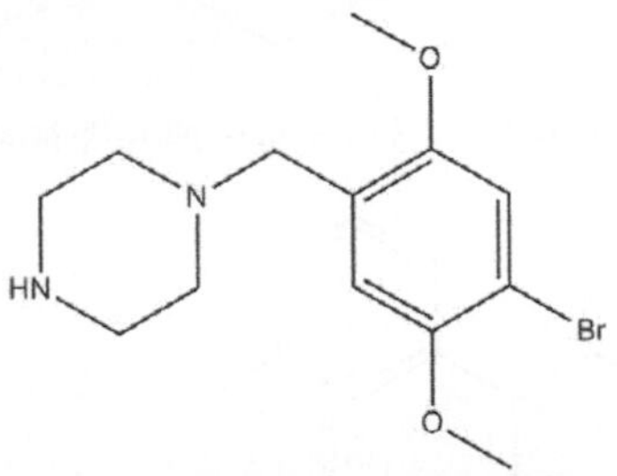

1-(4-Bromo-2,5-dimethoxybenzyl)piperazine

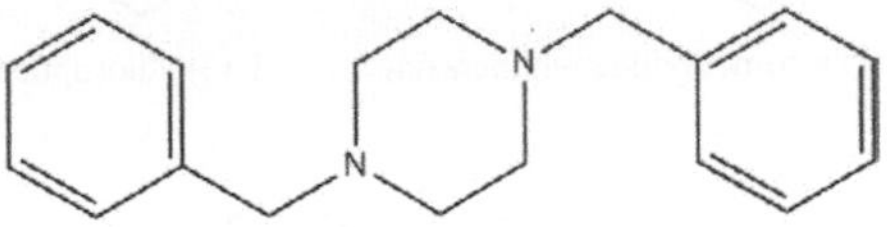

1,4-Dibenzylpiperazine

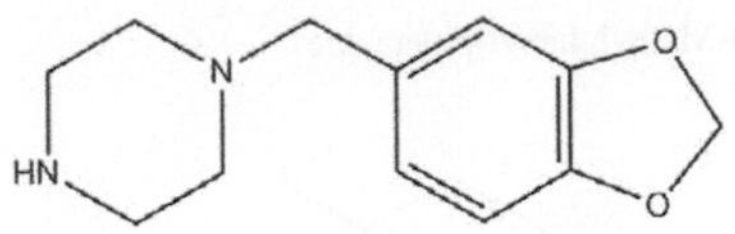

1-(3,4-Methylenedioxybenzyl)piperazine

FIGURE 8.2 Psychoactive benzyl-piperazine derivatives.

formula of $C_{10}H_{13}ClN_2$ with a molecular weight of 196.7g/mol. Structurally it is related to TFMPP with the chloro-moiety replacing the trifluoromethyl group. There are two positional isomers: 1-(4-chlorophenyl)piperazine (pCPP, para-CPP) and 1-(2-chlorophenyl piperazine, (oCPP, ortho-CPP). The positional isomer oCPP is an antagonist of the $5HT_{2C}$ receptor. Only limited investigations have been conducted on the properties of pCPP and oCPP. The isomers, predominantly mCPP, have been used as surreptitious substitutes for MDMA [29]. In Europe, their use has now increased, becoming more widespread than BZP [32].

A number of other piperazine derivatives have been reported as recreational drugs. Very little published information exists on the recreational use and associated toxicity of these substances. They are presented in Figs. 8.2 and 8.3 and a brief summary of their known properties is presented in Table 8.1.

Pharmacokinetics

The information available regarding the kinetics of piperazine analogues has been obtained predominantly from animal studies and three investigations using human volunteers. The human studies involved the oral administration of a single 200mg dose of BZP [33], a combination of BZP (100mg) and TFMPP (30mg) [34] and a single dose of 60mg TFMPP [35].

Absorption

Piperazine derivatives are usually found in capsule, pill and occasionally powder form. Use is usually by the oral route though intravenous [36,37], inhalation [38], nasal insufflations [39,40] and even rectal use [41] have been described.

Both BZP and TFMPP are rapidly absorbed following oral ingestion. Following the ingestion of 200mg BZP, a mean maximum plasma

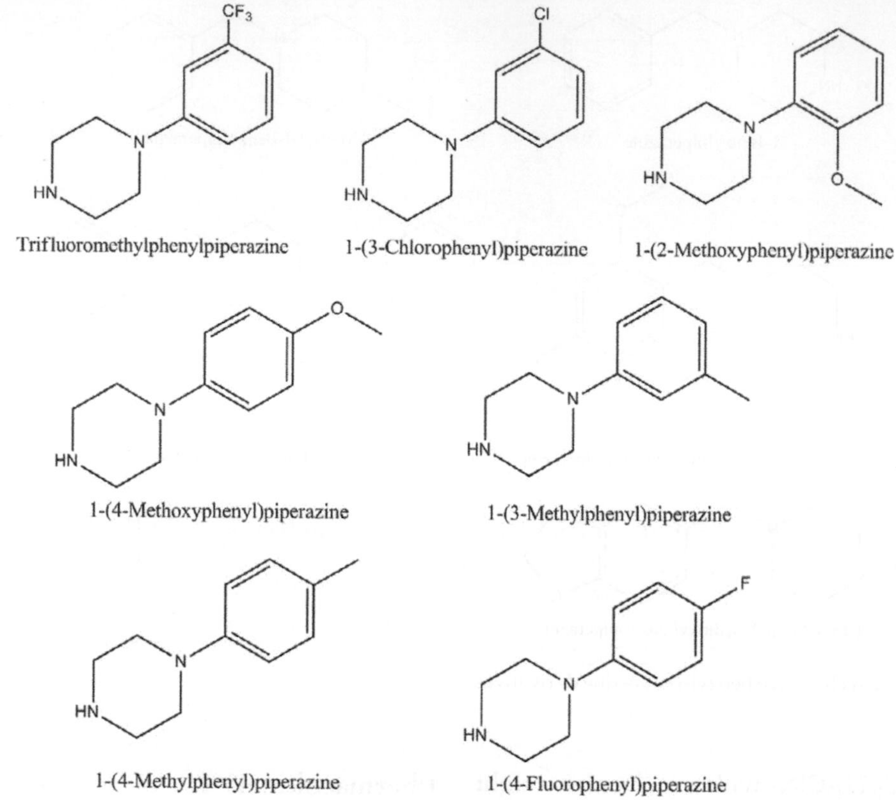

FIGURE 8.3   Psychoactive phenyl-piperazine derivatives.

concentration (T_{max}) occurred in 75min with a peak plasma concentration of 262ng/mL; [33] the calculated absorption half-life was 6.2 minutes (min). When 60mg TFMPP was administered, the T_{max} was achieved at 90min with an absorption half-life of 24.6min [35]. When BZP was co-administered with TFMPP, the demonstrated absorption half-lives were 6.0 and 13.3min, respectively [34]. The times to maximum concentration for BZP and TFMPP were 60min and 75min, respectively.

Following oral administration of 0.5mg/kg mCPP in 14 human volunteers, average peak levels occurred at 3.2 hours (h) with a maximum plasma concentration (C_{max}) of 54ng/mL; [42] the authors of this study did, however, report a large variability in their calculated

values (range 8–132ng/mL). Bioavailability as measured by AUC also varied eight-fold (64–2758h × ng/mL). C_{max} variability was also high in another study, varying eight-fold between individuals after oral administration [43].

Distribution

There is no information on the distribution of BZP. The apparent mean volume of distribution for TFMPP was calculated to be 891.34L, following the administration of 60mg to six human volunteers [35].

TFMPP has been shown to readily distribute across the blood–brain barrier; one animal study demonstrated peak concentrations in the brain within five minutes of intravenous administration [44].

TABLE 8.1 Summary of the Lesser Known Piperazine Derivatives

Name	Piperazine Class	Synonyms	CAS Number	Molecular Formula	Molecular Mass (g/mol)
1-(4-methoxyphenyl) piperazine	Phenyl	MeOPP, para-Methoxyphenylpiperazine, pMeOPP, Paraperazine	38212-30-5	$C_{11}H_{16}N_2O$	192.258
1-(2-methoxyphenyl) piperazine	Phenyl	pMeOPP	35386-24-4	$C_{11}H_{16}N_2O$	192.258
4-methyl-1-benzylpiperazine	Benzyl	MBZP	374898-00-7	$C_{12}H_{18}N_2$	190.294
1-(4-methylphenyl)piperazine	Phenyl	pMPP	39593-08-3	$C_{11}H_{16}N_2$	176.26
1-(3-methylphenyl)piperazine	Phenyl	mMPP	41186-03-2	$C_{11}H_{16}N_2$	176.26
1-(4-bromo-2,5-dimethoxybenzyl)piperazine	Benzyl	2C-B-BZP	–	$C_{13}H_{19}BrN_2O_2$	315.208
1,4-Dibenzylpiperazine	Benzyl	DBZP	2298-55-7	$C_{18}H_{22}N_2$	266.381
1-(3,4-Methylenedioxybenzyl) piperazine	Benzyl	MDBZP, 1-Piperonylpiperazine	32231-06-4	$C_{12}H_{16}N_2O_2$	220.268
1-(4-Fluorophenyl)piperazine	Phenyl	pFPP, 4-FPP; Fluoperazine, Flipiperazine	2252-63-3	$C_{10}H_{13}FN_2$	180.222

Metabolism

The metabolism of the piperazine derivatives BZP, TFMPP, mCPP and MeOPP occurs predominantly in the liver [45–48]. Summaries of the phase 1 metabolism for BZP, TFMPP, MeOPP and mCPP are presented in Figures 8.4, 8.5, 8.6 and 8.7, respectively.

Following the oral administration of BZP, 3–6% of the unchanged parent drug has been recovered from urine over the first 24 h [17,33]. Two hydroxylated species of metabolite have been identified: 4-OH BZP and 3-OH BZP [33]. However, these metabolites only account for 0.11% of the total dose. The phase 2 conjugated metabolites O-sulfate BZP and N-sulfate BZP were also recovered, making the total amount of BZP excretion 12.25%. This suggests that BZP may have low bioavailability, significant biliary excretion or strong protein or tissue binding. The protein binding of BZP in pooled human plasma was in the order of 12% [54]. Magyar reports recovering approximately 3.7% BZP in urine, following the administration of the parent drug EGYT-475 [17]. Animal studies, however, suggested that there may be other metabolites of BZP including benzylamine, N-benzylethylenediamine, piperazine [46] and glucuronide conjugates [46,49] nevertheless, as described above, the only formal study with humans detected only the hydroxyl BZP metabolites [33].

The major phase 1 metabolites of mCPP were found to be formed by hydroxylation of the aromatic ring to form 4-OH-1-(3-Chlorophenyl)piperazine, and degradation of the piperazine moiety to form N-(3-chlorophenyl)-1,2-ethanediamine, 3-chlorobenzenamine, 4-[(2-aminoethyl)amino]-2-chlorophenol and 4-amino-2-chlorophenol [53,55]. Phase 2 glucuronide and sulfate analogues have also been identified.

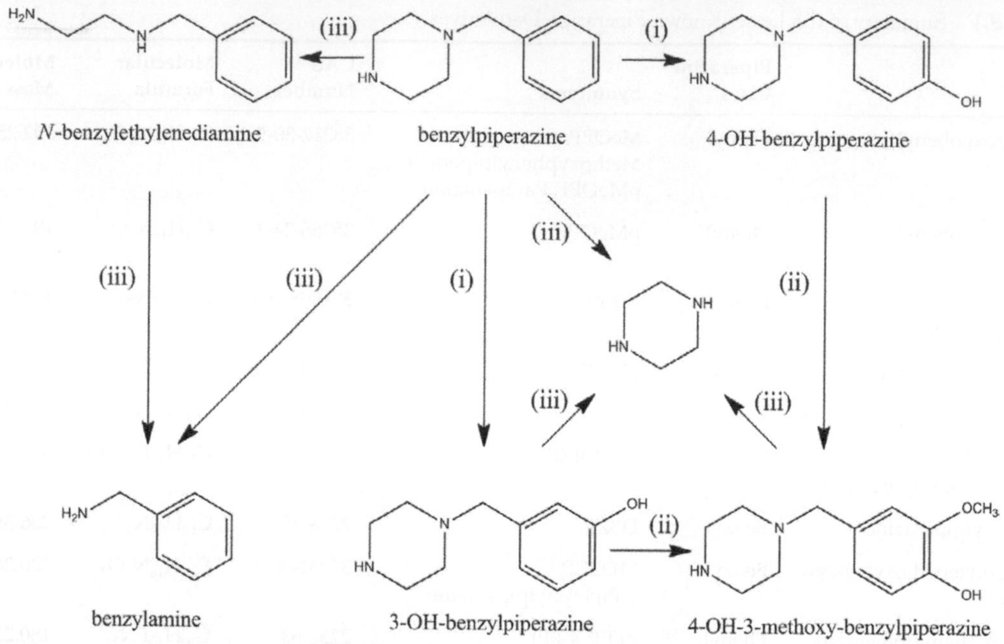

FIGURE 8.4 A proposed scheme for the phase I metabolic pathway of BZP in humans and rats [46,49] (i) aromatic hydroxylation, (ii) methylation, and (iii) dealkylation.

In vitro microsomal assays suggest the major isoenzymes involved in the metabolism of piperazine derivatives are CYP2D6, CYP1A2 and CYP3A4 [45,48,56,57]. Research has also shown BZP, TFMPP, pFPP, MeOPP, mCPP, MBZP and MDBP have an inhibitory effect on these and the isoenzymes CYP2C19 and CYP2C9 [56]. Further *in vitro* studies with human liver microsomes assessing BZP and TFMPP drug–drug interactions have suggested TFMPP causes more significant inhibition than BZP on test substrates (dextromethorphan and caffeine) [57]. Interactions with substrates can cause isoenzyme inhibition leading to unexpected high levels of either one or both. Various antidepressants (selective serotonin reuptake inhibitor and tricyclic antidepressants) and antipsychotics (phenothiazine and butyrophenone classes) are also recognised inhibitors of the isoenzymes responsible for the metabolism of piperazine derivatives [58]. They have been shown to increase plasma concentrations of mCPP, when metabolised with thioridazine [59] and haloperidol [60]. Such interactions not only cause unexpected elevations of the piperazine analogue, but also increase the risk of host toxicity. For example, chronic treatment with fluoxetine has been shown to cause a four-fold increase in plasma concentrations of mCPP in human subjects [61]. Non-prescribed recreational drugs can also influence piperazine derivative metabolism; cocaine has been shown to influence the metabolism of mCPP [62], and there is a theoretical suggestion BZP may inhibit CYP2D6 leading to elevated concentrations of MDMA, when both are co-ingested [57].

Elimination

In human studies approximately 12% of BZP and metabolites are recovered in urine over 24h [33]. This low recovery rate may be due, in part, to biliary excretion. The elimination half-life for BZP was calculated to be 5.5H [33]. Using

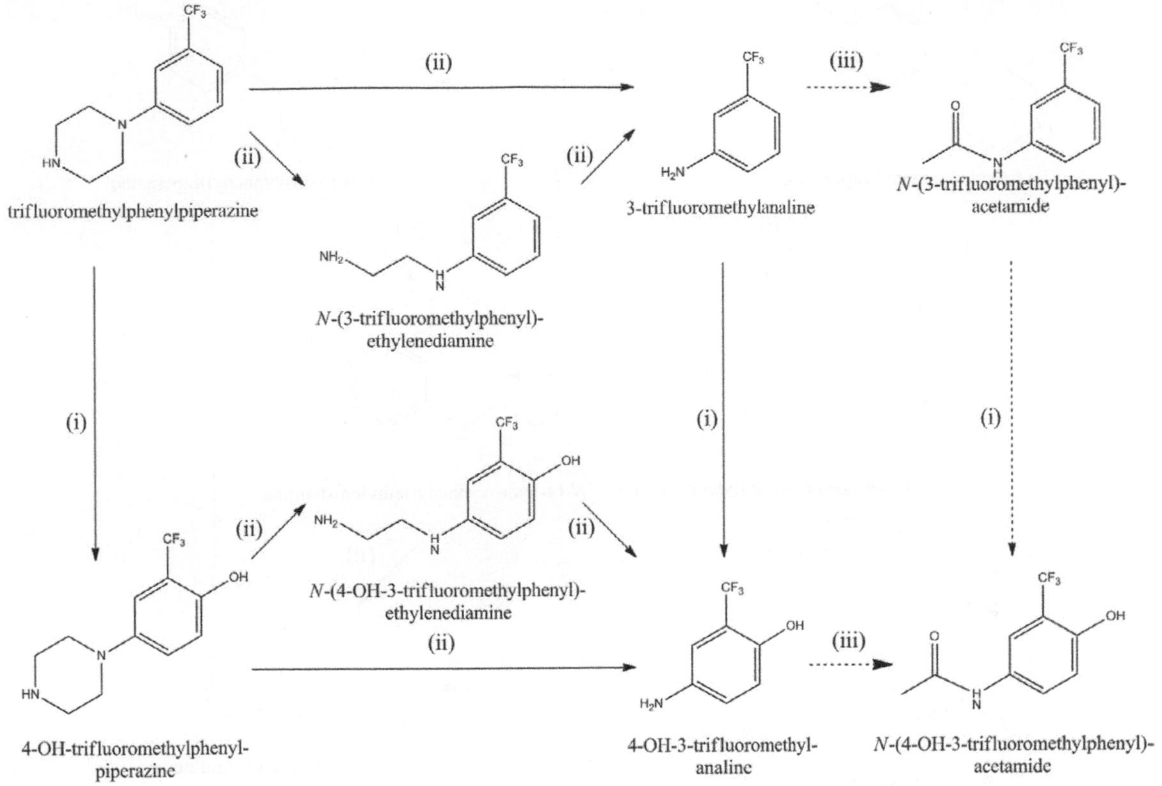

FIGURE 8.5 A proposed scheme for the phase I metabolic pathway of TFMPP in humans and rats [50,51] (i) aromatic hydroxylation, (ii) N-dealkylation and (iii) N-acetylation.

a detection limit of 5ng/mL, BZP could therefore theoretically be detected in urine up to 30h after an oral dose.

The elimination of TFMPP was determined to have two disposition phases; these measured elimination half-lives were 2.04 and 5.95h [35]. The apparent clearance of TFMPP in human volunteers is 384L/h [35]. Recovery rates of TFMPP from urine are less than 1% and suggest either very poor bioavailability or alternate excretion.

The mean rate of elimination of *m*CPP following oral and parenteral administration in human volunteers was 3.2 and 5.8h respectively [42]. Clearance following intravenous administration was 49.6mL/h.

Pharmacodynamics

Animal Data

BZP

BZP exerts its actions in a manner similar to amphetamine, causing an increase in the presynaptic release of dopamine, serotonin and noradrenaline. Research also suggests it inhibits the reuptake of dopamine [63,64] and other monoamine transmitters [65] by blocking the respective transport complexes thereby preventing the removal of transmitters from neural synapses [31]. It may also act by functioning as an agonist upon the postsynaptic dopamine receptors [66]. Studies assessing the nucleus

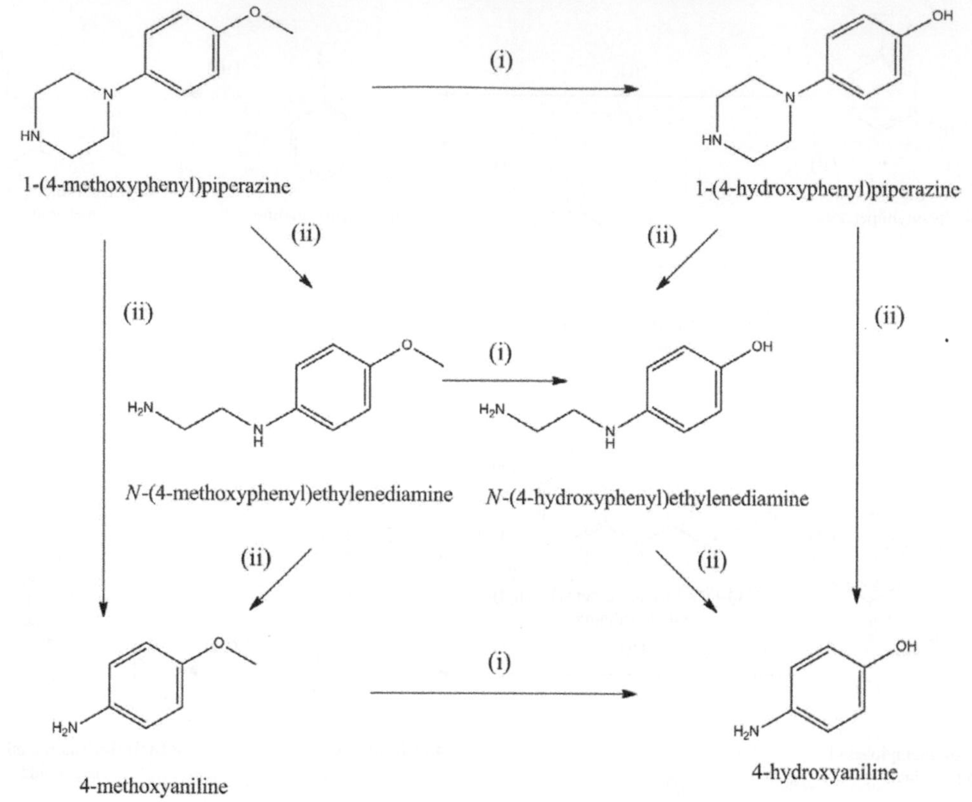

FIGURE 8.6   A summary of the proposed phase I metabolic pathway of MeOPP in rats [48] (i) Aromatic hydroxylation, (ii) N-dealkylation. Both 4-methoxyaniline and 4-hydroxyaniline can also undergo possible acetylation.

accumbens in rats demonstrated the parallel rise of dopamine and 5-HT following administration of BZP, but dopaminergic effects were more predominant [31].

In vitro studies have additionally shown BZP may also increase the release of norepinephrine in peripheral sympathetic nerve fibres [67]. Serial dilutions of BZP were applied to peripheral sympathetic nerve fibres (rabbit pulmonary artery preparations). Resting and stimulation evoked release of noradrenaline was significantly potentiated over a BZP concentration range of 10^{-7}M to 10^{-4}M, This occurred in a dose-dependent fashion. Elevations of smooth-muscle tone were also observed [67].

BZP acts as a locomotor stimulant producing hyperlocomotion and stereotypic behaviours [68]. Reinforcing effects were found in the place preference test with rats [63]. BZP given to adolescent rats produced higher levels of anxiety-like behaviour [69]. Further behavioural research is discussed below in the Dependence and Abuse section.

TFMPP

In contrast to BZP, TFMPP does not possess dopaminergic or adrenergic properties; [70] rather, it has non-selective agonist activity upon the 5-HT receptor classes [71,72], and also prevents the re-uptake of serotonin [73].

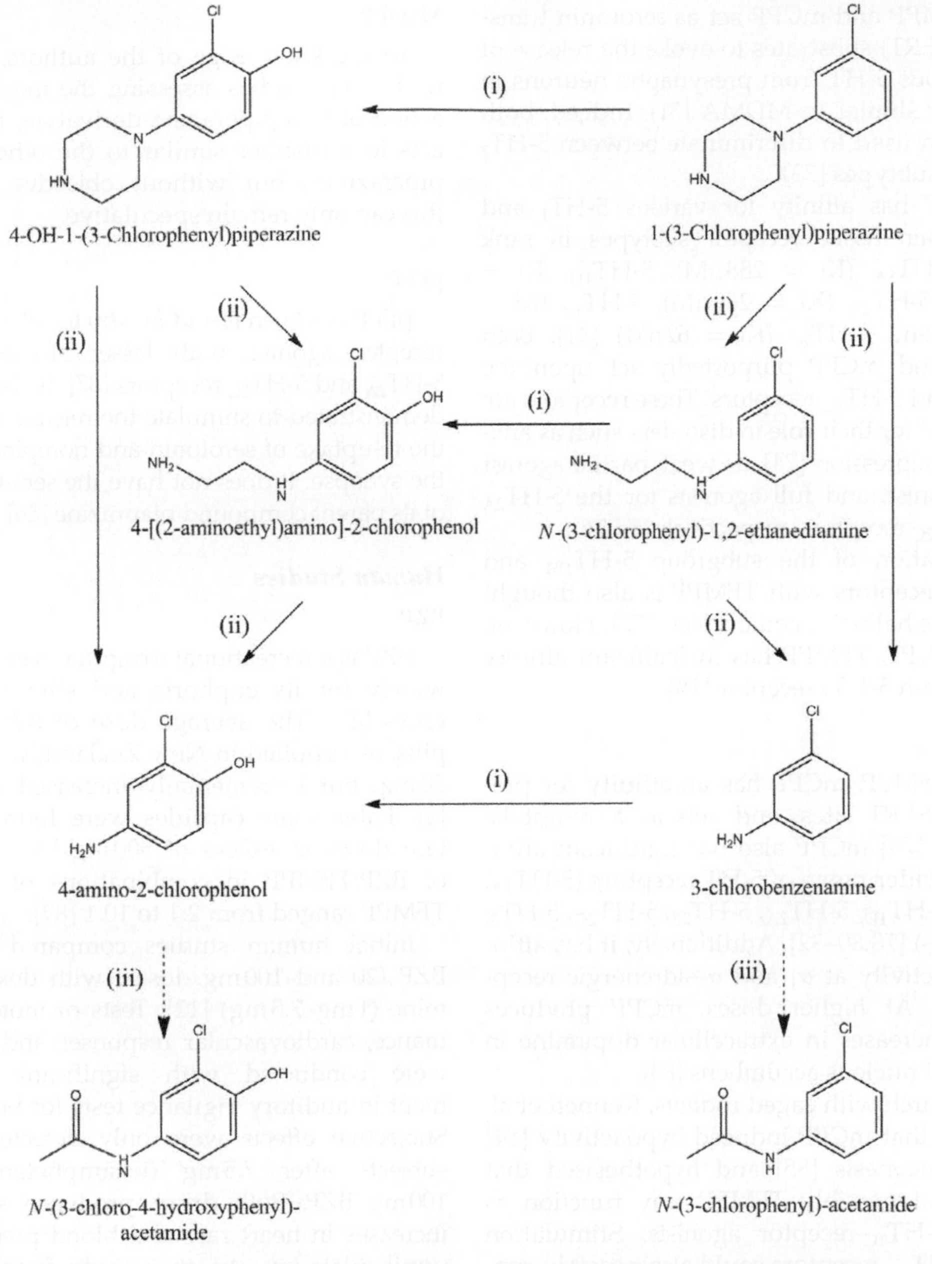

FIGURE 8.7 A summary of the proposed phase I metabolic pathway of mCPP in rats [52,53] (i) aromatic hydroxylation, (ii) N-dealkylation and (iii) N-acetylation.

Both TFMPP and mCPP act as serotonin transporter (SERT) substrates to evoke the release of endogenous 5-HT from presynaptic neurons in a manner similar to MDMA [31]. Indeed, both have been used to discriminate between 5-HT$_2$ receptor subtypes [73].

TFMPP has affinity for various 5-HT$_1$ and 5-HT$_2$ brain tissue receptor subtypes in rank order: 5-HT$_{1A}$ (Ki = 288nM), 5-HT$_{1D}$ (Ki = 282nM), 5-HT$_{2A}$ (Ki = 269nM), 5-HT$_{1B}$ (Ki = 132nM), and 5-HT$_{2C}$ (Ki = 62nM) [31]. Both TFMPP and mCPP purportedly act upon the 5-HT$_{2A}$ and 5-HT$_{2C}$ receptors. These receptors are recognised for their role in disorders such as anxiety and depression [74], as weak partial agonist or antagonist, and full agonists for the 5-HT$_{2A}$ and 5-HT$_{2C}$ receptors, respectively [75,76].

Stimulation of the subgroup 5-HT$_{1B1}$ and 5-HT$_{1B2}$ receptors with TFMPP is also thought to mediate hallucinogenic effects [77]. However, unlike mCPP, TFMPP has insignificant affinity for the brain 5-HT$_3$ receptor [78].

mCPP

Like TFMPP, mCPP has an affinity for presynaptic SERT sites, and acts as a re-uptake inhibitor; [79] mCPP also has significant affinity for a wider range of 5-HT receptors (5-HT$_{1A}$, 5-HT$_{1B}$, 5-HT$_{1D}$, 5-HT$_{2A}$, 5-HT$_{2B}$, 5-HT$_{2C}$, 5-HT$_3$, and 5-HT$_7$) [78,80–82]. Additionally, it has affinity and activity at α_1 and α_2-adrenergic receptors [82]. At higher doses, mCPP produces modest increases in extracellular dopamine in the rodent nucleus accumbens [83].

In research with caged rodents, Kennett et al. observed that mCPP induced hypoactivity [84] and anxiogenesis [85] and hypothesised that mCPP and possibly TFMPP may function as central 5-HT$_{1C}$-receptor agonists. Stimulation of the 5-HT$_{2A}$ receptors could also possibly contribute to depression-like behaviour. Increases in endogenous adrenocorticotropic hormone, cortisol, growth hormone and prolactin concentrations have also been shown to occur following the administration of mCPP [43,82,86].

MeOPP

To the knowledge of the authors, there are no known studies assessing the mechanism of action of this piperazine derivative; it possibly acts in a manner similar to the other phenylpiperazines, but without objective evidence, this can only remain speculative.

pFPP

pFPP has been found *in vitro* to act as a 5-HT$_{1A}$ receptor agonist, with lesser affinity for the 5-HT$_{2A}$ and 5-HT$_{2C}$ receptors [87]. It also has been demonstrated to stimulate the release and inhibit the re-uptake of serotonin and norepinephrine in the synapse. It does not have the sedative effects of its parent compound niaprazine [26].

Human Studies

BZP

BZP is a recreational drug that has been used widely for its euphoric and stimulant properties [88]. The average dose of BZP in party pills, as supplied in New Zealand, was initially 50mg, but incrementally increased to 250mg [4]. Later some capsules were found to contain doses in excess of 500mg [36]. The ratio of BZP:TFMPP in combinations of BZP and TFMPP ranged from 2:1 to 10:1 [89].

Initial human studies compared doses of BZP (20 and 100mg doses) with dexamphetamine (1mg–7.5mg) [12]. Tests of motor performance, cardiovascular responses and vigilance were conducted with significant improvement in auditory vigilance tests for both drugs. Subjective effects were only detected by the subjects after 7.5mg dexamphetamine and 100mg BZP. Both drugs produced significant increases in heart rate and blood pressure and pupil dilatation. Another study found a trend towards QT prolongation in 160 users but no recorded arrhythmias [36]. The study group of former amphetamine addicts were unable to distinguish between 100mg BZP and 7.5mg d-amphetamine [90].

TFMPP

Dexamphetamine-like effects, dysphoria, confusion and feelings of tension were recorded from 30 volunteers given 60 mg TFMPP [91]. Such symptoms, as noted by the authors, were consistent with other drugs that have significant serotonin activation such as mCPP and lysergic acid diethylamide. Volunteers in another investigation on the subjective and physiological effects of combined BZP/TFMPP reported dexamphetamine- and MDMA-like effects along with increased dysphoria and increased feelings of self-confidence [92]. They also experienced elevation of blood pressure and heart rate but no adverse effects were reported – this may have been due to the lower dose regime used.

mCPP

Despite having adrenoceptor affinity [82], mCPP does not appear to produce clinically significant sympathomimetic effects. There is, however, a risk of serotinergic overstimulation with reported clinical effects including anxiety, dizziness, hallucinations, nausea, warm and cold flushes, migraine and panic attacks [42,93,94]. This has tended to limit the use of mCPP as a recreational drug [8].

pFPP

The psychoactive effects of pFPP are likely to be due to $5-HT_{1A}$ agonism [87]. There are no peer reviewed papers reporting adverse effects associated with this piperazine derivative.

TOXICOLOGY

Animal Data

In mammals, the prodrug N-benzyl-piperazine-piconyl fumarate (EGYT-475) is metabolised to BZP. It was found to be toxic to beagle dogs at doses approaching 30 mg/kg and lethal at higher doses [16].

Baumann reported in his studies, comparing MDMA, BZP and TFMPP in rats, that animals exposed to doses of a 10 mg/kg combination of TFMPP and BZP developed seizures and ataxia [31]. Experiments at that dose were voluntarily stopped because of animal welfare concerns.

Human Data

User Reports

The majority of studies documenting the subjective effects of piperazine drugs have been following the use of BZP, TFMPP or mCPP. There has been little reported in scientific literature on the subjective effects of the other piperazine derivatives. The subjective effects reported following use of BZP, TFMPP and mCPP have been documented in clinical investigations using subjective ratings scales [88,91,92,95–98] and surveys or group discussions where recreational users were questioned about their experiences following use [99–103]. As the surveys are not controlled and are based on the self-reporting of symptoms they may not be particularly accurate. Users may or may not disclose the use of other coingestants. Additionally, following the recreational use of piperazine party drugs participants may have also experienced extended periods of exertion, lack of sleep or dehydration; these confounding factors may contribute to the effects displayed in some users, thereby potentially limiting the accuracy of survey questionnaires or group discussion data.

Three double blind and placebo controlled clinical studies investigations have been undertaken using BZP and TFMPP. The first investigated BZP in 27 healthy female volunteers (13 subjects given placebo, 14 subjects administered 200 mg BZP) [88], the second investigated TFMPP in 30 healthy males (15 in the placebo group, 15 administered 60 mg TFMPP) [91], while the third study investigated a combination of both BZP and TFMPP in 36 healthy male volunteers (16 in the placebo group, 20 administered 100 mg BZP/30 mg TFMPP combination) [92]. These studies used the addiction research centre inventory (ARCI) questionnaire, the profile of mood state (POMS) scales and visual analogue scales (VAS) to measure changes between, before and 120 minutes after drug or placebo administration. The results of these investigations showed BZP administered singularly produced statistically significant changes in a number of the ratings; the ARCI questionnaire showed increased scores in euphoria, dysphoria and dexamphetamine-like effects. The POMS scales showed decreases in depression/dejection, fatigue/inertia and confusion/bewilderment, and increases in vigour/activity and total mood disturbance. The results from VAS included increases in categories of drug effect, drug liking, being stimulated, high, anxious, talkative, and self-confident [88]. TFMPP administered by itself also showed changes in the ARCI, including increases in dysphoria and dexamphetamine-like effects. The POMS scale showed an increase in tension/anxiety while the VAS scores included increases in drug liking, and being high, and stimulated [91]. When BZP was given in combination with TFMPP, the ARCI questionnaire showed similar effects as when the drugs were administered individually, including increases in dysphoria and dexamphetamine-like effects. The only POMS scale to increase when subjects were given both drugs simultaneously was vigour/activity while the changes in the VAS included an increase in drug effect, good drug effect, drug liking, stimulation and self-confidence when the combination was administered [92].

mCPP is recognised as an effective serotonin agonist, and has therefore been used extensively as a clinical tool for investigating mechanisms of anxiety and other psychiatric disorders in humans [80]. A detailed review of these trials is, unfortunately, beyond the scope of this chapter, but a brief summary of reported subjective symptoms from a number of these papers are presented. Typically these studies have investigated the neuroendrocrine, physiological, and/or behavioural effects of mCPP, both in healthy volunteers and those suffering various psychiatric disorders. One study in 15 healthy volunteers given an oral 0.5 mg/kg dose of mCPP showed only a limited effect on mood with one-third to half of the participants not showing any changes [97]. There was, however, a small but significant increase in ratings on the activation-euphoria and anxiety subscales. Administered oral doses of 1.0 mg/kg mCPP, or greater, produced transient nausea and vomiting in 6 of 13 subjects [97]. Another study in which 0.1 mg/kg mCPP was administered both to healthy volunteers (n = 19) and patients suffering agrophobia and panic disorder (n = 23) showed a number of differences from baseline in the VAS scores in the healthy volunteer group [95]. There were significant increases in the drowsy, anxious, nervous and high categories with decreases in happy, calm, and energetic categories. A study of 56 healthy volunteers administered mCPP either orally (0.5 mg/kg) or IV (0.1 mg/kg) showed a trend towards increases in anxiety, activation-euphoria and functional deficit scales following oral mCPP administration and increases in anxiety, altered self-reality, functional deficit, depression, and dysphoria in the IV group [98]. In a further study, 10 healthy elderly volunteers and 12 patients with Alzheimer's disease were administered 0.1 mg/kg mCPP; the healthy volunteer group displayed significant increases in anxiety, functional deficit and dysphoria [96]. Schizophrenic patients experienced an exacerbation of psychopathology following the oral administration of 0.25 mg/kg mCPP [104].

Serotonin challenges using mCPP have also been performed on untreated patients with obsessive–compulsive disorder (OCD), causing marked though transient elevations of obsessive–compulsive symptoms [61,105–109].

In one study, for example, patients suffered an increase in subjective anxiety following both oral (0.5 mg/kg) and IV (0.1 mg/kg) mCPP administration [86].

Questionnaires and interviewing recreational users have been used to determine the subjective effects of BZP and TFMPP [101–103]. A telephone survey investigating their use was conducted in New Zealand in 2006. The survey sampled 2010 people aged between 13 and 45 years with 20% confirming they had previously used piperazine-based party pills. The results of this survey have shown a number of people reported adverse subjective effects following use, including strange thoughts (15.6% of subjects), mood swings (14.8%), confusion (12.1%), irritability (11.4%), short temper (10.9%), anxiety (10.0%), loss of sex urge (9.5%), visual hallucinations (8.8%), paranoia (8.4%), depression (8.2%), auditory hallucinations (6.9%), flashbacks (5.0%), panic attacks (3.2%), feelings of aggression (1.2%), and suicidal thoughts (0.8%) [101].

Another study in which subjective effects were documented in recreational users was conducted in New Zealand in 2006. It involved 58 people between the ages of 17 and 23 who were interviewed or participated in group discussions regarding their experiences with the use of these drugs. Some respondents documented both positive and negative subjective effects. The perceived positive effects included feeling confident, relaxed and wakeful with positive sensory and mind altering effects. Reported negative effects included feelings of tension, agitation, anxiety, and paranoia; the numbers of individuals experiencing these effects was not specified [99]. A further questionnaire was undertaken in patients who presented to a New Zealand emergency department; out of 1043 people who completed the survey, 125 reported using party pills in the past. The major subjective effects documented from this survey include extreme alertness and agitation [100].

Clinical Reports

Amongst the recreational piperazine derivatives, most of the available literature on their clinical effects is associated with BZP intoxication. Signs and symptoms following exposure to these derivatives are predominantly adrenergic, characteristic of sympathomimetic excess [110]. A full summary of the reported clinical effects of BZP is presented in Table 8.2.

The relationship between doses ingested and clinical effects appears to be limited. Recreational doses of BZP typically 'recommended' by the respective party drug distributors, have ranged from 75 to 250 mg [4] but some capsules on the market exceeded 500 mg [36]. Self-reported doses (by number of pills ingested) did not correlate with intensity of reported clinical signs and symptoms.

The presentation and severity of symptoms can also be exacerbated by the patient's co-ingestion of other substances, most typically alcohol but also cannabis, MDMA and other stimulants [36,99,100,103,112,115]. BZP and TFMPP are often ingested in combination, with the a risk of additive or synergistic toxicity. As measured by prevalence, females appear more susceptible to adverse effects from recreational use of BZP [36]. However, males predominate in cases of severe toxicity [111,117,120]. Dose responses may also vary widely due to inter-individual pharmacokinetic variations, such as CYP450-associated hepatic metabolism, that may be subject to genetic polymorphism [121].

The predominant symptoms experienced by 61 patients in a 5-month New Zealand ED-based study on BZP-related presentations were palpitations, vomiting and agitation [122]. Fifteen had toxic grand mal type seizures; most were self-limiting but two required intensive care for airway compromise and severe metabolic acidosis (pH 6.87 and 6.67). In contrast, a similar survey from another major centre in New Zealand only reported 26 cases over

TABLE 8.2 Summary of the Commonly Reported Adverse Effects Following Exposure to Benzylpiperazine (BZP) and Trifluoromtheylpiperazine (TFMPP)

Sign and Symptom	Reference	Sign and Symptom	Reference
Autonomic		**Neurological**	
Fever	[100,111]	Seizure	[36,100,101,111–113]
Sweating	[101,112]	Hallucination	[101,103]
Facial flushing	[12]	Psychosis	[114]
Nausea	[99,101,103,112,115,116]	Confusion	[36,91,101,111,112,115]
Vomiting	[36,89,99,101,103,115,116]	Anxiety	[89,101,112,114–116]
Difficulty urinating	[36,99,101,112]	Agitation	[36,99,100,112,115,116]
Mydriasis	[12,111,113,116]	Headache	[36,89,99,101,103,112,115]
		Insomnia	[36,99–101,103,112,115]
Neuromuscular		Dysphoria	[88,91,92]
Involuntary movement	[112,115]	Paranoia	[99,101,103]
Tremor	[36,89,99,101,112,115]	Paraesthesia	[101]
Bruxism	[116]	Decreased level of consciousness	[115]
Trismus	[112]		
Muscular/jointpains	[101]	**Other**	
		Severe hyperthermia	[111]
Renal		Hepatocellular injury	[111]
Loin pain	[117]	Disseminated intravascular coagulopathy	[111]
Proteinurea	[117]	Pruritis	[100]
Elevated creatinine	[117]	Acidosis (metabolic/respiratory)	[111,112]
Renal failure	[111,117,118]	Rhabdomyolosis	[111]
		Hypoglycaemia	[111]
		Collapse	[36,101,111–113,115]
		Chest pain	[36,100,101,103,112]
		Tachycardia	[88,92,99,111–113,116,119]
		Palpitations	[36,100,101,103,112,115]
		Hypertension	[88,92,111–113]
		Dyspnea	[100]
		Shortness of breath	[101]

(Continued)

TABLE 8.2 (Continued)

Sign and Symptom	Reference	Sign and Symptom	Reference
		Rapid breathing	[36,112]
		Dizziness	[36,99–101,103,112,115]
		Abdominal pain	[99,117]
		Vision problems	[101,118]

a 3-year interval (2002–2004) [115]. The latter study was conducted before New Zealand experienced a large increase in marketing and sales of BZP (during 2004). This may help explain the differing prevalence of adverse effects from the two centres.

A formal double blind placebo controlled study under laboratory conditions was conducted by Thompson and associates [89]. Thirty-five healthy volunteers were given either 300 mg BZP or 74 mg TFMPP alone, BZP plus TFMPP and 57.6 g (6 units) alcohol, placebo and alcohol, or double placebo. The primary outcome variable was a measure of driving performance as tested in a driving simulator. Other adverse events were measured as a secondary outcome. The study was terminated prematurely, short of the planned 64 subjects, because of severe adverse events experienced by the subjects; 41% of the participants who took the BZP/TFMPP combination reported agitation, anxiety, hallucinations, vomiting, insomnia and migraine. However, those that completed the study protocol demonstrated a significant improvement in their driving performance.

The largest study of BZP toxicity amongst recreational users to date [36] described 178 presentations to an Emergency Department after suffering adverse effects. Many of these symptoms persisted for more than 24 h after the BZP ingestion. Patients often co-ingested one (57%) or two other (13%) psychoactive substances. The most common co-ingestant was ethanol, with 95 patients (53.4%) admitting to taking this with BZP.

The age of patients in this study ranged from 15 to 42 years, with a median age of 19 years. Of the total, 53% were females and 47% were males. A higher proportion of females took BZP at a younger age than males. The average number of pills taken across both sexes was 3.9. Females took fewer pills than males on average, with 3.0 versus 5.0 taken respectively. BZP alone was taken by 30% of the patients.

Ninety-six patients had plasma BZP concentrations measured on admission, and these ranged from 0 to 6.29 μg/mL (mean 0.68 μg/mL). Logistic regression revealed a trend toward higher concentrations of BZP being associated with seizures. This was in the entire cohort of 96 patients whether or not they had co-ingested ethanol.

Seizures occurred in 33 patients (18.5% of presentations), typically within 2–6 hours of ingestion; two of these patients claimed to have used BZP intravenously. Most seizures terminated spontaneously followed by uncomplicated recoveries, though a few were prolonged or developed secondary complications. Two patients suffered severe sympathomimetic toxicity with extreme hyperthermia (core temperatures >40°C) culminating in reversible multi-organ failure. Two others had seizures associated with metabolic acidosis and two developed hyponatraemia (plasma sodium levels 118 and 125 μmol/L, respectively). There were no fatalities recorded during the study period. Three case reports are summarised in Box 8.1. Increasing plasma BZP concentrations were linked to a significant increase in

BOX 8.1

SELECTED CASE REPORTS [122] – SEIZURES

Patient 1

A 16-year-old female took 4 BZP tablets without alcohol. She suffered a tonic clonic seizure 2.5 h after her last tablet. Following further seizures at the emergency department, she was treated with diazepam (GCS 3/15 with intubation). Her initial vital signs were heart rate (HR) 149 b.p.m., blood pressure (BP) 70/55, temperature 36°C with blood glucose of 5.6 µmol/L. Following a further seizure, she developed a mixed metabolic and respiratory acidosis. She was transferred to the intensive care unit; extubation was possible 12 h later. (Laboratory analysis confirmed BZP and metabolites with no other recreational drugs. No apparent prolonged adverse effects were reported a week later.)

Patient 2

An 18-year-old female took an unknown quantity of BZP and subsequently had five seizures. She developed mixed metabolic and respiratory acidosis – pH 6.64. She was intubated and transferred to ICU but later recovered with no apparent long-term effects. Laboratory analysis confirmed the presence of BZP alone in the urine.

Patient 3

A 25-year-old male took two tablets with alcohol and a further two the following morning; he suffered a tonic seizure 3 h after the last tablet whilst driving a car. His HR was 170 b.p.m., BP 148/75, blood glucose 5.4 µmol/L. He was drowsy upon admission but had an uncomplicated recovery. Laboratory analysis of urine showed BZP metabolites and alcohol only.

the chances of having a seizure within the group that ingested BZP without co-ingestants. Logistic regression showed that increasing the plasma BZP concentration by 0.50 µg/mL produced an odds ratio of 1.80 (95% confidence interval 1.18–3.74). Although the correlation between elevated BZP plasma concentration and increased risk of seizures has not been demonstrated in other human studies, it has been reported in rats [30]. Several patients who had seizures had comparatively low plasma BZP concentrations. One patient had a seizure with a plasma BZP concentration of 0.15 µg/mL. By comparison, another patient who did not have a seizure had a plasma concentration of 6.29 µg/mL (over 40 times greater than that of the other patient). Delay in blood sampling may account for only some of this discrepancy.

With those that took ethanol and BZP, there was no association between BZP concentration and seizure risk (odds ratio was 1.00 with 95% CI 0.43–1.42). This unexpected finding suggests that ethanol may reduce (but not eliminate) the proconvulsant effect of BZP. Receptor studies show that ethanol is a CNS depressant, an effect mediated by reducing excitatory glutaminergic transmission as well as enhancing depressant gamma-aminobutyric acid

BOX 8.2

SELECTED CASE REPORTS [111] – SEROTONIN TOXICITY (SYNDROME)

Patient 1

A 19-year-old female took an unknown amount of BZP and was later arrested for intoxication. She had a seizure while in custody. She was transferred to hospital where she developed status epilepticus, hyperthermia, disseminated intravascular coagulation, rhabdomyolysis and renal failure. In hospital she was diaphoretic, her pupils were dilated at 7mm; she had dry mucous membranes and had no audible bowel sounds. Her temperature peaked at 40.2°C. Between ictal activity she had 3–4 beats of inducible ankle clonus raising the suspicion of serotinergic toxicity. Intravenous benzodiazepines did not stop seizure activity and she was intubated with thiopentone, fentanyl and suxamethonium. She had a significant metabolic acidosis with pH 7.14, PCO_2 40mmHg, PO_2 200mmHg, HCO_3 13.4µmol/L, base excess of 15 (on 100% oxygen).

An electrocardiogram showed a sinus tachycardia of 110 with QT interval of 413msec and QTc 559msec. A computed tomograph of the head showed no brain lesion or haemorrhage. Biochemistry showed rhabdomyolysis and an acute renal injury (creatinine kinase 27601IU/L, creatinine 0.27µmol/L, urea 16.5µmol/L). She developed disseminated intravascular coagulation with INR 2.9, APTT 82s, fibrinogen 0.5g/L, platelets 23×109/L and D-dimer >1000µg/L). Her creatinine peaked at 0.42µmol/L (normal range 0.05–0.11µmol/L) five days post-admission and later normalised without need for renal replacement therapy. A urine immunoassay screen revealed a strong colour change for amphetamine-like type compounds (cross-reaction from BZP). Her plasma concentration of BZP was 0.20mg/L, measured 10h after being placed in custody. Her exact time of ingestion was unknown. Benztropine metabolites (patient's medication), nicotine and caffeine were additionally detected in the urine.

The patient was haemodynamically unstable and showed no neurological response for seven days. Some improvement was then observed with flexion to painful stimuli. She remained febrile and tachycardic with a labile blood pressure for over 10 days, with intermittent muscle rigidity. She was unresponsive to boluses of benzodiazepines and orphenadrine. Clonidine was required to control surges of hypertension.

A total of 30 days hospitalization was required, including 11 days in the intensive care unit. After five months she underwent psychometric testing which showed significant cognitive disabilities with cognition and memory. The development of epilepsy prompted an MRI brain scan which revealed focal areas of encephalopathy.

Patient 2

A 22-year-old male who developed a similar pattern of toxicity from the combined use of BZP and MDMA. Concentrations measured in a blood sample collected 3h after admission found a BZP concentration of 2.23mg/L and an MDMA concentration of 1.05mg/L. A repeat BZP blood concentration 6h later was 0.1mg/L. He had a similar pattern of toxicity with hyperthermia (41.4°C), metabolic acidosis, rhabdomyolysis, hepatic injury and renal failure. His intensive care stay was longer due to respiratory complications but he made a good functional recovery. The contribution of MDMA was difficult to estimate in this case. Measured plasma levels do not correlate well with clinical toxicity; 1.05mg/L is within the range considered neurotoxic. Near fatal and fatal ingestion has been reported with levels between 0.11mg/L and 2.1mg/L. Survival has been reported after MDMA blood levels of 4.3mg/L [123].

TABLE 8.3 Effect of alcohol co-ingestion on the frequency of common symptoms experienced with BZP[36]

	BZP Ingestion only (%)	BZP & Alcohol Ingestion (%)	Relative Risk (CI)	Effect of Alcohol Co-ingestion
Confusion	16.7	26	1.56 (0.77–3.18)	Increased
Palpitation	32.4	31.5	1.55 (0.83–2.89)	No change
Agitation	29.6	38.4	1.29 (0.78–2.14)	Increased
Anxiety	38.4	40.7	0.94 (0.61–1.45)	No change
Vomiting	38.9	26	0.67 (0.40–1.12)	Decreased
Seizures	29.6	13.7	0.46 (0.23–0.94)	Decreased

(GABA)-A receptor activity [124,125]. Despite the possibility of offering some protection against seizures, co-ingestion of ethanol did lead to an increase in other BZP-related symptoms including anxiety, agitation, confusion and vomiting. A summary of clinical effects experienced by patients following BZP ingestion with and without ethanol is presented in Table 8.3.

Seven patients attending an Accident and Emergency Department in London after purportedly taking ecstasy had tablets subsequently identified as BZP. Two suffered grand mal seizures. Serum samples were analyzed in four of the patients, and BZP concentrations of 1.3, 1.9, 1.9 and 2.5 mg/L were measured [113,126].

A potential risk following the co-ingestion of TFMPP with BZP, and to a lesser extent BZP alone, is serotonin toxicity (syndrome) [111,116]. This syndrome has also been reported with mCPP ingestion [127]. Features typical of this syndrome include hyperthermia, autonomic instability and neuromuscular effects (such as tremor, rigidity, inducible clonus) [128,129]. Two case reports are summarised in Box 8.2.

Hyponatraemia does not appear to be common following BZP/TFMPP ingestion but may occur more commonly when BZP is taken with MDMA. This was first reported in the fatal case of a patient combining use of MDMA with BZP [130]. Three further cases arising from BZP use alone have been reported [36,122]. Acute hyponatremia is a rare but well recognised complication of MDMA use [131] and is possibly caused by the stimulated release of antidiuretic hormone and the excessive rehydration with water due to the profuse sweating and pyrexia. These mechanisms may also be at work with BZP. The three non-fatal cases recovered with fluid restriction; one had confusion but none had objective evidence of cerebral oedema.

As with other amphetamine-type stimulants, renal failure has been reported with BZP. Alansari and Hamilton reported that a 17-year-old male developed acute renal failure after consuming five BZP tablets with alcohol [117]. In the absence of rhabdomyolosis, the cause was postulated to be BZP-related arteriospasm. The patient received dialysis until intrinsic renal function recovered. There are also suggestions that either BZP or other unintended contaminants within the formulation may have a specific toxic effect producing acute renal injury [132,133]. Two additional cases of renal failure have also been reported, but these were due to complications associated with rhabdomyolosis [111]. The patients also developed multi-organ failure as a result of BZP-induced hyperthermia. Both had careful fluid management and regained normal renal function without requiring renal replacement therapy.

Adrenergic stimulants are also known to induce adverse psychiatric effects such as intense agitation and psychosis [134]. There are few reports of PZD-related psychosis to date. One case was reported of a 20-year-old man who developed a brief psychotic episode 12 h following the ingestion of BZP and TFMPP co-administered with cannabis and nitrous oxide [114]. He displayed persecutory delusional beliefs along with auditory and visual hallucinations, leading him to set fire to his residence. The man had no prior psychiatric history. In another case, Mohandas and Vecchio reported the precipitation of a first episode of mania in a patient with schizophrenia who started using BZP [135].

Methamphetamine is known to induce transient and prolonged psychosis by dysfunction of dopaminergic transmission; BZP may induce similar effects. Recent work has suggested that amphetamine users with low normal concentrations of catechol-O-methyl transferase (COMT) are more likely to develop both forms of induced psychosis [136].

Amongst the other piperazine derivatives, mCPP is the most widely used, accounting for up to 10% of illicit tablets sold in the EU as illicit ecstasy [29,137]. Despite its widespread use, there is only one case report documenting adverse effects. A 29-year-old woman developed anxiety, agitation, drowsiness, flushing, visual disturbances and tachycardia following the ingestion of three tablets [138]. mCPP was detected in plasma and urine at concentrations of 320 ng/mL and 2300 ng/mL respectively. Amphetamine (40 mg/mL), benzoylecgonine (47 ng/mL) and alcohol (0.7 g/L) were also detected. A remaining tablet was analyzed and found to contain 30 mg of mCPP [138].

There are no known published reports in the peer reviewed literature on the effects on humans (adverse or other) following use of MDBZP, pFPP or MeOPP.

TREATMENT

Management

Stabilization and Supportive Care

Most users presenting with adverse effects will have mild symptoms only. Agitation palpitations and anxiety usually resolve spontaneously or with a single dose of benzodiazepine [36,111]. An observation period of 4–6 hours from time of ingestion is recommended. Patients with significant toxicity may present with seizures, decreased level of consciousness or cardiovascular instability. Termination of seizures and airway control are treatment priorities. Intubation may be required to control the airway of obtunded patients. A core temperature above 38.5°C is a sign that serious sympathomimetic toxicity may be developing (discussed further below).

Gastrointestinal Decontamination

There is no evidence for the benefit of gastrointestinal decontamination following ingestion of piperazines. It is contentious whether gastric lavage or activated charcoal can confer any clinical benefit in most poisonings [139]. The vast majority of the piperazine poisonings have been sub-lethal and most patients typically recover with supportive measures only. There is also the additional risk of aspiration associated with decreased level of consciousness and seizures. Routine decontamination is therefore not recommended.

Agitation, Paranoia and Psychosis

Minor agitation can be managed by providing reassurance in a non-stimulating environment. If sedation is required, incremental doses of a benzodiazepine, such as midazolam, diazepam or lorazepam, are safe and effective

[36,111]. Severe agitation and psychosis may necessitate physical restraint and sedation with higher doses of benzodiazepines.

Seizures

Seizures usually occur within two hours post-ingestion, though they can be delayed for up to six hours [36,122]. Most are self-terminating and require temporary airway support, oxygen and a blood glucose assessment. Benzodiazepines are the first line of treatment if seizures are prolonged. A second benzodiazepine may be required if seizures recur. Persistent seizure activity may necessitate loading the patient with phenobarbitone (20 mg/kg) infused at a rate of not more than 100 mg/min [140].

Cardiovascular Effects

Recreational doses of BZP and other piperazine derivatives have been shown to cause tachycardia and mild elevation of blood pressure [12,13,92]. Patients presenting with only these symptoms should not typically require pharmaceutical intervention; however, they will respond to benzodiazepines that may have been given for other reasons. Unlike cases involving other adrenergic agents such as methamphetamine [134], there have been no recorded incidents of significant arrhythmias following the ingestion of piperazine derivatives. An initial electrocardiogram is recommended. If it is normal then further cardiac monitoring should not be required.

Severely ill patients with multisystem toxicity may have marked autonomic instability. Extreme hypertension and hypotension have been observed in these cases and will require invasive monitoring and intensive care management.

Hyperthermia and Serotonin Toxicity

Hyperthermia has been reported, typically in those suffering from severe toxicity [111], and is thought to result from disturbance of central thermoregulatory systems and increased muscle activity (such as seizures, clonus, rigidity and fasiculations) [141]. The combination of BZP and TFMPP has serotonergic effects, which may be exacerbated by other therapeutic or recreational serotonergic drugs, leading to an increased risk of hyperthermia and/or sertonin toxicity. A core temperature above 38.5°C may herald the onset of severe stimulant toxicity. This may progress to severe hyperthermia and multi-organ failure [142].

Initial treatment for hyperthermia and serotonin toxicity includes the cessation of any serotonergic drugs (including therapeutic drugs).

Control of agitation with benzodiazepines will reduce muscle activity and fluid replacement will correct dehydration. Active external cooling may be required and can be achieved with cooled intravenous fluids, ice packs or baths [143]. Refractory hyperthermia will additionally require neuromuscular paralysis with associated ventilatory support [143]. Those suffering hyperthermia must also be investigated for evidence of rhabdomyolysis, metabolic acidosis, hyponatraemia and renal failure. There is only one report of cyproheptadine used for BZP induced hyperthermia and no clinical effect was noted; the patient survived [111]. There are animal and human case reports of the successful treatment of serotonin-related hyperthermia with cyproheptadine and chlorpromazine but no trial data exists [144]. There is no data on the effectiveness of dantrolene in BZP toxicity. Controversy exists over the role and utility of dantrolene in severe stimulant and serotonin toxicity [145,146].

Electrolyte and Acid–Base Disturbances

Patients who develop hyponatraemia may present with confusion or convulsions. They will require careful fluid management. Electrolyte levels, renal function and osmolality should be monitored. Hyponatraemia (serum sodium below 135 μmol/L) will require an assessment of hydration and cerebral function for early signs of cerebral oedema. Restriction of oral fluid intake to 1000 mL per day will usually correct mild to moderate hyponatraemia (135–125 µmol/L). Clinically unwell patients, or those with evidence of cerebral oedema, may require more aggressive treatment with hypertonic saline, mannitol and diuretics. Correction of severe hyponatraemia (below 125 µmol/L) should be done slowly over 48–72 h in an intensive care setting [131,147].

Acidosis may develop secondary to seizures and/or inadequate tissue perfusion. Acid–base abnormalities should be treated by correcting the cause.

Rhabdomyolysis

Rhabdomyolysis may occur secondary to severe agitation, excessive muscular activity or hyperthermia [111,141]. Patients presenting with severe agitation should be monitored for rhabdomyolysis, including the assessment of serum creatine kinase (CK) and renal function; serum or urine myoglobin determinations can also be beneficial. Intravenous fluid administration is essential to maintain urine output. Some authors recommend urinary alkalinisation, however we would not recommend urinary alkalinisation in all patients with PZD-related rhabdomyolysis [148].

Renal Effects

Acute renal impairment or failure is not a common occurrence. It may develop, secondary to complications such as rhabdomyolysis, hyperthermia,or circulatory collapse [111,117]. A 17-year-old male developed bilateral loin pain after BZP use and was found to be in acute renal failure (serum creatinine 440 µmol/L, urea 10.6 mmol/L, sodium 140 mmol/L, potassium 4.5 mmol/L) without other signs of stimulant toxicity [117].

Urine output, serum urea and creatinine concentrations must therefore be monitored in patients with severe toxicity or unexplained back pain. Treatment should be directed at the underlying cause. Acute renal failure requires careful fluid management and also may include haemodialysis or haemofiltration using conventional indications.

Urinary retention has also been reported following use of BZP [36]. Physical examination and ultrasonography is required in those with a distended bladder and/or suffering suprapubic pain. Temporary catheterisation may be required to decompress the bladder.

Enhanced Elimination

There is no information on the usefulness of haemodialysis, haemoperfusion, haemodiafiltration, or haemofiltration to enhance removal of BZP and TFMPP or other PZDs.

Deaths

There has not been a confirmed death from direct piperazine derivative toxicity. A major problem in investigating the involvement of piperazines in fatalities is the lack of laboratory confirmation. Although numerous methods have been published, to date piperazine derivatives are not usually included in routine or specific toxicological analysis. Reference standards for these piperazines are not readily available for the calibration of mass spectrometers. A summary of fatalities where piperazine derivatives, notably BZP and TFMPP, were detected in the decedent's blood and urine is detailed in Table 8.4. There were a total of 26 cases, with ages ranging from 17 to 48 years (median age 30 years); the age of the deceased was not recorded for one case. All but one of the deaths involved a male.

BZP in combination with MDMA was associated with three fatalities; two in Sweden and one in Switzerland [4,130]. Wikstrom described two fatalities where BZP and TFMPP were detected in post-mortem blood; [4] additionally, MDMA and Δ^9-tetrahydrocannabinol (THC) were also detected. The role of BZP alone or in combination was not defined. Balmelli published a report of a 23-year-old female who ingested BZP followed by MDMA and consumed a large volume of fluid [130]. She became comatose with evidence of hyponatraemia (sodium

TABLE 8.4   Summary of Reported Deaths Associated with Benzylpiperazine (BZP) and Trifluoromtheylpiperazine (TFMPP) Intoxication

Alterative Cause of Death Determined	Piperazine(s) Detected	Sex	Coingestants Detected	Reference
None	BZP, TFMPP	M	Alcohol	[146]
None	BZP, TFMPP	M	Paracetamol, promethazine, venlafaxine	[146]
None	BZP, TFMPP	M	Lignocaine	[146]
None	BZP, TFMPP	M	Cannabis, paracetamol, venlafaxine	[146]
None	BZP, TFMPP	M	Alcohol, cannabis, cocaine, diazepam, levamisole, lignocaine	[146]
None	BZP, TFMPP	M	Amitriptyline, cannabis, diazepam, ketamine, methadone, morphine, sildenafil	[146]
None	BZP, TFMPP	M	Amphetamine, lansoprazole	[146]
Unknown	BZP	M	Amphetamine, MDMA	[3]
MDMA	BZP	–	MDMA, MDA, cannabis	[3]
MDMA	BZP	F	Benzodiazepines, caffeine, cocaine, MDMA, nicotine	[131]
Methadone	BZP, TFMPP	M	Diazepam, methadone	[146]
Heroin	BZP, TFMPP	M	Benzodiazepines, morphine	[146]
Heroin	BZP, TFMPP	M	Morphine	[146]
Heroin	BZP, TFMPP	M	Alcohol, cannabis, cocaine, morphine, trimethoprim	[146]
Heroin	BZP, TFMPP	M	Alcohol, cocaine, levamisole, morphine	[146]
Heroin	BZP, TFMPP	M	Amitriptyline, cannabis, diazepam, olanzapine, morphine	[146]
Trauma suicide	BZP, TFMPP	M	Alcohol, cannabis, citalopram, diazepam, methadone	[146]
Trauma s	BZP, TFMPP	M	Alcohol, cannabis, cocaine, levamisole, MDMA	[146]
Trauma suicide	BZP, TFMPP	M	Alcohol, ketamine	[146]
Trauma suicide	BZP, TFMPP	M	Not available	[146]
Trauma RTA	BZP, TFMPP	M	Alcohol, methadone	[146]
Trauma RTA	BZP, TFMPP	M	Alcohol, ketamine, cannabis, cocaine, ephedrine	[122]
Trauma RTA	BZP	M	Amphetamine, cannabis, cocaine, diltiazem, ketamine, MDMA	[122]
Trauma fall	BZP, TFMPP	M	Alcohol	[122]
Trauma gunshot	BZP, TFMPP	M	Alcohol, cannabis	[147]

115 µmol/L) and subsequently died from cerebral oedema and tonsillar herniation. MDMA was measured and was likely the major contributor to the fatal outcome. Elliott [120,149] has published reports where BZP has been detected in post-mortem samples from 22 deceased patients. In almost all cases the presence of BZP and TFMPP was confirmed in the urine. In six fatalities, heroin or methadone were presumed to be the primary toxic cause of death. Traumatic causes of death were seen in five (motor vehicle accidents and a fall) and suicide was listed as the primary cause in a further five. In seven BZP and TFMPP were detected and there were no other obvious causes of death. The contribution of these piperazines in these cases was unclear due to lack of toxicity data and in some the presence of other illicit and prescription drugs.

Cognitive impairment or psychosis from BZP may be an additional factor in trauma or apparent suicidal deaths. It is difficult to determine the role of PZDs in these cases. In one example a man who threatened police with a weapon and was fatally shot. A coronial inquest was conducted over the death. Witnesses reported he was agitated and behaving irrationally with features consistent with a state of paranoid psychosis. He had no psychiatric history, but BZP and THC were detected in a post-mortem urine sample [150].

The involvement of the piperazine derivatives may also be under-reported, possibly due to the lack of procedures to assay these piperazines, or alternatively they may be detected but not identified as being piperazines due to cross-reactivity with immunoflourescent assays for amphetamines [151].

CHRONIC TOXICITY

Animal Data

There is little data on chronic or accumulation toxicity of BZP in animals. One study

has been conducted where BZP was administered to peri-adolescent male and female rats, to determine the effects it may have on the development of the adolescent brain. Animals showed a heightened anxiety, which may possibly be due to the interference with mechanisms linked to anxiety-associated forebrain development and maturation [69].

Human Data

Presently there is no information on the long-term consequences of human use of BZP and other piperazine derivatives. It is unknown whether long-term use of the drug can lead to dependence, tolerance, or further neurological or cardiovascular sequelae, as has been reported with amphetamines such as MDMA [152] or methamphetamine [134,153]. With methamphetamine, for example, use in humans can produce cognitive and mood deficits that continue long after use of the drug has ceased [134]. A number of MRI studies have shown structural damage in users who regularly take a range of different amphetamines [154]. Although BZP and related piperazines are structurally different to the amphetamines, they similarly act as adrenergic agonists and there is a possibility similar anatomical changes may occur following long-term abuse. Future research is required

DEPENDENCE AND ABUSE POTENTIAL

Animal Data

BZP, amphetamine and methamphetamine had been shown in rats to induce dopamine release within the nigrostriatal dopamine pathway [66], in part responsible for reward and addiction [155]. Other research has shown that elevated doses of BZP led to increased ambulation, stereotypical movements and sniffing, which was not demonstrated with TFMPP alone [31]. Further investigations concluded that the rewarding properties of BZP are possibly attributable to D_1 and 5-HT$_3$ receptor activity [63,156]. Interestingly, in rhesus monkeys TFMPP does not appear to reinforce self-administration and when used in combination with BZP reduces the self-administration rate associated with BZP alone [157]. Overall, these results suggest that BZP has the propensity for stimulant abuse and that repeated exposure can lead to sensitisation [158].

Human Data

Early research by Campbell and associates [13], studying the subjective effects of BZP in 18 former addicts noted its perceived similarity to dexamphetamine. They concluded that BZP necessitated statutory controls because of its potential for abuse.

User reports mention frequent adverse symptoms such as anxiety, nausea, vomiting, insomnia, headache and significant hangover effects [37,99,101]. There are also reports of the development of short to medium term drug tolerance. It has been suggested that these adverse effects may mitigate the development of addiction [101].

Dependency on BZP has also been investigated by use of surveys and self-reporting. Although some studies have shown that this form of investigation may not necessarily be reliable [159,160], it may provide a simple overview on addiction. Extended periods of exertion, lack of sleep, or dehydration can however contribute to symptoms displayed in some users [99].

Wilkins and associates conducted a survey of BZP users. The responses of many participants were very suggestive of regular use despite adverse health and social consequences [37]. These findings appear indicative of drug dependence. Participants reported adverse financial consequences (8%), decreased energy and vitality (19.3%), adverse health

consequences (14.6%) and adverse effects on personal relationships (4.0%) and work/study (2.9%). Respondents self-reported that they sometimes wished they could stop (14.5%) and they 'sometimes' felt their legal party pill use was out of control (9.7%). On applying the Short Dependency Scale screen for stimulant addiction, 2.2% of respondents met the criteria.

CONCLUSION

The piperazine derivatives include 1-benzylpiperazine (BZP), 1,3-trifluoromethylphenylpiperazine (TFMPP), 1-(3-chlorophenyl) piperazine (mCPP) and 1-(4-methoxyphenyl) piperazine (MeOPP). These novel psychoactive substances are classed as 'designer drugs'. They possess structural similarity to amphetamines. They are primarily stimulants and potentiate central dopamine, serotonin and noradrenaline neurotransmission. BZP has the greatest effect on dopamine release and is a moderately potent stimulant. It possesses rewarding stimulant properties which are tempered by adverse symptoms such as nausea and prolonged hangover effects. In contrast, TFMPP acts more directly as a serotonin agonist. Piperazine derivatives are increasingly being substituted for 3,4-methylenedioxymethamphetamine (MDMA), either as single constituents or more commonly as mixtures of piperazine derivatives. Toxicity is mostly due to excessive sympathomimetic effects; typical symptoms include palpitations, agitation, anxiety, dizziness, confusion, tremor, headache, mydriasis, insomnia and vomiting. Features of serotonergic toxicity may also be present. The most serious complications from recreational use are seizures, hyperthermia, renal failure, hyponatraemia and disseminated intravascular coagulation. There are case reports of precipitation or exacerbation of psychiatric illness. Based on animal studies and human reports piperazine derivatives appear to have mild to moderate potential for abuse and dependency.

Most users experience only mild adverse effects. They can usually be managed with observation and reassurance. The features of more severe toxicity include confusion, seizures, hyperthermia, cardiovascular instability, hyponatraemia or coma. Early recognition of severe toxicity and good supportive treatment are the key elements of effective management.

ACKNOWLEDGMENT

The authors wish to acknowledge Robin Slaughter.

REFERENCES

[1] DEA Office of domestic intelligence, Domestic strategic intelligence unit. BZP and TFMPP: chemicals used to mimic MDMA's effects. Microgram 2002;35: 123.

[2] Couch RAF, Moore G. Detection of 1-benzylpiperazine, a designer recreational drug marketed in New Zealand. Clin Exp Pharmacol Physiol 2004;31 A227–A227.

[3] de Boer D, Bosman IJ, Hidvégi E, Manzoni C, Benkö AA, dos Reys LJ, et al. Piperazine-like compounds: a new group of designer drugs-of-abuse on the European market. Forensic Sci Int 2001;121:47–56.

[4] Wikström M, Holmgren P, Ahlner J. A2 (N-benzylpiperazine) a new drug of abuse in Sweden. J Anal Toxicol 2004;28:67–70.

[5] Joint formulary committee. British national formulary 59. London: Pharmaceutical Press; 2010.

[6] Dessouky YM, Ismaiel SA. Colorimetric determination of piperazine in pharmaceutical formulations. Analyst 1974;99:482–6.

[7] King LA, Nutt D. Seizures in a night club. Lancet 2007;370:220.

[8] Nikolova I, Danchev N. Piperazine based substances of abuse: a new party pills on Bulgarian drug market. Biotechnol Biotechnol Equip 2008;22:652–5.

[9] Baltzly R, Buck SJ, Lorz E, Schön W. The Preparation of N-Mono-substituted and unsymmetrically disubstituted piperazines. J Am Chem Soc 1944;66:263–6.

[10] EMCDDA Report on the risk assessment of BZP in the framework of the council decision on new psychoactive substances. Lisbon: EMCDDA; 2009. Available: <http://www.emcdda.europa.eu/attachements.cfm/att_70975_EN_EMCDDA_risk_assessment_8.pdf> [accessed 01.03.13].

[11] Barrett PA, Caldwell AG, Walls LP. Treatment of depression with n-benzylpiperazine. United States: Search USPTO Assignment Database; 1968.

[12] Bye C, Munro-Faure AD, Peck AW, Young PA. A comparison of the effects of 1-benzylpiperazine and dexamphetamine on human performance tests. Eur J Clin Pharmacol 1973;6:163–9.

[13] Campbell H, Cline W, Evans M, Lloyd J, Peck AW. Comparison of effects of dexamphetamine and 1-benzylpiperazine in former addicts. Eur J Clin Pharmacol 1973;6:170–6.

[14] Shulgin A, Shulgin A. PiHKAL: a chemical love story. Berkley, CA: Transform; 1991.

[15] Feher L, Fonyo M, Magyar K. Computer analysis of the pharmacokinetic parameters of EGYT-475, a new antidepressant agent. Pol J Pharmacol Pharm 1987;39:167–71.

[16] Fonyo M, Guttmann A, Magyar K. Species differences in the in vitro metabolism of EGYT-475, a new antidepressant agent. Pol J Pharmacol Pharm 1987;39:129–34.

[17] Magyar K. Pharmacokinetic aspects of the mode of action of Egyt-475, a new antidepressant agent. Pol J Pharmacol Pharm 1987;39:107–12.

[18] Malomvolgyi B, Tothfalusi L, Tekes K, Magyar K. Comparison of serotonin agonistic and antagonistic activities of a new Antidepressant agent trelibet (Egyt-475) and its metabolite EGYT-2760 On isolated rat fundus. Acta Physiol Hung 1991;78:201–9.

[19] Szucks Z, Szentendrei T, Fekete MIK. The effect of EGYT-475 (trelibet) and its metabolites on the potassium-stimulated H-3 noradrenaline release from cortical slices of rat-brain. Pol J Pharmacol Pharm 1987;39:185–93.

[20] Kast RE. Trazodone generates m-CPP: in 2008 risks from m-CPP might outweigh benefits of trazodone. World J Biol Psychiatry 2009;10:682–5.

[21] Rotzinger S, Fang J, Baker GB. Trazodone is metabolized to m-chlorophenylpiperazine by CYP3A4 from human sources. Drug Metab Dispos 1998;26:572–5.

[22] Staack RF, Maurer HH. Studies on the metabolism and the toxicological analysis of the nootropic drug fipexide in rat urine using gas chromatography-mass spectrometry. J Chromatogr B Analyt Technol Biomed Life Sci 2004;804:337–43.

[23] Milne GWA. Drugs: synonyms and propertie. Aldershot, UK: Ashgate Publishing; 2002.

[24] Guy C, Blay N, Rousset H, Fardeau V, Ollagnier M. Fever caused by fipexide. Evaluation of the national pharmacovigilance survey. Therapie 1990;45:429–31.

[25] Durand F, Samuel D, Bernuau J, Saliba F, Pariente EA, Marion S, et al. Fipexide-induced fulminant hepatitis. Report of three cases with emergency liver transplantation. J Hepatol 1992;15:144–6.

[26] Keane PE, Strolin Benedetti M, Dow J. The effect of niaprazine on the turnover of 5-hydroxytryptamine in the rat brain. Neuropharmacology 1982;21:163–9.

[27] Division of pharmacovigilance. Core safety profile niazaprine. Oslo, Norway: The Norwegian Medicines Agency Available: <http://www.legemiddelverket.no/upload/Coresafetyprofiles/CSPniaprazine200903.pdf >; 2009 [accessed 01.03.13].

[28] Baron M, Elie M, Elie L. An analysis of legal highs: do they contain what it says on the tin?. Drug Test Anal 2011;3:576–81.

[29] Vogels N, Brunt TM, Rigter S, van Dijk P, Vervaeke H, Niesink RJ. Content of ecstasy in the Netherlands: 1993-2008. Addiction 2009;104:2057–66.

[30] Baumann MH, Clark RD, Budzynski AG, Partilla JS, Blough BE, Rothman RB. Effects of 'Legal X' piperazine analogs on dopamine and serotonin release in rat brain. Ann N Y Acad Sci 2004;1025:189–97.

[31] Baumann MH, Clark RD, Budzynski AG, Partilla JS, Blough BE, Rothman RB. N-substituted piperazines abused by humans mimic the molecular mechanism of 3,4-methylenedioxymethamphetamine (MDMA, or 'Ecstasy'). Neuropsychopharmacology 2005;30:550–60.

[32] Bossong MG, Brunt TM, Van Dijk JP, Rigter SM, Hoek J, Goldschmidt HM, et al. mCPP: an undesired addition to the ecstasy market. J Psychopharmacol 2010;24:1395–401.

[33] Antia U, Lee HS, Kydd RR, Tingle MD, Russell BR. Pharmacokinetics of 'party pill' drug N-benzylpiperazine (BZP) in healthy human participants. Forensic Sci Int 2009;186:63–7.

[34] Antia U, Tingle MD, Russell BR. In vivo interactions between BZP and TFMPP (party pill drugs). N Z Med J 2009;122:29–38.

[35] Antia U, Tingle MD, Russell BR. Validation of an LC-MS method for the detection and quantification of BZP and TFMPP and their hydroxylated metabolites in human plasma and its application to the pharmacokinetic study of TFMPP in humans. J Forensic Sci 2010;55:1311–8.

[36] Gee P, Gilbert M, Richardson S, Moore G, Paterson S, Graham P. Toxicity from the recreational use of 1-benzylpiperazine. Clin Toxicol (Phila) 2008;46:802–7.

[37] Wilkins C, Girling M, Sweetsur P, Huckle T, Huakau J. Legal party pill use in New Zealand: prevalence of use, availability, health harms and 'gateway effects' of benzylpiperazine (BZP) and triflourophenylmethyl-piperazine (TFMPP). Auckland: Centre for Social and Health Outcomes Research and Evaluation (SHORE); 2006. Available: <http://www.ndp.govt.nz/moh.nsf/pagescm/993/$File/legalpartypillusenz.pdf>; [accessed 01.03.13].

[38] Pipes of piperazines: experience with cannabis, BZP, TFMPP (ID 14942). 2004. Available: <http://www.erowid.org/experiences/exp.php?ID=6138014942>.

[39] Surprisingly close to perfection: experience with Piperazines – mCPP , MeOPP (ID 57182). Available: <http://www.erowid.org/experiences/exp.php?ID=57182>; 2006 [accessed 04.03.13].

[40] A burning nose: experience with piperazines – MeOPP (ID 61380). Available: <http://www.erowid.org/experiences/exp.php?ID=61380>; 2007 [accessed 04.03.13].

[41] Wormwood (hash)+Piperazines + Methylphenidate (rectally)+Leonotis leonorus – Exp. Available: <http://www.bluelight.ru/vb/threads/189411>; 2005 [accessed 04.03.13].

[42] Gijsman HJ, Van Gerven JM, Tieleman MC, Schoemaker RC, Pieters MS, Ferrari MD, et al. Pharmacokinetic and pharmacodynamic profile of oral and intravenous meta-chlorophenylpiperazine in healthy volunteers. J Clin Psychopharmacol 1998;18:289–95.

[43] Feuchtl A, Bagli M, Stephan R, Frahnert C, Kolsch H, Kuhn KU, et al. Pharmacokinetics of m-chlorophenylpiperazine after intravenous and oral administration in healthy male volunteers: implication for the pharmacodynamic profile. Pharmacopsychiatry 2004;37:180–8.

[44] Caccia S, Fong MH, Urso R. Ionization constants and partition coefficients of 1-arylpiperazine derivatives. J Pharm Pharmacol 1985;37:567–70.

[45] Rotzinger S, Fang J, Coutts RT, Baker GB. Human CYP2D6 and metabolism of m-chlorophenylpiperazine. Biol Psychiatry 1998;44:1185–91.

[46] Staack RF, Fritschi G, Maurer HH. Studies on the metabolism and toxicological detection of the new designer drug N-benzylpiperazine in urine using gas chromatography-mass spectrometry. J Chromatogr B Analyt Technol Biomed Life Sci 2002;773:35–46.

[47] Staack RF, Paul LD, Springer D, Kraemer T, Maurer HH. Cytochrome P450 dependent metabolism of the new designer drug 1-(3-trifluoromethylphenyl)piperazine (TFMPP). In vivo studies in Wistar and Dark Agouti rats as well as in vitro studies in human liver microsomes. Biochem Pharmacol 2004;67:235–44.

[48] Staack RF, Theobald DS, Paul LD, Springer D, Kraemer T, Maurer HH. In vivo metabolism of the new designer drug 1-(4-methoxyphenyl)piperazine (MeOPP) in rat and identification of the human cytochrome P450 enzymes responsible for the major metabolic step. Xenobiotica 2004;34:179–92.

[49] Tsutsumi H, Katagi M, Miki A, Shima N, Kamata T, Nakajima K, et al. Metabolism and the urinary excretion profile of the recently scheduled designer drug

N-benzylpiperazine (BZP) in the rat. J Anal Toxicol 2006;30:38–43.

[50] Staack RF, Fritschi G, Maurer HH. New designer drug 1-(3-trifluoromethylphenyl) piperazine (TFMPP): gas chromatography/mass spectrometry and liquid chromatography/mass spectrometry studies on its phase I and II metabolism and on its toxicological detection in rat urine. J Mass Spectrom 2003;38:971–81.

[51] Tsutsumi H, Katagi M, Miki A, Shima N, Kamata T, Nakajima K, et al. Isolation, identification and excretion profile of the principal urinary metabolite of the recently banned designer drug 1-(3-trifluoromethylphenyl)piperazine (TFMPP) in rats. Xenobiotica 2005;35:107–16.

[52] Mayol RF, Cole CA, Colson KE, Kerns EH. Isolation and identification of the major urinary metabolite of m-chlorophenylpiperazine in the rat. Drug Metab Dispos 1994;22:171–4.

[53] Staack RF, Maurer HH. Piperazine-derived designer drug 1-(3-chlorophenyl)piperazine (mCPP): GC-MS studies on its metabolism and its toxicological detection in rat urine including analytical differentiation from its precursor drugs trazodone and nefazodone. J Anal Toxicol 2003;27:560–8.

[54] Szoko E, Kalasz H, Kerecsen L, Magyar K. Studies on serum binding of some drugs. Pol J Pharmacol Pharm 1987;39:177–83.

[55] Maurer HH, Kraemer T, Springer D, Staack RF. Chemistry, pharmacology, toxicology, and hepatic metabolism of designer drugs of the amphetamine (ecstasy), piperazine, and pyrrolidinophenone types– A synopsis. Ther Drug Monit 2004;26:127–31.

[56] Antia U, Tingle MD, Russell BR. Metabolic interactions with piperazine-based 'party pill' drugs. J Pharm Pharmacol 2009;61:877–82.

[57] Murphy M, Antia U, Chang HY, Han JY, Ibrahim U, Tingle MD, et al. Party pills and drug-drug interactions. N Z Med J 2009;122:3564.

[58] Sandson NB, Armstrong SC, Cozza KL. An overview of psychotropic drug-drug interactions. Psychosomatics 2005;46:464–94.

[59] Yasui N, Otani K, Kaneko S, Ohkubo T, Osanai T, Ishida M, et al. Inhibition of trazodone metabolism by thioridazine in humans. Ther Drug Monit 1995;17:333–5.

[60] Mihara K, Otani K, Ishida M, Yasui N, Suzuki A, Ohkubo T, et al. Increases in plasma concentration of m-chlorophenylpiperazine, but not trazodone, with low-dose haloperidol. Ther Drug Monit 1997;19:43–5.

[61] Hollander E, Decaria C, Gully R, Nitescu A, Suckow RF, Gorman JM, et al. Effects of chronic fluoxetine treatment on behavioral and neuroendocrine responses to meta-chlorophenylpiperazine

in obsessive-compulsive disorder. Psychiatry Res 1991;36:1–17.

[62] Staack RF, Paul LD, Schmid D, Roider G, Rolf B. Proof of a 1-(3-chlorophenyl)piperazine (mCPP) intake: use as adulterant of cocaine resulting in drug-drug interactions?. J Chromatogr B Analyt Technol Biomed Life Sci 2007;855:127–33.

[63] Meririnne E, Kajos M, Kankaanpaa A, Seppala T. Rewarding properties of 1-benzylpiperazine, a new drug of abuse, in rats. Basic Clin Pharmacol Toxicol 2006;98:346–50.

[64] Tekes K, Tothfalusi L, Malomvolgyi B, Herman F, Magyar K. Studies on the biochemical-mode of action of EGYT-475, a new antidepressant. Pol J Pharmacol Pharm 1987;39:203–11.

[65] Nagai F, Nonaka R, Kamimura KSH. The effects of non-medically used psychoactive drugs on monoamine neurotransmission in rat brain. Eur J Pharmacol 2007;559:132–7.

[66] Oberlander C, Euvrard C, Dumont C, Boissier JR. Circling behavior induced by dopamine releasers and-or uptake inhibitors during degeneration of the nigrostriatal pathway. Eur J Pharmacol 1979;60:163–70.

[67] Magyar K, Fekete MIK, Tekes K, Torok TL. The action of trelibet, a new antidepressive agent on [H-3] noradrenaline release from rabbit pulmonary artery. Eur J Pharmacol 1986;130:219–27.

[68] Yarosh HL, Katz EB, Coop A, Fantegrossi WE. MDMA-like behavioral effects of N-substituted piperazines in the mouse. Pharmacol Biochem Behav 2007;88:18–27.

[69] Aitchison LK, Hughes RN. Treatment of adolescent rats with 1-benzylpiperazine: a preliminary study of subsequent behavioral effects. Neurotoxicol Teratol 2006;28:453–8.

[70] Herndon JL, Pierson ME, Glennon RA. Mechanistic investigation of the stimulus properties of 1-(3-trifluoromethylphenyl)piperazine. Pharmacol Biochem Behav 1992;43:739–48.

[71] Cunningham KA, Appel JB. Possible 5-hydroxytryptamine 1 (5HT1) receptor involvement in the stimulus properties of 1-(m-trifluoromethylphenyl)piperazine (TFMPP). J Pharmacol Exp Ther 1986;237:369–77.

[72] Miranda F, Orozco G, Velazquez-Martinez DN. Full substitution of the discriminative cue of a 5-HT(1A/1B/2C) agonist with the combined administration of a 5-HT(1B/2C) and a 5-HT(1A) agonist. Behav Pharmacol 2002;13:303–11.

[73] Auerbach SB, Kamalakannan N, Rutter JJ. TFMPP and RU24969 enhance serotonin release from rat hippocampus. Eur J Pharmacol 1990;190:51–7.

[74] Berg KA, Harvey JA, Spampinato U, Clarke WP. Physiological and therapeutic relevance of

constitutive activity of 5-HT 2A and 5-HT 2C receptors for the treatment of depression. Prog Brain Res 2008;172:287–304.

[75] Conn PJ, Sanders-Bush E. Relative efficacies of piperazines at the phosphoinositide hydrolysis-linked serotonergic (5-HT-2 and 5-HT-1c) receptors. J Pharmacol Exp Ther 1987;242:552–777.

[76] Grotewiel MS, Chu H, Sanders-Bush E. m-chlorophenylpiperazine and m-trifluoromethylphenylpiperazine are partial agonists at cloned 5-HT2A receptors expressed in fibroblasts. J Pharmacol Exp Ther 1994;271:1122–6.

[77] Appel JB, Callahan PM. Involvement of 5-HT receptor subtypes in the discriminative stimulus properties of mescaline. Eur J Pharmacol 1989;159:41–6.

[78] Robertson DW, Bloomquist W, Wong DT, Cohen ML. mCPP but not TFMPP is an antagonist at cardiac 5HT3 receptors. Life Sci 1992;50:599–605.

[79] Pettibone DJ, Williams M. Serotonin-releasing effects of substituted piperazines in vitro. Biochem Pharmacol 1984;33:1531–5.

[80] Gatch MB. Discriminative stimulus effects of m-chlorophenylpiperazine as a model of the role of serotonin receptors in anxiety. Life Sci 2003;73:1347–67.

[81] Rajkumar R, Pandey DK, Mahesh R, Radha R. 1-(m-Chlorophenyl)piperazine induces depressogenic-like behaviour in rodents by stimulating the neuronal 5-HT(2A) receptors: proposal of a modified rodent antidepressant assay. Eur J Pharmacol 2009;608:32–41.

[82] Silverstone PH, Rue JE, Franklin M, Hallis K, Camplin G, Laver D, et al. The effects of administration of mCPP on psychological, cognitive, cardiovascular, hormonal and MHPG measurements in human volunteers. Int Clin Psychopharmacol 1994;9:173–8.

[83] Eriksson E, Engberg G, Bing O, Nissbrandt H. Effects of mCPP on the extracellular concentrations of serotonin and dopamine in rat brain. Neuropsychopharmacology 1999;20:287–96.

[84] Kennett GA, Curzon G. Evidence that mCPP may have behavioural effects mediated by central 5-HT1C receptors. Br J Pharmacol 1988;94:137–47.

[85] Kennett GA, Whitton P, Shah K, Curzon G. Anxiogenic-like effects of mCPP and TFMPP in animal models are opposed by 5-HT1C receptor antagonists. Eur J Pharmacol 1989;164:445–54.

[86] Goodman WK, McDougle CJ, Price LH, Barr LC, Hills OF, Caplik JF, et al. m-Chlorophenylpiperazine in patients with obsessive-compulsive disorder: absence of symptom exacerbation. Biol Psychiatry 1995;38:138–49.

[87] Scherman D, Hamon M, Gozlan H, Henry JP, Lesage A, Masson M, et al. Molecular pharmacology of niaprazine. Prog Neuropsychopharmacol Biol Psychiatry 1988;12:989–1001.

[88] Lin JC, Bangs N, Lee H, Kydd RR, Russell BR. Determining the subjective and physiological effects of BZP on human females. Psychopharmacology (Berl) 2009;207:439–46.

[89] Thompson I, Williams G, Caldwell B, Aldington S, Dickson S, Lucas N, et al. Randomised double-blind, placebo-controlled trial of the effects of the 'party pills' BZP/TFMPP alone and in combination with alcohol. J Psychopharmacol 2010;24:1299–308.

[90] Campbell H, Peck AW, Lloyd J, Clane W, Evans M. Comparison of effects of dexamphetamine and 1-benzylpiperazine in former addicts. Br J Pharmacol 1972;44:P369–70.

[91] Jan RK, Lin JC, Lee H, Sheridan JL, Kydd RR, Kirk IJ, et al. Determining the subjective effects of TFMPP in human males. Psychopharmacology (Berl) 2010;211:347–53.

[92] Lin JC, Jan RK, Lee H, Jensen MA, Kydd RR, Russell BR. Determining the subjective and physiological effects of BZP combined with TFMPP in human males. Psychopharmacology (Berl) 2011;214:761–8.

[93] Tancer M, Johanson CE. Reinforcing, subjective, and physiological effects of MDMA in humans: a comparison with d-amphetamine and mCPP. Drug Alcohol Depend 2003;72:33–44.

[94] Tancer ME, Johanson CE. The subjective effects of MDMA and mCPP in moderate MDMA users. Drug Alcohol Depend 2001;65:97–101.

[95] Charney DS, Woods SW, Goodman WK, Heninger GR. Serotonin function in anxiety. II. Effects of the serotonin agonist MCPP in panic disorder patients and healthy subjects. Psychopharmacology (Berl) 1987;92:14–24.

[96] Lawlor BA, Sunderland T, Mellow AM, Hill JL, Molchan SE, Murphy DL. Hyperresponsivity to the serotonin agonist m-chlorophenylpiperazine in Alzheimer's disease. A controlled study. Arch Gen Psychiatry 1989;46:542–9.

[97] Mueller EA, Murphy DL, Sunderland T. Neuroendocrine effects of M-chlorophenylpiperazine, a serotonin agonist, in humans. J Clin Endocrinol Metab 1985;61:1179–84.

[98] Murphy DL, Mueller EA, Hill JL, Tolliver TJ, Jacobsen FM. Comparative anxiogenic, neuroendocrine, and other physiologic effects of m-chlorophenylpiperazine given intravenously or orally to healthy volunteers. Psychopharmacology (Berl) 1989;98:275–82.

[99] Butler RA, Sheridan JL. Highs and lows: patterns of use, positive and negative effects of benzylpiperazine-containing party pills (BZP-party pills) amongst young people in New Zealand. Harm Reduct J 2007;4:18.

[100] Nicholson TC. Prevalence of use, epidemiology and toxicity of 'herbal party pills' among those presenting to the emergency department. Emerg Med Australas 2006;18:180–4.

[101] Wilkins C, Girling M, Sweetsur P. The prevalence of use, dependency and harms of legal 'party pills' containing benzylpiperazine (BZP) and trifluorophenylmethylpiperazine (TFMPP) in New Zealand. J Subst Use 2007;12:213–24.

[102] Wilkins C, Sweetsur P. Differences in harm from legal BZP/TFMPP party pills between North Island and South Island users in New Zealand: a case of effective industry self-regulation?. Int J Drug Policy 2010;21:86–90.

[103] Wilkins C, Sweetsur P, Girling M. Patterns of benzylpiperazine/trifluoromethylphenylpiperazine party pill use and adverse effects in a population sample in New Zealand. Drug Alcohol Rev 2008;27:633–9.

[104] Iqbal N, Asnis GM, Wetzler S, Kahn RS, Kay SR, van Praag HM. The MCPP challenge test in schizophrenia: hormonal and behavioral responses. Biol Psychiatry 1991;30:770–8.

[105] Gross-Isseroff R, Cohen R, Sasson Y, Voet H, Zohar J. Serotonergic dissection of obsessive compulsive symptoms: a challenge study with m-chlorophenylpiperazine and sumatriptan. Neuropsychobiology 2004;50:200–5.

[106] Hollander E, DeCaria CM, Nitescu A, Gully R, Suckow RF, Cooper TB, et al. Serotonergic function in obsessive-compulsive disorder. Behavioral and neuroendocrine responses to oral m-chlorophenylpiperazine and fenfluramine in patients and healthy volunteers. Arch Gen Psychiatry 1992;49:21–8.

[107] Hollander E, Fay M, Cohen B, Campeas R, Gorman JM, Liebowitz MR. Serotonergic and noradrenergic sensitivity in obsessive-compulsive disorder: behavioral findings. Am J Psychiatry 1988;145:1015–7.

[108] Pigott TA, Hill JL, Grady TA, L'Heureux F, Bernstein S, Rubenstein CS, et al. A comparison of the behavioral effects of oral versus intravenous mCPP administration in OCD patients and the effect of metergoline prior to i.v. mCPP. Biol Psychiatry 1993;33:3–14.

[109] Zohar J, Mueller EA, Insel TR, Zohar-Kadouch RC, Murphy DL. Serotonergic responsivity in obsessive-compulsive disorder. Comparison of patients and healthy controls. Arch Gen Psychiatry 1987;44:946–51.

[110] Schep LJ, Slaughter RJ, Vale JA, Beasley DM, Gee P. The clinical toxicology of the designer 'party pills' benzylpiperazine and trifluoromethylphenylpiperazine. Clin Toxicol (Phila) 2011;49:131–41.

[111] Gee P, Jerram T, Bowie D. Multiorgan failure from 1-benzylpiperazine ingestion – legal high or lethal high?. Clin Toxicol (Phila) 2010;48:230–3.

[112] Gee P, Richardson S. Researching the toxicity of party pills. Nurs N Z 2005;11:12–13.

[113] Wood DM, Dargan PI, Button J, Holt DW, Ovaska H, Ramsey J, et al. Collapse, reported seizure – and an unexpected pill. Lancet 2007;369 1490-1490.

[114] Austin H, Monasterio E. Acute psychosis following ingestion of 'Rapture'. Australas Psychiatry 2004;12:406–8.

[115] Theron Y, Jansen K, Miles J. Benzylpiperizine-based party pills' impact on the Auckland city hospital emergency department overdose database (2002-2004) compared with ecstasy (MDMA or methylene dioxymethamphetamine), gamma hydroxybutyrate (GHB), amphetamines, cocaine, and alcohol. N Z Med J 2007;120:U2416.

[116] Wood DM, Button J, Lidder S, Ramsey J, Holt DW, Dargan PI. Dissociative and sympathomimetic toxicity associated with recreational use of 1-(3-trifluoromethylphenyl) piperazine (TFMPP) and 1-benzylpiperizine (BZP). J Med Toxicol 2008;4:254–7.

[117] Alansari M, Hamilton D. Nephrotoxicity of BZP-based herbal party pills: a New Zealand case report. N Z Med J 2006;119:U1959.

[118] Harnett MA. Piperazine-based party drugs: case series of 73 poisonings [Conference abstract]. Clin Toxicol (Phila) 2007;45:373.

[119] Poon WT, Lai CF, Lui MC, Chan AY, Mak TW. Piperazines: a new class of drug of abuse has landed in Hong Kong. Hong Kong Med J 2010;16:76–7.

[120] Elliott S, Smith C. Investigation of the first deaths in the United Kingdom involving the detection and quantitation of the piperazines BZP and 3-TFMPP. J Anal Toxicol 2008;32:172–7.

[121] Staack RF, Maurer HH. Metabolism of designer drugs of abuse. Curr Drug Metab 2005;6:259–74.

[122] Gee P, Richardson S, Woltersdorf W, Moore G. Toxic effects of BZP-based herbal party pills in humans: a prospective study in Christchurch, New Zealand. N Z Med J 2005;118:U1784.

[123] Lynton RC, Albertson TE. Dart RC, editors. Medical toxicology. Philadelphia: Lippincott Williams and Wilkins; 2004.

[124] Nie Z, Yuan X, Madamba SG, Siggins GR. Ethanol decreases glutamatergic synaptic transmission in rat nucleus accumbens in vitro: naloxone reversal. J Pharmacol Exp Ther 1993;266:1705–12.

[125] Sanna E, Serra M, Cossu A, Colombo G, Follesa P, Cuccheddu T, et al. Chronic ethanol intoxication induces differential effects on GABAA and NMDA receptor function in the rat brain. Alcohol Clin Exp Res 1993;17:115–23.

[126] Button J, Wood DM, Dargan PI, Ovaska H, Jones AL, Ramsey J, et al. A gas chromatography

mass-spectrometric method for the quantitative analysis of the recreational drug n-benzylpiperazine in serum. Ther Drug Monit 2007;29:496.

[127] Klaassen T, Pian KLH, Westenberg HGM, den Boer JA, van Praag HM. Serotonin syndrome after challenge with the 5-HT agonist meta-chlorophenylpiperazine. Psychiatry Res 1998;79:207–12.

[128] Dunkley EJ, Isbister GK, Sibbritt D, Dawson AH, Whyte IM. The hunter serotonin toxicity criteria: simple and accurate diagnostic decision rules for serotonin toxicity. Q J M 2003;96:635–42.

[129] Isbister GK, Buckley NA, Whyte IM. Serotonin toxicity: a practical approach to diagnosis and treatment. Med J Aust 2007;187:361–5.

[130] Balmelli C, Kupferschmidt H, Rentsch K, Schneemann M. [Fatal brain oedema after ingestion of ecstasy and benzylpiperazine] German. Dtsch Med Wochenschr 2001;126:809–11.

[131] Hartung TK, Schofield E, Short AI, Parr MJ, Henry JA. Hyponatraemic states following 3,4-methylenedioxymethamphetamine (MDMA, 'ecstasy') ingestion. Q J M 2002;95:431–7.

[132] Berney-Meyer L, Putt T, Schollum J, Walker R. Nephrotoxicity of recreational party drugs. Nephrology (Carlton) 2012;17:99–103.

[133] Cole M. Poison in party pills is too much to swallow. Nature 2011;474:253.

[134] Schep LJ, Slaughter RJ, Beasley DMG. The clinical toxicology of metamfetamine. Clin Toxicol (Phila) 2010;48:675–94.

[135] Mohandas A, Vecchio D. A case report of Benzylpiperazine induced new onset affective symptoms in a patient with schizophrenia. Eur Psychiatry 2008;23:S315–6.

[136] Suzuki A, Nakamura K, Sekine Y, Minabe Y, Takei N, Suzuki K, et al. An association study between catechol-O-methyl transferase gene polymorphism and methamphetamine psychotic disorder. Psychiat Genet 2006;16:133–8.

[137] EMCDDA. Europol–EMCDDA Joint report on a new psychoactive substance: 1-(3-chlorophenyl) piperazine (mCPP). Available: <http://www.emcdda.europa.eu/attachements.cfm/att_132213_EN_Final%20Joint%20Report%20mCPP.pdf>; 2005 [accessed 04.03.03].

[138] Kovaleva J, Devuyst E, De Paepe P, Verstraete A. Acute chlorophenylpiperazine overdose: a case report and review of the literature. Ther Drug Monit 2008;30:394–8.

[139] Bailey B. Gastrointestinal decontamination triangle. Clin Toxicol (Phila) 2005;43:59–60.

[140] Shah AS, Eddleston M. Should phenytoin or barbiturates be used as second-line anticonvulsant therapy for toxicological seizures?. Clin. Toxicol. (Phila) 2010;48:800–5.

[141] Callaway CW, Clark RF. Hyperthermia in psychostimulant overdose. Ann Emerg Med 1994;24:68–76.

[142] Rusyniak DE, Sprague JE. Hyperthermic syndromes induced by toxins. Clin Lab Med 2006;26:165–84. ix.

[143] Greene SL, Kerr F, Braitberg G. Review article: amphetamines and related drugs of abuse. Emerg Med Australas 2008;20:391–402.

[144] Sun-Edelstein C, Tepper SJ, Shapiro RE. Drug-induced serotonin syndrome: a review. Expert Opin Drug Saf 2008;7:587–96.

[145] Hall AP, Henry JA. Acute toxic effects of 'Ecstasy' (MDMA) and related compounds: overview of pathophysiology and clinical management. Br J Anaesth 2006;96:678–85.

[146] Isbister GK, Buckley NA. The pathophysiology of serotonin toxicity in animals and humans: implications for diagnosis and treatment. Clin Neuropharmacol 2005;28:205–14.

[147] Zietse R, van der Lubbe N, Hoorn EJ. Current and future treatment options in SIADH. NDT Plus 2009;2:iii12–iii19.

[148] Bosch X, Poch E, Grau JM. Current concepts: rhabdomyolysis and acute kidney injury. N Engl J Med 2009;361:62–72.

[149] Elliott S. Current awareness of piperazines: pharmacology and toxicology. Drug Test Anal 2011;3:430–8.

[150] Bellingham SJ. Findings of Inquest; Coronial Court at Christchurch, New Zealand. 15 December 2009. 2011; COR REF: CSU-2008-WHG-000273.

[151] Smith-Kielland A, Olsen KM, Christophersen AS. False-positive results with Emit II amphetamine/methamphetamine assay in users of common psychotropic drugs. Clin. Chem. 1995;41:951–2.

[152] Capela JP, Carmo H, Remiao F, Bastos ML, Meisel A, Carvalho F. Molecular and cellular mechanisms of ecstasy-induced neurotoxicity: an overview. Mol Neurobiol 2009;39:210–71.

[153] Fatovich DM, McCoubrie DL, Song SJ, Rosen DM, Lawn ND, Daly FF. Brain abnormalities detected on magnetic resonance imaging of amphetamine users presenting to an emergency department: a pilot study. Med J Aust 2010;193:266–8.

[154] Berman S, O'Neill J, Fears S, Bartzokis G, London ED. Abuse of amphetamines and structural abnormalities in the brain. Ann N Y Acad Sci 2008;1141:195–220.

[155] Wise RA. Roles for nigrostriatal–not just mesocorticolimbic–dopamine in reward and addiction. Trends Neurosci 2009;32:517–24.

[156] Johnstone AC, Lea RA, Brennan KA, Schenk S, Kennedy MA, Fitzmaurice PS. Benzylpiperazine: a drug of abuse?. J Psychopharmacol 2007;21:888–94.

[157] Fantegrossi WE, Winger G, Woods JH, Woolverton WL, Coop A. Reinforcing and discriminative stimulus effects of 1-benzylpiperazine and trifluoromethylphenylpiperazine in rhesus monkeys. Drug Alcohol Depend 2005;77:161–8.

[158] Brennan K, Johnstone A, Fitzmaurice P, Lea R, Schenk S. Chronic benzylpiperazine (BZP) exposure produces behavioral sensitization and cross-sensitization to methamphetamine (MA). Drug Alcohol Depend 2007;88:204–13.

[159] Kim MT, Hill MN. Validity of self-report of illicit drug use in young hypertensive urban African American males. Addict Behav 2003;28:795–802.

[160] Lee MO, Vivier PM, Diercks DB. Is the self-report of recent cocaine or methamphetamine use reliable in illicit stimulant drug users who present to the emergency department with chest pain? J Emerg Med 2009;37:237–40.

This page intentionally left blank

Mephedrone

David M. Wood and Paul I. Dargan

Guy's and St Thomas' NHS Foundation Trust and King's Health Partners and King's College London, London, UK

INTRODUCTION

The cathinones are synthetic ring-substituted phenylethylamines with substitution of a keto group at the beta carbon [1,2]. Mephedrone (4-methylmethcathinone) is one of a number of synthetic cathinones that have emerged as novel psychoactive substances (NPS) since 2005/6 [1–3]. A total of 34 cathinones have been reported to the EMCDDA Early Warning System over the period 2005–2011 (this is the second largest single drug group after the synthetic cannabinoid receptor agonists) [3]. There is limited data available on the pharmacology, epidemiology of use and potential for acute/chronic toxicity associated with the use of many of these cathinones. This chapter will focus on mephedrone as it is the most widely used cathinone and the one for which there is the most published literature. In May 2010, mephedrone was risk assessed by the EMCDDA extended scientific committee and based on this risk assessment, the Council of the European Union decided that mephedrone should be controlled in all European Member States [4]; details of the control status of mephedrone in other countries are within Chapter 1 of this book.

PHARMACOLOGY

Physical and Chemical Description

'Mephedrone' is the common name for the synthetic cathinone 4-methylmethcathinone (see Fig. 9.1 for the chemical structure of mephedrone). The International Union of Pure and Applied Chemistry (IUPAC) name for mephedrone is 2-methylamino-1-(4-methylphenyl)propan-1-one and its molecular formula is $C_{11}H_{15}NO$. Other names for mephedrone include β-keto-(4,N-dimethylamphetamine), 4,N-dimethylcathinone, N-methylephedrone, p-methyl-methcathinone and 2-aminomethyl-1-tolyl-propan-1-one.

Mephedrone was first synthesised in 1929 [4]. The synthesis of mephedrone is relatively straightforward and can be carried out with similar equipment and knowledge as that required for the synthesis of amphetamines [5].

Mephedrone is most commonly encountered in powder form [4,6]. Mephedrone powder is typically a white crystalline powder which sometimes has a light yellow or light brown colour [4,7,8]. Mephedrone is also available as tablets (tableting machines for pressing powder into tablets have been seized in Europe) and

Novel Psychoactive Substances.
DOI: http://dx.doi.org/10.1016/B978-0-12-415816-0.00009-2

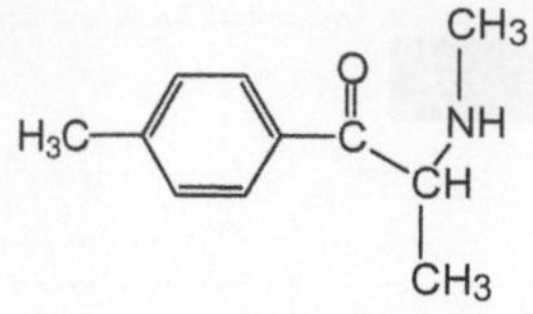

FIGURE 9.1 Chemical structure of mephedrone.

in capsule form; although these preparations appear to be less common [4,7,8]. Mephedrone is highly soluble in water and so the powder can be dissolved into water for oral use; this mephedrone solution may also be used rectally or injected intravenously or intramuscularly [4,9].

Mephedrone powder, like many other NPS, is often sold to users in small sealed ('zip locked') bags labelled as 'research chemical', 'not for human consumption' or 'not tested for hazards or toxicity' [4,6,7,8]. It is also sold under a number of other names including 'plant food' and 'bath salts', although it has no legitimate use as a plant food or cosmetic product [4]. In Europe, particularly prior to its control, mephedrone was often sold under its own name, or as 'plant food'; in the USA, mephedrone, alongside many other NPS is more commonly sold as 'bath salts' [4,6–11].

There are reports of the use of mephedrone by nasal insufflation of powder; oral ingestion of dissolved powder, powder wrapped in paper ('bombing'), or powder contained in capsules or tablets; intramuscular injection; intravenous injection and rectal insertion [6,7,12–15].

The most common routes of use of mephedrone appear to be nasal insufflation and oral ingestion. In the 2010 MixMag survey, 70% of mephedrone users reported use of mephedrone by nasal insufflation and 30% oral ingestion [16]. In a follow up to this study, 100 mephedrone users were questioned more closely and 79% reported nasal insufflation of powder, 11.1% oral use of mephedrone powder dissolved in a drink and 9.9% oral use of mephedrone

powder wrapped in cigarette paper [17]. There are numerous reports of individuals using mixed routes during a single session (oral and nasal, oral and rectal, oral and intramuscular) [6,7,9,12,14]. Some intranasal users report nasal irritation associated with nasal insufflation and as a result some users move from nasal insufflation to oral use of mephedrone [4,12].

Data is available on the doses used by users from a variety of sources including Internet drug user fora, subpopulation surveys, focus groups and reports of acute toxicity. Single use doses reported on Internet user fora vary from 5–125 mg for nasal insufflation and 15–250 mg for oral ingestion [18]. In a focus group study of mephedrone users, there were reports of users starting with lower doses of mephedrone (50–75 mg) and then increasing the doses used to 200–400 mg [6]. Doses of mephedrone users in individuals presenting to hospital with acute toxicity range from 0.3–7.0 g [14,19–22]. Users commonly report re-dosing during a single session with total doses typically being between 0.5 and 2.0 g [7,9,17,18]. In a follow up study to the 2010 MixMag survey, the 100 mephedrone users surveyed, reported that their first session lasted for a median of 6 hours (interquartile range (IQR) 4–10 hours) during which it was reported that a median of four doses were taken (IQR 2–7), with a median total dose during this session of 500 mg mephedrone (IQR 200–1000 mg) [17]. At the time of the follow up study, users had been using mephedrone for a mean ± standard deviation of 6.1 ± 3.1 months. Typical using sessions last for a median of 10 (IQR 6–16) hours, during which a median of 5.5 (IQR 3–10) doses were taken with a median total dose per session of 1000 (IQR 250–1275) mg. During a typical user session none of the users reported using mephedrone alone, it was used together with alcohol (n = 82), cannabis (36), ketamine (35), cocaine (26) and ecstasy (23). In the 4 weeks prior to the study, the median total dose of mephedrone users was 1500 mg (500–4000) mg.

Pharmacokinetics

There is limited information available on the pharmacokinetics of mephedrone. Three studies have investigated mephedrone metabolism and one animal study focusing predominantly on pharmacodynamics has limited pharmacokinetic data, including time to maximum concentration and area under the curve [23–26]. Therefore, information on other pharmacokinetic parameters has to be extrapolated from user reports on Internet discussion fora and data collected in user surveys.

User reports on Internet discussion fora suggest onset of desired effects within a few minutes of intravenous injection or nasal insufflation and within 15–45 minutes of oral ingestion [12,27]. Some users report a slower onset of action after oral use of mephedrone on a full stomach [27]. The duration of desired effects is approximately 2–4 hours and users often report use of repeated doses during a single session to prolong the duration of the desired effects [6,15,17,27]. Users report a much shorter duration (30 minutes) of desired effects after intravenous use of mephedrone [27].

The first study to investigate the metabolism of mephedrone was a study in rats administered a single 20 mg/kg dose of mephedrone by gastric intubation [23]. Urine was collected over a 24-hour period after mephedrone administration. In addition to mephedrone, the following metabolites were detected: nor-mephedrone, nor-dihydro mephedrone, hydroxytolyl mephedrone and nor-hydroxytolyl mephedrone. No data was provided in this study on the time course of these metabolites as all urine collected over a 24-hour period after mephedrone administration was pooled as a single sample. In this study a human urine sample submitted by a mephedrone user was also analyzed and a further metabolite, 4-carboxy-dihydro mephedrone was detected in addition to the metabolites found in the animal study. The authors

postulated that the overlapping metabolic pathways that were thought to be responsible for these metabolites were as follows:

1. N-demethylation to the primary amine (metabolites nor-mephedrone, nor-dihydro mephedrone and nor-hydroxytolyl mephedrone);
2. reduction of the keto moiety to the respective alcohol (metabolites nor-dihydro mephedrone and 4-carboxy-dihydro mephedrone); and
3. oxidation of the tolyl moiety to the corresponding alcohol (metabolites hydroxytolyl mephedrone and nor-hydroxytolyl mephedrone) – it is thought that the hydroxytolyl mephedrone and nor-hydroxytolyl mephedrone metabolites are partly excreted as glucuronides and sulphates.

In a more recent *in vitro* study using Sprague-Dawley rat liver hepatocytes, the half-life of mephedrone was 61.9 minutes and a total of 17 metabolites formed by Phase I and Phase II metabolism were elucidated [24]. Only nor-mephedrone was formed in 'large amounts'; concentrations of the 4-(carboxy)methcathinone metabolite increased after 2 hours whilst the formation of the 4-(hydroxymethyl)methcathinone and 4-methylephedrine metabolites began to decrease after 30 minutes.

In an *in vitro* study using human liver microsomal preparations and cDNA-expressed cytochrome-P450 enzymes, the main isoenzyme responsible for Phase I metabolism of mephedrone was cytochrome P450 2D6 (CYP2D6), with a lesser contribution from NAPDH-dependent enzymes [25]. The hydroxytolyl-mephedrone and nor-mephedrone metabolites were purified and used for analysis of four samples from forensic traffic cases where mephedrone was detected. Both of these metabolites were identified in these human cases and the 4-carboxy-dihydro-mephedrone metabolite was also detected along with two other new metabolites (dihydro-mephedrone and 4-carboxy-mephedrone). The blood concentrations in the forensic traffic cases ranged from 1–51 µg/kg for mephedrone, and from not detected to 9 µg/kg for hydroxytolyl-mephedrone.

The only study to investigate other pharmacokinetic parameters was a study in Wistar (n = 3) and Sprague-Dawley (n = 3) rats [26]. The peak plasma concentration of mephedrone following subcutaneous injection of 5.6 mg/kg mephedrone was 1206 ng/mL in Sprague-Dawley rats and 868 ng/mL in Wistar rats, with the same time to peak plasma concentration of 0.25 hours (15 minutes). In both animal models there was rapid reduction in plasma concentrations with an area under the curve of 1170 ng-hr/mL and 870 ng-hr/mL for Sprague-Dawley and Wistar rats respectively.

Pharmacodynamics

A number of animal studies have investigated the effects of mephedrone at a receptor level and on neurological, thermoregulatory and cardiovascular parameters. We have summarised the data from these studies below which suggest that overall mephedrone has similar actions both at a cellular and whole animal model level to other stimulant recreational drugs. It should be noted that some of the expected effects such as increased temperature are not always seen in all animal models and therefore extrapolation of some of this data to humans should be undertaken with caution.

Mephedrone effects on serotonin (5-HT) and dopamine uptake were investigated using isolated rat synaptosomes [28]. Mephedrone inhibited the uptake of both serotonin (IC50 0.31 ± 0.08 µM) and dopamine (IC50 0.97 ± 0.05 µM). This study also demonstrated that mephedrone has affinity for both serotonin and dopamine membrane transporters and receptors (5-HT2 and D2 receptors) and that the profile of these effects was similar to amphetamines. In a study using microdialysis of the nucleus accumbens in a rat model, mephedrone was shown to increase extracellular serotonin to a greater extent than extracellular dopamine [29]. A further microdialysis study in the rat nucleus accumbens, demonstrated that mephedrone and MDMA had similar effects on serotonin concentrations (941% and 911% respectively) compared to amphetamine (165% increase) [30]. However, mephedrone and amphetamine had similar effects on increasing dopamine concentrations (496% increase and 412% increase respectively) and this effect was greater than the effect of MDMA on dopamine concentrations (235% increase).

The effects of mephedrone, 3,4-methylenedioxypyrovalerone (MDPV), methamphetamine, methcathinone and dopamine were studied by patch clamp studies on *Xenopus laevis* oocytes expressing human dopamine transporters [31]. Mephedrone, similar to methamphetamine, methcathinone and dopamine produced an inward (depolarising) current when applied to the *Xenopus laevis* oocytes; in addition there was a similar 'shelf current' following removal of the mephedrone as seen with the other stimulant drugs (methamphetamine and methcathinone) that was not seen following dopamine administration. In dose-response studies, mephedrone was less potent than both methcathinone and methamphetamine, with an EC50 of 0.84 ± 0.14 µM (methamphetamine EC50 0.64 ± 0.15 µM, methcathinone EC50 0.23 ± 0.03 µM). Interestingly, using the same model, application of MDPV was associated with an outward (hyper-polarising) current, similar to that seen with cocaine.

In an invertebrate model, using a planarian assay, mephedrone (500–1000 µM) produced stereotypical movements that were decreased by a dopamine receptor antagonist (SCH23390 at a dose of 0.3 µM) [32]. In a rat model, the effect of intraperitoneal mephedrone (15 mg/kg and 30 mg/kg) was compared to methamphetamine [33]. Mephedrone caused significant locomotor hyperactivity at both doses and reduced social preference. Subsequent histological analysis of the rat brains using Fos immunocytochemistry demonstrated that the effects of mephedrone were similar to the combined pattern seen with methampethamine and MDMA, with strong Fos expression in the cortex, dorsal and ventral striatum, ventral tegmental area (typical of both MDMA and methamphetamine) and supraoptic nucleus (typical of MDMA).

In another study investigating locomotor effects, young adult male Lister hooded rats were administered mephedrone, MDMA and cathinone in an intermittent 'binge use' style, where the drugs were administered on two consecutive days for three weeks (which would simulate the use of a recreational drug on two days at the weekend (e.g. Friday and Saturday) [34]. Low doses of mephedrone (1 mg/kg) had no significant impact on locomotor activity in rats following either the first or sixth administration of mephedrone, whereas higher doses (10 mg/kg) were associated with significantly increased locomotor activity following both the first and sixth dose. Both cathinone (1 mg/kg and 4 mg/kg) and MDMA (10 mg/kg) increased locomotor activity following the first and sixth doses. The locomotor response to mephedrone was short-lived, lasting less than or equal to 15 minutes before returning to baseline following the first dose, whereas following the sixth dose there was a more prolonged increase in the duration of the locomotor response which was significantly elevated above baseline for 55 minutes following administration. The response to both cathinone and MDMA was sustained and remained significantly elevated above baseline for at least an hour (recording of locomotor activity stopped at one hour following administration) for both the first and sixth dose. Following a familiarisation trial, the impact of mephedrone, MDMA and cathinone on novel object discrimination compared to saline was determined; all three drugs significantly impaired ability to discriminate a novel object

compared to the saline vehicle. The concentrations of dopamine, serotonin (5-HT) and their major metabolites were measured in the striatum, frontal cortex and hippocampus 2 hours after a single dose of drug (defined as 'acute') and following six doses as described above ('chronic'). There were no changes in dopamine, 5-HT or their major metabolites in the striatum and frontal cortex for any of the drugs for either acute or chronic administration or in the hippocampus following a single acute dose of mephedrone or cathinone. MDMA, however, was associated with a significant decrease in both serotonin (5-HT) and its major metabolite 5-Hydroxyindoleacetic acid (5-HIAA) concentration in the hippocampus. Following chronic administration there was no change in hippocampal concentrations of 5-HT or 5-HIAA for any of the three substances. Finally, whilst administration of cathinone, mephedrone (10 mg/kg) and MDMA were all associated with a decrease in both hippocampal concentrations of dopamine and its major metabolite 3,4-dihydroxyphenylacetic acid (DOPAC), this was only statistically significant compared to saline vehicle for mephedrone. Interestingly administration of mephedrone at a lower concentration (4 mg/kg) was associated with an increase in hippocampal concentrations of both dopamine and DOPAC.

In a study investigating both neurochemical and behavioural effects of mephedrone, male adolescent Wistar rats received single or repeated (once daily for 10 days) intraperitoneal mephedrone 30 mg/kg, methamphetamine 2.5 mg/kg or vehicle [35]. Rats administered mephedrone for 10 days gained significantly less weight than single dose mephedrone and single/repeated methamphetamine or vehicle ($p < 0.05$). There was increased locomotor activity in both the methamphetamine and mephedrone treated animals ($p < 0.001$), with no difference in locomotor activity in animals who received single or 10-day repeated drug administration. In the biochemical studies, striatal dopamine was elevated ($p < 0.05$)

and striatal serotonin depressed (p < 0.001) in rats treated with mephedrone or methamphetamine; these changes were seen in animals treated with both single and repeated doses. Striatal homovanillic acid (HVA) concentrations were significantly higher in rats who received single dose compared to the repeat dose treated rats (p < 0.01). Striatal serotonin metabolism (5-HIAA/5-HT) was increased in rats treated with single and repeated doses of mephedrone (p < 0.001); this was greater after single dose mephedrone administration (p < 0.01). Autoradiographic studies using a [(125)I]CLINDE ligand used as a marker for translocator protein expression suggested that there was no inflammation associated with either single or repeated mephedrone administration. In the second part of this study rats were given vehicle or 7.5 mg/kg, 15 mg/kg or 30 mg/kg intraperitoneal mephedrone for 10 days. Behavioural studies were carried out 5 weeks later and impaired novel object recognition was seen only in the high dose (30 mg/kg) mephedrone group (p < 0.01). However, there were no differences in any of the mephedrone treated groups in other tests including elevated plus maze or social preference tests. There were also no differences in striatal or hippocampal dopamine, serotonin, 5-HT/5-HIAA or HVA at 5 weeks in any of the mephedrone treated groups.

The effects of binge-like regimen of mephedrone (30 mg/kg twice daily) for four days were investigated in C57BL/J6 mice (behavioural and biochemical testing) and Wistar rats (biochemical testing) [36]. In mice, body temperature increased by approximately 2°C (p < 0.0005) and there was no effect of treatment day suggesting that there is no tolerance or sensitisation to this hyperthermic response; there was no increase in body temperature associated with mephedrone administration to rats. Mephedrone treated mice had a significantly lower rate of spontaneous alternations compared to saline treated controls in a T-maze used to test spontaneous alternation. In the biochemical experiments, two weeks

following the last drug administration, there were no changes seen in mouse or rat brain frontal cortex, striatal or hippocampal serotonin (5-HT), 5-Hydroxyindoleacetic acid (5-HIAA), dopamine, 3,4-dihydroxyphenylacetic acid (DOPAC) or norepinephrine related to mephedrone administration. There was a 22% decrease (p < 0.05) in striatal homovanillic acid (HVA) in the mephedrone treated mice, there was no difference in striatal HVA in mephedrone treated rats.

The effects of mephedrone on rectal and tail temperature in individually housed Lister hooded rats was compared to MDMA, cathinone and methacathinone [37]. All drugs were administered to rats by intraperitoneal injection at a doses of both 4 mg/kg and 10 mg/kg. Administration of MDMA resulted in a significant reduction in both rectal and tail temperature, with the rectal temperature being significantly lower for two hours following administration, whereas the tail temperature was only significantly lower for the first hour following administration. Both cathinone and methacathinone at 4 mg/kg and 10 mg/kg had no effect on tail temperature and 4 mg/kg had no effect on rectal temperature compared to vehicle control; administration of 10 mg/kg significantly increased rectal temperature for 2 hours following administration of methcathinone and 80 minutes following administration of cathinone. The effects of mephedrone on temperature were similar to that seen with MDMA, although the decrease in rectal temperature was only short lived (20 minutes for 4 mg/kg and 40 minutes for 10 mg/kg) whereas the decrease in tail temperature was more prolonged and was significantly lower for the 2 hours following administration. Prazosin (an α-1 adrenoreceptor antagonist), BRL 44408 (an α-2a adrenoreceptor antagonist), SCH 23390 (a dopamine-D1 receptor antagonist) and L-741626 (dopamine-D2 receptor antagonist) all increased not only the duration of mephedrone-related decreases in rectal temperature compared to those rats treatment with mephedrone alone, but also the degree of the decrease in temperature. Finally, the changes in tail and rectal temperature following administration of mephedrone seen in individually housed rats were not seen when the rats were group housed (three rats per cage). Administration of 10 mg/kg mephedrone was associated with a statistically significant increase in plasma noradrenaline concentrations, which could be completely abolished by pre-treatment with prazosin, BRL 44408 and SCH 23390. It appeared that pre-treatment with L-741626 was not associated with this abolishment of the increase in plasma noradrenaline concentrations seen with mephedrone. The authors reported that mephedrone was associated with a statistically significant increase in plasma adrenaline concentrations; however, the effect of the individual antagonists on this change in plasma adrenaline concentrations was not reported in full in this paper.

Female C57BL/6 mice were treated with four doses of mephedrone (20 mg/kg or 40 mg/kg), with doses given every two hours, as this is representative of the pattern of use of mephedrone in humans and also because this pattern of administration in animal models of substituted amphetamines and cathinones has been shown to cause dopaminergic nerve ending damage [38]. The authors in this study reported that 'lower doses' of mephedrone (5–10 mg/kg) were not neurotoxic and did not cause changes in body temperature, although the data for this was not presented in the paper. Striatal concentrations of dopamine, tyrosine hydroxylase and the dopamine transporter were compared to control concentrations by immunblot. Two days following administration of mephedrone, there was no change in striatal concentrations of dopamine and tyrosine hydroxylase, whereas there was a small but statistically significant increase in dopamine transporter concentrations in animals treated with higher concentrations of mephedrone (40 mg/kg) which was not seen in those treated with the lower concentration (20 mg/kg). When the concentrations were measured seven days after administration, there was no difference in striatal tyrosine hydroxylase or dopamine transporter concentrations, whereas there was a small but statistically significant increase in striatal dopamine concentrations in those treated with 40 mg/kg mephedrone. In this study, intraperitoneal injection of 20 mg/kg mephedrone to female C57BL/6 mice was associated with an increase in core body temperature, which remained elevated following subsequent intraperitoneal injections throughout the study period. Interestingly, intraperitoneal injection of 40 mg/kg mephedrone had a different response, in that the first and each subsequent injection of mephedrone was associated with an immediate drop in core body temperature, but the drop in core body temperature increased in magnitude with each successive injection of mephedrone. Within 40 minutes of injection after this initial immediate drop in core body temperature, the animal's core body temperature had then risen above that of the control animals and remained elevated until the next injection of mephedrone. Administration of 20 mg/kg and 40 mg/kg of mephedrone to female C57BL/6 mice led to an increase in locomotor activity, distance travelled, movement time and stereotypical episodes compared to controls. Interestingly, when compared to methamphetamine which led to a constant increase in locomotor activity following administration, mephedrone was associated with 'cyclic bouts of explosive and stereotyped activity followed by short periods of inactivity'.

Thermoregulatory and locomotor stimulant effects of mephedrone were studied using both Wistar and Sprague-Dawley rats [26]. Following administration of 3.2 mg/kg, 5.6 mg/kg and 10 mg/kg mephedrone there was a decrease in body temperature in the Wistar animals when maintained at an ambient temperature of either 23°C or 27°C, whereas for the Sprague-Dawley rats there appeared to be only minimal effect on the body temperature at both 23°C and 27°C. For both Wistar and Sprague-Dawley rats there was an increase in locomotor activity in the hour following administration of mephedrone at both 23°C and 27°C, although there was a greater increase in activity seen in the Sprague-Dawley rats.

The cardiovascular effects of mephedrone have been studied both *in vitro* and *in vivo* [39,40]. Application of mephedrone up to 30μM to isolated guinea pig cardiac myocytes, demonstrated that mephedrone had no effects on the recorded action potential or on major voltage-dependent cardiac ion channels [39]. In the second component of this study, mephedrone was administered subcutaneously (3mg/kg and 15mg/kg) conscious telemetry implanted rats and effects compared to saline control. Heart rate increased after subcutaneous mephedrone administration peaking at 30% higher than pre-mephedrone administration 25 minutes after injection of 3mg/kg mephedrone and 39% higher at 45 minutes after 15mg/kg mephedrone. Heart rate remained higher for 105 minutes after the 3mg/kg dose and three hours after the 15mg/kg mephedrone dose. Subcutaneous mephedrone administration also resulted in an increase in mean, systolic and diastolic blood pressure, with an increase of approximately 10–15% above pre-mephedrone administration blood pressure for the 3mg/kg mephedrone treatment group and 20–25% for the 15mg/kg mephedrone treatment group. Peak blood pressure effects were seen at 1–2 hours after mephedrone administration and blood pressure gradually decreased to baseline over the next three hours. Pre-treatment with reserpine had a minimal effect on these heart rate and blood pressure changes. The final component of this study involved echocardiography after intravenous administration of mephedrone (1mg/kg) to anaesthetised rats. Mephedrone administration was associated with an increase in cardiac output (81.3 ± 3.0ml/min to 95.0 ± 5.6ml/min, $p < 0.05$), ejection fraction ($74.3 \pm 2.5\%$ to $86.1 \pm 2.5\%$, $p < 0.05$) and fraction shortening ($44.8 \pm 2.4\%$ to $57.8 \pm 3.3\%$, $p < 0.05$). The echocardiographic changes seen after

intravenous methamphetamine 0.3mg/kg were similar qualitatively and quantitatively to those seen after intravenous 1mg/kg mephedrone. Similar, but lower magnitude findings were seen after lower dose (0.3mg) intravenous mephedrone administration.

In a more recent study, the cardiovascular effects of mephedrone were compared to methamphetamine in male Sprague-Dawley rats [40]. In both groups, the baseline heart rate and mean blood pressure were comparable (mephedrone 379 ± 10b.p.m., 118 ± 4mmHg; methamphetamine: 381 ± 12b.p.m., 116 ± 3mmHg). After intravenous administration of either mephedrone or methamphetamine (0.01–9mg/kg of both drugs), there were comparable changes in both heart rate and blood pressure. There was a continual trend in the increase in blood pressure over the doses used (0.1–9mg/Kg) with approximately 10mmHg increase in blood pressure with 0.1mg/kg administration and approximately 20mmHg increase in blood pressure with 9mg/kg administration. The increases in mean arterial pressure peaked within 2–3 minutes of mephedrone administration and lasted from 0.09 ± 0.02 hours for 0.1mg/kg administered to 1.53 ± 0.26 hours for 9mg/kg administered. In terms of heart rate responses, there appeared to be less of a dose-response increase for methamphetamine (maximal increase with 0.1mg/kg administration), whereas for mephedrone there appeared to be a more of a dose-response to increasing doses of mephedrone (approximate 50–60b.p.m. increase with 0.1mg/kg compared to approximate 90–100b.p.m. increase for 9mg/kg). Similar to that seen with the blood pressure changes, the tachycardic responses peaked within 2–5 minutes after mephedrone administration and lasted between 0.3 ± 0.23 hours following 0.1mg/kg administration and 1.27 ± 0.57 hours following 9mg/kg administration. Pre-treatment with the β1-adrenergic antagonist atenolol (1mg/kg) had no impact on the blood pressure effects seen following the administration of 3mg/kg of mephedrone, whereas pre-treatment with the α-adrenergic antagonist phentolamine (3 mg/kg) significantly inhibited the increase in blood pressure seen with mephedrone. Both atenolol and phentolamine reduced the heart rate responses seen following mephedrone administration, although atenolol had a greater impact compared to phentolamine – the increase in heart rate with mephedrone following: 1) no pre-treatment approximately 80 b.p.m.; 2) atenolol approximately 15–20 b.p.m.; and 3) phentolamine approximately 40 b.p.m.

PREVALENCE OF USE

Information on the prevalence of use of mephedrone is available from a number of different data sources. These include information from law enforcement and border agency seizures, along with population and subpopulation surveys.

The first report of the detection of mephedrone with the European Union to the European Monitoring Centre for Drugs and Drug Addiction (EMCDDA) through its Early Warning System was in 2007. This related to a seizure of capsules in Finland that were found to contain mephedrone [4,41]. There were further detections by border and law enforcement agencies other Scandinavian countries (Sweden, Norway and Denmark), as well as the UK in 2008 [4,41,42]. By the time of the EMCDDA risk assessment of mephedrone in 2010, it had been detected by law enforcement agencies in 31 European and neighbouring countries suggesting that its availability, and therefore probably its use had become widespread across Europe [4]. Some countries had reported significant seizures of many kilograms, noting that production and export took place in Asia and in particular China [4]. Data in the United Nations Office on Drugs and Crime (UNODC) 2012 World Drug Report suggested that mephedrone was the most frequently seized substance in Hungary and that 286 out of 3564 drug seizures in North Ireland were mephedrone

[43]. Mephedrone has subsequently also been detected in law enforcement and border agency seizures in other areas of the world, including North America and Australasia [43].

Population Level Surveys

International population level reports, such as the World Drug Report from the United Nations Office on Drugs and Crime (UNODC) and the EMCDDA Annual Report typically do not contain detailed data on the prevalence of use NPS such as mephedrone when they first appear on the market as it takes time for new drugs to enter national drugs surveys.

To our knowledge, currently only England and Wales collect data at a population level on the use of mephedrone at a national level [44,45]. In the 2010/2011 British Crime Survey there was inclusion of questions about the use of mephedrone for the first time. Last year use of mephedrone in those aged 16–59 years old was 1.4% (the same as last year use of ecstasy). There was a higher reported prevalence of last year use amongst younger individuals: 4.4% in 16–24-year-olds and 0.6% in 25–59-year-olds. The 2011/2012 survey was renamed the Crime Survey England and Wales and showed that last year use of mephedrone had fallen both in those aged 16–59 years and 16–24 years (15–59-year-olds: 1.4% in 2010/2011 compared to 1.1% in 2011/2012; 15–24-year-olds: 4.4% in 2010/2011 compared to 3.3% in 2011/2012) [44,45]. Despite this reduction in last year use of mephedrone, it was the fourth most prevalently used drug. There was variation in the prevalence of last year use across a number of demographic factors:

1. sex: males (1.5%) compared to females (0.7%);
2. martial status: single (2.7%) compared to married/civil partnership (0.1%); and
3. employment status: employed (0.9%) compared to unemployed (2.2%) and student (2.7%) [45].

Subpopulation Surveys

There have been a number of subpopulation surveys which have reported on the use of mephedrone amongst students, clubbers and those who frequent other night-time economy venues and the men who have sex with men (MSM or gay) community. The majority of these surveys are from the UK, but there are also surveys from elsewhere in Europe and other areas of the world. One of the difficulties with the subpopulation level surveys, particularly from the USA, is that some ask about the use of NPS using colloquial terms such as 'bath salts' rather than by individual drug name such as mephedrone.

There is evidence of use of mephedrone amongst younger age groups. In a survey of 1006 Scottish school and college/university students, 20.3% reported that they had tried mephedrone on at least one occasion previously [46]. In a survey of 154 students aged 14/15 years undertaken in Northern Ireland in May 2010, although the students reported that '70% of their friends had tried mephedrone', only 40% of respondents reported that they themselves had tried mephedrone [4,47].

A number of studies have reported that mephedrone use is higher in those who frequent clubs and other night-time economy venues. In the 2011/2012 Crime Survey England and Wales, data was published comparing the last year use of mephedrone amongst those who visited bars and nightclubs compared to those who did not [45]. Last year use of mephedrone was associated with a higher frequency of visiting nightclubs: 0.5% for no visits to a nightclub in the last month, compared to 2.6% for between one and three nightclub visits in the last month and 9.7% for four or more nightclub visits in the last month. There was a similar association for visits to other night-time economy venues such as wine bars and pubs (bars): 0.3%, 0.7%, 2.0% and 4.7% last year use for 0, 1–3, 4–8 and more than nine visits to wine

bars or pubs in the last month, respectively. Finally there was also an association between alcohol intake and the last year use of mephedrone: 0.2% for those with no alcohol intake increasing to 1.9% for those drinking three or more days per week.

The 2009/10 MixMag survey reported life-time and last month mephedrone use rates of 41.7% and 33.6% respectively [16]. Further analysis of the 2295 UK respondents to this survey, reported a life-time use rate of 41.3%, with 38.7% and 33.2% reporting use within the last year and last month, respectively [48]. In the 2010/11 MixMag survey, life-time use of mephedrone had increased to 61% and use of mephedrone within the last year had also increased to 51% [49]. The recent (last year) use of mephedrone decreased with increasing age, with last year use rates of 58%, 53% and 37% for those aged 18–20 years old, 21–30 years old and older than 30 years, respectively. Finally, in the expanded 2011/2012 Global Drugs Survey, life-time prevalence of mephedrone use amongst UK respondents had fallen 42.7%, probably due to differing sampling methodology [50]. There was geographical variation in use, with last year use of 19.5% amongst all UK respondents compared to only 2% for US respondents. There was also higher use amongst those who were regular clubbers in the UK (30%) compared to the total respondent population. However, in a convenience sample of 207 bar-goers in Lancashire, UK, only 5% of those surveyed reported use of mephedrone within the last month [51]. This suggests that, even within a country, there is likely to be regional variation in the use of mephedrone.

There have been several studies that have reported on the use of mephedrone, along with other NPS in the MSM population. In a survey of 308 attendees in South London 'gay friendly' nightclubs in the summer of 2010, 54% reported life-time use of mephedrone and 52% reported use within the last year [52]. A subsequent survey in the same nightclubs one year after the

UK ban showed that there was ongoing significant use of mephedrone with 41% reporting that on the night of the survey they had already taken mephedrone and/or were planning on taking mephedrone later that evening; it was also the most commonly reported 'favourite drug' (20.4%) by those surveyed [53]. In the 'same-sex attracted adults' survey of 572 individuals aged 18–25 years old living or spending time in Sydney, Australia, the lifetime prevalence of use (4.0%) of mephedrone was similar to that in the British Crime Survey 2010/2011 and the Crime Survey England and Wales 2011/2012 [44,45,54].

In a further study from Australia of 693 individuals who self-reported regular 'ecstasy' use, 28% reported that they had used an emerging (novel) psychoactive substance in the last six months [55]. Mephedrone was the most commonly used, with 21% reporting that they had used it ever and 17% reporting that they had used it in the last six months.

ACUTE TOXICITY

Animal Data

The whole animal studies that have investigated the acute toxicity of mephedrone, in particular cardiovascular and thermoregulatory effects, are summarised within the pharmacodynamic section of this chapter.

Human Data

Similar to other NPS, there are numerous sources of information that can be combined together to build a profile of the acute toxicity associated with the use of mephedrone. These include user reports on Internet discussion fora and subpopulation level surveys, data from calls to poisons information services and published case reports and case series. The main limitation of these different data sources is that typically they are based on self-reported drug(s) used, rather than analytical confirmation of the substance use (either of the product itself or from biological samples from the individual(s) involved). However, despite this limitation, since this process, more commonly known as 'data triangulation', draws on a variety of different sources of information it allows a more robust profile of the acute toxicity to be described [56]. In this next section, we will utilise this process of data triangulation to draw on the information sources in relation to mephedrone to describe the pattern of acute toxicity associated with its use. In addition, for mephedrone, there is one series of analytically confirmed acute toxicity which means that reliable data on the pattern of acute toxicity in humans is available [21].

The majority of data from clinical sources (case reports/series and poisons information services/centres) on the acute unwanted effects is from the UK and Europe. There is limited data from the USA on the acute toxicity of mephedrone, since the published information often relates to the use of 'bath salts', a colloquial term typically used in the USA for a range of NPS [11,57–59]. Therefore it is not possible to utilise information from these US data sources as part of an assessment of the acute toxicity of mephedrone, since the majority of these cases will relate to other NPS rather than to mephedrone itself.

User Reports

There are numerous reports on Internet discussion fora such as Erowid and Drugs Forum [7,12,13,27]. These reports provide detailed physiological and psychological information or more descriptive information of the effects experienced after use. Some users may have used mephedrone as part of a cocktail of recreational drugs used and therefore it is not always possible to determine which of the effects seen are related to mephedrone rather than the other drugs. Despite these caveats, the unwanted symptoms described by users appear to be similar to the adverse or unwanted effects seen following these use of other stimulant recreational drugs such as amphetamine, MDMA or cocaine. Symptoms described include loss of appetite, insomnia and nightmares, increased body temperature (more commonly known as 'mephedrone sweat'), chest pain, nausea, vomiting and abdominal pain, painful joints, discolouration or mottling of extremities and joints (particularly the lower limbs), post-use depression and/or craving for mephedrone, anxiety and agitation, paranoia and hallucinations. Additionally, following nasal insufflation of mephedrone, a significant proportion of individuals complain of nasal irritation and/or bleeding, which can lead to switching to oral ingestion of mephedrone [27]. A survey of 1506 previous and/or current UK mephedrone users conducted over the Internet reported that 20% had suffered at least one 'significant negative reaction' related to the use of mephedrone [60]. Interestingly, the rate of significant negative reactions was higher amongst 'friends' of users (28%), suggesting that users may not acknowledge or self-report some of the significant adverse effects/reactions that they experience. In this Internet survey, only skin discolouration/blotches were specifically asked about and 21% of respondents reported experiencing this in relation to using mephedrone.

In the 2009 MixMag survey of over 2000 UK clubbers, data was collected specifically on the incidence of a number of pre-determined adverse effects from the cohort of 900 individuals who self-reported previous mephedrone use [48]. The frequency of these effects were: sweating (67% of those who had previously used mephedrone), headaches (51%), palpitations (43%), nausea (27%) and cold or blue fingers (15%) [16,48]. Subsequent telephone follow-up interviews were conducted with those individuals who had self-reported mephedrone use [17]. From a cohort of 218 individuals, 100 were recruited to provide qualitative information on both the positive and negative effects of

mephedrone. The study used the product of the 'frequency of effect' combined with the 'severity of effect' (maximum score 9 to enable comparison of the impact between infrequent but severe effects and frequent but less severe effects). Bruxism had the highest frequency-intensity effect product score (5.1). In decreasing product score, other physical effects related to mephedrone included: body sweats (4.4), heart racing (3.8), overheating (2.8), tremor (2.6), shortness of breath (1.9), headache (1.4) and chest pain (0.8). Using this system, 'forgetting things' had the highest neuro-psychiatric frequency-intensity effect product (3.5); other commonly reported neuropsychiatric effects included: restlessness/anxiety (3.3), paranoia (1.4), panic (1.2), agitation (1.4), visual hallucinations (0.8), auditory hallucinations (0.5) and aggression (0.2). There was a very low frequency-intensity effect product for unwanted effects such as cold/numb extremities (0.9), blue/red skin (0.5), skin rashes (0.3); this suggests that these less severe effects were occurring infrequently.

In a Scottish survey of school and college/university students, of those who self-reported using mephedrone on at least one occasion, 56% reported that they had suffered at least one adverse effect related to its use [46]. The frequency of adverse effects described was: bruxism (28.3%), paranoia (24.9%), sore nasal passages (24.4%), hot flushes (23.4%), sore mouth/throat (22.9%), nose bleeds (22.4%), suppressed appetite (21.5%), blurred vision (21.0%), palpitations (20.5%), insomnia (19.5%), hallucinations (18.0%), addiction/dependence (17.6%), nausea/vomiting (17.1%), burns (17.1%) and blue/cold extremities (14.6%). Similar to the Internet discussion fora, there was a high reported rate of local irritant effects related to the use of mephedrone. This probably reflects that mephedrone is used by nasal insufflation in a similar manner to the way that cocaine is used; however, unlike cocaine, mephedrone does not have any local anaesthetic effects.

Poisons Information Service Data

The first published data on enquiries to poisons information services was of 150 calls to the Swedish Poisons Centre relating to cathinones, of which 100 related to the use of mephedrone [61]. Clinical features were reported for all of the calls, so it is not possible to separate out the frequency of unwanted effects related to mephedrone itself. The unwanted effects reported by clinicians contacting the Swedish Poisons Centre included tachycardia (54% of cases), restlessness (37%), mydriasis (25%), hypertension (14%) and anxiety (14%).

The first enquiry to the UK National Poisons Information Service (NPIS) relating to mephedrone was in May 2009 [4]. Detailed information has been published on 131 calls to the UK NPIS relating to self-reported use of mephedrone or combined mephedrone/alcohol use [19]. Although 26 further calls were excluded as the individual contacting NPIS mentioned one or more co-used substance, it is possible that in the cohort of 131 reported, that the individual contacting may have either not asked about or reported to the NPIS other co-used substances. The most commonly reported unwanted effects were: agitation/aggression (24% of enquiries), tachycardia (22%), anxiety (15%), confusion or psychosis (14%), chest pain (13%), palpitations (11%) and nausea (11%). Less frequently occurring effects (<10% of enquiries) included dizziness, 'peripheral vasoconstriction' and 'skin changes/rashes', fever and/or sweating, headache, abdominal pain, insomnia, dizziness, hypertension, convulsions, myoclonus and reduced level of consciousness. It appears from this case series, that the duration of unwanted effects were prolonged, with 45% having symptoms continuing for more than 24 hours and 30% having symptoms for more than 48 hours after the use of mephedrone.

In addition to the information on the range and frequency of unwanted effects, data from the poisons information service can provide information on the changes of accesses/enquiries over time. Looking at the UK National Poisons Information Service data, it appears that the number of accesses to TOXBASE and enquiries to the NPIS telephone service peaked in March 2010, and subsequently declined after the control of mephedrone in the UK [4]. This apparent reduction in the number of cases of acute toxicity relating to the use of mephedrone following its control, may not only be the impact of control; clinicians were more likely to be aware of mephedrone and its toxicity from the general media interest and therefore less likely to call as they already had awareness of the compound. Additionally, studies have shown that the purity of mephedrone reduced after control [48,62], suggesting that total exposure to mephedrone in a single use session could potentially be less and therefore it may have been less likely to be associated with adverse effects. This trend was also seen in presentations to a London (UK) hospital Emergency Department with acute mephedrone toxicity [63]. There was a peak in presentations prior to the control of mephedrone, with 31 presentations in the two-month period prior to control; the frequency of presentations then fell significantly to between three and five presentations per two-month period.

Emergency Department Case Reports and Case Series

The first analytically confirmed case of acute toxicity related to the recreational use of mephedrone was in an individual who used mephedrone by both oral ingestion and intramuscular injection who developed typical features of sympathomimetic toxicity [14]. In addition, there have been a number of individual case reports of acute toxicity related to the use of mephedrone, including severe agitation and aggression [64], diabetic ketoacidosis in an individual with underlying Type 1 diabetes mellitus using mephedrone [65], myocarditis, myopericarditis and/or other ECG abnormalities [66] and a case of presumed 'mephedrone-induced euvolaemic hypo-osmotic hyponatraemia with encephalopathy and raised intra-cranial pressure' [67]. Some of these individual case reports include analytical confirmation of mephedrone use from screening of biological samples drawn from the patients involved, although some are based only on self-reported use or analysis of the powder taken.

In addition to these individual case reports, there have been published case series of presentations to the Emergency Department describing the clinical features of acute mephedrone toxicity [4,20–22].

We have previously published data on 72 presentations to our central London Emergency Department with acute toxicity related to the self-reported use of mephedrone between 1st January 2009 and 15th June 2010 [4,21,42]. In the reporting of these patients, information was extracted from the medical and nursing notes rather than using a proforma with pre-determined unwanted adverse effects. The majority of patients were male (81.9%) with a mean ± SD age of 27.8 ± 8.7 years (range 16–54 years). Mephedrone was typically used by nasal insufflation (19 (54.3%)) where route of use was specified) and oral ingestion (12 (34.3%)). Patients were included if they self-reported co-used other substances and/or ethanol; 63 of the 72 had used one or more other substance. Therefore, it is possible that some of the clinical effects seen may relate to the other co-used substances rather than to the mephedrone itself. The commonest adverse effects documented in the medical and nursing notes were: agitation (38.9% of patients), palpitations (25.0%), vomiting (13.9%), chest pain (12.5%), self-limiting pre-hospital seizures (6.9%) and headache (7.2%). We were not able to identify any cases of skin mottling, cool limbs or blue extremities as described in some of the Internet discussion fora. We had pre-defined a number of physiological parameters which we felt were indicators of clinically significant sympathomimetic (stimulant) toxicity:

1. significant hypertension – systolic BP ≥160 mmHg;
2. significant tachycardia – heart rate ≥140 b.p.m.; and
3. hyperpyrexia – temperature >38.5°C.

In this case series, 13.9% had significant hypertension and 8.3% had significant tachycardia, but no patients had hyperpyrexia. This lack of hyperpyrexia may reflect that the majority of cases presented during the autumn/winter/spring months, where ambient air temperature in the UK is lower and therefore there is likely to be a reduced risk of hyperpyrexia developing.

A subset of nine individuals from this cohort of patients underwent detailed toxicological screening of blood or urine collected at the time of presentation to confirm the use of mephedrone [22]. Of these nine individuals, mephedrone was detected in seven; in the remaining two it was thought that mephedrone was not detected in the biological matrix screened (serum/blood) as they presented more than 24 hours after use. Of those where mephedrone was detected, no other substances were detected in four, and the others had used cocaine (two patients) and a combination of butylone and MDPV (one patient). The frequency of unwanted effects seen in this subset of patients was comparable to that seen in the whole case series of patients: agitation (57.1% of patients), palpitations (28.6%), chest pain (28.6%), self-limiting pre-hospital seizure (14.3%) and headache (14.3%). In terms of clinically significant sympathomimetic toxicity, 42.9% had significant hypertension, and 14.3% had a significant tachycardia.

The second case series is of 89 patients who presented to the Emergency Department in Aberdeen, Scotland [20]. Of these 89 patients, 30 self-reported lone mephedrone use, 27 self-reported use of both mephedrone and alcohol within the same drug use session, and the remaining 32 reported use of one or more other recreational drug at the same time as the mephedrone. The authors of this case series only reported data on the acute toxicity from the 57 patients with either self-reported lone mephedrone use or co-used mephedrone and alcohol. Commonly reported clinical features seen in these individuals were anxiety or agitation (40.4%), chest pain (24.6%), parasthesiae (24.6%), palpitations (21.1%), dyspnoea (17.5%), confusion (17.5%), collapse (14.0%) and 'oral symptoms' (not defined; 12.3%). It is not possible the determine the proportion of individuals who had clinically significant hypertension or tachycardia, since only the range of systolic blood pressure (88–184 mmHg) and heart rate (68–184 b.p.m.) was reported. However, based on this data there is evidence that a proportion of them would have had clinically significant cardiovascular features of toxicity on presentation to the Emergency Department.

MEPHEDRONE-RELATED DEATHS

At the time of the EMCDDA Risk Assessment of mephedrone in 2010, there were reports of deaths related to mephedrone from Sweden and the UK, and deaths in which mephedrone had been detected but not contributed to death from the UK and Ireland [4]. In addition, there were many deaths in which either mephedrone had been detected in post-mortem samples or the individual had anecdotally been using mephedrone, but the inquests into death had not been completed and so the role of mephedrone in the deaths was not clear [4]. These potential mephedrone-related deaths (many of which subsequently turned out either not to involve mephedrone or to have an alternative cause of death) received significant coverage in the UK popular press [4,68]. It has been postulated that this media coverage helped to increase interest in mephedrone amongst drug users and potential users; a recent study has shown an association between peaks in media coverage

concerning potential mephedrone related deaths and searches on the Internet investigating purchase of mephedrone [68].

The first death related to mephedrone was in an 18-year-old female in Sweden [69]. She reported use of mephedrone and cannabis and had an out of hospital cardiorespiratory arrest. She was resuscitated in the Emergency Department and found to have a metabolic acidosis, cerebral oedema and hyponatraemia (serum sodium 120 mmol/L); no data was provided in the report to be able to determine whether tests were undertaken to investigate the cause of the hyponatraemia. The patient was declared brain dead on the intensive care unit 36 hours later. Qualitative analysis of blood and urine samples revealed the presence of mephedrone only, with no other drugs or alcohol detected.

The second published case in which mephedrone was detected in post-mortem samples was from Maryland, USA [70]. This was a 22-year-old male found collapsed and unresponsive at home; resuscitation was unsuccessful and he was pronounced dead in hospital. In addition to mephedrone, urine screening was positive for 6-acetylmorphine, codeine, morphine and doxylamine. The post-mortem blood mephedrone concentration was 0.5 mg/L. Death was recorded at the inquest as due to 'accidental multiple drug toxicity' and it is not possible to determine from the published report the role that mephedrone played in this death.

In the UK the National Programme on Substance Abuse Deaths (np-SAD) collates data on drug-related deaths. In 2012, this group reported data on cases reported to them up to the end of October 2011 in which there had been suspicion of mephedrone involvement [71]. There were a total of 128 potential mephedrone associated deaths. Mephedrone was not found in post-mortem analysis in 25 cases and analytical results were still pending in a further 10 cases. In the 90 cases in which mephedrone was detected, inquests into the cause of death had been completed in 69 cases and data was available for 62 of these. The mean ± SD age of these 62 cases was 28.8 ± 11.3 years (range 14–64 years). Mephedrone was the only drug detected in 8 cases, other substances detected included alcohol (n = 26), stimulants e.g. amphetamine, cocaine, MDMA (22), sedative/hypnotics (22), opioids (13), piperazeins (13), other acthinones (13), antidepressants (8), antipsychotics (5), gamma-hydroxybutyrate (GHB, 5). In a number of cases other factors were recorded (e.g. cardiovascular disease, bronchopneumonia) and self-harm (18 cases) including hanging (11 cases) was also commonly recorded. Mephedrone was included in the cause of death in 36 cases, from the data available in the paper it is not possible to determine whether mephedrone was the cause of death in these 36 cases.

In 2011, a case series of four deaths in Scotland in which mephedrone was detected in post-mortem samples was published [72]. Death was reported to be due to mephedrone in one of these cases (blood mephedrone concentration 0.98 mg/L); other factors played a role in the other three cases (other drugs in two cases and trauma in a road traffic accident in the third case).

In a recent report from Northern Ireland, post-mortem toxicology analyses undertaken by the Forensic Science Service in Northern Ireland from late 2009 to the end of 2010 were reported. Mephedrone was detected in 12 deaths and was felt to be the cause of two deaths (blood mephedrone concentration were 2.1 and 1.94 mg/L in these two cases) [73].

There are also reported deaths related to mephedrone from Poland [74], Italy [75] and the Netherlands [64].

CHRONIC TOXICITY

Animal Data

Animal models of mephedrone toxicity and unwanted effects have focused on single dose or short-term (up to 1–3 weeks) administration of mephedrone, rather than longer-term administration of mephedrone to determine its chronic toxicity and therefore currently there is no data available from animal studies on the chronic toxicity of mephedrone.

Human Data

There are no published cases of long-term mephedrone use being associated with specific chronic toxicity. There are anecdotal user reports that long-term mephedrone use is associated with post-use 'depression' [12,13,27]. In the MixMag survey telephone follow-up survey, the frequency-intensity product for depression, which was reported by 56 individuals, was 2.3 [17]. However, there are no experimental or clinical data to support users' hypotheses that this relates to depletion of serotonin or dopamine.

One study assessed working memory, phonological and semantic fluency, psychomotor speed and executive control in 20 mephedrone users whilst intoxicated and then when drug free and compared this with 20 controls [76]. In addition, psychological wellbeing, schizotypy and depression scores were compared between controls and mephedrone users. The mephedrone group had used mephedrone for a mean ± SD of 1.35 ± 0.37 years for 4.05 ± 1.70 days per month (range 2–8). Mephedrone users had higher depression scores than controls (p = 0.01), this remained significant when other factors were controlled including use of other drugs. Prose recall, both immediate and delayed, was poorer in the mephedrone users (p < 0.001) and the mephedrone group have higher scores in schizotypy (p < 0.001). There was no difference in verbal fluency between the mephedrone group and controls.

Finally, acute toxicity of mephedrone can be associated with long-term consequences related to secondary complications similar to that seen with other stimulant recreational drugs; an example of this would be mephedrone-induced convulsions leading to cerebral hypoxia and long-term neurological consequences.

DEPENDENCE AND ABUSE POTENTIAL

Animal Data

There are no published animal studies or models that have investigated the development of mephedrone-related dependence. However, there have been studies which have reported on whether mephedrone is associated with self-administration. This could be considered to be a surrogate marker of dependence and abuse potential, since drug (and food) can be used in animal models to induce self-administration secondary to the associated reward of being delivered either a drug with beneficial effects or food.

In these animal models of self-administration there is evidence that mephedrone is associated with increased rates of self-administration and total doses administered, suggesting that it may be associated with a risk of abuse potential [77,78]. In a model of intravenous mephedrone self-administration, both Sprague-Dawley and Wistar rats had increasing self-administration of mephedrone over a 10 session study when a dose of 1 mg/kg/infusion was used, whereas there was a decrease to a relatively stable plateau of rate of self-administration when a dose of mephedrone of 0.5 mg/kg/infusion was used [78]. Additionally, in further experiments, there appeared to be a difference in self-administration between the two strains of rats, with Wistar rats appearing to have greater self-administration than Sprague-Dawley rats.

In a rat model, mephedrone self-administration was compared to self-administration of methamphetamine and saline [77]. The mean total amount of mephedrone self-administered increased from 1.77 ± 0.15 mg on Day 1 to 6.78 ± 1 mg on Day 8. In this model, there was an increasing number of 'active lever' presses

from Day 1 throughout the whole study. In addition the discrimination between the mephedrone delivery lever and the control inactive lever increased from 2.65:1 (mephedrone active:inactive) presses on Day 1 to 10.71:1 on Day 8, suggesting that over time the rats were able to differentiate which lever would lead to administration, and also increased their overall administration of mephedrone. In the saline control rats, there was a decrease in the rate of active lever pressing rapidly from day 1 of the study. Finally, when methamphetamine was used for self-administration, rats rapidly developed a steady state of active lever pressing over time and had a much lower mean total amount of methamphetamine administered per day $(2.55 \pm 0.06$ mg).

Human Data

There is anecdotal evidence from users of dependence-like symptoms related to the use of mephedrone [4,8,12,13,15,42,79,80]. In a survey of Scottish school and college/university students, 17.6% of the 205 who self-reported mephedrone use reported 'addiction and/or dependence type symptoms' [42]. From Internet discussion fora, those individuals with high and/or frequent use of mephedrone report cravings for mephedrone between use sessions [8,12,13,15,80]. Data compiled by the Slovenian organisation DrogArt from their outreach work at dance events and nightclubs, Internet/telephone counselling and over 6000 users of an Internet discussion forum suggested that 'craving' was the main problem associated with the use of mephedrone and that was more severe that that seen with cocaine, methamphetamine or speed [4,79].

At the time of the EMCDDA risk assessment of mephedrone in 2010, there were anecdotal reports of mephedrone dependence reported to the UK National Drug Treatment Monitoring system [4]. There was a suggestion that the 300% increase in referrals to Forum for Action on Substance Abuse and Suicide Awareness (FASA) in Belfast, Northern Ireland between January 2009 and January 2010 was related to the use of mephedrone [4]. Similarly the Dublin Youth Drug and Alcohol Service in Ireland reported that 11% of their assessments in the first six months of 2010 were related to the use of 'head shop' drugs, which would have included mephedrone [4].

Questions relating to mephedrone-related dependence were also included in the telephone follow-up survey of 100 regular mephedrone users recruited through the 2010/2011 MixMag survey [17]. These questions included the DSM-IV dependence criteria, and frequency of positive responses amongst the mephedrone users were: 1) usual dose no longer has desired effect – 54.1%; 2) taken mephedrone or other stimulant to relieve withdrawal effects – 12.2%; 3) taken for longer or in larger amounts that intended – 62.2%; 4) wanted to cut down or stop but had not been successful – 14.3%; 5) much time obtaining, taking or recovering from – 20.4%; 6) important social, occupational or recreational activities given up – 7.1%; and 7) continued to take in spite of physical or psychological problems – 24.5%. Overall, 29.5% of individuals reported three or more of these criteria, and therefore met the criteria for a 'possible clinical diagnosis of stimulant dependence'. It is possible that some of the dependence criteria could be related to one or more co-used substances, since 10 of the 14 individuals who submitted urine samples for screening had one or more other substance detected in addition to mephedrone.

To date we are aware of only one published case of formally diagnosed mephedrone-related dependence published in the medical literature [81]. A young professional male presented to psychiatry services in Scotland, UK with dependence symptoms following 18 months use of oral, nasal and rectal mephedrone. Prior to presentation he was bingeing on 4–5 g of mephedrone twice per week. He fulfilled the International Classification of Diseases (ICD) 10 criteria for dependence syndrome; he was managed with changing his fluoxetine to olanzapine and was discharged after four weeks inpatient treatment.

Based on this information, it appears that the dependence related to the use of mephedrone is similar to that seen with other stimulant recreational drugs such as MDMA and cocaine. In particular, it appears that this is a psychological dependence, rather than a physical dependence. Therefore treatment of this is likely to involve appropriate psychological support (e.g. motivational interviewing, cognitive behavioural therapy) rather than requiring the need of specific pharmacological treatment for an acute physical withdrawal and/or maintaining abstinence.

REFERENCES

[1] Gibbons S. 'Legal highs' – novel and emerging psychoactive drugs: a chemical overview for the toxicologist. Clin Toxicol (Phila) 2012;50:15–24.

[2] Gibbons S, Zloh M. An analysis of the 'legal high' mephedrone. Bioorg Med Chem Lett 2010;20:4135–9.

[3] EMCDDA Annual report 2012: the state of the drugs problem in Europe. Available:<http://www.emcdda.europa.eu/attachements.cfm/att_190854_EN_TDAC12001ENC_.pdf> [accessed 18.02.13].

[4] EMCDDA Risk assessments: report on the risk assessment of mephedrone in the framework of the council decision on new psychoactive substances. Available: <http://www.emcdda.europa.eu/attachements.cfm/att_116646_EN_TDAK11001ENC_WEB-OPTIMISED%20FILE.pdf> [accessed 18.02.13].

[5] Camilleri A, Johnston MR, Brennan M, Davis S, Caldicott DG. Chemical analysis of four capsules containing the controlled substance analogues 4-methylmethcathinone, 2-fluoromethamphetamine, alpha-phthalimidopropiophenone and N-ethylcathinone. Forensic Sci Int 2010;197:59–66.

[6] Newcombe R. Mephedrone. The use of mephedrone (M-cat, Meow) in Middlesbrough. Manchester, UK: Lifeline Publications and Research; Available: <http://ewsd.wiv-isp.be/Publications%20on%20new%20psychoactive%20substances/Mephedrone/M-cat%20report%20small.pdf> [accessed 18.02.13].

[7] Psychonaut WebMapping Research Group Mephedrone report. London: Institute of Psychiatry, King's College London; 2009.

[8] Schifano F, Albanese A, Fergus S, Psychonaut Web Mapping ReDNet Research Groups Mephedrone (4-methylmethcathinone; 'meow meow'): chemical, pharmacological and clinical issues. Psychopharmacology (Berl) 2011;214:593–602.

[9] 4-Methylmethcathinone basics. Available: <http://www.erowid.org/chemicals/4_methylmethcathinone/4_methylmethcathinone_basics.shtml> [accessed 18.02.13].

[10] Prosser JM, Nelson LS. The toxicology of bath salts: a review of synthetic cathinones. J Med Toxicol 2012;8:33–42.

[11] Spiller HA, Ryan ML, Weston RG, Jansen J. Clinical experience with and analytical confirmation of 'bath salts' and 'legal highs' (synthetic cathinones) in the United States. Clin Toxicol (Phila) 2011;49:499–505.

[12] 4-Methylmethcathinone reports. Available: <http://www.erowid.org/experiences/subs/exp_4Methylmethcathinone.shtml> [accessed 18.02.13].

[13] Mephedrone and Beta-Ketones. Available: <http://www.drugs-forum.com/forum/forumdisplay.php?f=377> [accessed 18.02.13].

[14] Wood DM, Davies S, Puchnarewicz M, Button J, Archer R, Ovaska H, et al. Recreational use of mephedrone (4-methylmethcathinone, 4-MMC) with associated sympathomimetic toxicity. J Med Toxicol 2010;6:327–30.

[15] Measham F, Moore K, Newcombe R, Welch Z. Tweaking, bombing, dabbing and stockpiling: the emergence of mephedrone and the perversity of prohibition. Drugs Alcohol Today 2010;10:14–21.

[16] Dick D, Torrance C. MixMag drugs survey. MixMag 2010;225:44–53.

[17] Winstock A, Mitcheson L, Ramsey J, Davies S, Puchanarewicz Marsden J. Mephedrone: use, subjective effects and health risks. Addiction 2011;106:1991–6.

[18] 4-Methylmethcathinone dose. Available: <http://www.erowid.org/chemicals/4_methylmethcathinone/4_methylmethcathinone_dose.shtml> [accessed 18.02.13].

[19] James D, Adams RD, Spears R, National Poisons Information Service. Clinical characteristics of mephedrone toxicity reported to the UK National Poisons Information Service. Emerg Med J 2011;28:686–9.

[20] Regan L, Mitchelson M, Macdonald C. Mephedrone toxicity in a Scottish emergency department. EMJ 2011;28:1055–8.

[21] Wood DM, Davies S, Greene SL, Button J, Holt DW, Ramsey J, et al. Case series of individuals with analytically confirmed acute mephedrone toxicity. Clin Toxicol (Phila) 2010;48:924–7.

[22] Wood DM, Greene SL, Dargan PI. Clinical pattern of toxicity associated with the novel synthetic cathinone mephedrone. Emerg Med J 2011;28:280–2.

[23] Meyer MR, Wilhelm J, Peters FT, Maurer HH. Beta-keto amphetamines: studies on the metabolism of the designer drug mephedrone and toxicological detection of mephedrone, butylone, and methylone in urine using gas chromatography-mass spectrometry. Anal Bioanal Chem 2010;397:1225–33.

[24] Khreit OI, Grant MH, Zhang T, Henderson C, Watson DG, Sutcliffe OB. Elucidation of the Phase I and Phase II metabolic pathways of (±)-4'-methylmethcathinone (4-MMC) and (±)-4'-(trifluoromethyl)methcathinone (4-TFMMC) in rat liver hepatocytes using LC-MS and LC-MS^2. J Pharm Biomed Anal 2013;72:177–85.

[25] Pedersen AJ, Reitzel LA, Johansen SS, Linnet K. In vitro metabolism studies on mephedrone and analysis of forensic cases. Drug Test Anal 2012 May 9 [Epub ahead of print].

[26] Wright Jr MJ, Angrish D, Aarde SM, Barlow DJ, Houseknecht KL, Dickerson TJ, et al. Effect of ambient temperature on the thermoregulatory and locomotor stimulant effects of 4-methylmethcathinone in Wistar and Sprague-Dawley rats. PLoS One 2012;7:e44652.

[27] 4-Methylmethcathinone effects. Available: <http://www.erowid.org/chemicals/4_methylmethcathinone/4_methylmethcathinone_effects.shtml> [accessed 18.02.13].

[28] Martínez-Clemente J, Escubedo E, Pubill D, Camarasa J. Interaction of mephedrone with dopamine and serotonin targets in rats. Eur Neuropsychopharmacol 2012;22:231–6.

[29] Baumann MH, Ayestas Jr MA, Partilla JS, Sink JR, Shulgin AT, Daley PF, et al. The designer methcathinone analogs, mephedrone and methylone, are substrates for monoamine transporters in brain tissue. Neuropsychopharmacology 2012;37:1192–203.

[30] Kehr J, Ichinose F, Yoshitake S, Goiny M, Sievertsson T, Myberg F, et al. Mephedrone, compared with MDMA (ecstasy) and amphetamine, rapidly increases both dopamine and 5-HT levels in nucleus accumbens of awake rats. Br J Pharmacol 2011;164:1949–58.

[31] Cameron K, Kolanos R, Verkariya R, De Felice L, Glennon RA. Mephedrone and methylenedioxypyrovalerone (MDPV), major constituents of 'bath salts', produce opposite effects at the human dopamine transporter. Psychopharmacology (Berl) 2013 [Epub ahead of print].

[32] Ramoz L, Lodi S, Bhatt P, Reitz AB, Tallarida C, Tallardia RJ, et al. Mephedrone ('bath salt') pharmacology: insights from invertebrates. Neuroscience 2012;208:79–84.

[33] Motbey CP, Hunt GE, Bowen MT, Artiss S, McGregor IS. Mephedrone (4-methylmethcathinone, 'meow'): acute behavioural effects and distribution of Fos expression in adolescent rats. Addict Biol 2012;17:409–22.

[34] Shortall SE, Macerola AE, Swaby RT, Jayson R, Korsah C, Pilledge KE, et al. Behavioural and neurochemical comparison of chronic intermittent cathinone, mephedrone and MDMA administration to the rat. Eur Neuropsychopharmacol 2012 Oct 7 doi:10.1016/j.euroneuro.2012.09.005.

[35] Motbey CP, Karanges E, Li KM, Wilkinson S, Winstock AR, Ramsay J, et al. Mephedrone in adolescent rats: residual memory impairment and acute but not lasting 5-HT depletion. PLoS One 2012;7:e45473.

[36] den Hollander B, Rozov S, Linden AM, Uusi-Oukari M, Ojanperä I, Korpi ER. Long-term cognitive and neurochemical effects of 'bath salt' designer drugs methylone and mephedrone. Pharmacol Biochem Behav 2013;103:501–9.

[37] Shortall SE, Green AR, Swift KM, Fone KCF, King MV. Lost in translation: preclinical studies on 3,4-methylenedioxymethamphetamine provide information on mechanisms of action, but do not allow accurate prediction of adverse events in humans. Br J Pharmacol 2012;166:1523–36.

[38] Angoa-Pérez M, Kane MJ, Francescutti DM, Sykes KE, Shah MM, Mohammed AM, et al. Mephedrone, an abused psychoactive component of 'bath salts' and methamphetamine congener, does not cause neurotoxicity to dopamine nerve endings of the striatum. J Neurochem 2012;120:1097–107.

[39] Meng H, Cao J, Kang J, Ying X, Ji J, Reynolds W, et al. Mephedrone, a new designer drug of abuse, produces acute hemodynamic effects in the rat. Toxicol Lett 2012;208:62–8.

[40] Varner KJ, Daigle K, Weed PF, Lewis PB, Mahne SE, Sankaranarayanan A, et al. Comparison of the behavioral and cardiovascular effects of mephedrone with other drugs of abuse in rats. Psychopharmacology (Berl) 2013;225:675–85.

[41] Europol–EMCDDA Joint report on a new psychoactive substance: 4-methylmethcathinone (mephedrone). Available: <http://www.emcdda.europa.eu/attachements.cfm/att_132203_EN_2010_Mephedrone_Joint%20report.pdf> [accessed 18.02.13].

[42] Dargan PI, Sedefov R, Gallegos A, Wood DM. The pharmacology and toxicology of the synthetic cathinone mephedrone (4-methylmethcathinone). Drug Test Anal 2011;3:454–63.

[43] United Nations Office on Drugs and Crime (UNODC) World Drug report 2012 <http://www.unodc.org/documents/data-and-analysis/WDR2012/WDR_2012_web_small.pdf> [accessed 18.02.13].

[44] Drug Misuse Declared: Findings from the 2010/11 British Crime Survey England and Wales. Available: <http://www.homeoffice.gov.uk/publications/science-research-statistics/research-statistics/crime-research/hosb1211/hosb1211?view=Binary> [accessed 18.02.13].

[45] Drug Misuse Declared: Findings from the 2011/12 Crime Survey for England and Wales. 2nd ed. July 2012. <http://www.homeoffice.gov.uk/publications/science-research-statistics/research-statistics/crime-research/drugs-misuse-dec-1112/drugs-misuse-dec-1112-pdf?view=Binary> [accessed 18.02.13].

[46] Dargan PI, Albert S, Wood DM. Mephedrone use and associated adverse effects in school and college/university students before the UK legislation change. QJM 2010;103:875–9.

[47] Meehan C. Doctoral study on adolescent drug taking in Northern Ireland. University of Ulster.

[48] Winstock AR, Mitcheson LR, Deluca P, Davey Z, Corazza O, Schifano F. Mephedrone, new kid for the chop? Addiction 2011;106:154–61.

[49] Winstock A. The 2011 MixMag drugs survey. MixMag 2011;238:49–59.

[50] Winstock A. MixMag global drugs survey. MixMag 2012;251:68–73.

[51] Measham F, Moore K, Østergaard J. Mephedrone, 'Bubble' and unidentified white powders: the contested identities of synthetic 'legal highs'. Drugs Alcohol Today 2011;11:137–47.

[52] Measham F, Wood DM, Dargan PI, Moore K. The Rise in Legal Highs: prevalence and patterns in the use of illegal drugs and first and second generation 'legal highs' in south London gay dance clubs. J Subs Use 2011;16:263–72.

[53] Wood DM, Measham F, Dargan PI. 'Our favourite drug': prevalence of use and preference for mephedrone in the London night time economy one year after control. J Subs Use 2011;68:853–6.

[54] Lea T, Reynolds R, De Wit J. Mephedrone use among same-sex attracted young people in Sydney, Australia. Drug Alcohol Rev 2011;30:438–40.

[55] Bruno R, Matthews AJ, Dunn M, Alali R, McIlwraith F, Hickey S, et al. Emerging psychoactive substance use among regular ecstasy users in Australia. Drug Alcohol Depend 2012;124:19–25.

[56] Wood DM, Dargan PI. Understanding how data triangulation identifies acute toxicity of novel psychoactive drugs. J Med Toxicol 2012;8:300–3.

[57] Adebamiro A, Perazella MA. Recurrent acute kidney injury following bath salts intoxication. Am J Kidney Dis 2012;59:273–5.

[58] Centers for Disease Control, Prevention (CDC) Emergency department visits after use of a drug sold as 'bath salts'– Michigan, November 13, 2010– March 31, 2011. MMWR Morb Mortal Wkly Rep 2011;60:624–7.

[59] Kasick DP, McKnight CA, Klisovic E. 'Bath salt' ingestion leading to severe intoxication delirium: two cases and a brief review of the emergence of mephedrone use. Am J Drug Alcohol Abuse 2012;38:176–80.

[60] Carhart-Harris RL, King LA, Nutt DJ. A web-based survey on mephedrone. Drug Alcohol Depend 2011;118:19–22.

[61] Hägerkvist R, Hultén P, Personne M. Increasing abuse of new cathinone derivatives in Sweden – a poisons centre study for the years 2008–2009. Clin Toxicol (Phila) 2010;48:291–2.

[62] Winstock A, Mitcheson L, Marsden J. Mephedrone: still available and twice the price. Lancet 2010;376:1537.

[63] Wood DM, Greene SL, Dargan PI. Emergency department presentations in determining the effectiveness of drug control in the United Kingdom: mephedrone (4-methylmethcathinone) control appears to be effective using this model. Emerg Med J 2013;30:70–1.

[64] Lusthof KJ, Oosting R, Maes A, Verschraagen M, Dijkhuizen A, Sprong AG. A case of extreme agitation and death after the use of mephedrone in The Netherlands. Forensic Sci Int 2011;206:e93–5.

[65] Wong ML, Holt RI. The potential dangers of mephedrone in people with diabetes: a case report. Drug Test Anal 2011;3:464–5.

[66] Nicholson PJ, Quinn MJ, Dodd JD. Headshop heartache: acute mephedrone 'meow' myocarditis. Heart 2010;96:2051–2.

[67] Sammler EM, Foley PL, Lauder GD, Wilson SJ, Goudie AR, O'Riordan JI. A harmless high?. Lancet 2010;376:742.

[68] Forsyth AJ. Virtually a drug scare: mephedrone and the impact of the internet on drug news transmission. Int J Drug Policy 2012;23:198–209.

[69] Gustavsson D, Escher C. Mephedrone – Internet drug which seems to have come and stay. Fatal cases in Sweden have drawn attention to previously unknown substance. Lakartidningen 2009;106:2769–71.

[70] Dickson AJ, Vorce SP, Levine B, Past MR. Multiple-drug toxicity caused by coadministration of 4-methylmethcathinon (mephedrone) and heroin. J Anal Toxicol 2010;34:162–8.

[71] Schifano F, Corkery J, Ghodse AH. Suspected and confirmed fatalities associated with mephedrone (4-methylmethcathinone, 'meow meow') in the United Kingdom. J Clin Psychopharmacol 2012;32:710–4.

[72] Maskell PD, De Paoli G, Seneviratne C, Pounder DJ. Mephedrone (4-methylmethcathinone)-related deaths. J Anal Toxicol 2011;35:188–91.

[73] Cosbey SH, Peters KL, Quinn A, Bentley A. Mephedrone (methylmethcathinone) in toxicology casework: a Northern Ireland perspective. J Anal Toxicol 2013;37:74–82.

[74] Adamowicz P, Tokarczyk B, Stanaszek R, Slopianka M. Fatal mephedrone intoxication – a case report. J Anal Toxicol 2013;37:37–42.

[75] Aromatario M, Bottoni E, Santoni M, Ciallella C. New 'lethal highs': a case of a deadly cocktail of GHB and Mephedrone. Forensic Sci Int 2012;223:e38–41.

[76] Freeman TP, Morgan CJ, Vaughn-Jones J, Hussain N, Karimi K, Curran HV. Cognitive and subjective effects of mephedrone and factors influencing use of a 'new legal high'. Addiction 2012;107:792–800.

[77] Hadlock GC, Webb KM, McFadden LM, et al. 4-Methylmethcathinone (mephedrone): neuropharmacological effects of a designer stimulant of abuse. J Pharmacol Exp Ther 2011;339:530–6.

[78] Aarde SM, Angrish D, Barlow DJ, et al. Mephedrone (4-methylmethcathinone) supports intravenous self-administration in Sprague-Dawley and Wistar rats. Addict Biol 2013 Jan 30 doi:10.1111/adb.12038.

[79] Pas M. Mephedrone in Slovenia. DrogArt report. May 2010. In: EMCDDA – Appendix 2 to Annex 1 (Technical report on mephedrone): Mephedrone. Additional studies, overview of prevalence, use patterns, effects. Available: <http://www.ofdt.fr/BDD/publications/docs/rarOEDTmephA1A2.pdf> [accessed 18.02.13].

[80] Deluca P, Davey Z, Corazza O, et al. Identifying emerging trends in recreational drug use; outcomes from the Psychonaut Web Mapping Project. Prog Neuropsychopharmacol Biol Psychiatry 2012;39:221–6.

[81] Bajaj N, Mullen D, Wylie S. Dependence and psychosis with 4-methylmethcathinone (mephedrone) use. BMJ Case Rep 2010;2010 Nov 3. doi: pii: bcr0220102780.

This page intentionally left blank

Pipradrol and Pipradrol Derivatives

Michael W. White* and John R.H. Archer†

*Formerly of the Forensic Science Service Ltd, London, UK
†Guy's and St Thomas' NHS Foundation Trust, London, UK

PHARMACOLOGY

Physical and Chemical Description

Pipradrol and pipradrol derivatives are a group of amphetamine type stimulants structurally related to methylamphetamine (Fig. 10.1) and are characterised by the presence of a large hydrophobic diphenylmethyl substituent attached to the α-carbon atom of a cyclic amine.

Various pipradrol derivatives have been investigated and found to have stimulant properties [1–19]. Of particular interest is desoxypipradrol, a 'legal high' which appeared on the United Kingdom (UK) market in 2009 in a collected sample of 'Ivory Wave' [20], and the

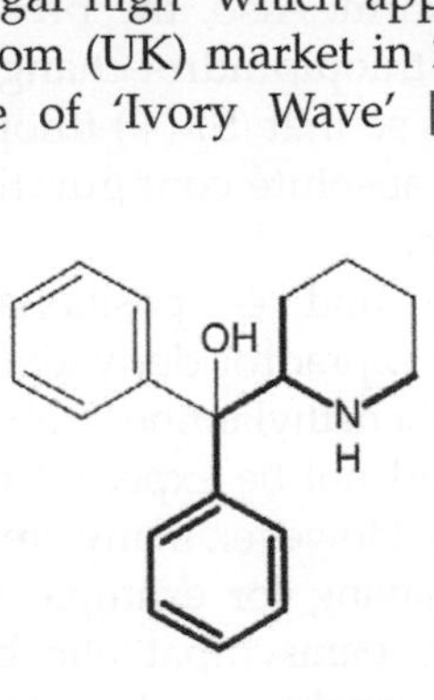

FIGURE 10.1 Structure of pipradrol showing the methylamphetamine substructure in bold.

pyrrolidine derivatives, diphenylprolinol and 2-(diphenylmethyl)pyrrolidine (Fig. 10.2), which according to UK Early Warning System seizure data [21], have since replaced desoxypipradrol with a total of 16 seizure reports in 2011 compared to none for desoxypipradrol, other than a test purchase sample from a head shop in Edinburgh [22].

Other simple modifications to the pipradrol structure (Fig. 10.3), such as addition of halo, alkyl, hydroxy or alkoxy groups to one or both phenyl rings or substitution on the ring nitrogen atom with alkyl, alkylenyl, haloalkyl or hydroxyalkyl groups have also been reported in the literature and many are claimed to have a stimulant effect on the central nervous system (CNS) [1,2,7,9,10,12,15]. Replacement of the hydroxyl group of pipradrol with an alkyl or hydroxyalkyl group also produces compounds with stimulant properties [5,10]. The piperidine ring may also be modified by substitution on a ring carbon atom with an hydroxy group [13,18].

Other pipradrol derivatives with stimulant properties reported in the literature include compounds in which the piperidine ring has been replaced with another cyclic amine,

Novel Psychoactive Substances.
DOI: http://dx.doi.org/10.1016/B978-0-12-415816-0.00010-9

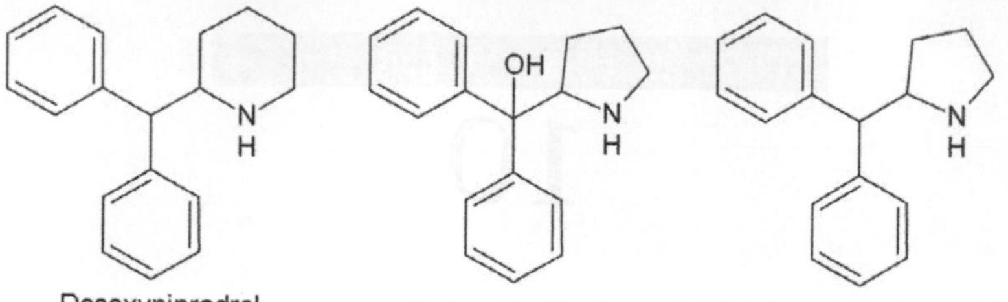

Desoxypipradrol   Diphenylprolinol   2-(Diphenylmethyl)pyrrolidine

FIGURE 10.2 Structures of pipradrol, desoxypipradrol, D2PM and desoxy-D2PM.

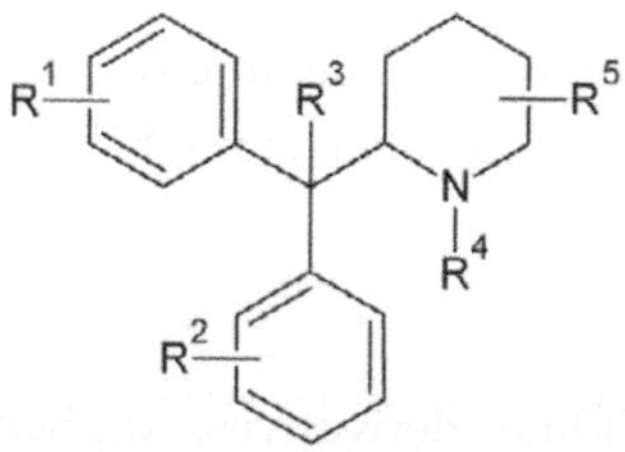

R^1 and R^2 = H, alkyl, alkoxy or halide

R^3 = H, hydroxy, alkyl, or hydroxyalkyl

R^4 = H, alkyl, alkylenyl, haloalkyl, or hydroxyalkyl

R^5 = H or hydroxy

FIGURE 10.3 Structures of other pipradrol derivatives.

including azepane, morpholine, thiomorpholine, or a pyridine ring (Fig. 10.4) [11,14,16,19].

Pipradrol derivatives in which one of the phenyl groups is replaced by hydrogen [23], alkyl [24], hydroxyl and ester derivatives (e.g. levophacetoperane) [25], carbomethoxy (e.g. methylphenidate) or another ring system [5,7], have been investigated, as well as acyclic amine analogues; [26] however, the structures of these compounds do not possess the characteristic cyclic amine and hydrophobic diphenylmethyl group found in pipradrol, and are therefore not considered in detail in this chapter.

Like amphetamine, pipradrol and most of the pipradrol derivatives have an asymmetric carbon atom at the 2-position of the piperidine ring so that each compound exists as a pair of stereoisomers, each a mirror image of the other. The absolute configurations of pipradrol, desoxypipradrol and the thiomorpholine analogue of pipradrol ('thiopipradrol') have been determined and compared to those of amphetamine [27]. Interestingly, the more pharmacologically active enantiomers of pipradrol, desoxypipradrol and thiopipradrol are stereochemically superimposable, but are not superimposable on the more active (S)-(+)-enantiomer of amphetamine (Fig. 10.5). The nomenclature denoting stereoisomers can lead to confusion as the hydrochloride salts of pipradrol, desoxypipradrol and thiopipradrol rotate the plane of polarised light in the opposite direction to that for the base form. Also, the presence of a sulphur atom in thiopipradrol changes the substituent priorities so that (S)-(+)-thiopipradrol base has the same absolute configuration as (R)-(+)-pipradrol base.

The 1-, 3- and 4- positional isomers of pipradrol and pipradrol derivatives do not contain the β-phenethylamine substructure and therefore would not be expected to have stimulant properties. However, many are pharmacologically active having for example antihistaminic, antispasmodic, parasympathetic blocking (antiacetylcholine), analgesic, depressant or antipsychotic activity [6,28–31]. The routine forensic

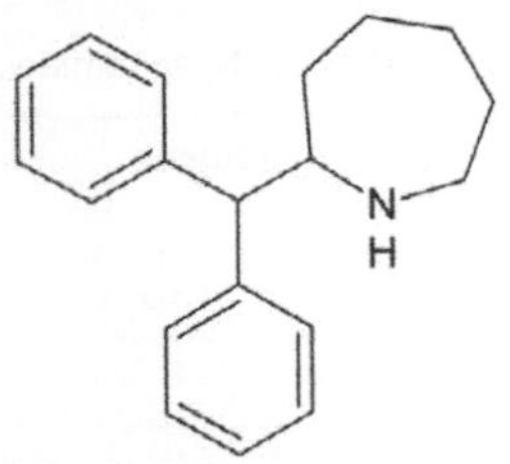

2-(Diphenylmethyl)azepane

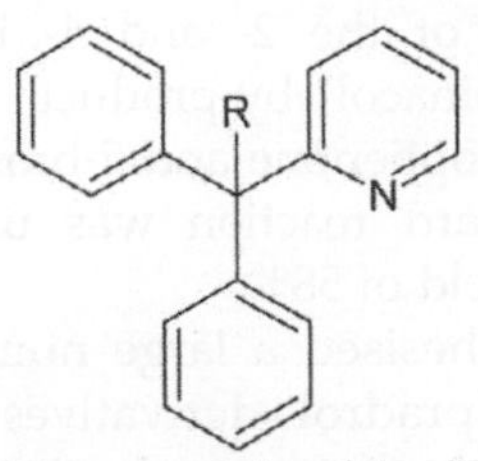

2-(Diphenylmethyl)pyridine (R=H)
α,α-Diphenyl-2-pyridinemethanol (R=OH)

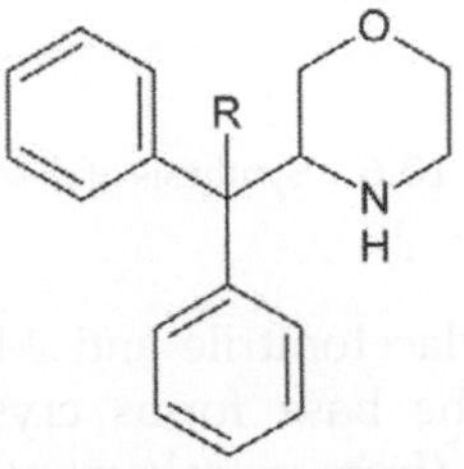

3-(Diphenylmethyl)morpholine (R=H)
α,α-Diphenyl-3-morpholinemethanol (R=OH)

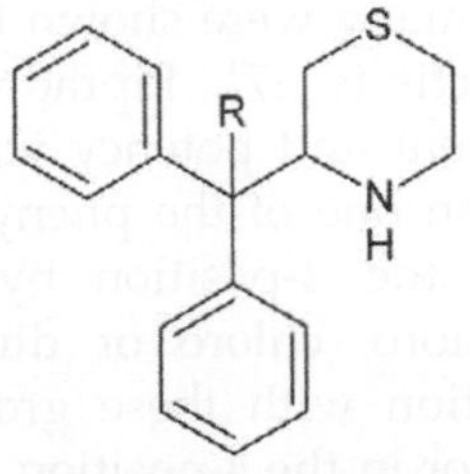

3-(Diphenylmethyl)thiomorpholine (R=H)
'Thiopipradrol' (R=OH)

FIGURE 10.4   Structures of other heterocyclic pipradrol derivatives.

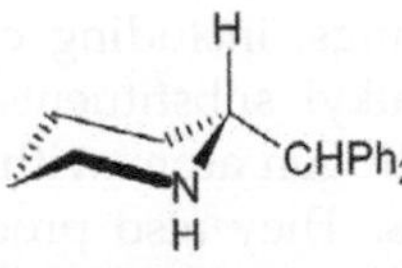

(R)-(+)-Desoxypipradrol 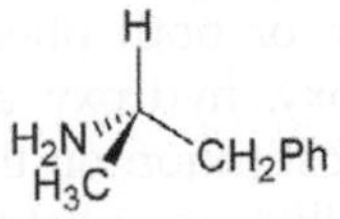   (S)-(+)-Amphetamine

FIGURE 10.5   Stereochemistry of desoxypipradrol in relation to amphetamine.

analysis of drugs seized by law enforcement agencies is usually performed by gas chromatography–mass spectrometry (GC-MS). However, positional isomers, such as the 2-, 3- and 4- isomers of desoxypipradrol, are likely to have similar mass spectra and therefore, in the absence of reference standards for each of the isomers, it is necessary to use specialist analytical techniques such as NMR to confirm the positional isomer. Therefore, in order to avoid any unnecessary burden on forensic laboratories, the recommendation to control pipradrol and pipradrol derivatives in the UK under the Misuse of Drugs Act, 1971,

includes positional isomers within the proposed generic definition [32].

Pipradrol

Pipradrol (also known as α,α-diphenyl-2-piperidinemethanol, α-(2-piperidyl)-benzhydrol, and MRD-108) is a mild stimulant with a similar psychopharmacological action to methylphenidate. It was developed in the 1940s in the USA and patented in 1953 along with several derivatives, including the N-alkyl compounds, which were claimed to be CNS stimulants [1] and was first marketed in the USA in the mid 1950s as the antidepressant, Meratran [33]. Pipradrol, a white powder (melting point 308–309°C), was synthesised by catalytic hydrogenation of the corresponding pyridine compound, α,α-diphenyl-2-pyridinemethanol, which was synthesised using the Emmert reaction in which pyridine is reacted with benzophenone in the presence of magnesium or aluminium and mercuric chloride. However, the reaction product includes a mixture of the 2- and 4- isomers together with a pinacol by-product [34,35]. Synthesis from benzophenone and 2-bromopyridine by the Grignard reaction was used by Tilford [34] with a yield of 58%.

McCarty [7] synthesised a large number of ring substituted pipradrol derivatives using several different methods for production of the intermediate α,α-diphenyl-2-pyridinemethanol. The compounds were tested by oral administration to mice and many were shown to have central stimulant activity [7]. Pipradrol was one of the most potent and potency was usually maintained when one of the phenyl rings was substituted at the 4-position by alkyl, alkoxy, hydroxy, fluoro, chloro or dimethylamino [7]. Substitution with these groups in the 2- or 3-position, or in the 4-position of both phenyl rings generally reduced the potency by a considerable amount [7]. Replacement of one of the phenyl rings with another ring system (2-piperidyl, 2-furyl or 2-tetrahydrofuryl rings) also decreased the potency slightly [7].

An improved method for the synthesis of the α,α-diphenyl-2-pyridinemethanol intermediate was patented in 1996 [35] and employs the reaction of 2-cyanopyridine with benzophenone in the presence of sodium, which produces the desired product in 70% yield.

The (R) and (S) stereoisomers of pipradrol were synthesised by hydrogenation of picolinic acid to pipecolic acid (piperidine-2-carboxylic acid) followed by resolution of the optical isomers, esterification and reaction of the resulting ester with the Grignard reagent phenylmagnesium bromide [36,37].

Desoxypipradrol

Desoxypipradrol (also known as deoxypipradrol, 2-benzhydrylpiperidine, 2-diphenylmethylpiperidine, 2-DPMP) was discovered by CIBA in the 1950s whilst conducting research on cyclic amine derivatives [3] and was patented in 1958 as a potential psychomotor stimulant and narcoleptic [8]. It was synthesised from

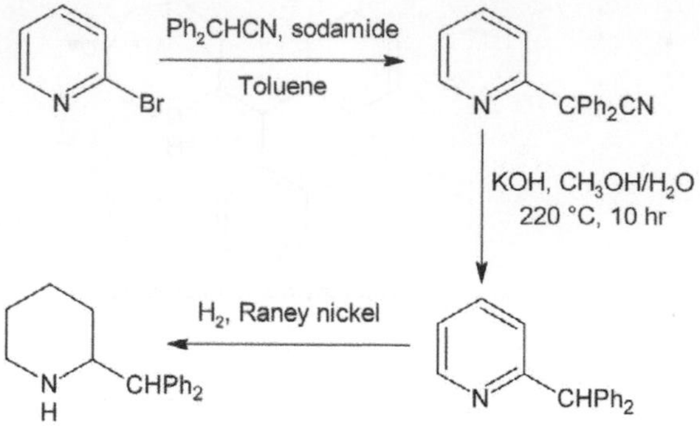

FIGURE 10.6   Synthesis of desoxypipradrol.

diphenylacetonitrile and 2-bromopyridine (Fig. 10.6). The base forms crystals, melting point 65–67°C (from petroleum ether), and the hydrochloride salt forms colourless needles, melting point 286–287°C.

The diphenylacetonitrile route was also used by Sury and Hoffmann [2] to synthesise a series of desoxypipradrol derivatives with substituents in one or both phenyl rings, including chloro, methoxy, hydroxy and alkyl substituents, and by substitution on the nitrogen atom with alkyl and alkylene substituents. They also produced a compound in which the two phenyl groups were linked to form 2-(9-fluorenyl)-piperidine. Most of the compounds were reported to have central stimulating effects in mice at doses as low as 1 mg/kg, although desoxypipradrol was found to be the most effective [2].

Pipradrol derivatives containing the benzylic hydroxyl group, and the corresponding pyridine precursors, can also be converted to the desoxy form by reduction of the hydroxyl group. Heer used sodium in butanol to reduce the thiophene analogue, α-phenyl-α-(2-thienyl)-2-piperidinemethanol, to the corresponding desoxy compound [5]. Shafi'ee used lithium and ammonia in the Birch method to reduce pipradrol to desoxypipradrol [27]. Benzyl alcohols can also be reduced with hydriodic acid as in the conversion of ephedrine to methylamphetamine. Krumkalns

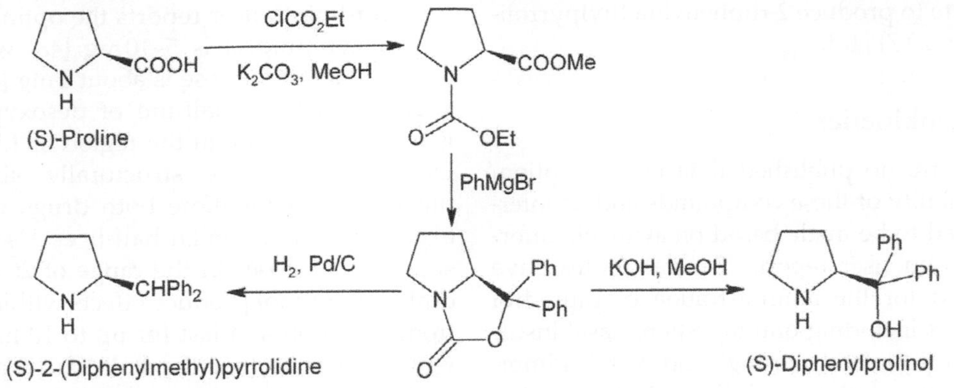

FIGURE 10.7   Synthesis of diphenylprolinol and 2-diphenylmethylpyrrolidine.

used hydriodic acid to reduce the fluorene analogue of pipradrol, 9-(2-pyridyl)-9-fluorenol, to 9-(2-pyridyl)fluorene, which was claimed to be useful for controlling the growth of aquatic weeds [38]. This method was also used to convert diphenylprolinol to 2-(diphenylmethyl) pyrrolidine [16]. The N-alkyl derivatives of the 3- and 4- isomers of pipradrol can be converted to the desoxy derivatives by dehydration with sulphuric acid followed by hydrogenation of the double bond [30,39].

Diphenylprolinol and 2-(Diphenylmethyl) pyrrolidine

Diphenylprolinol (also known as diphenyl-2-pyrrolidinylmethanol, D2PM, α,α-diphenyl-2-pyrrolidinemethanol, and GPI-2089 [40]) and 2-(diphenylmethyl)pyrrolidine (also known as desoxy-D2PM, 2-benzhydrylpyrrolidine) are the pyrrolidine analogues of pipradrol and desoxypipradrol, respectively. A detailed review of the methods of synthesis for 2-(diphenylmethyl)pyrrolidine and diphenylprolinol was published on the former Rhodium website and is currently archived on the Erowid website [41]. Diphenylprolinol was first synthesised in 1933 from proline ethyl ester by reaction with phenylmagnesium bromide. However, it was not until 1961, when racemic diphenylprolinol was synthesised using this method for potential

clinical applications, that this compound was shown to have central stimulant activity when administered parentally to rats [14]. A few years later, the reduction of diphenylprolinol using hydriodic acid was reported in the patent literature [16] for the production of (S)-2-diphenylmethylpyrrolidine as a potential central stimulant.

Interestingly, pipradrol, diphenylprolinol and 2-(diphenylmethyl)pyrrolidine later found use as chiral organic catalysts [37,42–44], which subsequently led to an explosion of research on organocatalysts [44]. Kanth reported a convenient synthesis of diphenylprolinol from (S)-proline using ethyl chloroformate to simultaneously produce the (S)-proline ethyl ester with an N-protecting group. This intermediate was then reacted with phenylmagnesium bromide to produce a cyclic carbamate which was then hydrolyzed with potassium hydroxide (Fig. 10.7) [45]. A nitrogen-protecting group has also been utilised for the synthesis of diphenylprolinol from pyrrolidine and benzophenone [41]. The chiral properties of 2-(diphenylmethyl)pyrrolidine have also been exploited in analytical chemistry as a chiral solvating agent for the determination of enantiomeric composition of chiral carboxylic acids by NMR analysis [46]. For this application, Bailey modified the method of Kanth by hydrogenating the cyclic carbamate to produce 2-diphenylmethylpyrrolidine (Fig. 10.7) [46].

Pharmacokinetics

There are no published data on absorption/bioavailability of these compounds and so inferences need to be made based on available information from user reports. Several routes have been used for the administration of pipradrol derivatives including oral ingestion, nasal insufflation of powder (snorting) and rectal administration of a solution [20,47]. Oral ingestion appears to be the preferred route for the use of desoxypipradrol, whereas rectal administration appears to be the usual mode for diphenylprolinol [20]. The rate of absorption of desoxypipradrol following oral ingestion appears to be slower than that for pipradrol with effects being described within 60 minutes of ingestion of 1–10 mg of desoxypipradrol [48,49] compared to a delay of 30–40 minutes for oral ingestion of 3 mg of pipradrol [50]. After oral administration of 20–50 mg of diphenylprolinol the desired effects were described within about two hours, and re-dosing appears to occur quite often [51].

There is a dearth of published information on the metabolism of pipradrol and pipradrol derivatives. Vree investigated the renal excretion of pipradrol in man and the use of sodium bicarbonate to suppress excretion as a potential method of circumventing doping controls. The excretion half-life of pipradrol was 25 hours [52] and when used in combination with sodium bicarbonate renal excretion was only partially suppressed, but the stimulant effect was greatly enhanced [52]. The half-life of desoxypipradrol has been quoted as 16–20 hours [53], but there is no original reference for the source of this data. The half-life of desoxypipradrol is likely to be much longer than this, as the duration of effects, from 3–7 days [53,54], is considerably longer than that for pipradrol, 12–14 hours [50], partly due to the absence of a hydroxyl group which can form excretable conjugates.

According to user reports the optimum dose of desoxypipradrol is 5–10 mg [48] whereas a clinically effective dose is about 1 mg [4], which suggests that the half-life of desoxypipradrol is more likely to be in the region of 1.5–2 days. Diphenylprolinol is structurally similar to pipradrol and therefore both drugs would be expected to have similar half-lives. User reports suggest that doses in the range of 20–50 mg of diphenylprolinol produce effects within 2 hours from ingestion and last for up to 10 hours [51], which suggests that the half-life of diphenylprolinol is comparable to or slightly less than that of pipradrol. This is in contrast to cases of acute toxicity associated with the use of diphenylprolinol in which individuals used much larger doses, typically 1 g of powder or 5 capsules, and presented to emergency departments with symptoms that had been ongoing for 24–96 hours [55].

Piperidine and pyrrolidine compounds are most susceptible to metabolism by oxidation at a ring carbon atom, usually adjacent to the nitrogen atom, to produce the corresponding hydroxyl derivative, which can form excretable conjugates or oxidise further to the lactam followed by ring-opening hydrolysis, e.g. N-benzylpiperidine is metabolised *in vitro* to the α-, β- and γ-hydroxy metabolites [56]. The N-alkylpiperidine derivatives may also be converted to the corresponding N-oxides [56]. Studies on the metabolism of other drugs containing the diphenylmethyl group such as cinnarizine, fenoctimine, diphenhydramine and modafinil, have identified aromatic hydroxylation as one of the metabolic pathways [57–60].

Pharmacodynamics

Pipradrol

Pipradrol and amphetamine have the same stimulating actions on the higher functions of the CNS but pipradrol is without sympathomimetic properties [61]. It has no effect on blood pressure or appetite, seldom interferes with nocturnal sleep and any anxiety side reactions are less severe than those encountered with amphetamine [33]. Experiments on cats, however, showed that pipradrol reduced food consumption and delayed the eating response, and there was no measurable difference between the anti-appetite action of pipradrol and that of amphetamine, methylamphetamine and methylphenidate [62]. It has been used in the treatment of obesity, but only as an adjunct in the dietary management of obesity [63]. In mice given racemic pipradrol orally, motor stimulation was evident at doses above 3 mg/kg and motor activity increased with increasing dose until it reached a peak at 17 mg/kg. The CNS activity of pipradrol resides in the (R) isomer [36]. The (S) isomer is devoid of stimulant activity and does not significantly antagonise the effect of the (R) isomer. It is interesting that although the (S)-(+)-pipradrol isomer is inactive as a stimulant it has anticonvulsant properties. There is no apparent antagonism between (R)- and (S)-pipradrol, which suggests that the two stereoisomers act at different receptor sites [36]. In larger animals, such as dogs, pipradrol significantly increased purposeful activity which was more intense than that produce by amphetamine but without the side-effects produced by the latter. There was no change in blood pressure, heart rate or respiratory function and there was a wide safety margin [64].

Therapeutically, pipradrol was found to benefit reactive depression, chronic fatigue states and narcolepsy. However, the drug has a tendency to exacerbate pre-existing anxiety and therefore is contraindicated in psychotic patients [65] and in conditions where there is severe tension, agitation or anxiety [50]. In geriatric patients it has been used to improve psychomotor reactions and produce mood elevation [66]. In a study of 99 healthy elderly volunteers, treatment for one week with a pipradrol-vitamin product (Alertonic®) produce no significant side-effects, although there were no significant effects on mood, memory or appetite [67].

The neurochemical effects of pipradrol have not been extensively investigated. It is a dopamine and norepinephrine reuptake inhibitor, but is generally less potent than (+)-amphetamine in its effects on *in vitro* release and uptake inhibition of dopamine and norepinephrine [68]. It releases dopamine from the reserpine-sensitive dopamine pool [69], but in normal and reserpinised rat striatum pipradrol is most potent as an uptake inhibitor [70].

Desoxypipradrol

Desoxypipradrol was tested on non-anaesthetised animals [3]. The drug produced a marked central arousal effect, initially as general unrest and later by a large increase in coordinated motility, enhanced reflexes, forced movements and relatively low respiratory stimulation. Depending on the animal and route of administration, the effects were observed with doses from 1/50 to 1/20 of the LD_{100} dose. The effects continued for several hours and then subsided with signs of tiredness. The psychomotor effects were quantified by determination of the effective dose required to produce 2000 'single' cage movements in the first hour. Using this this method desoxypipradrol was equally as effective as methylamphetamine ($ED_{2000} = 1$ mg/kg). However, when administered to rats anaesthetised with 200 mg/kg of barbital, the wake-up effect of desoxypipradrol was considerably stronger than that of methylamphetamine. The blood pressure of anaesthetised cats and rabbits at low doses (0.1–1 mg/kg i.v.) was slightly increased by about 5 mm of Hg; however, at high doses (3–6 mg/kg i.v.) it decreased slightly by about 15 mm of Hg [3,71]. Peripheral sympathomimetic effects in the whole animal and in isolated organs were scarce and no specific effects were described [3].

Tests on animals at higher doses (1/20–1/10 of the LD_{100}) showed manifest signs of restlessness, agitation and later a marked increase in coordinated motility. The effects continued for a few hours and then subsided with signs of fatigue [4].

Desoxypipradrol was tested on both animals and humans as a drug for reducing recovery time from anaesthesia [4]. The awakening effect of desoxypipradrol was studied on a group of mice anaesthetised with 0.2–0.3 mg/kg of pentothal. Administration of 2–4 mg/kg desoxypipradrol significantly shortened the time of anaesthesia. However, the drug did not significantly raise the lethal dose of barbiturates; tolerance to pentothal was only increased by 3–4 mg compared to controls. In rabbits an intravenous dose of 0.1–1 mg/kg produced only mild short-term increase in blood pressure, while 3–6 mg/kg resulted in a mild hypertension action. The lower doses produced an increased frequency of breathing in both awake and anaesthetised animals and even in humans there was an increased frequency and depth of breathing. In humans the drug causes a mild increase in metabolic rate 30 minutes after administration of a 2 mg dose, but no changes in body temperature were observed in either humans or animals [4]. Mild leucocytosis was detected, but no further information was provided on white blood cell counts or the type of white blood cells responsible for the increase in counts [4], although the original report by CIBA (now Novartis) states that slight leucocytosis and a weak eosinophilia was observed in chronically treated rabbits [71].

In a clinical trial of desoxypipradrol, the drug was used to accelerate the awakening of patients from barbiturate anaesthesia following about 50 different interventions, including abdominal surgery, urology, and orthopaedics [4]. A few minutes after intravenous administration the appearance of reflexes was observed, there was an increased frequency and depth of breathing, and a return of motility and pain sensitivity. There was little or no effect on pulse rate or blood pressure, but no data were provided. The initial dose administered was 1 mg in 5 ml saline injected slowly into the drip tube or directly into a vein. A subsequent administration of an equal amount was given once the first dose was shown to be having an effect.

Desoxypipradrol is a norepinephrine-dopamine re-uptake inhibitor, which causes increased extracellular concentrations of norepinephrine and dopamine and therefore, an increase in adrenergic and dopaminergic neurotransmission [23,72,73].

Maxwell et al. investigated the structure and conformational possibilities of desoxypipradrol, methylphenidate, cocaine and tricyclic antidepressants (desmethylimipramine, protriptyline and nortriptyline), in relation to their activity as competitive inhibitors of the uptake of norepinephrine in the peripheral adrenergic nerves of rabbit aorta [72]. The potencies of these compounds fall into the following order: tricyclic antidepressants > desoxypipradrol > methylphenidate > cocaine. These compounds can all adopt the *trans*-staggered confirmation of β-phenethylamine so that one phenyl ring and a positively charged nitrogen atom can be exactly superimposed on their counterparts in β-phenethylamine. This suggests that they bind to the same 'amine pump' receptor site for norepinephrine. The relative binding affinities of these compounds are shown in (Table 10.1) and can be rationalised by structural and conformational differences [72]. These compounds have a second phenyl group or a carbomethoxy group that projects above the plane of the phenethylamine and aids in binding to the pump receptor. There is also a marked difference between the potencies of the two stereoisomers of desoxypipradrol with the (R)-(−)isomer being the more potent inhibitor [72]. None of the compounds were found to release norepinephrine. One reason for the high potency of the tricyclic antidepressants is probably due to the two carbon bridge between the two phenyl groups which restricts the rotation of the phenyl group which projects above the phenethylamine plane. In desoxypipradrol the second phenyl group is free to rotate which is less favourable for bonding to the receptor [72].

Although pipradrol and its derivatives, like amphetamine, all contain the β-phenethylamine

TABLE 10.1 Relative Potencies of Norepinephrine Uptake Inhibitors in Rabbit Aorta Strips [72]

Compound	Relative Potency[a]
Desmethylimiprimine HCl	3000
Prototriptyline HCl	280
Nortriptyline	120
(R)-(−)-Desoxypipradrol	63
(S)-(+)-Desoxypipradrol	1
(2 R:2′R)-(+)-*threo*-Methylphenidate HCl	38
Cocaine HCl	8

[a]*Molar concentrations of inhibitors required to reduce NE uptake from a 10^{-7} M solution by 50% (ID50) relative to the ID50 of a reference standard.*

sub-structure and cause an increase in the extracellular concentrations of norepinephrine and dopamine, their mechanism of action differs from that of amphetamine due to differences in the stereo-chemical requirements of the receptor sites of the different phenethylamine pumps [23,73]. The uptake of tritium labelled dopamine and norepinephrine into synaptosomal preparations of three regions of the rat brain indicated that (R)-(−)desoxypipradrol is more potent than the (S)-(+)-desoxypipradrol isomer as an inhibitor of the 'phenethylamine pump', whereas the stereoisomers of amphetamine are equally effective and less potent inhibitors than (R)-(−)-desoxypipradrol [73]. The results suggested that the receptor sites on all the 'phenethylamine pumps' studied are not sensitive to the orientation of the small α-methyl group of amphetamine but are sensitive to the orientation of the larger hydrophobic piperidine ring of desoxypipradrol. In contrast to the results obtained with peripheral adrenergic nerves of the rabbit aorta, the tricyclic antidepressants are less potent inhibitors of catecholamine uptake into the rat striatum than desoxypipradrol [73]. Desoxypipradrol did not release norepinephrine or dopamine from the synaptosomes of rat cortex and hypothalamus

but both stereoisomers produced a small but comparable release of dopamine from the rat striatum. The results indicated that the phenethylamine pump in the rat striatum is different to those in which the tricyclic antidepressants are potent inhibitors of catecholamine uptake [73].

In addition to inhibition of the phenethylamine pumps present in neuronal membranes, desoxypipradrol also inhibits the uptake of norepinephrine and dopamine into the synaptic vesicles used to store neurotransmitters [23]. In contrast to the effects in neuronal membranes, the (S)-(+)-isomer of amphetamine was 10 times more potent than the (R)-(−)-isomer as an inhibitor of norepinephrine and dopamine uptake into synaptic vesicles isolated from whole rat brain, rat striatum and rat hypothalamus, whereas the stereoisomers of desoxypipradrol were equally effective and less potent inhibitors [23].

Experiments *in vitro* using carbon fibre electrodes attached to rat brain slices demonstrated that desoxypipradrol was more potent than cocaine as a stimulator of dopamine release from the region of the nucleus accumbens, considered to be a target for psychostimulant drugs, which could account for its psychotogenic effects [32,74].

Not all DAT inhibitors have the reinforcing effects of cocaine, which suggests that DAT inhibitors may induce conformational changes in the DAT or may bind at different sites than cocaine. The binding affinities of a number DAT inhibitors were determined using DAT mutations (W84L, D313N), which have a much greater affinity for cocaine compared to the wild-type transporter, but have a significantly lower affinity for the DAT inhibitor benztropine. Methylphenidate and a derivative of desoxypipradrol (D-254; Fig. 10.8) exhibited a cocaine-like binding profile in that they were potent inhibitors of both wild-type and mutant DATs, whereas compounds possessing the diphenylmethoxy moiety of benztropine exhibited a binding profile dissimilar to cocaine-like

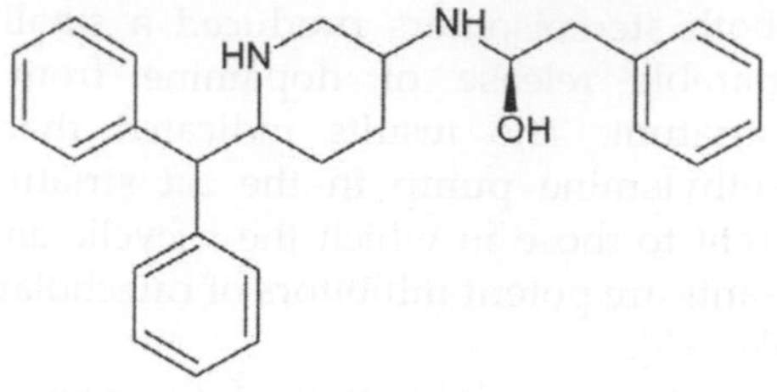

FIGURE 10.8   Structure of D-254.

compounds [75]. This suggests that other des-oxypipradrol derivatives may also have a cocaine-like binding profile.

Diphenylprolinol

Cloned human DAT protein has been shown to have several binding sites, including a binding site for cocaine which is distinct from that of dopamine [40]. It has therefore been suggested that compounds which bind selectively to the cocaine site without inhibiting dopamine uptake may be useful in the treatment of cocaine addiction or overdose. The idea being that such compounds antagonise cocaine's binding to the DAT without inhibiting the normal DAT function [40,76]. Potentially useful compounds would therefore have a DAT uptake to cocaine binding ratio (Ki_{uptake}/$Ki_{binding}$) greater than that for cocaine. The (R)-(+) and (S)-(−) stereoisomers of diphenylprolinol have binding ratios of 4.25 and 4.12, respectively, compared to 1.67 for cocaine. The (R)-(+) isomer is about 10 times more active than the (S)-(−) isomer at both sites and therefore the uptake to binding ratios for the two stereoisomers are about the same. The N-alkyl and the bis(4-fluoro-3-methylphenyl) derivatives of diphenylprolinol also had uptake to binding ratios greater than that for cocaine [76]. However, pharmacotherapies for cocaine addiction that typically target the receptor sites, have as yet been unsuccessful, and often generate adverse side effects [77].

Conversely, it has also been claimed that diphenylprolinol and its N-alkyl derivatives may be useful for the treatment of attention deficit disorder (ADD), now known as attention deficit hyperactivity disorder (ADHD), in that like methylphenidate they actually inhibit normal DAT function by binding to the cocaine site of the dopamine transporter [76].

2-(Diphenylmethyl)pyrrolidine

2-(Diphenylmethyl)pyrrolidine was patented in 1964 and shown to stimulate the activity of rats at doses ranging from 1/20 to 1/5 of LD_{50}(i.p.), whilst at the higher doses the drug also increased aggressiveness of the animals in response to an electrical stimulus [16]. The French patent claims that the drug can be used for the treatment of depressive states, psychasthenia, convalescence and mental and physical fatigue and is active at doses from 25 to 100mg in adults with a daily dose of 100 to 750mg, depending on the route of administration. This indicates that desoxypipradrol is about 100 times more potent than the pyrrolidine derivative.

Other Pipradrol Derivatives

Likhosherstov [17] compared the pharmacological activity of pipradrol with the five and seven membered ring analogues and their N-alkyl derivatives, and found that the most active compounds were those containing the piperidine ring.

Chemical modifications of pipradrol were investigated with a view to increasing the selectivity of action. Belleau [11] reasoned that replacing the piperidine ring with a thiomorpholine ring should retain the stimulant activity while the sulphide group should minimise toxicity by virtue of the marked susceptibility of sulphides to detoxification by enzymes. When administered to mice the (+) isomer of α,α-diphenyl-3-thiomorphinylmethanol ('thiopipradrol') produces a pronounced increase in motor activity at a minimum dose of 5mg/kg (i.p.) and with an LD_{50} (mice) (i.p.) of 127 mg/kg was less toxic than pipradrol. There was also a virtual absence of undesirable side effects. Studies on cats and dogs also produced an increase in motor activity together with increases in alertness, reaction times and responsiveness to stimuli [78].

The effects of introducing of a 3-hydroxy substituent into the piperidine ring have also been investigated. In studies on mice, the 3-hydroxypiperidine derivative of pipradrol was shown to have a stimulant effect with a potency similar to that of pipradrol with a lowest effect dose of 10 mg/kg (i.p.), whereas the 3-hydroxypiperidine derivative of desoxypipradrol produced stimulant effects at a dose of 0.3 mg/kg (i.p.) [18].

A clinical investigation of the 3-hydroxy-N-methylpiperidine derivative of pipradrol, 2-diphenylmethyl-3-hydroxy-1-methylpiperidine (SCH 5472), showed that a daily dose of 0.25 to 3 mg produced a moderate to marked improvement in 76% of patients with exhaustion syndromes, neurotic depression and hypersomnia; however, the variation in the individual patient response for a given dose was too marked [13].

The activity of desoxypipradrol has been compared to the other three positional isomers and to an N-alkyl derivative of the 4-isomer. The effective doses required to produce spontaneous motility in mice (ED_{2000}) showed that desoxypipradrol (0.001 mg/kg) is 100 times more potent than the 1- and 4-isomers (0.1 mg/kg), six times more potent than the 3- isomer (0.015 mg/kg) and twice as potent as pipradrol (0.002 mg/kg) [2].

Enyedy [19] employed a pharmacophore-based 3D-database to search for novel DAT inhibitors which would potentially function as cocaine antagonists. Using a refined pharmacophore model derived from several known DAT inhibitors, the National Cancer Institute 3D-database was searched and a total 1104 potential DAT inhibitors were identified. Of these, 4-diphenylmethylpyridine was found to be a significant inhibitor of dopamine reuptake.

The 2- and 3- isomers were then tested to gain insights into the structure activity relationship. All three isomers showed quite potent activities in binding and uptake assays with K_i values of 0.742, 0.780 and 0.079 µM in binding and 1.064, 0.860 and 0.255 µM in inhibition of DA reuptake, for the 2-, 3- and 4- isomers respectively. The 2-diphenylmethylpyridine isomer was as potent as cocaine in the DA uptake assay.

Most positional isomers of pipradrol and pipradrol derivatives are not stimulants but have other pharmacological properties. In animal studies, the 4-isomer of pipradrol, azacyclonol (marketed as Frenquel), has been shown to have some effects opposite to those of pipradrol [6]. In mice and rats, azacyclonol depresses activity when administer in small to medium doses, but at large doses it has a stimulant effect. In some animal studies the drug was found to antagonise the effects of other stimulant drugs, including pipradrol [79].

PREVALENCE OF USE

Prevalence data on the use of pipradrol derivatives are not collected by any of the major surveys or reports, such as the British Crime Survey, the United Nations Office on Drug and Crime (UNODC) World Drug Report or the MixMag Global Drugs Survey. However, the EMCDDA collects data on seizure reports and collected samples of new psychoactive substances through the Early Warning System (EWS). The data are not published, but are available to registered users on the European Database of New Drugs (EDND) [21] and have been included in this review to give some indication of the prevalence of use of pipradrol derivatives. Data on the use of 'pipradrols' was collected in a recent survey of 315 individuals attending gay-friendly nightclubs in South East London in July 2011. The results indicated a relatively low level of use of pipradrols in comparison to other drugs such as mephedrone: 1.6% reported ever using a 'pipradrol', 1.0% had used within the last month and 0.6% were using or planning to use on the night compared to 63.8%, 53.2% and 41.0% for mephedrone respectively [80].

Pipradrol

Pipradrol is a previously licensed medicine in the UK [32] and is listed as the active ingredient of several pharmaceutical products [81], including Alertonic Elixir®, which is used in the treatment of patients with functional or psychogenic fatigue and as a dietary supplement in nutritional fatigue. There are no recent published reports of misuse of pipradrol or of proprietary products containing pipradrol, although in 2007 a user report from Australia described an attempt to get high by oral administration of Alertonic without any apparent effects [48].

Desoxypipradrol

In 1975, Shulgin attempted to anticipate the future direction for drugs of abuse and predicted that structural manipulation of pipradrol (with and without the benzylic hydroxyl group) would likely lead to new and more potent drugs [82]. In 1990, a similar article by Cooper from the DEA highlighted a number of pipradrol derivatives that would be suitable candidates for clandestine production with 2-diphenylmethylpiperidine being the best bet as it was the most potent adrenomimetic compound in the series [83]. Arguably, such articles may encourage experimentation with new drugs, including pipradrol derivatives. However, the first report of users experimenting with desoxypipradrol did not appear until early 2007 [84].

The first seizure of desoxypipradrol was reported to the EMCDDA in 2009 following a seizure in Finland in December 2008 [85]. Since then the EMCDDA has received further reports of seizures in Finland and also in the UK and Hungary, although the total number of seizures has remained relatively low compared to other

new psychoactive substances such as 4-methylmethcathinone (mephedrone) [21]. In the UK in 2009, the drug was identified in a packet labelled 'Desoxypipradol' (sic) '20mg' analyzed by TICTAC Communications Ltd, and in a collected sample of a product labelled 'Ivory Wave' submitted to a forensic science laboratory in Scotland [86]. The composition of 'Ivory Wave' is known to be variable and some samples have previously been found to contain methylenedioxypyrovalerone (MDPV) rather than desoxypipradrol [87–89]. Furthermore, a seizure of powder made in the UK by Lincolnshire Police in December 2009 contained desoxypipradrol mixed with N,N-dimethylcathinone (main component), 4-methylmethcathinone and MDPV [21]. A collected sample of 'Ivory Wave' examined by the Glasgow Forensic Science Laboratory in 2010 only contained about 20% desoxypipradrol together with unidentified carbohydrate material [21]. In Ireland in June 2010, analyses of test purchases from head shops of a product named 'Whack' revealed that it contained a mixture of desoxypipradrol and fluorotropacocaine [89,90].

In 2010 the number of seizures reported to the EMCDDA increased, with 47 seizure reports from Finland, mostly powders with some tablets and liquids, one seizure report from Hungary and five seizure reports from the UK [21].

Following a series of hospital admissions in the summer of 2010 and 3 deaths [86], all believed, but not conclusively confirmed, to be linked to the use of desoxypipradrol, there were no further seizures in the UK in 2011, although a test purchase from a head shop in Edinburgh showed that the drug was still available under the product name 'Lunar Wave' [22]. The number of seizures in Finland also fell with only five seizures reported to the EMCDDA during the first half of 2011 [21]. Importation of the drug into the UK was banned from 4 November 2010 under an Open General Import Licence [91], and the UK Advisory Council on the Misuse of Drugs (ACMD) subsequently recommended that desoxypipradrol and related pipradrol derivatives should be controlled under the UK Misuse of Drugs Act, 1971 as Class B drugs [32], although it appears that the poor reputation of desoxypipradrol amongst users has now self-regulated its availability. Desoxypipradrol and other pipradrol derivatives subsequently became Class B drugs in the UK under the Misuse of Drugs Act, 1971 on 13 June 2012 [92].

Diphenylprolinol (D2PM)

In the second quarter of 2007, the Drugs Group at the ESR laboratory in New Zealand reported receiving an unusual case containing fake blue ecstasy tablets with a 'Playboy Bunny' logo. The tablets were found to contain diphenylprolinol and benzophenone, together with traces of 1-benzylpiperazine (BZP) and 1-[3-(trifluoromethyl)phenyl]piperazine (TFMPP) [93]. Later in 2007 a man in New Zealand was hospitalised after taking part in a trial of 'Neuro Blast' pills, a non-BZP product designed to beat the impending Government ban on BZP. The media report stated that scientific testing showed that the pills contained diphenylprolinol [94]. In the UK, diphenylprolinol was first identified in March 2007 in tablets purchased from a website [86]. More recently, data from the Home Office Forensic Early Warning System showed that the drug continued to be available in the UK in 2011 [95] although the number of police seizures examined by the Forensic Science Service (FSS) in 2011 was still low compared to the new psychoactive substance mephedrone.

Diphenylprolinol and benzophenone were also detected in the urine of two individuals who had used 'legal highs' which contained diphenylprolinol [96]. The presence of benzophenone was thought to have originated from a precursor residue in the 'legal high' product suggesting that the diphenylprolinol was not commercially produced. However, it was later suspected that the detection of benzophenone may have been due to thermal degradation of diphenylprolinol in the GC-MS injection port.

Subsequent analysis by liquid chromatography tandem mass-spectrometry confirmed the presence of diphenylprolinol but no benzophenone was detected. This confirmed that the detection of benzophenone by GC-MS was an analytical artifact [55]. Since GC-MS is routinely used in forensic analyses, the detection of benzophenone in a diphenylprolinol tablet by the New Zealand laboratory may also be due to an artefact.

Further cases of acute toxicity due to recreational use of diphenylprolinol prompted the UK ACMD to recommend an import ban in the UK on both diphenylprolinol and 2-(diphenylmethyl)pyrrolidine [95]. The importation ban subsequently came into force on 15 November 2011 [97] and diphenylprolinol was controlled under the UK Misuse of Drugs Act, 1971 as a Class B drug on 13 June 2012 [92].

2-(Diphenylmethyl)pyrrolidine (Desoxy-D2PM)

2-(Diphenylmethyl)pyrrolidine was found to be an active ingredient of 'Slim Xtreme', a fat burning product marketed by Athletic Xtreme (AX) in 2009 [98]. The product was also being used as a 'legal high' [99,100] and was subsequently removed from the company's website [101]. The drug has also been identified in a 'legal high' product marketed as 'A3A Methano' [85] and data from the UK Home Office Forensic Early Warning System showed that the drug was available in the UK in 2011. However, only one seizure was encountered by the FSS in 2011 and in this case the drug was identified in a mixture with 4-fluoromethcathinone (4-FMC), methylenedioxy-α-pyrrolidinobutiophenone (MDPBP), methylenedioxypyrovalerone (MDPV) and p-methoxymethylamphetmine (PMMA) [21].

ACUTE TOXICITY

There is limited information available regarding the human toxicity profiles of pipradrol and pipradrol derivatives. Predominately, information relating to the toxicity of pipradrol comes from animal studies and early therapeutic human clinical trials. However, with their increasing prevalence of use as recreational drugs over recent years, several case reports describing their toxicity in humans have been reported [53–55,86,94,102–105]. Despite this the safety profile of these compounds remains unknown and the concentrations of desoxypiradrol, diphenylprolinol, and 2-(diphenylmethyl)pyrrolidine associated with toxicity or death in humans have not been determined.

Animal Studies

Experimental animal studies have shown pipradrol to have a stimulant effect on the central nervous system, inducing behavioural activation and incoordinated motor activity and ataxia [36]. Pipradrol administered orally to mice causes an increase in motor activity with increasing doses until it reaches a peak at approximately 17 mg/kg. With doses beyond this, tremor and clonic convulsions are seen [36]. In a rabbit model, pipradrol was observed to induce anxiety and gnawing behaviour, indicating stimulation of striatal dopaminergic receptors [106]. It also evokes a dose-dependent hyperthermic response [107], which is significantly antagonised by drugs with prominent α-adrenergic receptor blocking capability, e.g. chlorpromazine, but enhanced through the blockade of seretonergic receptors such as with cyproheptadine, suggesting an important role for α-adrenergic receptors in this drug response.

The LD_{50} values for pipradrol, diphenylprolinol, desoxypipradrol, and 2-(diphenylmethyl)pyrrolidine are compared with amphetamine and methylamphetamine and listed in Table 10.2. The LD_{50} of desoxypipradrol indicates that it is potentially more toxic than pipradrol in animals [4,6,32,64,108,109]. In addition its LD_{50} is lower than amphetamine and methylamphetamine in the mouse and rabbit but higher than these

compounds in the rat [4,6] Desoxypipradrol is more potent than methylphenidate (Ritalin) as a competitive inhibitor of noradrenaline uptake in the peripheral adrenergic nerves of the rabbit [23]. In adult and young monkeys the LD_{50} for methylphenidate is 15–20 mg/kg and 5 mg/kg respectively [110]. In humans, adult deaths have been reported after methylphenidate ingestion at doses as low as 1.3 mg/kg [110]. There are very little animal toxicity data on the other pipradrol derivatives. The data in Table 10.2 indicate that the smaller ring size of diphenylprolinol reduces toxicity in animals, but further modification of diphenylprolinol by introduction of an N-alkyl substituent increases toxicity and eliminates most of the stimulant properties [17]. The introduction of a sulphur atom also reduces the toxicity of pipradrol. The LD_{50} values for racemic, (+) and (−)-α,α-diphenyl-3-thiomorphinylmethanol in mice (intraperitoneal) have been reported to be 361 mg/kg, 127 mg/kg and 292 mg/kg, respectively [11].

4-(Diphenylmethyl)piperidine (4-DPMP) and its N-methyl derivative produced vacuolisation of pancreatic islet cells and a reduction in pancreatic insulin levels when administered orally to rats for 14 days with a daily dose of 130 and 260 μmoles/kg. The 2- and 3-diphenylmethylpiperidine isomers caused no significant changes in the endocrine function of the pancreas. However, the N-methyl derivative of 3-diphenylmethylpiperidine, produced an increase in pancreatic insulin levels at a dose of 130 μmoles/kg, but the reason for the increase was not known. The toxicity of the 4-(diphenylmethyl)-1-methylpiperidine may be due to demethylation in vivo to 4-DPMP [111]. In RINm5F rat insulinoma, 4-DPMP inhibited the production of insulin, but the 2-DPMP isomers had no activity [112].

Human Data

Pipradrol

Adverse clinical effects of pipradrol in humans were evident in early clinical trials investigating

TABLE 10.2 Animal Studies Indicating Toxicity, as Measured by LD_{50} (mg/kg), of Pipradrol, Diphenylprolinol, Desoxypipradrol, 2-(Diphenylmethyl) pyrrolidine and other Compounds.

Animal/ Administration	Pipradrol	Diphenyl- prolinol	Desoxy- pipradrol	2-(Diphenylmethyl)- pyrrolidine	Amphetamine [Reference32]	d-Methylamphetamine [Reference32]
Mouse						
IV	20[108]		20[32]	28[16]	50	20
SC	147[6]		47[32]		60	80
	240[109]					
PO	120[6]	300[a]	50[32]	76[16]	70	150
	74[109]					
IP	74[109]	152[17]		77[16]		
	90[108]					
RAT						
IV	30[64]		15[4]		12	23
SC	240[64]		30[4]		12	15
PO	180[64]	980[a]	80[32]		13	25
RABBIT						
IV	15[64]		6[32]		40	30
SC			7[32]		45	20
PO		3200[a]	80[32]		170	200
GUINEA PIG						
EP			10[4]			

IV, intravenous; SC, subcutaneous; PO, oral; IP, intraperitoneal; EP, extraperitoneal

[a](S)-Diphenylprolinol [Material Safety Data Sheet]. Clearsynth Labs Pvt. Ltd., India; [Accessed on 13 July 2012]. Available from: http://www.clearsynth.com/docs/MSD-CS-IS-21627.pdf.

its therapeutic effects in patients with depression. In a control matched study, doses of up to 3mg per day given orally to 24 normal healthy participants caused an insidious elevation of mood, with heightened confidence, greater ability to concentrate, and an increased work output. The elevation of mood was not of sufficient degree to be characterised as euphoria. It reached a subjectively detectable level 30–40 minutes after taking the drug and persisted for up to 12–14 hours afterwards. Two participants (8%) experienced the occurrence of undesirable after-effects

(not specified), while an unspecified number of others reported interference with sleep [50]. In patients with depression given a higher dose of up to 7.5mg per day, the main adverse effect was that pipradrol had a tendency to exacerbate pre-existing anxiety and even on occasions cause an unpredictable and marked exacerbation of underlying psychotic symptoms [50]. However, the drug manufacturer has claimed that these adverse effects are mainly associated with the improper use of pipradrol in patients for whom pipradrol is contraindicated [65]. In a later trial with 111 mildly to moderately depressed outpatients, a daily dose of 5.0–7.5 mg of pipradrol was found to have only limited efficacy and caused significant side effects (excitement and insomnia) in addition to anorexic effects and weight loss compared to placebo [113].

Due to these adverse effects and the development of alternative antidepressant drugs, pipradrol was predominately removed from the global market and it has been controlled under international drugs legislation since the early 1970s due to its potential for abuse [114].

Pipradrol is the main active constituent of Alertonic® and although not licensed in the UK, is still available in Canada and South Africa, and until recently was available in Australia, but now appears to have been discontinued or is no longer actively marketed [115]. A web-based product information sheet [116] states that overdoses of Alertonic® cause nausea, anxiety, insomnia and abdominal pain, but symptoms generally disappear when the drug is withdrawn. In severe cases convulsions may occur. It also describes two cases of overdose: one in an adult after a single oral dose of 250 mg and the other in a 2.5 year-old after ingestion of 15 mg. While both cases were reported to have survived, no details on the clinical features were described [116].

Desoxypipradrol (2-DPMP)

The desoxy form of pipradrol, desoxypipradrol, appears to have similar clinical effects to amphetamines and, as a dopamine re-uptake inhibitor, it may have a similar action to cocaine [74]. However, the pipradrol derivatives, particularly the desoxy forms, are highly lipophilic molecules, lacking polar functional groups that are typical targets for metabolic enzymes [117]. As such this characteristic gives them a prolonged duration of action (increased half-life) when compared to other stimulants e.g. cocaine [3,23], and this is a likely contributory factor in reports of related adverse effects.

Since 2007 the popularity of desoxypipradrol use as a recreational drug has increased [32,118].

User reports are now available on web-based recreational drug forums which contain first-hand accounts, identify factors such as administration routes, drug dosing and related effects [48]. It is evident that oral ingestion appears to be the usual route of administration; [48] however, intravenous, inhalation (smoking), nasal insufflation (snorting) and rectal routes have also been described [48]. Dosing of desoxypipradrol appears to be in the range of 1–10 mg with a typical dose being 1–2 mg and an optimum dose of 5–10 mg [48], which is higher than the divided 2 mg dose used in clinical trials as an awaking drug from anaesthesia [4]. These reports suggest that the effects of desoxypipradrol are felt within 60 minutes from oral administration and may last up to 24–48 hours [32,48,49].

There is little published data describing the adverse effects of desoxypipradrol in humans, although during early clinical testing the cardiovascular effects of desoxypipradrol were demonstrated. After administration of 1–2 mg doses a reduction in circulation time (defined as the time that it takes an injected indicator to reach a detectable concentration at the place of detection) was seen accompanied by a decrease in blood pressure amplitude in approximately 50% of cases. This was then followed by a relative increase in blood pressure amplitude. At doses ≥ 3 mg headache, nausea and palpitations were experienced [119]. More recently two case series have been published that describe the acute toxicity associated with the recreational use of desoxypipradrol. From May to June 2010 a cluster of patients who had suffered adverse effects after taking a substance called 'Whack' were identified by the National Poisons Information Centre in the Republic of Ireland [54]. In total 49 patients had presented to hospital with signs compatible with sympathomimetic toxicity including tachycardia, hypertension, agitation and psychosis. Those with psychosis had severe delusions of parasitosis and hallucinations, some of which persisted up to five days after ingestion. A sample of 'Whack' was later analyzed and found to contain two active ingredients. One was fluorotropacocaine, a synthetic cocaine, and the other, desoxypipradrol; [54] however, no analysis of biological samples was undertaken so it cannot be confirmed that these drugs were responsible for the clinical symptoms of those affected. The second cluster of cases was seen during August 2010 in Edinburgh [53]. Thirty-four patients with acute toxicity after using 'Ivory Wave' presented to the Emergency Department (ED). Of these 19 (56%) required admission for in-patient hospital care while a further 12 cases were admitted directly to a psychiatric hospital with a predominance of psychiatric rather than physical symptoms. In those with physical symptoms, patients demonstrated a toxidrome characterised by tachycardia (>100/min), tachypnoea (>16/min), dystonia, leucocytosis ($>11 \times 10^9$/L), rhabdomyolysis (CK >170IU/L), agitation, aggression, insomnia, intense paranoia and hallucinations which were commonly auditory and tactile with formication and itch. Patients did not present until after a mean of 2.7 days ±2.2 (standard deviation) and their symptoms lasted several days. A sample of 'Ivory Wave' was later analyzed with the active constituent being identified as desoxypipradrol. Adequate biological samples (blood or urine) were only available for analysis from five patients who reported 'Ivory Wave' use. Desoxypipradrol was detected in four of these samples, despite the delay from ingestion being on average 5 days (range 3–7) [53]. This suggests that desoxypipradrol may have a prolonged half-life and may help to explain the extended duration of the symptoms experienced. However, while its half-life has been quoted in the literature to be 16–20 hours [53,120] there is no original source data to support this. In both case series described here analytical testing of biological specimens could not be completed in all patients suspected of toxicity. Given the possibility that the composition of the drugs may have varied, it is not unequivocal that desoxypipradrol was the causative agent for all presentations seen. However, the comparable clusters of hospital attendances, the detection of the agent in locally obtained samples of both 'Whack' and 'Ivory Wave' and the similarities of symptoms reported through user reports [48] supports desoxypipradrol as being the causative factor.

Diphenylprolinol (D2PM)

Diphenylprolinol, another norepinephrine-dopamine reuptake inhibitor, was first identified in March 2007 in tablets bought online and analyzed by the toxicology services at St George's, University of London [21,86]. Insufflation, oral and rectal administration are described [51,86] and user reports suggest that oral doses range from 20–50mg, causing the desired euphoria (similar to that seen with amphetamines) within two hours from ingestion and lasting for up to 10 hours [51]. Associated effects included jaw clenching, babbling speech and dilated pupils, with prolonged use causing craving and the need to re-dose [51]. The first analytically confirmed case of acute toxicity from the recreational use of diphenylprolinol was reported in 2008 [102,103], in which a 21-year-old male presented to a London ED with chest pain several hours after ingesting a compound named 'Head Candy' bought from a local street shop. At admission he had evidence of sympathomimetic toxicity, consisting of significant hypertension (blood pressure of >200/100mmHg), sinus tachycardia (heart rate of 126b.p.m.) with normal ECG parameters, and agitation. He was treated with 20mg of oral diazepam given in divided doses over a 12-hour period. During this time he required hospital admission for observation until his symptoms had settled. Blood samples collected at admission were later analyzed and diphenylprolinol and glaucine were detected at estimated concentrations of 0.17mg/L and 0.10mg/L respectively. No previous reports of glaucine toxicity have described sympathomimetic features and the authors felt the features were predominantly due to diphenylprolinol [102].

Further detailed information on the acute toxicity related to the use of diphenylprolinol comes from a series of five cases [55]. These involved male patients aged 17–33 years, who presented to a London ED on unrelated occasions having used a range of novel psychoactive substances, none of which were known to contain diphenylprolinol. They presented with ongoing and prolonged neuropsychiatric symptoms including agitation, paranoia, anxiety, visual hallucination and insomnia lasting 24–96 hours post-ingestion. At the time of presentation none had evidence of other features of sympathomimetic toxicity (hypertension or tachycardia). Two patients required treatment with 5 mg of oral diazepam while another, despite being given 5 mg of lorazepam, had ongoing agitation and neuropsychiatric symptoms. This patient required further treatment with 5 mg olanzapine and was transferred to a local psychiatric hospital for ongoing management. The remaining patients all required a period of observation in hospital before being discharged home. Urine samples collected at the time of presentation to the ED were subsequently analyzed and all tested positive for diphenylprolinol. Mephedrone was also identified in one sample and MDMA and amphetamine in another [55]. The authors concluded that the initial pattern of acute toxicity seen after diphenylprolinol is similar to that exhibited by other sympathomimetic drugs, with users describing a 'high' or 'rush' [51,55]. However, the notable consistent pattern observed in the case series of the presence of neuropsychiatric symptoms lasting for up to and longer than 24–72 hours after use of diphenylprolinol differentiates it from many other sympathomimetic drugs.

2-(Diphenylmethyl)pyrrolidine

2-(Diphenylmethyl)pyrrolidine (desoxy-D2PM, DPMP) is a stimulant substance which is structurally related to diphenylprolinol and desoxypipradrol. To date there are no animal or human studies demonstrating the pharmacokinetic or pharmacodynamic properties of 2-(diphenylmethyl)pyrrolidine. However the chemical similarities with the other pipradrol derivatives suggest it could have the same pharmacological activity [117]. Despite a paper from 1965 describing 2-(diphenylmethyl)pyrrolidine as inducing aggressiveness and 'psychotonic' properties in a rat model [16] reports of acute or chronic toxicity in humans remain limited. User reports on Internet forums indicate that 2-(diphenylmethyl)pyrrolidine may have been available as an alternative 'legal high' or fat burning supplement [99,100]. At an oral dose of 5 mg, a mild euphoric effect was noted within 45–60 minutes from ingestion, with a stimulant effect that lasted for 10–12 hours. Other effects include appetite suppression, vasoconstriction of the peripheries and improved motivation [100]. In 2010 a Drug Expert Witness for Cambridgeshire Constabulary in the UK submitted a report to the Early Warning System for the EMCDDA, describing 2-(diphenylmethyl) pyrrolidine toxicity in a 42-year-old woman [104,105]. It was established that she had been taking 'A3A Methano' bought over the Internet and was admitted to hospital with neuropsychiatric symptoms of extreme agitation, violent behaviour and hallucinations, and sympathomimetic toxicity involving hypertension and tachycardia (heart rate 120 b.p.m.). Although there was no analysis of biological samples taken from the patient, the EMCDDA later reported that 2-(diphenylmethyl)pyrrolidine was analytically proven to be present from test purchases in the Internet products 'A3A Methano' and 'Green Powder' and was suspected to be associated with a number of adverse clinical effects as described [32,85].

Mortality from Pipradrol Derivatives

Fatality involving pipradrol derivatives appears to be uncommon. Desoxypipradrol was implicated in the report of three deaths that occurred in the UK in August 2010 [86]. All three cases occurred in young adults who had no significant co-morbidity. The first was a 34-year-old male who was found dead in a fetal position inside a locker on a yacht that he had been repairing. The direct cause of death was thought to be suffocation; however, an open packet of 'Ivory Wave' was found at the scene and a post-mortem toxicology screen found desoxypipradrol at a concentration of 1.16 mg/L in his blood. The coroner gave a cause of death as 'Unascertained' as it is not known if this concentration of desoxypipradrol could have killed the decedent and an 'Open' verdict was given. In the second case a 24-year-old male was seen running along the edge of a cliff with his arms outstretched. The following day his body was found floating in the sea with widespread head and body injuries, thought to have occurred secondary to a fall from the cliff. During the preceding days he had required voluntary mental health counselling for paranoia and auditory and visual hallucinations and been advised he had had a psychotic reaction to 'Ivory Wave', alcohol and mephedrone. Post-mortem toxicological screening found desoxypipradrol at a concentration of 0.79 mg/L in his blood. The coroner returned an 'Open' verdict and indicated that the ingestion of this 'legal high' may have been a very strong contributory factor to the deceased's 'out of character' behaviour. In the final case a 35-year-old woman had been purchasing 'Ivory Wave' from the Internet 10 months prior to her death and was thought to have been taking this in an effort to lose weight. Over four months her dress size dropped from size 16 to size 6 (UK) and she developed paranoia, aggressive agoraphobia, insomnia and auditory hallucinations. One night she lost consciousness and was hospitalised but died 12 days later on the intensive care unit with cerebral oedema and heart failure. The post-mortem toxicology screen found a blood desoxypipradrol concentration of 0.025 mg/L. The cause of death was recorded as 'desoxypipradrol overdose' and the coroner returned a verdict of 'Accidental Death'. Due to the lack of comparative data it is not possible to comment on whether the desoxypipradrol concentrations found in the post-mortem samples had toxicological significance. However, the neuropsychiatric symptoms displayed by the deceased are consistent with the case reports of desoxypipradrol toxicity and it is likely to have been a contributory factor.

CHRONIC TOXICITY

There are no published data on the chronic toxicological effects of pipradrol or its derivatives in humans. However, results from an *in vitro* study implicate that diphenylprolinol may lead to an impairment of neuronal development [121]. When added to PC12 cells (a rat phaeochromocytoma cell line used as a modelling system for neuronal differentiation) diphenylprolinol induced a dose-dependent cell death. The (R) and (S) isomers of diphenylprolinol were equally toxic and both were more toxic than MDMA, which is known to produce a degeneration of serotonergic nerve terminals. When treated with nerve growth factor (NGF) most PC12 cells produce neurites; however, in the presence of diphenylprolinol the number of NGF-induced neurite positive cells and the length of the neuritis was significantly decreased. These results suggest that diphenylprolinol causes impaired neuronal development, which may be particularly relevant in a drug-taking teenager and in the fetus of a mother who is abusing the drug [121].

There are no user reports relating to the prolonged use of pipradrol derivatives. However, some users have re-dosed on desoxypipradrol, in one case staying awake for up to eight days, experiencing hallucinations due to sleep deprivation and manic repetitive OCD-like (obsessive–compulsive disorder) physical activity, although all symptoms cleared up after two weeks of rest [48].

DEPENDENCE AND ABUSE POTENTIAL

There are no published studies relating to the dependence and abuse potential of pipradrol derivatives in animals or humans, although there is some evidence for the abuse potential of pipradrol.

By 1960, pipradrol (Meratran) and methylphenidate (Ritalin) abuse and illicit trafficking in Sweden had become a significant problem and therefore these drugs were subject to control in Sweden from 1 January 1961 [122]. In the UK there was an increase in the misuse of amphetamine and related stimulants and therefore the Drugs (Prevention of Misuse) Act (DPMA) 1964 was introduced to control the import and possession of these substances. Pipradrol and methylphenidate fell within the scope of the generic definition in Clause 3 of the schedule to the DPMA 1964; however, in 1970 a DPMA Modification Order replaced the generic definition with a list of named stimulants, which included pipradrol and methylphenidate [123]. In the same year the Advisory Committee on Drug Dependence advised that pipradrol should not be used for the treatment of depression as there was a possibility that anxiety, schizophrenic manifestations, and hysterical symptoms may be exacerbated [124]. Following a review of drug efficacy from 1966 to 1968, pipradrol and a large number of other drugs were withdrawn from the US market by the FDA [125].

The World Health Organisation (WHO) rated the abuse potential of pipradrol as low [126]. There had been a few cases of abuse of pipradrol a number of years earlier, but the references cited [127,128] refer to amphetamine-like stimulants in general and do not provide any specific evidence of pipradrol abuse. Nevertheless, it was decided that pipradrol must be judged as having a dependence potential, albeit less than that of methylphenidate. The drug was therefore included in Group (c) *'drugs whose liability to abuse constitutes a small but still significant risk*

to health and having a therapeutic usefulness ranging from little to great' and was recommended for international control [129]. Pipradrol became subject to international control in 1971 [114] and in the UK was included in the Misuse of Drugs Act 1971 as a Class C drug. The drug is still available in some countries, but the product information includes a warning that it should not be administered for long periods or given to patients known to be prone to drug dependence.

Desoxypipradrol has received several negative comments from drug users and is not considered by some users to be a good drug for recreational purposes, primarily due to its long duration of action and dangers from sleep deprivation [48]. Desoxypipradrol still has the potential to be abused as a stimulant to stay awake for long periods and user reports refer to its use to stay awake whilst operating hazardous machinery or driving a motorcycle [48]. However, it is the acute toxicity associated with the recreational use of pipradrol derivatives and the deaths associated with these drugs that have highlighted their potential for abuse and led to a ban on the importation of desoxypipradrol into the UK [91]. A UK ACMD report considered that use of these substances has potential social harms, particularly in relation to impairment of function, loss of relationships and potential harm to others and recommended that they should be controlled under the Misuse of Drugs Act, 1971 as Class B drugs [32]. The UK ban on importation of desoxypipradrol on 4 November 2010 [91], together with the media reports and negative comments from some users, appears to have effectively reduced its availability in the UK as there were no seizures in 2011. However, diphenylprolinol and to a lesser extent 2-(diphenylmethyl)pyrrolidine continued to be encountered in seizures and in accident and emergency presentations. The UK ACMD therefore provided further advice on these drugs on 10 November 2011 [95] and a ban on their importation into the UK was implemented on 15 November 2011 [97]. Desoxypipradrol and other pipradrol derivatives eventually became Class B drugs in the UK under the Misuse of Drugs Act, 1971 on 13 June 2012 [92]. The drugs were added to Schedule 2 of the Act in the form of a generic definition designed to include not only desoxypipradrol, diphenylprolinol and 2-(diphenylmethyl)pyrrolidine, but also other pipradrol derivatives that might be produced as alternative 'legal highs' [32]. Additionally, at this time, pipradrol was changed from a Class C drug to a Class B drug [92].

CONCLUSIONS

Desoxypipradrol, diphenylprolinol and 2-(diphenylmethyl)pyrrolidine are examples of failed or redundant pharmaceuticals that have since been 'rediscovered' in the scientific literature and used by 'researchers' to explore their potential as legal recreational drugs. There are also many other pipradrol derivatives with stimulant properties that have been synthesised by simple modifications to the chemical structure of pipradrol and manufacturers of legal highs might also design new, untested, pipradrol derivatives that have not previously been investigated by pharmaceutical companies.

Reports on the use of desoxypipradrol and diphenylprolinol as recreational drugs first appeared on the Internet in 2007 [48,51]. There is limited published literature on the pharmacology, recreational dose, toxicity and associated harm caused by the use of these compounds. However, user reports describing the experimental use of these substances published on Internet drug forums provide some detail as to their effects. Several deaths associated with the use of pipradrol derivatives have also been described in the literature. Pipradrol derivatives are central nervous system stimulants, but the doses used for recreational purposes are usually in excess of those originally intended for medicinal purposes. The initial pattern of acute toxicity is similar to that for other recreational drugs such as MDMA and other amphetamines, but the key clinical feature distinguishing the pipradrol derivatives from other recreational drugs is their long duration of action often associated with prolonged neuropsychiatric symptoms.

Cases of acute toxicity of pipradrol derivatives requiring hospitalisation appear to have been a result of the wider marketing of these drugs as 'legal highs' on the Internet and from 'head shops', with product names such as 'Ivory Wave', 'Whack', 'Head Candy', 'NRG', and 'A3A Methano', in packages bearing no details of the active ingredients or purity. Therefore, users of these products do not know exactly what drug they are taking and the content of some of these products are known to be variable. It has therefore been suggested that cases of acute toxicity relating to the use of such products occur when users are not intentionally using the drug that was subsequently identified in biological samples [47]. Localised clusters of poisonings are also thought to occur when the active constituents of a product are changed, often to circumvent new drugs legislation [53].

The outbreaks of poisonings from pipradrol derivatives appear to have been short-lived, probably due to their undesirable adverse effects (neuropsychiatric symptoms) and long duration of action. Whilst new legislation may have helped to reduce the supply of these substances from 'head shops' and UK-based websites, and might deter the production of new pipradrol derivatives, legislation has not been effective at reducing the availability of other new psychoactive substances such as mephedrone. From a harm reduction point of view, the risk of poisoning from pipradrol derivatives might have been reduced if information on the ingredients of 'legal high' products and warnings relating to their harmful effects had been more widely available at the earliest opportunity. Early identification of new psychoactive substances is an essential first step in the information gathering process and therefore it is recommended that the monitoring activities of the EMCDDA Early Warning System and the Home Office Forensic Early Warning System [130] should be continued and the prevalence of emerging novel psychoactive drugs should be monitored through surveys and seizure data so that appropriate action can be taken to minimise harms. If cases of acute toxicity do occur, clinicians should be encouraged to collect biological samples so that the causative agents can be identified.

REFERENCES

[1] Werner HW, Tilford CH, inventors; Wm. S. Merrell Co., assignee. Alpha,alpha diaryl piperidino methanols. US Patent 2624739; 1953.

[2] Sury E, Hoffmann K. Über alkylenimin-derivate. Piperidin-Derivate mit zentralerregender Wirkung I [About alkyleneimine derivatives. Piperidine derivatives with a central stimulant effect I]. Helv Chim Acta 1954;37:2133–45. [German].

[3] Tripod J, Sury E, Hoffmann K. Zentralerregende wirkung eines neuen piperidinderivates [Central stimulant effect of a new piperidine derivative]. Experientia 1954;10(6):261–2. [German].

[4] Bellucci G. Il 2-difenilmetil-piperidin-idrocloruro e l'estere metilico dell'acido α-fenil-α-piperidil-(2)-cloroacetico, farmaci dotati di azione risvegliante [2-Diphenylmethyl-piperidine hydrochloride and the methyl ester of α-phenyl-α-(2-piperidyl)-chloroacetic acid: two drugs possessing an awakening action]. Minerva Anestesiol 1955;21(6):125–8. [Italian].

[5] Heer J, Sury E, Hoffmann K. Über alkylenimin-derivate. Piperidin-derivate mit zentralerregender Wirkung II [Piperidine derivatives with central stimulant effects]. Helv Chim Acta 1955;38:134–40. [German].

[6] Brown BB, Braun DL, Feldman RG. The pharmacologic activity of α-(4-piperidyl)-benzhydrol hydrochloride (Azacyclonol hydrochloride); an ataractive agent. J Pharmacol Exp Ther 1956;118(2):153–61.

[7] McCarty FJ, Tilford CH, Van Campen Jr MG. Central stimulants. α,α-Disubstituted 2-piperidinemethanols and 1,1-disubstituted heptahydroöxazolo[3,4-a]pyridines. J Am Chem Soc 1957;79:472–80.

[8] Hoffmann K, Heer J, Sury E, Urech E, inventors; CIBA Pharmaceutical Products, Inc., assignee. 2-Diphenylmethyl-piperidine. US Patent 2820038; 1958.

[9] Hoffmann K, Heer J, Sury E, Urech E, inventors; CIBA Pharmaceutical Products, Inc., assignee. 2-Diphenylmethyl-piperidine compounds. US Patent 2826583; 1958.

[10] Hoffmann K, Heer J, Sury E, Urech E, inventors; CIBA Pharmaceutical Products, Inc., assignee. Piperidines and their manufacture. US Patent 2849453; 1958.

[11] Belleau B. The synthesis of (±),(+) and (−) α-(3-thiamorpholinyl)-benzhydrol, a new selective stimulant of the central nervous system. J Med Pharm Chem 1960;2(5):553–62.

[12] Hoffmann K, Heer J, Sury E, Urech E, inventors; CIBA Pharmaceutical Products, Inc., assignee. Substituted 2-diphenylmethyl-piperidine compounds. US Patent 2957879; 1960.

[13] Nodine JH, Bodi T, Slap J, Levy HA, Siegler PE. Preliminary trial of a new stimulant SCH 5472 in ambulatory patients with depression, exhaustion, or hypersomnia syndromes. Antibiotic Med Clin Ther 1960;7:771–6.

[14] Winthrop SO, Humber LG. Central stimulants. Cyclized diphenylisopropylamines. J Org Chem 1961;26:2834–6.

[15] Hoffmann K, Sury E, inventors; CIBA Corporation, assignee. 1-Ethyl-2-diphenylmethyl-piperidines. US Patent 3048594; 1962.

[16] Roussel-UCLAF, inventor; Nouveau médicament actif sur le systéme nerveux central et doué notamment d'effet psychotonique [New drug active in the central nervous system with particular psychotonic effect]. French Patent [Fr. M 3638]; 1964. [French].

[17] Likhosherstov AM, Raevskii KS, Lebedeva AS, Kritsyn AM, Skoldinov AP. Azacycloalkanes III. Tertiary α-azacycloalkyl carbinols. Synthesis and pharmacological activity. Pharm Chem J 1967;1:27–31.

[18] Walter LA, Springer CK, Kenney J, Galen SK, Sperber N. Derivatives of 3-piperidinol as central stimulants. J Med Chem 1968;11(4):792–6.

[19] Enyedy IJ, Sakamuri S, Zaman WA, Johnson KM, Wang S. Pharmacophore-based discovery of substituted pyridines as novel dopamine transporter inhibitors. Bioorg Med Chem Lett 2003;13:513–7.

[20] Corkery JM, Elliot S, Forbes Forsyth VA, Schifano F, Corazza O, Ghodse AH. First deaths involving 2-DPMP (desoxypipradrol, 2-benzhydrylpiperidine, 2-diphenylmethylpiperidine); 2012. In preparation.

[21] European monitoring centre for drugs and drug addiction. EDND – European information system and database on new drugs; 2012. [Access restricted to registered users].

[22] ACMD Consideration of the Novel Psychoactive Substances ('Legal Highs'). London: Advisory Council on the Misuse of Drugs; 2011. Available: <http://www.homeoffice.gov.uk/publications/agencies-public-bodies/acmd1/acmdnps2011> [accessed 18.07.12].

[23] Ferris RM, Tang FLM. Comparison of the effects of the isomers of amphetamine, methylphenidate

and deoxypipradrol on the uptake of l-[^{3}H]norepinephrine and [^{3}H]dopamine by synaptic vesicles from rat whole brain, striatum and hypothalamus. J Pharmacol Exp Ther 1979;210(3):422–8.

[24] Hoffman K, Sury E, inventors; CIBA Pharmaceutical Products, Inc., assignee. 2(1'-Phenyl-lower alkyl) piperidines and their intermediates. US Patent 2830057; 1958.

[25] Jacob RM, Joseph NM, inventors; Rhone-Poulenc, assignee. New esters. US Patent 2928835; 1960.

[26] Schmidt D, Hüller H, Amon I, Peters R. Zur pharmakologie einer serie von diphenylaminopropanderivaten [A series of diphenylaminopropane derivatives for pharmacology]. Pharmazie 1973;28(10):677–80. [German].

[27] Shafi'ee A, Hite G. The absolute configurations of the pheniramines, methyl phenidates, and pipradrols. J Med Chem 1969;12:266–70.

[28] Sperber N, Papa D, inventors; Schering Corporation, assignee. Substituted alkyl piperidines. US Patent 2739969; 1956.

[29] Forster W, Henderson AL. A clinical study of 'Frenquel' (alpha (4-piperidyl) benzhydrol hydrochloride) in chronic schizophrenia. Can Med Assoc J 1957;76:97–101.

[30] Sperber N, Villani FJ, Papa D, inventors; Schering Corporation, assignee. Substituted piperidines. US Patent 2739968; 1956.

[31] Schumann EL, Van Campen, Jr MG, Pogge RC, inventors; Wm. S. Merrell Co., assignee. Tranquilizing composition comprising alpha phenyl, alpha-(4-piperidyl)-benzyl alcohol and method of using same. US Patent 2804422; 1957.

[32] ACMD Consideration of desoxypipradrol (2-DPMP) and related pipradrol compounds. London: Advisory Council on the Misuse of Drugs; 2011. Available: <http://www.homeoffice.gov.uk/publications/agencies-public-bodies/acmd1/desoxypipradrol-report> [accessed 18.07.12].

[33] Fabing HD, Hawkins JR, Moulton JAL. Clinical studies on α-(2-piperidyl)benzhydrol hydrochloride, a new antidepressant drug. Am J Psychiatry 1955;111(11):832–6.

[34] Tilford CH, Shelton RS, Van Campen Jr MG. Histamine antagonists. Basically substituted pyridine derivatives. J Am Chem Soc 1948;70:4001–9.

[35] Murugan R, Goe GL, Scriven EFV, inventors; Reilly Industries, Inc., assignee. Processes for producing α-pyridyl carbinols. US Patent 5541331; 1996.

[36] Portoghese PS, Pazdernik TL, Kuhn WL, Hite G, Shafi'ee A. Stereochemical studies on medicinal agents. V. Synthesis, configuration, and pharmacological activity of pipradrol enantiomers. J Med Chem 1968;11(1):12–15.

[37] Rao AVR, Gurjar MK, Sharma PA, Kaiwar V. Entantioselective reductions of ketones with oxaborolidines derived from (R) and (S)-α,α-diphenyl-2-piperidine methanol. Tetrahedron Lett 1990;31(16):2341–4.

[38] Krumkalns EV, inventor; Eli Lilly and Company, assignee. Method of regulating the growth of aquatic weeds with pyridine derivatives. US Patent 4043790; 1977.

[39] Berger L, Lee J, inventors; Hoffman-La Roche Inc. and Nutley, assignee. 1-Alkyl-3-benzohydryl piperidines. US Patent 2599364; 1952.

[40] Jackson PF, Slusher BS, inventors; Guilford Pharmaceuticals Inc., assignee. Pharmaceutical compositions and methods for treating compulsive disorders using pyrrolidine derivatives. US Patent 5925666; 1999.

[41] Synthesis of diphenyl-2-pyrrolidinyl-methanol and diphenyl-2-pyrrolidinyl-methane [Review article]. Rhodium archive hosted by Erowid. Available from: <http://www.erowid.org/archive/rhodium/chemistry/pyrrolidinyl.html>; August 2004 [accessed 08.11.11].

[42] Enders D, Pieter R, Renger B, Seebach D. Nucleophilic α-sec-aminoalkylation: 2-(diphenylhydroxymethyl)pyrrolidine. Org Syntheses Coll 1988;6:542. Available: <http://www.orgsyn.org/orgsyn/prep.asp?prep=cv6p0542> [accessed 18.07.12].

[43] Corey EJ, Shibata T, Lee TW. Asymmetric Diels-Alder reactions catalysed by a triflic acid activated chiral oxazaborolidine. J Am Chem Soc 2002;124(15):3808–9.

[44] Carlone A. Enantioselective aminocatalysis: new reactions and new directions. [PhD Thesis]. Bologna: Università di Bologna; 2008. p. 5. Available: <http://amsdottorato.cib.unibo.it/1040/1/Tesi_Carlone_Armando.pdf> [accessed 18.07.12].

[45] Kanth JVB, Periasamy M. Convenient method for the synthesis of chiral α,α-diphenyl-2-pyrrolidinemethanol. Tetrahedron 1993;49(23):5127–32.

[46] Bailey DJ, O'Hagan D, Tavasli M. A short synthesis of (S)-2-(diphenylmethyl)pyrrolidine, a chiral solvating agent for NMR analysis. Tetrahedron Asymmetry 1997;8(1):149–53.

[47] Wood DM, Dargan PI. Use and acute toxicity associated with the novel psychoactive substances diphenylprolinol (D2PM) and desoxypipradrol (2-DPMP). Clin Toxicol (Phila.) 2012;50(8):727–32.

[48] Drugs info: 2-DPMP/desoxypipradrol [Internet forum]. Drugs-Forum. Available: <http://www.drugsforum.com/forum/showthread.php?t=30973>; 2007 [accessed 09.07.12].

[49] Desoxypipradrol research – preliminary conclusions [Internet forum]. Bluelight. Available: <http://www.bluelight.ru/vb/threads/305531-Desoxypipradrol-research-preliminary-conclusions>; 2007 [accessed 10.07.12].

[50] Begg WGA, Reid AA. 'Meratran'; a new stimulant drug. BMJ 1956 April;1(4973):946–9.

[51] Experiences – Diphenyl-2-Pyrrolidinyl-Methanol (diphenylprolinol) trip reports [Internet forum]. Drugs-Forum. Available: http://www.drugs-forum.com/forum/showthread.php?t=41534>; 2007 [accessed 12.07.12].

[52] Vree TB, Van Rossum JM. Suppression of renal excretion of fencamfamine in man. Eur J Pharmacol 1969;7(2):227–30.

[53] Murray DB, Potts S, Haxton C, Jackson G, Sandilands EA, Ramsey J, et al. 'Ivory wave' toxicity in recreational drug users; integration of clinical and poisons information services to manage legal high poisoning. Clin Toxicol (Phila.) 2012;50(2):108–13.

[54] Herbert JX, Daly F, Tracey JA. Whacked! BMJ Lett. 341. Available: <http://www.bmj.com/rapid-response/2011/11/02/whacked>; 2010 [accessed 18.07.12].

[55] Wood DM, Puchnarewicz M, Johnston A, Dargan PI. A case series of individuals with analytically confirmed acute diphenyl-2-pyrrolidinemethanol (D2PM) toxicity. Eur J Clin Pharmacol 2012 April;68(4):349–53. [Epub 2011 October 29].

[56] WHO Food Additives Series: 56. Safety evaluation of certain food additives. Prepared by the sixty-fifth meeting of the Joint FAO/WHO Expert Committee on Food Additives (JECFA). Genève: World Health Organization. p. 358-9. Available: <http://whqlibdoc.who.int/publications/2006/9241660562_part2_g_eng.pdf>; 2006 [accessed 18.07.12].

[57] Provigil (modafinil) Tablets. United States: Cephalon, Inc.; Revised October 2010. [Medication Guide]. Available: <www.provigil.com/media/PDFs/prescribing_info.pdf> [accessed 12.08.12].

[58] Brandenberger H. Part 3.9 Antihistamines Brandenberger H, Maes RAA, editors. Analytical toxicology: for clinical, forensic, and pharmaceutical chemists. Berlin: Walter de Gruyter; 1997. p. 546. Available: <http://books.google.co.uk/books?id=ZhYtynyC4kAC> [accessed 12.08.12].

[59] Kariya S, Isozaki S, Narimatsu S, Suzuki T. Oxidative metabolism of cinnarizine in rat liver microsomes. Biochem Pharmacol 1992;44(7):1471–4.

[60] Wu WN, Hills JF, Chang SY. In vivo biotransformation of fenoctimine in rat, dog and man. Xenobiotica 1994;24(11):1133–48.

[61] Jacobsen E. The comparative pharmacology of some psychotropic drugs. Bull World Health Organ 1959;21:411–93. Available: <www.ncbi.nlm.nih.gov/pmc/articles/PMC2537982/> [accessed 25.08.12].

[62] Karczmar AG, Howard Jr JH. Anorexigenic action of methylphenidate (Ritalin) and pipradrol (Meratran). Proc Soc Exp Biol Med 1959;102:163–7.

[63] Gelvin EP, McGavack TH, Kenigsberg S. Alpha-(2-piperidyl) benzhydrol hydrochloride (pipradrol) as an adjunct in the dietary management of obesity. NY State J Med 1955;55(16):2336–8.

[64] Brown BB, Werner HW. Pharmacologic studies on a new central stimulant, α-(2-piperidyl)benzhydrol hydrochloride (MRD-108). J Pharmacol Exp Ther 1954;110(2):180–7.

[65] Pogge RC. Meratran. BMJ 1956;2(4988):361.

[66] Council on Drugs New drugs and developments in therapeutics. JAMA 1962 October;182(4):346–7. Available: jama.ama-assn.org.

[67] Shader RI, Harmatz JS, Kochansky GE, Cole JO. Psychopharmacologic investigations in healthy elderly volunteers: effects of pipradrol-vitamin (Alertonic) elixir and placebo in relation to research design. J Am Geriatr Soc 1975;23(6):277–9.

[68] Robbins TW, Watson BA, Gaskin M, Ennis C. Contrasting interactions of pipradrol, d-amphetamine, cocaine, cocaine analogues, apomorphine and other drugs with conditioned reinforcement. Psychopharmacology 1983;80:113–9.

[69] White NM, Hiroi N. Pipradrol conditioned place preference is blocked by SCH23390. Pharmacol Biochem Behav 1992;43:377–80.

[70] Ross SB, Kelder D. Inhibition of 3H-dopamine accumulation in reserpinized and normal rat striatum. Acta Pharmacol Toxicol (Copenh.) 1979;44:329–35.

[71] Information on Prep. No. 14'469 (desoxypipradrol), a new synthetic stimulant with central point of application. Novartis; 1955. [Unpublished report].

[72] Maxwell RA, Chaplin E, Batmanglidj Eckhardt S, Soares JR, Hite G. Conformational similarities between molecular models of phenethylamine and of potent inhibitors of the uptake of tritiated norepinephrine by adrenergic nerves in rabbit aorta. J Pharmacol Exp Ther 1970;173(1):158–65.

[73] Ferris RM, Tang FLM, Maxwell RA. A comparison of the capacities of isomers of amphetamine, deoxypipradrol and methylphenidate to inhibit the uptake of tritiated catecholamines into rat cerebral cortex slices, synaptosomal preparations of rat cerebral cortex, hypothalamus and striatum and into adrenergic nerves of rabbit aorta. J Pharmacol Exp Ther 1972;181(3):407–16.

[74] Davidson C, Ramsey J. Desoxypipradrol is more potent than cocaine on evoked dopamine efflux in the nucleus accumbens. J Psychopharmacol 2012;26(7):1036–41.

[75] Schmitt KC, Zhen J, Kharkar P, Mishra M, Chen N, Dutta AK, et al. Interaction of cocaine-, benztropine-, and GBR12909-like compounds with wild-type and mutant human dopamine transporters: molecular features that differentially determine antagonist-binding properties. J Neurochem 2008;107:928–40.

[76] Jackson PF, inventor; Guilford Pharmaceuticals Inc., assignee. Pyrrolidine derivatives. US Patent 5650521; 1997.

[77] Dickerson TJ, Janda KD. Recent advances for the treatment of cocaine abuse: central nervous system immunopharmacotherapy. AAPS J 2005;7(3):E579–86. Available: <http://link.springer.com/content/pdf/10.1208%2Faapsj070359.pdf> [accessed 06.05.13].

[78] Pindell MH, Doran KM, Frenzel KC, Tisch DE. Pharmacological studies on 3-(α,α-diphenyl-α-hydroxymethyl)thiomorpholine HCl, a CNS stimulant. Fed Roc Soc 1959;78:433.

[79] Fabing HD. New blocking agent against the development of LSD-25 psychosis. Science 1955;121:208–10.

[80] Wood DM, Hunter L, Measham F, Dargan PI. Limited use of novel psychoactive substances in South London nightclubs. QJM 2012;Oct;105(10):959–64.

[81] Nordegren T. The A-Z encyclopedia of alcohol and drug abuse. Boca Raton, FL: Brown Walker Press; 2002. Available: <http://books.google.co.uk/books?id=4yaGePenGKgC&dq=Gato+Diablo&redir_esc=y> [accessed 18.07.12].

[82] Shulgin AT. Drugs of abuse in the future. Clin Toxicol 1975;8(4):405–56. Available: <http://www.erowid.org/archive/rhodium/chemistry/shulgin.futuredrugs.html> [accessed 18.07.12].

[83] Cooper DA. Future synthetic drugs of abuse. In: Proceedings of the International Symposium on the forensic aspects of controlled substances. Washington, DC: Laboratory Division, FBI, US Government Printing Office. p. 79–103. Available: <http://www.erowid.org/archive/rhodium/chemistry/future_drugs.html>; 1990 [accessed 03.02.12].

[84] Desoxypipradrol information request [Internet forum]. Bluelight; Postings from February 2007 to June 2009. Available: <http://www.bluelight.ru/vb/threads/292451-desoxypipradol-information-request> [accessed 03.02.12].

[85] EMCDDA-Europol 2010 Annual Report on the implementation of Council Decision 2005/387/JHA. Lisbon: European Monitoring Centre for Drugs and Drug Addiction. Available: <www.emcdda.europa.eu/attachements.cfm/att_132857_EN_EMCDDA-Europol%20Annual%20Report%202010A.pdf>; 2011 [accessed 12.07.12].

[86] Corkery JM, Elliott S, Schifano F, Corazza O, Ghodse AH. 2-DPMP (desoxypipradrol, 2-benzhydrylpiperidine, 2-phenylmethylpiperidine) and D2PM (diphenyl-2-pyrrolidin-2-yl-methanol, diphenylprolinol): a preliminary review. Prog Neuropsychopharmacol Biol Psychiatry 2012;June [Epub ahead of print].

[87] Kim HS, Aftab A, Shah M, Nayar J. Physical and psychological effects of the new legal high 'Ivory Wave': a case report. BJMP 2010;3(4):a343.

[88] Durham M. Ivory Wave: the next mephedrone? Emerg Med J 2011;28(12):1059–60.

[89] Kelleher C, Christie R, Lalor K, Fox J, Bowden M, O'Donnell C. An overview of new psychoactive substances and the outlets supplying them. Dublin: National Advisory Committee on Drugs; 2011. Available: <http://www.nacd.ie/publications/Head_Report2011_overview.pdf> [accessed 18.07.12].

[90] Kavanagh PV, Sharma J, McNamara S, et al. Head shop 'legal highs' active constituents identification chart (June/July 2010, '511' – '714'). Dublin: Department of Pharmacology and Therapeutics, School of Medicine, Trinity Centre for Health Sciences, St James's Hospital; 2010. Available: <www.medicine.tcd.ie/bulletin/oct-nov-2010/HS%20ID%20Poster%20511-714.pdf> [accessed 18.07.12].

[91] Home Office. Imports of Desoxypipradrol (2-DPMP, 2-Benzhydrylpiperidine, 2-Diphenylmethylpiperidine) [Home Office notice]. London: Home Office. Available: <http://www.homeoffice.gov.uk/publications/alcohol-drugs/drugs/drug-licences/desoxypipradrol/>; 4 November 2010 [accessed 15.07.12].

[92] Dangerous Drugs: The Misuse of Drugs Act 1971 (Amendment) Order 2012 [Statutory Instrument]. 30 May 2012[SI 1390].

[93] Intelligence alert – Recent, unusual drug submissions in New Zealand. In: Microgram Bulletin. Washington, DC: Drug Enforcement Administration, Office of Forensic Sciences. Available: <http://www.justice.gov/dea/programs/forensicsci/microgram/mg1207/mg1207.html>; 2007 [accessed 18.07.12].

[94] Gower P. Herald investigation: party pill pushers beat the ban [Media report]. Available: <www.nzherald.co.nz/nz/news/article.cfm?c_id=1&objectid=10473827>; 2007 November 3 [accessed 30.08.12].

[95] ACMD Further advice on diphenylprolinol (D2PM) and diphenylmethylpyrrolidine. London: Advisory Council on the Misuse of Drugs; 10 November 2011. Available: <http://www.homeoffice.gov.uk/publications/agencies-public-bodies/acmd1/acmd-d2pm> [accessed 16.07.12].

[96] Wood DM, Puchnarewicz M, Holt DW, Ramsey J, Dargan PI. Detection of the precursor benzophenone in individuals who have used legal highs containing diphenyl-2-pyrrolidinemethanol (D2PM). Basic Clin Pharmacol Toxicol 2011;109(Suppl.1):86.

[97] Home Office. Import ban on diphenylprolinol and diphenylmethylpyrrolidine [Home Office Notice]. London: Home Office; 15 November 2011 Available: <www.homeoffice.gov.uk/publications/alcohol-drugs/drugs/import-ban-d2pm> [accessed 16.07.12].

[98] Mysterious stimulant in Slim Xtreme dietary supplement [Internet forum]. Bluelight Available: <http://www.bluelight.ru/vb/threads/462408-Mysterious-stimulant-in-Slim-Xtreme-dietary-supplement>; 2009 [accessed 07.02.12].

[99] Effects: Difference: 2-(diphenylmethyl)pyrrolidine vs diphenyl-2-pyrrolidinyl-methanol (D2PM) [Internet forum]. Drugs-Forum Available: <http://www.drugs-forum.com/forum/showthread.php?t=143109>; 2010 [accessed 12.07.12].

[100] Experiences: 2-Diphenylmethylpyrrolidine   aka 2-Benzylhydrylpyrrolidin [Internet forum]. Drugs-Forum   Available: <http://www.drugs-forum.com/forum/showthread.php?t=144908>.

[101] Toxic designer stim update: confirmed. [Web newspaper]. Steroid Times. Available: <http://www.steroidtimes.com/toxic-designer-stim-update-confirmed/2009>; 2009 [accessed 07.02.12].

[102] Lidder S, Dargan PI, Sexton M, et al. Cardiovascular toxicity associated with recreational use of diphenyl-prolinol (diphenyl-2-pyrrolidinemethanol [D2PM]). J Med Toxicol 2008;4(3):167–9.

[103] Wood DM, Button J, Lidder S, Ovaska H, Ramsey J, Holt DW, et al. Detection of the novel recreational drug diphenyl-2-pyrrolinemethanol (D2PM) sold 'legally' in combination with glaucine. Clin Toxicol 2008;46(5):393.

[104] Sistema nazionale di allerta precoce per le droghe [National Early Warning System for Drugs]. 2-(Diphenylmethyl)pyrrolidine (desoxy-D2PM). Rome: Dipartimento Politiche Antidroga. [Italian]. Available: <http://www.droganews.it/pubdownload.php?id=2472>; 2010 [accessed 12.07.12].

[105] Nayar S. Green Powder called A3A Methano.: Cambridgeshire Constabulary. Available: <http://ewsd.wiv-isp.be/Other%20information%20on%20new%20psychoactive%20substances/Desoxy-D2PM/Report_UK_Dec%202010_Green%20Powder%20called%20A3A%20Methano.pdf>; 2010 [accessed 12.07.12].

[106] Di Chirara G, Gessa GL. Pharmacology and neurochemistry of apomorphine. Adv Pharmacol Chemother 1978;15:87–160.

[107] Small SF, Quock RM, Malone MH. Pipradrol-induced hyperthermia in the rabbit. Pharmacol Res Commun 1984;16(9):923–31.

[108] Pipradrol hydrochloride toxicity data [Datasheet]. LookChem Available: <http://www.lookchem.com/TempWithNoSign.aspx?cas=71-78-3&key=pipradrol+hydrochloride>; 2012 [accessed 13.01.12].

[109] ChemIDplus Advanced [Database]. United States: National Library of Medicine. Available: <chem.sis.nlm.nih.gov/chemidplus>; 2012 [accessed 13.01.12].

[110] Inchem. Methylphenidate hydrochloride [Poisons Information Monographs]. International Programme on Chemical Safety (Revised August 1997). Available: <http://www.inchem.org/documents/pims/pharm/pim344.htm>; 1992 [accessed 09.07.12].

[111] Hintze KL, Aboul-Enein HY, Fischer LJ. Isomeric specificity of diphenylmethylpiperidine in the production of rat pancreatic islet cell toxicity. Toxicology 1977;7:133–40.

[112] Miller CP, Reape TJ, Fischer LJ. Inhibition of insulin production by cyproheptadine in RINm5F rat insulinoma cells. J Biochem Toxicol 1993;8(3):127–34.

[113] Rickels K, Schneider B, Pereira-Organ JA, Perloff MM, Segal A, Vandervort W. Pipradrol in mild depression: a controlled study. J Clin Pharmacol 1974;14:127–33.

[114] Convention on Psychotropic Substances. Vienna: International Narcotics Control Board. Available: <http://www.incb.org/incb/convention_1971.html>; 1971 [accessed 20.07.12].

[115] Martindale: the complete drug reference [online pharmacopoeia]. Medicines Complete. Available: www.medicinescomplete.com/mc/martindale/current/login.htm>; 2012 [accessed 10.09.12].

[116] Alertonic elixir [Information sheet]. South African electronic package inserts.; Published 26/2/75 and updated April 2004. Available: <http://home.intekom.com/pharm/adcock/alerton.html> [accessed 06.01.12].

[117] Coppola M, Mondola R. Research chemicals marketed as legal highs: the case of pipradrol derivatives. Toxicol Lett 2012;212:57–60.

[118] Psychonaut.   Desoxypipradrol   technical   paper. London: Psychonaut Web Mapping Research Group. Institute of Psychiatry, King's College London. Available: <www.rednetproject.eu/groups/2dpmp>; 2009 [accessed 14.08.12].

[119] Drassdo A, Schmidt M. Effect of 2-diphenylmethylpiperidine hydrochloride on blood circulation, mental capacity and metabolism. Medizinische Monatsschrift fur Pharmazeuten 1956;10(11):738–43. [German].

[120] LoGiCal. Analytical monograph: Desoxypipradrol. LGC Standards; Last revised 23 July 2012. Available: <www.logical-standards.com/uploads/pdfs/english/Desoxypipradol_with%20copyright.pdf> [accessed 30.07.12].

[121] Kaizaki A, Tanaka S, Tsujikawa K, Numazawa S, Yoshida T. Recreational drugs, 3,4-methylenedioxymethamphetamine (MDMA), 3,4-methylenedioxyamphetamine (MDA) and diphenylprolinol, inhibit neurite outgrowth in PC12 cells. J Toxicol Sci 2010;35(3):375–81.

[122] Bejerot N. Narkotikafrågan och samhället [Drug issues in the community], 2nd ed. Volume 28 of Aldus aktuellt. Stockholm: Bokförlaget aldus/Bonnier. [Swedish]. Available: <www.nilsbejerot.se/samhallet.pdf>; 1968 [accessed 18.07.12].

[123] King LA. Chapter 5: Drug legislation in the UK Forensic chemistry of substance misuse. Cambridge: Royal Society of Chemistry; 2009. p. 31.

[124] Department of Health and Social Security The amphetamines and lysergic acid diethylamide (LSD): report by the Advisory Committee on Drug Dependence. London: HMSO; 1970.

[125] Shorter E. The drug efficacy study in 1966–1968 Ban TA, Healy D, Shorter E, editors. Reflections on twentieth century psychopharmacology. CINP; 2004. p. 75–6. Available : <http://cinp.org/fileadmin/documents/history/books/VOL4_opt.pdf> [accessed 18.07.12].

[126] Isbell H, Chruáciel TL. Dependence liability of 'non-narcotic' drugs. Geneva: World Health Organization; 1970. Available: <http://whqlibdoc.who.int/bulletin/1970/Vol43/supplement/bulletin_1970_43%28supp%29.pdf> [accessed 18.07.12].

[127] Goldberg L. Drug abuse in Sweden (I). Bull Narcot 1968;20(1):1–31.

[128] Goldberg L. Drug abuse in Sweden (II). Bull Narcot 1968;20(2):9–36.

[129] World Health Organization. Technical Report Series No. 437. WHO Expert Committee on Drug Dependence, Seventeenth Report. Geneva: World Health Organization. Available: <http://whqlibdoc.who.int/trs/WHO_TRS_437.pdf>; 1970 [accessed 18.07.12].

[130] Annual report on the Home Office forensic early warning system (FEWS). Home Office. Available: <www.homeoffice.gov.uk/publications/alcohol-drugs/drugs/drug-strategy/fews>; 2012 [accessed 07.09.12].

This page intentionally left blank

Aminoindane Analogues

Simon D. Brandt[*], Robin A. Braithwaite[†], Michael Evans-Brown[**] and Andrew T. Kicman[†]

[*]School of Pharmacy and Biomolecular Sciences, Liverpool John Moores University, Liverpool, UK
[†]Drug Control Centre, Department of Forensic Science, Division of Analytical and Environmental Sciences, King's College London, London, UK
[**]European Monitoring Centre for Drugs and Drug Addiction, Lisbon, Portugal

INTRODUCTION

While a wide range of central nervous system (CNS)-active substances are developed for clinical and therapeutic purposes, others are used experimentally to probe neurobiological and pharmacological properties and functions within a research setting. The implementation of structural modification of chemical entities lies at the heart of drug discovery, but the interest in the synthesis and availability of novel psychoactive substances (NPS) extends well beyond members of the scientific community and the pharmaceutical industry. Individual users of novel drugs are often very interested in exploring their novel effects, while entrepreneurs and criminal organisations show a commercial interest. Within the context of communities interested in recreational drugs, the Internet impacts in various ways, ranging from information exchange about the effects of unexplored substances and marketing of commercially available products to the aim of providing various forms of harm reduction advice [1]. Many of the emerging NPS have been described to some extent in the scientific literature and drug patents, whereas others are the result of structural modifications that derive from those candidates. One of the classes of the NPS is represented by analogues and derivatives that are based on the 2-aminoindane nucleus, i.e. 2,3-dihydro-1H-inden-2-amine (**1**, 2-AI; Fig. 11.1). The present chapter will provide an account of selected aminoindanes regarding their chemistry, pharmacology and toxicology, expanding on a review published in 2011 [2].

PHARMACOLOGY

Physical and Chemical Description

This section predominantly focuses solely on routes of synthesis and laboratory identification of the aminoindanes.

Novel Psychoactive Substances.
DOI: http://dx.doi.org/10.1016/B978-0-12-415816-0.00011-0

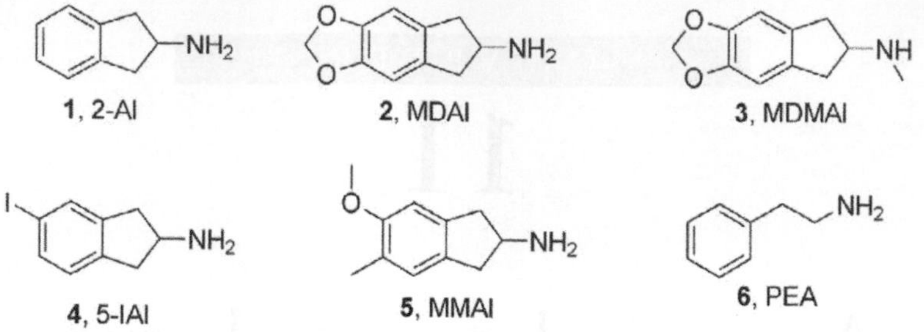

FIGURE 11.1   (1) 2-Aminoindane; (2) 5,6-methylenedioxy-2-aminoindane; (3) 5,6-methylenedioxy-N-methyl-2-aminoindane; (4) 5-iodo-2-aminoindane; (5) 5-methoxy-6-methyl-2-aminoindane; (6) β-phenethylamine.

Synthesis

The 2-AI nucleus serves as a core structure for many pharmacologically active substances and some have been the subject of attention regarding their occurrence as NPS, having been found to be available for sale on the Internet and some retail outlets. From a thematic viewpoint, an emphasis is placed on 5,6-methylenedioxy-2-aminoindane (2, MDAI), 5,6-methylenedioxy-N-methyl-2-aminoindane (3, MDMAI), 5-Iodo-2-aminoindane (4, 5-IAI) and 5-methoxy-6-methyl-2-aminoindane (5, MMAI), respectively (Fig. 11.1). Some additional examples are provided in order to add some historical comments, which are based on the early recognition of the structural similarity between 2-AI and β-phenethylamine (6, PEA) [3]. The early interest in the exploration of aminoindane derivatives derived from the realisation that these would be considered as rigid analogues of various catecholamines, phenethylamines and amphetamines that have already been under investigation for their bioactive properties, including CNS-stimulation, bronchodilation and analgesic activity.

A representative example for the synthesis of aminoindanes is shown in Figure 11.2, which depicts the preparation of MDAI following the approach published by the Nichols group [4] (with the exception of steps (vi) and (vii)). Piperonal, i.e. 3,4-(methylenedioxy)benzaldehyde (a), was heated with malonic acid

(pyridine/piperidine) to undergo Knoevenagel condensation which yielded 3,4-(methylenedioxy)cinnamic acid (b). This particular route can be adopted to a variety of aminoindane products depending on the substituted benzaldehyde precursor. Hydrogenation was carried out with Pd/C (in ethanol) until hydrogen was fully absorbed which then gave the propanoic acid product, namely 3,4-(methylenedioxy)dihydrocinnamic acid (c). Interestingly, a microwave-accelerated alternative was investigated by Quinn and co-workers [5] who prepared (c) and a variety of additional dihydrocinnamic acid derivatives. Under these conditions, 1,4-cyclohexadiene served as the hydrogen transfer source and product formation was usually observed within 5 minutes [5]. 5,6-Methylenedioxy-1-indanone (d) resulted from an intramolecular cyclisation reaction via formation of an acid chloride intermediate under the conditions reported by Nichols et al. [4]. In this particular example, thionyl chloride (in benzene) was used, although phosphorus pentoxide was also evaluated but with poor yields. The crude acid chloride was re-dissolved in dichloromethane and exposed to tin(IV) chloride to give indanone (d). Alternatively, oxalyl chloride may also be used as demonstrated by Johnson et al. during the preparation of MMAI [6]. Another alternative used for this cyclodehydration reaction was offered by Koo who studied a variety of cyclisations with polyphosphoric acid (PPA) which included the preparation

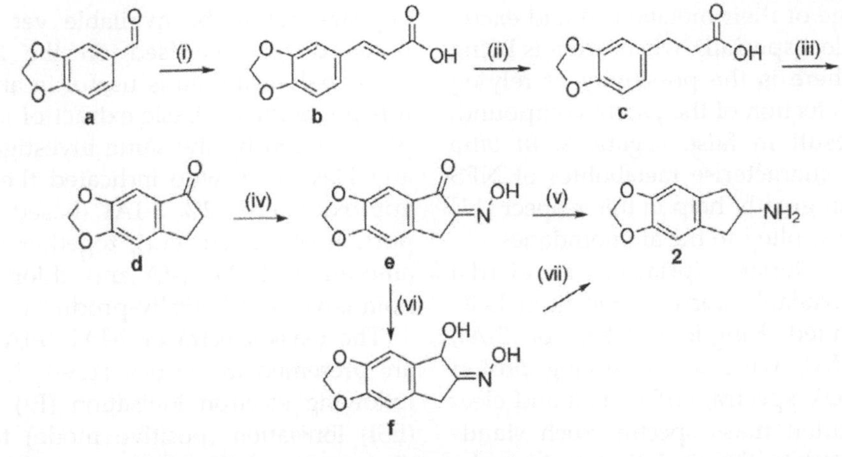

FIGURE 11.2 Synthesis of MDAI. (**1**) (i) malonic acid, pyridine, piperidine, HCl, steam bath; (ii) H₂, 10% Pd/C, ethanol; (iii) thionyl chloride, benzene, reflux, then tin(IV) chloride, dichloromethane; (iv) isoamyl nitrite, HCl, methanol; (v) H₂, 10% Pd/C, acetic acid, sulfuric acid; [4] (vi) sodium borohydride, ethanol; (vii) H₂, Pd/C, ethanol, HCl [12].

of several di-and trimethoxy-1-indanones [7]. Examples of PPA use were also provided by Nichols et al. during the synthesis of DOM-AI [8] and Coutts and Malicky who prepared various indanones, indanols and aminoindanes and also attempted a cyclisation with phosphorous oxychloride but reported low yields [9]. Oxime (**e**), or in other words the hydroxyimino ketone, was prepared from the indanone precursor (in methanol) using isoamyl nitrite and concentrated hydrochloric acid [4]. A variation with butyl nitrite (in absolute ethanol) was employed by Nichols et al. when making the precursor to DOM-AI [8] whereas Cannon and colleagues used butyl nitrite for the 4-methoxy-substituted indanone [10]. Furthermore, ethyl nitrite [9] and pentyl nitrite have also been employed [11]. Catalytic hydrogenation in acetic acid and concentrated sulphuric acid then afforded MDAI [4]. One variation was reported with the use of sodium borohydride that gave the indanol (**f**) [12,13] which was then converted to MDAI [12]. Another variation of the theme, using lithium aluminium hydride (LAH), was presented by Coutts and Malicky [9] who prepared the 1-aminoindane version of DOM-AI. The 1-indanone

precursor was first converted to the oxime with hydroxylamine and then reduced by LAH. The reaction was heated at reflux for 30 hours (in diethyl ether) and a 12-h alternative was also successfully developed using a hydrogenation (in ethanol) at normal pressure catalyzed by platinum dioxide [9].

Chemical Analysis

This is discussed in more detail in Chapter 6. Analytical techniques. Currently, there appears to be a lack of validated procedures in the peer-review literature regarding detecting the presence of the 2-aminoindanes in biological matrices and thus analytical toxicologists are faced with developing their own methodology, invariably employing chromatography-mass spectrometry, given current limitations concerning cross-reactivity in immunoassay screens. LC-MS/MS has been recently applied to the analysis of pooled urine collected from a portable urinal at a single nightclub, which showed the presence of 2-AI amongst other novel recreational drugs [14]. Targeting the parent compound is usually the first step for detecting the presence of these substances, but then knowledge of their metabolism and excretion is desirable, especially when urine is being analysed, as there is the possibility of relying solely on the detection of the parent compound, which may result in false negatives. *In vitro* approaches to characterise metabolites of NPS compounds can greatly help in this respect [15] and need to be applied to the aminoindanes.

Reference materials (primary standards) are currently available, for example from LGC Standards (United Kingdom; UK), for 2-AI, MDAI and 5-IAI, with accompanying proton NMR, IR and UV spectra, HPLC data and electrospray ionisation mass spectra. Such standards are essential to the analytical toxicologist for the incorporation of the aminoindanes into their drug screens and for matching purposes in confirmatory analysis. To the best of our knowledge, at the time of writing, reference materials for MMAI and MDMAI do not appear to be available, although they are under current development. Without standards or authentic biological samples (from a controlled study), drug intelligence regarding the potential use of MMAI and MDMAI is likely to be limited to the analysis of seizures, where the amounts present are conducive to analysis by spectroscopy (NMR, IR, UV and Raman). Even so, high resolution mass spectrometry is becoming increasingly employed and the determination of the elemental composition of NPS can be particularly useful when they are present as mixtures. However, unambiguous identification will often fail when it is attempted to differentiate between isomers without chromatographic comparison so an over-reliance on high resolution mass spectrometry alone should be avoided.

5-IAI has been synthesised and then extensively characterised by electron ionisation MS, NMR and IR spectroscopy, as has MDAI. 4-Iodo-2-aminoindan (4-IAI) and the biologically inactive 4,5-methylenedioxy-2-aminoindane were also investigated for analytical delineation from 5-IAI and MDAI, respectively, because of positional isomerism [16,17]. 4-IAI appears not to be available yet as a recreational drug advertised on the Internet, but such analytical data is useful in anticipation of this possibility. A basic extract of an illicit sample analysed by the same investigators, Casale and Hays [17], who indicated the presence of approximately 92% 5-IAI (based on the proportion of ion current), together with a small amount of 2-IAI, 4-IA and chloro-iodo-2-aminoindan as synthetic by-products.

The mass spectra of 2-IAI, 5-IAI and MDAI are presented in Figures 11.3–11.5, respectively, following electron ionisation (EI); electrospray (ESI) ionisation (positive mode) for MDAI is also presented. The ESI spectrum was obtained from a Certificate of Analysis provided by LGC Standards (UK) and the EI spectra were obtained from the *Microgram Journal* [16,17], published by the Drug Enforcement Administration (USA). The proposed structures of the fragments have been inserted on to the Figures by the authors of this chapter with the caveat that without further studies, these are provisional.

With respect to the EI spectra of MDAI (Fig. 11.5), the TMS derivatives fragment considerably differently under EI conditions compared to the underivatised compounds, presumably due to charge retention in other regions of the molecule (in this case at the nitrogen), leading to a combination of charge-driven and charge-remote fragmentation. It may be suggested that the m/z 150 (Fig. 11.5c) results from the consecutive loss of a methyl radical followed by ring cleavage (one of several possibly ring cleavages is depicted in the figure). Under hydrogen rearrangements, the formation of a stable conjugated ion is conceivable, which would be in agreement with the high abundance of this fragment ion.

Pharmacokinetics

There are currently no published studies on the pharmacokinetics of the aminoindane class of compounds in laboratory animals or in humans.

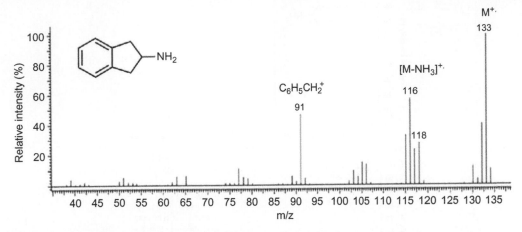

FIGURE 11.3 Electron ionisation (EI) mass spectrum of 2-aminoindane (2-AI, **1**). The proposed assignments to the ions have been added by the authors of this chapter.

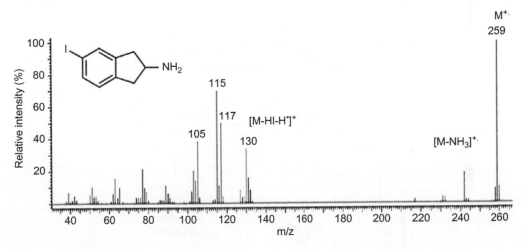

FIGURE 11.4 Electron ionisation (EI) mass spectrum of 5-iodo-2-aminoindane (5-IAI, **4**). The proposed assignments to the ions have been added by the authors of this chapter.

Pharmacodynamics

There are extensive studies of the pharmacological action of a number of aminoindanes in various animal and *in vitro* models; these are summarised below. However, there are currently no published human studies that have involved the administration of this class of compounds to human volunteers.

MDAI and 5-IAI-Related 2-Aminoindanes

5,6-Methylenedioxy-2-aminoindane (**2**, MDAI) was found to show a number of pharmacological similarities to MDMA-type entactogens (e.g. **7**, 3,4-methylenedioxymethamphetamine, Fig. 11.6). The first indication was reported in 1990 by the Nichols group when testing for stimulus generalisation in rats using a drug discrimination

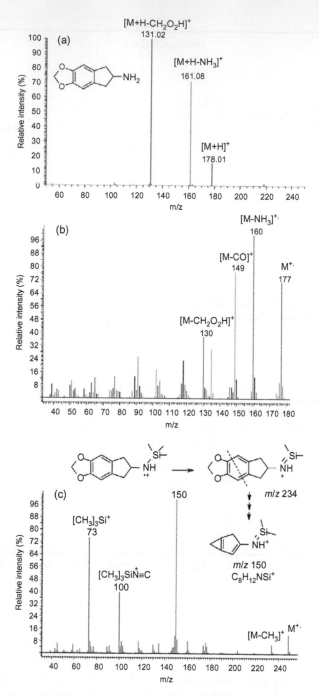

FIGURE 11.5 Mass spectra of 5,6-methylenedioxy-2-aminoindane (MDAI, **2**) following a) positive electrospray ionisation (courtesy of LGC Standards) and b) electron ionisation (EI) of the underivatised and c) trimethylsilyl derivative with proposed mechanism for formation of the m/z 150 fragment (only one of at least two possible ring cleavages depicted). Both b) and c) adapted from [16]. The proposed assignments to the ions (and the mechanism) have been added by the authors of this chapter. The authors are grateful to Dr Mario Thevis, Institute of Biochemistry, Center for Preventive Doping Research, German Sports University Cologne, for providing this proposed mechanism.

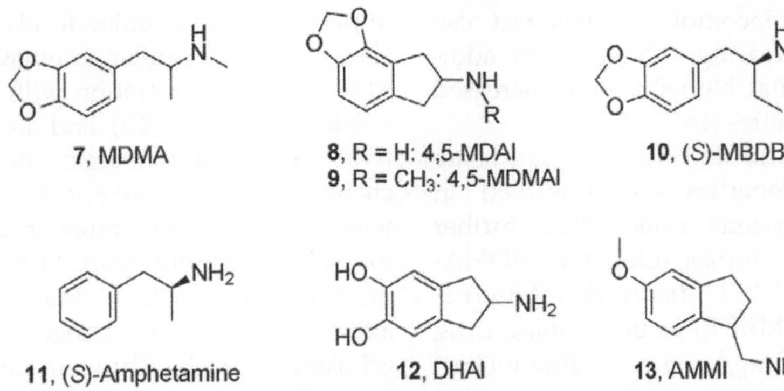

FIGURE 11.6 (**7**) 3,4-Methylenedioxymethamphetamine; (**8**) 4,5-methylenedioxy-2-aminoindane; (**9**) 4,5-methylene-dioxy-N-methyl-2-aminoindane; (**10**) (S)-(+)-N-methyl-1-(1,3-benzodioxol-5-yl)-2-butanamine; (**11**) (S)-amphetamine; (**12**) 5,6-dihydroxy-2-aminoindane; (**13**) 1-aminomethyl-5-methoxyindane.

paradigm [4]. Under these conditions, the ability of a test drug to substitute for a training drug (lever responding) was determined, including by dose-response evaluations and comparison of potency. While MDAI did not substitute in d-lysergic acid diethylamide (LSD)-trained rats (intraperitoneal injection, i.p.) it was found to produce an MDMA-like discriminative stimulus with comparable potency (MDAI ED_{50} (95% CI) 0.58 mg/kg, 2.66 μmol/kg) using (R/S)-MDMA as the training drug. However, given that racemic MDMA was used, it was pointed out that MDAI potency would have to be considered about half that of the more active (S)-MDMA enantiomer, since MDAI is not a chiral molecule. The suggestion that MDAI could serve as a non-neurotoxic MDMA replacement came from within the framework of serotonergic neurotoxicity and 5-HT depletion studies using analytical and radioligand binding techniques. Following a subcutaneous injection of 40 mg/kg racemic MDMA HCl, MDAI HCl or saline the impact on rat brain monoamine and metabolite levels was determined in the cortex and hippocampus region one week afterwards. High-performance liquid chromatography (HPLC) coupled with electrochemical detection (ECD) was employed to confirm significant reduction in serotonergic markers (5-HT and its metabolite 5-hydroxyindoleacetic

acid, 5-HIAA) when comparing MDMA with saline. As far as 5-HIAA was concerned, a cortical reduction of 71% and a hippocampal reduction of 53% were observed but no significant difference in noradrenaline and dopamine levels were found. By contrast, MDAI did not appear to show reduced serotonergic and noradrenergic levels which indicated a lack of 5-HT neurotoxicity. The second set of parameters tested included the determination of dissociation constants (K_d) and B_{max} values based on [^{3}H]paroxetine binding using rat brain cortical or hippocampal homogenates. Consistent with the concept of serotonergic toxicity it was found that (R/S)-MDMA caused a significant reduction in 5-HT binding sites when compared with saline treatment. MDAI, on the other hand, did not lead to a significant reduction in [^{3}H]paroxetine-labelled 5-HT binding sites. The 4,5-methylenedioxy-2-aminoindane isomer (**8**, 4,5-MDAI) did not appear to display MDMA-like properties in rats since it lacked the corresponding discriminative stimulus properties [4]. In a separate study, 4,5-methylenedioxy-N-methyl-2-aminoindane (**9**, 4,5-MDMAI) and 4,5-MDAI were also investigated. While 4,5-MDAI resulted in only partial MDMA-stimulus generalisation, a significant depression in locomotor activity and rearing frequency was noted. In comparison, 4,5-MDMAI was found to be inactive in terms of locomotor activity but also showed depressed rearing behaviour. In addition, it was found that N-methylation increased 5-HT$_{2A}$ receptor affinities [18].

Supporting evidence that MDAI might share pharmacological properties also observed in MDMA-type entactogens came from further drug discrimination studies using the MDMA-like (S)-(+)-N-methyl-1-(1,3-benzodioxol-5-yl)-2-butanamine (**10**, (S)-MBDB) as the training drug (training dose 1.75 mg/kg i.p. (7.18 μmol/kg)) and complete substitution was observed for the hydrochloride salts of (R/S)-MDMA and its single enantiomers, (R/S)-MDA and its single enantiomers and MDAI. The same was also found for its N-methyl derivative, i.e. MDMAI, that fully substituted for (S)-MBDB (ED$_{50}$ 3.01 μmol/kg). MDAI showed a slightly lower ED$_{50}$ value (2.04 μmol/kg) than MDMAI which indicated a slightly higher potency [19].

Drug discrimination studies showed that MDAI abolished (S)-(+)amphetamine like discriminative stimulus (i.p., (S)-amphetamine sulphate), which provided further support that it differed pharmacologically from (S)-amphetamine-like stimulants [20]. A separate study investigated the co-administration of MDAI with a variety of substances, including (S)-amphetamine and other dopaminergic agents and monoamine oxidase inhibitors (MAOIs). Short-term changes (3h post-drug) of monoamines in rat cortex, striatum and hippocampus were assessed by HPLC-ECD after a single MDAI HCl administration (s.c., 40mg/kg). For example, a substantial decrease in 5-HT was noted (e.g. $7 \pm 1\%$ frontal cortex, $11 \pm 1\%$ striatum and $21 \pm 3\%$ hippocampus relative to saline set at 100%), whereas dopamine, DOPAC and homovanillic acid (HVA) levels were significantly increased in the frontal cortex and striatum. 5-HT levels were found to return to around 80% of control after 8 hours and [³H]paroxetine binding studies (cortex, hippocampus and striatum) carried out one week after single and multiple MDAI treatment confirmed the absence of impact on catecholamine levels and 5-HT uptake sites. However, similar to what was reported on MMAI, co-administration with (S)-amphetamine resulted in short (3h) and long-term (1 week) reduction in serotonergic markers. Long-term changes (1 week) were not observed when MDAI was combined with a monoamine oxidase inhibitor MAO-A (chlorgyline), MAO-B ((R)-deprenyl) or dopamine uptake (GBR-12909) inhibitors although short-term changes in 5-HT and catecholamine levels (3h) were noted [21]. Sprague and colleagues studied whether MAO inhibition would potentiate the neurotoxicity of MDAI. A pre-treatment regimen included administration of MAO-A inhibitor clorgyline and MAO-B inhibitor (S)-deprenyl. Each pre-treatment was followed by MDAI application 15 minutes later. Analysis of frontal cortex and striatum tissue one week afterwards, by HPLC-ECD, showed that dopamine levels were only significantly increased in the striatum, but not in the frontal cortex. As expected, MDAI injection without pre-treatment did not show any significant effect on 5-HT, 5-HIAA or dopamine levels. However, it was commented that multiple MDAI doses produced slight long-term deficits in serotonin markers [22]. Malmusi et al. [18] confirmed earlier findings that MDAI was equipotent with MDMA and produced MDMA-like stimulus effects in rats. In contrast to MDMA, locomotor activity studies showed that MDAI was only weakly active in mice which indicated weaker central stimulant activity as judged by distance traversed in cm and rearing frequency. MDMAI, however, was considered inactive up to a dose of 30mg/kg (slight increase but not statistically significant) but, on the other hand, it was equally able to produce MDMA-stimulus generalisation, although it was less potent that MDAI. Both MDAI counterparts showed very low 5-HT$_{2A}$ receptor affinities (K$_i$ < 10000nM, [³H]ketanserin and [³H]DOB as radioligands), but modest affinity (K$_i$ 4600 ± 300nM) was observed with MDMAI using [³H]ketanserin in contrast to when [³H]DOB was used (K$_i$ < 10000nM), respectively [18].

A potential metabolite of MDAI, i.e. 5,6-dihydroxy-2-aminoindane (**12**, DHAI) displayed moderate selectivity for [³H]NE uptake, which was relatively similar to MDAI as far as this particular reuptake transporter was concerned [23].

5-Iodo-2-aminoindane (**4**, 5-IAI) was behaviourally active in rats (ED$_{50}$ 0.65 mg/kg, 2.19 µmol/kg using (R/S)-MDMA and 0.79 mg/kg, 2.67 µmol/kg using (S)-MBDB) based on drug discrimination studies and substituted fully for (R/S)-MDMA HCl and (S)-MBDB HCl [24]. Subcutaneous application of 40 mg/kg 5-IAI HCl (0.13 mM/kg) was followed by HPLC-ECD analysis one week later and revealed slight, but statistically significant, decreases of hippocampal 5-HT levels (15%) which contrasted with 5-HIAA (cortical and hippocampal) values where this was not the case. Catecholamine levels were not affected. In addition, 5-IAI acted as an inhibitor of [³H]5-HT in rat brain cortical synaptosomes (IC$_{50}$ 241 ± 21 nM) and was shown to slightly reduce cortical uptake sites ([³H]paroxetine) but not hippocampal uptake. Overall, this led to the conclusion that 5-IAI would also be considered to show limited potential to cause serotonergic deficits in the rat [24].

5-Methoxy-6-Methyl-2-Aminoindane (MMAI)

MMAI fully substituted in (R/S)-MDMA- and (S)-MBDB-trained rats (ED$_{50}$ 0.81 mg/kg, 3.77 µmol/kg using racemic MDMA and ED$_{50}$ 0.56 mg/kg, 2.63 µmol/kg based on (S)-MBDB) but, as observed with MDAI, LSD substitution did not occur [6]. Furthermore, MMAI was not found to substitute for (S)-amphetamine either at doses tested between 0.50 and 2.0 mg/kg, which also pointed towards entactogen-like properties of MMAI. Short-term, i.e. 3 h post-drug, cortical and hippocampal monoamine changes were observed using HPLC-ECD procedures following a single s.c. administration of 20 mg/kg MMAI HCl. For example, cortical dopamine levels increased, while both cortical and hippocampal 5-HT and 5-HIAA levels decreased significantly.

Interestingly, monoamines and their metabolite levels remained unchanged when determined one and two weeks after single and multiple dosing regimens using either 10 or 20 mg/kg MMAI HCl. In addition, [³H]paroxetine binding studies (frontal cortex and hippocampus) two weeks after sub-acute dosing showed the absence of changes in binding density [6].

[³H]Monoamine uptake inhibition studies into synaptosomes from rats confirmed that MDAI, MMAI and 5-IAI were selective for 5-HT over DA and NE – this was particularly the case with MMAI. In addition, uptake inhibition studies were carried out with reserpinised synaptosomes (leading to particularly lower [³H]5-HT IC$_{50}$ values), presumably to study the effects of blocked uptake into storage vesicles [23]. It was furthermore confirmed [25] that MMAI inhibited [³H]5-HT uptake into platelet plasma membrane vesicles and [³H]imipramine binding to the corresponding 5-HT transporters. In addition, MMAI-mediated [³H]5-HT efflux via the serotonin transporter was confirmed, which meant that it acted as a substrate for the transporter. For this reason, a decreased [³H]5-HT efflux was observed in the absence of external Na⁺ ions. A second serotonergic mechanism shared by both MMAI and MDMA was the ability to inhibit [³H]5-HT accumulation by chromaffin granule membrane vesicles which appeared to be a model system for H⁺ dependent storage within synaptic vesicles. What differed from MDMA, however, was the inability of MMAI to both inhibit dopamine transport (lack of displacement of the cocaine analogue β-[¹²⁵I]CIT) and stimulation of [³H] dopamine efflux [25], which was consistent with the idea of high selectivity for 5-HT over dopamine. More support for the ability of MMAI HCl (and 4-MTA HCl) to trigger 5-HT release *in vivo* was obtained from microdialysis studies in anaesthetised rats. HPLC-ECD was used to determine the increase in extracellular 5-HT concentration in the dorsal hippocampus following MMAI and 4-MTA (**14**, 4-methylthioamphetamine, Fig. 11.7) (i.p., 1 and 5 mg/kg) and fluoxetine (10 mg/kg)

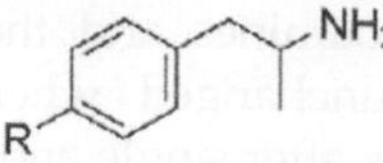

14, R = SCH₃: 4-MTA
15, R = F: 4-FA
16, R = Cl: 4-CA
17, R = I: 4-IA

FIGURE 11.7 (**14**) 4-Methylthio-; (**15**) 4-fluoro-; (**16**) 4-chloro-; (**17**) 4-iodoamphetamine.

administration. While fluoxetine caused an increase in 5-HT output of around 180% after 40 min, MMAI application resulted in an increase of around 1350% at 40 min post-injection and after 100 min 5-HT output returned to around 450%. 4-MTA, on the other hand, triggered a 2000% increase which declined much slower to around 1500% after 100 min. MMAI proved to be a weak MAO-A inhibitor but lacked MAO-B inhibiting properties [26]. The third transmembrane helix (TMH III) of the serotonin re-uptake protein (hSERT) has been implicated in the recognition of substrates and antagonists. As part of a study to determine the impact of selected amino acid residues on recognition, five hSERT TMH III mutants have been prepared by selected mutagenesis. A range of structurally diverse substances, including three aminoindanes, have been tested for [³H]5-HT uptake inhibition in comparison with the wild-type hSERT. The potency values of MMAI remained relatively similar apart from the S174M mutant that caused a fivefold drop in potency (hSERT Ki 600 ± 200 nM vs S174M Ki 3100 ± 300 nM). 1-Aminomethyl-5-methoxyindane (**13**, AMMI), on the other hand, displayed a particularly decreased Ki value for the A169D mutant [27].

While MMAI lacked long-term reduction in serotonergic markers in rat brain, persistent changes, however, were induced by co-administration with (S)-amphetamine. MMAI administration alone, i.e. 10 mg/kg or 20 mg/kg injections every 12 h for four days, revealed a lack of reduction of cortical 5-HT and 5-HIAA (HPLC-ECD) and [³H]paroxetine binding sites

one week afterwards [28]. The combination with (S)-amphetamine, on the other hand, led to a significant reduction in these cortical parameters. In addition, the impact was found to be dose-dependent, i.e. (S)-amphetamine + MMAI (10 mg/kg) resulted in an approximate 20% decrease whereas (S)-amphetamine and MMAI (20 mg/kg) decreased serotonergic marker levels by around 50–60%. Hippocampal results were similar. Changes in cortical noradrenaline or dopamine levels were not observed, although an approximately 55% decrease was observed in caudate 3,4-dihydroxyphenylacetic acid (DOPAC) levels following the 20 mg/kg MMAI combination [21]. The accumulation of (S)-3,4-dihydroxyphenylalanine (DOPA) was determined in the rat striatum by HPLC-ECD measurements 45 min after s.c. administration of MMAI HCl alone (i.p., 20 mg/kg) and in combination with (S)-amphetamine (i.p., 2.5 mg/kg). While MMAI alone did not yield increased DOPA levels, the combination with (S)-amphetamine, however, resulted in a DOPA increase to close to 200% which was consistent with the hypothesis that 5-HT might be involved in the potentiation of dopamine synthesis [29].

MMAI was used as a training drug for drug discrimination studies (ED₅₀ 0.56 mg/kg, 2.64 µmol/kg) and it was revealed that 4-MTA fully substituted for it (ED₅₀ 0.21 mg/kg, 0.96 µmol/kg) and full substitution was also observed in MDMA- and (S)-MBDB-trained rats [30]. Monte and co-workers employed MMAI, including (S)-MBDB and (S)-amphetamine, for additional studies in rats using a number of benzofuran, indan and tetrahydronaphthalene analogues of 3,4-methylenedioxyamphetamine and found that some of them showed comparable potencies [31]. MMAI HCl, together with (S)-amphetamine sulphate (S)-MBDB HCl, were also used as a training drug in order to assess the properties of 4-fluoroamphetamine (**15**, 4-FA). Interestingly, in contrast to 4-chloroamphetamine (**16**, 4-CA), 4-FA only fully substituted in (S)-amphetamine-trained rats (ED₅₀ 0.43 mg/kg,

2.11 µmol/kg) whereas 4-CA did not mimic (S)-amphetamine at all, but substituted for the serotonergic training drugs MMAI (ED$_{50}$ 0.14 mg/kg, 0.69 µmol/kg) and (S)-MBDB (ED$_{50}$ 0.17 mg/kg, 0.82 µmol/kg), respectively. In addition to monoamine uptake studies in whole brain synaptosomes and microdialysis studies of striatal extracellular levels of dopamine, DOPAC and HVA it was concluded that 4-FA showed psychostimulant-like effects rather than serotonergic, e.g. 5-HT-releasing, properties [32]. Interestingly, 4-iodoamphetamine (**17**, 4-IA) showed similar potency to 5-IAI in rats trained to discriminate (S)-MBDB from saline (ED$_{50}$ 0.54 mg/kg, 1.81 µM/kg) although both were less potent than 4-CA (ED$_{50}$ 0.17 mg/kg, 0.82 µM/kg) [24].

The impact of administration of MBDB, 4-MTA and MMAI, on hormone secretion in rats (adrenal corticotrophin (ACTH), corticosterone, prolactin, oxytocin and renin), was also studied. As far as MMAI was concerned, the secretion of plasma ACTH, corticosterone, prolactin, plasma oxytocin and renin increased in a dose-dependent manner and vasopressin secretion was not stimulated. Blood pressure, on the other hand, reduced significantly after 15 min and heart rate also reduced significantly from 5 min after MMAI administration. Pre-treatment with fluoxetine, however, led to a significant reduction in ACTH, prolactin and oxytocin responses, although renin responses were not significantly inhibited. Even though all three test drugs were classified as 5-HT-releasing agents, some differences were observed in terms of hormone secretion and response to fluoxetine pre-treatment which indicated that serotonin release was important but that additional mechanisms may have been involved in the changes seen in hormonal secretion [33]. Stress-induced reduction in consumption of sucrose solutions, i.e. preference deficit, was used as a chronic mild stress model in Sprague-Dawley rats to compare the antidepressant properties of MMAI HCl, 4-MTA HCl and sertraline HCl, respectively. Chronic administration with MMAI and 4-MTA revealed

significant antidepressant effects, i.e. restoration of sucrose drinking, within 3 weeks and improved onset and magnitude when compared with the serotonin specific re-uptake inhibitor (SSRI) sertraline [34]. Another in-depth investigation of MMAI behavioural pharmacology was provided by Marona-Lewicka and Nichols [34] who confirmed symmetrical substitution in drug discrimination studies (MMAI-trained rats) between MMAI (ED$_{50}$ 0.56 mg/kg, 2.64 µmol/kg), (R/S)-MDMA (ED$_{50}$ 0.70 mg/kg, 3.03 µmol/kg) and (S)-MBDB (ED$_{50}$ 0.36 mg/kg, 1.44 µmol/kg) [35]. 5-HT-releasing compounds, such as (S)-(+)-fenfluramine (ED$_{50}$ 0.997 mg/kg, 3.72 µmol/kg) and 4-CA (ED$_{50}$ 0.14 mg/kg, 0.69 µmol/kg) produced full substitution in MMAI-trained rats whereas d-LSD tartrate (ED$_{50}$ 0.08 mg/kg, 0.186 µmol/kg) produced partial substitution (75%) but also 47% disruption. (S)-Amphetamine, on the other hand, did not produce substitution but caused potent disruption (ED$_{50}$ 0.54 mg/kg, 2.97 µmol/kg). Antagonism tests with the moderately selective 5-HT$_{2A}$ receptor antagonist ketanserin showed that the MMAI was not blocked, although discriminability was attenuated and this was also observed with fluoxetine and paroxetine. On the other hand, pindolol, methiothepin and yohimbine were unable to block the MMAI cue, although disruption was observed. Additional support for 5-HT involvement in MMAI stimulus generalisation came from pre-treatment with the 5-HT biosynthesis inhibitor p-chlorophenylalanine (PCPA) which led the subjects to show saline-like responses and attenuation was still detectable after eight days. Radioligand studies with a variety of adrenoceptor bindings sites also confirmed relatively weak affinities (α_1: [^{3}H]prazosin, IC$_{50}$ > 10 µM; α_2: [^{3}H]RX781094, IC$_{50}$ 0.41 µM; α_2: [^{3}H]yohimbine, IC$_{50}$ 1.58 ± 0.21 µM; α_2: [^{3}H]clonidine, IC$_{50}$ 2.45 ± 0.35 µM; β_1/β_2: [^{3}H]dihydroalprenolol, IC$_{50}$ > 10 µM) [34]. Behavioural effects in rats indicated hypolocomotion which was consistent with previous findings on sedative activity of MMAI which led to reduced investigatory behaviours that could not be blocked by fluoxetine [36].

MMAI, but not fluoxetine, substituted for the 5-HT$_{2A}$ receptor agonist quipazine and antagonism tests carried out with the selective 5-HT$_{2A}$ receptor antagonist MDL 100907 (0.5 mg/kg) confirmed blockage of MMAI substitution (1.25 mg/kg) for quipazine [37]. This was in contrast to the study mentioned above that showed that 5-HT antagonists such as ketanserin and methiothepin were unable to block stimulus effects of MMAI [35]. Furthermore, Smith et al. induced 5-HT depletion (i.e. reduction of 5-HT and 5-HIAA levels) by pre-treatment with p-chlorophenylalanine (PCPA) (300 mg/kg for 3 days) and it was found that MMAI's substitution (1.25 mg/kg) for quipazine was reduced, whereas this was not the case with fenfluramine [37]. Moreover, phosphoinositide hydrolysis studies, carried out in NIH 3T3 fibroblasts stably transfected with rat brain 5-HT$_{2A}$ receptor, revealed that MMAI was much less effective than quipazine in formation of inositol monophosphate ([^{3}H]IP) which indicated low intrinsic activity as a direct 5-HT$_{2A}$ agonist. This provided support for the notion that MMAI substitution was facilitated by 5-HT release [37].

It was also shown that racemic fenfluramine HCl and MMAI HCl differed from MDMA when using the conditioned place preference (CPP) paradigm [38]. In contrast to MDMA HCl, that showed significant CPP effects at 2.5, 5.0, 10 and 20 mg/kg (i.p.), fenfluramine (all doses tested) and MMAI (only with 10 and 20 mg/kg) displayed place aversion which supported the idea that MMAI may show a human psychopharmacological profile similar to fenfluramine rather than MDMA. In addition, microdialysis studies with freely moving rats showed that dopamine and DOPAC levels in nucleus accumbens remained unchanged after MMAI (6.2 mg/kg) and fenfluramine (7.7 mg/kg) injection whereas MDMA (6.3 mg/kg) caused a significant increase in dopamine levels [38]. In order to evaluate the extent to which 5-HT-releasing effects of MMAI are blocked by inhibition of 5-HT uptake, a number of SSRIs and tricyclic antidepressants were used in drug discrimination studies [39]. Rats trained to discriminate MMAI HCl from saline showed partial substitution with citalopram and sertraline but also produced a relatively high percentage of disruption, especially at the highest dose tested. Fluoxetine did not satisfy the criteria for partial generalisation. In comparison, MMAI produced partial substitution in sertraline-trained rats and fully substituted for a citalopram cue, respectively, which indicated potentially similar mechanisms involved in discriminative stimulus effects. A series of antagonism tests (pre-treatment 60 min before test drug) were also carried out in MMAI-trained rats and it was observed that fluoxetine (at 10 mg/kg) caused significant disruption, whereas neither sertraline nor citalopram (at 2.5, 5.0 and 10 mg/kg) could completely block the MMAI cue. Partial substitution was also observed with the tricyclic antidepressants imipramine and clomipramine, but not desipramine in MMAI-trained rats [39].

Miscellaneous Aminoindane Examples

The pharmacological characterisation of N-ethyl-5-trifluoromethyl-2-aminoindan (**18**, ETAI, Fig. 11.8) and 5-trifluoromethyl-2-aminoindan (**19**, TAI) was reported by Cozzi et al. who compared them with their phenethylamine counterparts, i.e. fenfluramine (**20**, 3-trifluoromethyl-N-ethylamphetamine, FEN) and one of its metabolites, namely norfenfluramine (**21**, NoFEN), respectively [40]. Drug discrimination studies were based on (S)-amphetamine sulphate, (S)-MBDB HCl, MMAI HCl and LSD tartrate as training drugs. TAI was the only candidate that showed full substitution for (S)-MBDB (ED$_{50}$ 0.56 mg/kg) whereas partial substitution was reported for ETAI, which also partially substituted for LSD. On the other hand, MMAI and (S)-amphetamine effects were not mimicked. Inhibition of [^{3}H]neurotransmitter uptake was investigated using crude synaptosomes. ETAI (IC$_{50}$ 2.61 ± 0.34 µM) was the least potent inhibitor of [^{3}H]5-HT accumulation into synaptosomes and the corresponding

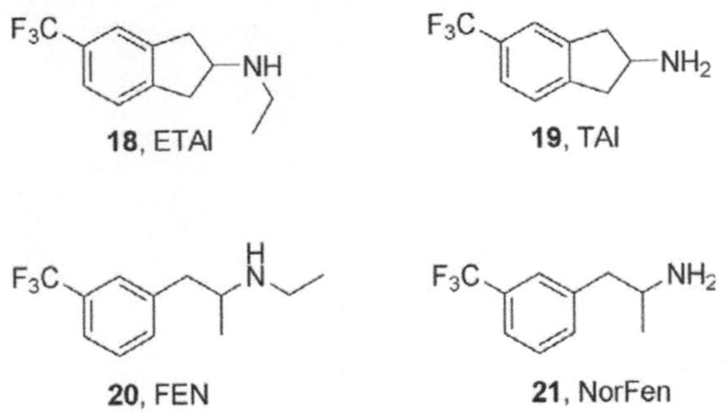

18, ETAI 19, TAI 20, FEN 21, NorFen

FIGURE 11.8 (18) N-Ethyl-5-trifluoromethyl-2-aminoindane; (19) 5-trifluoromethyl-2-aminoindane; (20) fenfluramine; (21): norfenfluramine.

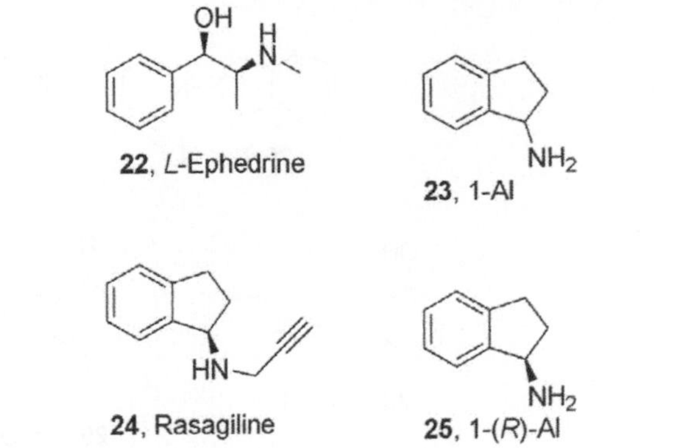

22, *L*-Ephedrine 23, 1-AI 24, Rasagiline 25, 1-(*R*)-AI

FIGURE 11.9 (22): *L*-Ephedrine; (23): 1-aminoindane; (24): *N*-propargyl-1-(*R*)-aminoindane; (25): 1-(*R*)-aminoindane.

IC_{50} values were $0.490 \pm 0.017\,\mu M$ (FEN), $0.590 \pm 0.062\,\mu M$ (NorFEN) and $0.604 \pm 0.066\,\mu M$ (TAI), respectively. Both FEN and NorFEN were generally found to be more potent at inhibiting [^{3}H]DA and [^{3}H]NE uptake than ETAI and TAI. Binding densities were determined at rat cortex, hippocampus and neostriatum after 10 days using [^{3}H]paroxetine following a 4-day drug treatment pattern of 2×10 mg/kg/day. In comparison with saline it was found that both FEN and NorFEN administrations resulted in a 60–70% reduction in binding sites whereas ETAI and TAI showed a less severe reduction of about 30–35%. In addition, it was also reported that the same treatment regimen led to significant reduction in rat body weight (10–15%) for all four substances [40].

Early published studies indicated that a number of aminoindanes showed both bronchodilating and analgesic properties [41]. 2-Aminoindane itself was also known to show a variety of bioactive properties. For example, Kenner and Matthews mentioned a personal communication with H.H. Dale that intravenous 2-AI HCl injection of 'a few milligrams' would lead to blood pressure increase similar to 'β-phenylethylamine', although details were not provided [3]. 2-AI, as well as its *N*-benzyl and *N*-methyl HCl salts, have been reported (in a summarised form and no further details) to be more effective bronchodilators than *L*-ephedrine (**22**, Fig.11.9) based on a bronchial perfusion method using isolated rabbit lungs. Intravenous injection studies carried out in white rabbits revealed less toxicity than amphetamine HCl. Interestingly, it was also pointed out that their 2-amino-1-indanone and 1-amino-2-indanol derivatives would also show bronchodilator activity [42]. Witkin et al. carried out an extensive studies on animals into the analgesic properties and other bioactive effects (e.g. central stimulation and gastrointestinal motility) of 2-AI [43]. It was concluded that 2-AI showed an analgesic potency and therapeutic index comparable with morphine sulphate (tail-flick test) but without respiratory depression. In addition, it was not antagonised by nalorphine (*N*-allylnormorphine) and sub-threshold doses potentiated the effects of morphine [43]. [^{3}H]DA and [^{3}H]NE uptake inhibition studies into rat brain synaptosomes confirmed that 2-AI was 28 times more potent than 1-aminoindane (**23**, 1-AI) in inhibiting catecholamine uptake in the hypothalamus ([^{3}H]NE) whereas a 300-fold higher potency was observed in the striatum preparations ([^{3}H]DA), respectively. In comparison, however, racemic amphetamine appeared to be more potent than both [44]. 1- AI HCl and 2-AI HCl have also been tested

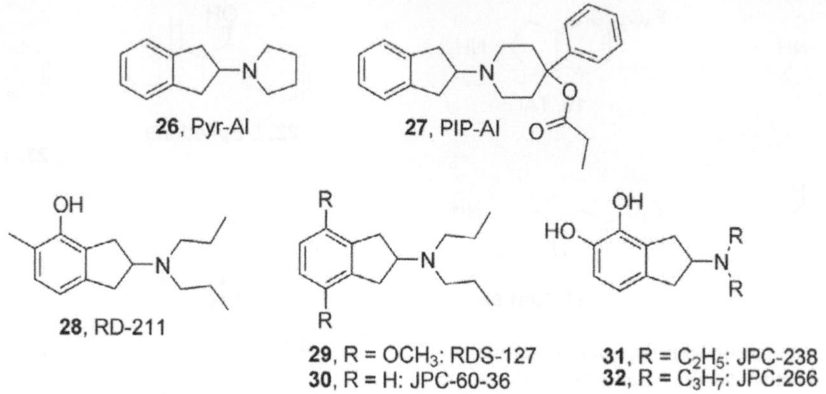

26, Pyr-AI   **27**, PIP-AI

28, RD-211

29, R = OCH$_3$: RDS-127   **31**, R = C$_2$H$_5$: JPC-238
30, R = H: JPC-60-36   **32**, R = C$_3$H$_7$: JPC-266

FIGURE 11.10   (**26**) 1-(2-Indanyl)pyrrolidine; (**27**) 1-(2-indanyl)-4-phenyl-4-propionoxypiperidine; (**28**) 4-hydroxy-5-methyl-2-N,N-dipropylaminoindane; (**29**) 4,7-dimethoxy-2-N,N-dipropylaminoindane; (**30**) 2-N,N-dipropylaminoindane; (**31**) 4,5-dihydroxy-2-N,N-diethylaminoindane; (**32**) 4,5-dihydroxy-2-N,N-dipropylaminoindane.

for their anorectic properties and motor activities in rats. Whereas the 2-substituted isomer led to increased motor activity (149% compared to control) and the 1-isomer produced depressed activity (−69%). Food intake reduction of starved, drug-treated rats was observed for 1h, starting 15min after injection; the 50% reduction in food intake (anorexia RD$_{50}$) was also determined. 2-AI was significantly more anorectic than 1-AI and comparable with (S)-amphetamine sulphate and fenfluramine HCl. However, since 2-AI did produce the least dramatic impact on motor activity it was considered more selective. Interestingly, N-ethyl-2-AI and N-isopropyl-2-AI were inactive in both paradigms tested [45]. 2-AI induced locomotor stimulation in rodents although it was shown to be less active than amphetamine [46]. Another drug discrimination study was carried out using (S)-amphetamine as the training drug and the results appeared to be somewhat comparable. In this particular case, partial substitution was observed with a higher 2-AI dose, although a higher rate of disruption was reported [20]. N-propargyl-1-(R)-aminoindan (**24**, rasagiline) is a selective and irreversible MAO-B inhibitor used for the treatment of parkinsonism and it is

also thought to have neuroprotective properties. Metabolism, affected by the CYP1A2 isoform, leads to the formation of 1-(R)-aminondan (**25**, 1-(R)-AI). Interestingly, although it is not known to be a potent MAO-B inhibitor itself, evidence appears to be mounting that it might show neuroprotective effects by changing pro- and anti-antiapoptotic markers, increase of antioxidant activity and reduction of the GAPDH-MAO-B death cascade [47].

A 'strong amphetamine-type activity' has been reported for the pyrrolidine derivative of 2-AI (**26**, Fig. 11.10) [41] and 'a long duration' of activity was observed at the 10mg/kg dose in mice. Details about the procedure were not provided but neuropharmacological screenings were based on observational methods [48]. A piperidine derivative (**27**, PIP-AI), on the other hand, showed analgesic properties comparable to meperidine [41]. One of the many examples that show simple modifications on both phenyl ring and the nitrogen atom is 4-hydroxy-5-methyl-2-N,N-dipropylaminoindan (**28**, RD-211) that has been shown to produce bradycardia and hypotension in cats. The fact that sulpiride antagonised cardiovascular effects induced by RD-211 indicated involvement of DA$_2$-receptor stimulation [49]. This was in agreement with previous studies where RD-211 HBr was used to induce rotational behaviour in rats that have been exposed to unilateral lesions of the nigrostriatal pathway by 6-hydroxydopamine pre-treatment. Dopamine receptor interactions were indicated by antagonism by haloperidol [50]. Interestingly, racemic RD-211 also showed considerable affinity towards the rat 5-HT$_{1A}$ receptor. Rat brain cortex homogenates were employed using [^{3}H]8-hydroxy-2-(dipropylamino)tetralin ([^{3}H]8-OH-DPAT) as the radioligand. The values obtained for the inhibition constants from both stereoisomers also differed significantly from each other ((R)-RD-211: 10nM vs (S)-RD-211: 340nM). In addition, the 4,7-dimethoxy derivative of RD-211, i.e. RDS-127, was shown to be equipotent to apomorphine in inhibiting artificially stimulated heart rates in cats [49]. Fascinatingly, both RDS-127 and its non-methoxylated analogue (30, JPC-60-36) also stimulated rat locomotor activity when administered subcutaneously while the N,N-diethyl- (31, JPC-238) and N,N-dipropyl-4,5-dihydroxy counterparts (32, JPC-266) also showed inhibitory effects at stimulated heart rates in cats [49] and strong emetic activities were also observed in adult mongrel dogs [51]. In addition to long-lasting locomotor activity in rats, RDS-127 significantly decreased food-intake in rats with little dose-dependency and was more marked than reduction observed with (S)-amphetamine. Both locomotor activity increase and food intake reduction were antagonised by pre-treatment with pimozide, a potent dopamine antagonist. Interestingly, food intake reduction was produced at lower dosage levels whereas effects on locomotor activity were observed at higher dosage levels [52].

One of the many classic substitution patterns found in the phenethylamine (2C-X class) and amphetamine (DO-X class) series that has been explored by Shulgin and Shulgin is summarised in PIHKAL [53]. In this particular series the 2,5-dimethoxy blueprint is left unchanged

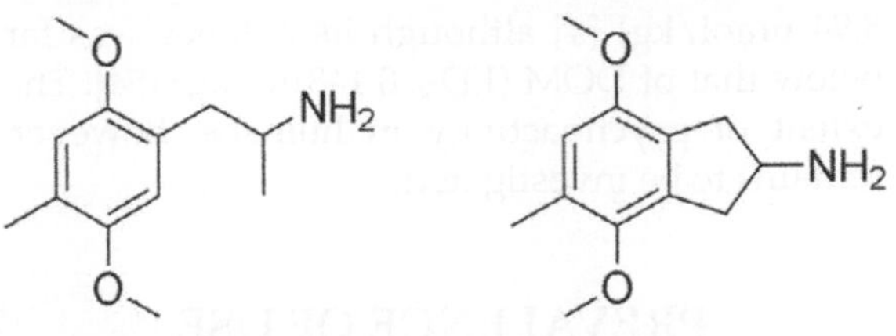

33, DOM   **34, DOM-AI**

FIGURE 11.11   (33) 2,5-Dimethoxy-4-methylamphetamine; (34) 4,7-dimethoxy-5-methyl-2-aminoindane.

whereas the substituent at the 4-position can be modified to a large extent while maintaining psychoactive properties [53]. What follows is the question as to whether their rigid aminoindane analogues would provide any activity in animals or humans. Indeed, the PHIKALisation of aminoindanes, i.e. the approach to adopt substitution patterns found to provide highly active derivatives reported in PIHKAL, might be of interest to scientific studies. One such example may be found in the comparatively potent and long-lasting hallucinogen 2,5-dimethoxy-4-methylamphetamine (33, DOM, Fig. 11.11) that has been around since the 1960s [54]. The corresponding aminoindane counterpart is 4,7-dimethoxy-5-methyl-2-aminoindan (34, DOM-AI) and it has been subject to some investigations but with mixed results. Behavioural studies with rats using the conditioned avoidance response test did not reveal mescaline or DOM-like responses for DOM-AI (40–160μmol/kg) although some form of sedation was observed. At higher dosage levels salivation was also observed [8]. An additional report stated that orally administered DOM-AI showed DOM-like responses in rats and an active intraperitoneal dose of 30mg/kg was also claimed to give DOM-like responses with the 5-bromo analogue, i.e. DOB-AI, respectively [9]. In rats trained to discriminate LSD tartrate (ED$_{50}$ 0.021mg/kg, 0.048μmol/kg) from saline, however, DOM-AI produced full substitution (ED$_{50}$ 2.18mg/kg,

8.94 µmol/kg) [4] although its potency was far below that of DOM (ED$_{50}$ 0.148 mg/kg) [54]. The extent of psychoactivity in humans, however, remains to be investigated.

PREVALENCE OF USE

There is limited information about the use of the aminoindanes as recreational drugs. Despite predictions in the UK media in 2010 that MDAI was likely to rapidly replace the newly controlled stimulant mephedrone (4-methylmethcathinone) [56–59], information suggests that neither MDAI, nor the other aminoindanes have become widely available or are being used extensively to date. In fact, most use appears to be limited to a small number of 'innovators' and 'early adopters' [60] that believe that they are using MDAI or 5-IAI [56,61–63].

Similar to the early stages of diffusion [60] of other types of NPS, much of what is currently known about the use and harms of the aminoindanes comes from triangulation [1,64–67] of limited information from a range of informal and formal sources. These include: laboratory detection of the substances from seizures by law enforcement agencies, biological samples, or test purchases; [16,17,68–79] monitoring online shops that claim to sell the substances; [80] anecdotes from users such as posts on Internet drug discussion forums [81–85], media reports; [56–59] and a small number of queries to poison control centres [86], case reports of acute toxicity [87], and surveys in populations which are expected to contain 'early adopters' of such substances, such as clubbers or self-reported users of NPS [61–63].

Since 2006, a few aminoindanes have been detected in Member States of the European Union and formally notified through its early-warning system for NPS [65,87,88]. There is some indication that 2-AI was detected in 2006, but the presence of the 1-aminoindane isomer could not be excluded at that time; subsequently, 2-AI was confirmed in 2009 and 2011 by other Member States; MDAI was detected in 2010; and 5-IAI in 2011. Subsequently a few other Member States have reported the detection of these substances. These were largely as a result of seizures by law enforcement agencies, although in at least one case it was from a test purchase. The circumstances of their seizure or test purchase, such as being sold as 'research chemicals', and/or the co-presence of other new substances known to be used recreationally or the dosage forms that were seized (such as tablets, capsules), suggest that they were intended to be used for their psychoactive effects [68–73]. It appears that, neither MDMAI nor MMAI have so far been detected, although some online retailers claim to sell these substances.

Within the UK, there is currently a Home Office funded Forensic Early Warning System (FEWS), which is a ministerial led project aimed at identifying NPS encountered in the UK and running from December 2010 to March 2014. FEWS has involved a number of collection plans to gather samples which are then analysed by forensic providers initially by GC-MS. The results are then considered by a team of experts who review the data and assess whether any further analytical technique is required, these can include high resolution MS and NMR (nuclear magnetic resonance). The collection plans have included samples from: 1) test purchases (both from the Internet and from 'head shops'); 2) the police; and 3) festivals (amnesty bins, test purchasing and any sample that the police may encounter). FEWS has encountered aminoindanes in all collection plans; often these are in mixtures, which can include one or more aminoindanes, or aminoindanes with other NPS or mild stimulants, such as ephedrine or caffeine. The aminoindanes encountered were 2-AI, 5-IAI and MDAI. In the 18 months, only 12 aminoindanes samples have been encountered in over 1300 samples [79].

There have only been a few subpopulation surveys that have examined the prevalence of use of the aminoindanes as recreational substances [61–63]. The surveys used convenience samples of clubbers, or self-reported users of NPS, and were conducted between 2010 and 2012. By their very nature these types of populations are likely to contain 'innovators' and 'early adopters' and they generally do not reflect the level of use in the wider population [1].

Measham et al., in their survey of clubbers in two 'gay-friendly' clubs in South London, UK during July 2010 (n = 308; mean age 30; 82% male) found that 6% reported lifetime use of MDAI, while use in the last year and last month was 6% and 4%, respectively [61]. Two percent reported use on the night of the fieldwork [61]. Similarly, the annual survey conducted online by the dance magazine *Mixmag* in 2011 (n = 2560; mean age 25; 69% male) found that 6.7% of respondents reported lifetime use of the substance; while use in the last year was 4.7% [62]. Finally, an online survey that recruited self-reported users of NPS resident in Ireland (n = 329; mean age 25; 67% male) found that 1.6% had used powders or products labelled as '2-Ai' (n = 2), while 2.7% (n = 5) had used MDAI and 1% had used '5-iAi' [63]. It is important to note that analytical data has shown that while some products do contain 2-AI, MDAI or 5-IAI, others are 'misbranded'. For example products sold as containing MDAI or 5-IAI (and/or labelled as such) have been found to contain other substances, such as mephedrone, methylone, 1-benzylpiperazine (BZP) and 1-(3-(trifluoromethyl)phenyl)piperazine (3-TFMPP) (Table 11.1) [75–78], which in some countries are controlled substances. This may place retailers and consumers of such substances at risk of prosecution although in the UK, Section 28 of the Misuse of Drugs Act, 1971 may provide a defence for such circumstances [90]. Overall, despite the limited data, it appears that the prevalence of

TABLE 11.1 Claimed Substances Versus Actual Substances from Test Purchases of Products Sold Online as MDAI or 5-IAI

Claimed Substance	Actual Substance	Reference
MDAI	Mephedrone	[75]
MDAI	Methylone	[76]
MDAI	BZP, 3-TFMPP, caffeine	[77,78]
5-IAI	BZP, 3-TFMPP, caffeine	[77,78]

use of the aminoindanes is still relatively low. Ongoing surveillance, monitoring and research into the use of these substances will be required in order to get a better picture of trends in their use, including whether they diffuse to broader populations of illicit drug users, and the harms that they may pose.

REGULATION OF THE AMINOINDANES

A review of the legal status of the aminoindanes is beyond the scope of this chapter, but it is worth noting that these substances are not under international control [91]. Furthermore, it appears that few countries currently regulate these substances under their national drug control laws; although this is less clear in some countries that have legal provisions that extend to analogues of controlled substances. At the time of writing they are not controlled in either the UK [92] or USA; [16] while The Czech Republic controls MDAI [93] and Switzerland controls 2-AI, MDAI, 5-IAI [94]. Of note, a growing number of Member States of the European Union are also using other, existing, legal measures (such as consumer protection laws or medicine laws) or introducing new ones to help regulate such substances [95,96].

ACUTE AND CHRONIC TOXICITY

Animal Data

We are not aware of any formal phase 1 studies on aminoindanes, and hence there is a paucity of information regarding their potential acute and chronic toxicity, e.g. LD_{50} investigations, mutagenicity and teratogenicity. The majority of animal studies have focused on comparing the effects of aminoindanes with the amphetamines and related drugs to elucidate their pharmacological effects, particularly with regard to dopamine and serotonin pathways (this is covered in more detail in Pharmacology above).

Human Data

Deaths

At the time of writing, data on any confirmed deaths linked to the use of the aminoindane group of compounds have not been received.

Clinical Reports

There is only one published case report of acute toxicity associated with the use of the aminoindanes. This describes multi-organ failure associated with the use of 5 g of what was thought to be MDAI. However, analytical confirmation of MDAI and/or other substances was not undertaken and so it is not possible to be certain that the effects were due to MDAI [87]. The patient made a slow recovery and was subsequently transferred to a psychiatric hospital where he remained for three months after exposure.

It is worth noting that, from the small number of user experiences published on Erowid (www.erowid.org), the common doses reported were between 100 and 200 mg (with some taking larger doses, split over time). Data from the National Poisons Information Service (NPIS) 2010/2011 annual report in the UK indicated that there were less than 40 queries related to MDAI during the reporting period; once again these cases were not analytically confirmed [86].

DEPENDENCE AND ABUSE POTENTIAL

Animal and Human Data

Detailed studies specifically evaluating abuse potential and dependence have, to date, not been performed.

CONCLUSIONS

It currently appears that serotonergic aminoindane representatives such as MDAI, MMAI or 5-IAI play a relatively minor role on the novel psychoactive substance market for recreational drugs. However, detailed insights into the prevalence of use of aminoindanes to be certain about this are absent at the moment. The appearance of previously unknown substances that are used for recreational purposes has been noted since the 1960s followed by the 'designer-drugs' wave in the 1980s/90s that was particularly driven by phenethylamine and amphetamine-type substances [53] with MDMA being one of the best known examples. The unique psychopharmacological profile of MDMA, traditionally regarded as the 'gold standard' by experienced users of stimulant entactogens will continue to attract the attention of drug aficionados (scientific researchers included) within all communities that deal with psychoactive substances in one form or another. The changing nature of drug markets appears to guarantee a constant supply of NPS.

It appears based on currently available data that some of the properties of serotonergic aminoindanes are similar to MDMA, for example when examining some of the drug discrimination studies. But there are differences, and so far, the pharmacological literature indicates a reduced potential to cause serotonergic deficits which may differ from MDMA. There also appears to be a reduced impact on dopamine and noradrenaline accumulation which is a feature that MDMA does not appear to show either. If one considers a potentially reduced capacity of aminoindane analogues currently observed on the NPS market (e.g. MDAI) to mimic the dopaminergic component found within MDMA, then it might be speculated that the effects in humans may not be comparable to this classic entactogen. Future studies in humans are required to address this issue in order to obtain data on both psychostimulant and entactogenic properties. Dopamine release has been discussed as a contributing factor to the formation of serotonergic deficits and co-administration studies of aminoindanes with (S)-amphetamine were found to potentiate the reduction of serotonergic markers. Whether, as speculated, substances such as MMAI show a human psychopharmacological profile similar to fenfluramine rather than MDMA remains to be investigated, but, on the other hand, fenfluramine has also been shown to cause long-term 5-HT depletion whereas the impact of MMAI may be less pronounced. Future studies will hopefully shed more light on the properties of aminoindanes in humans. However, there exists a wealth of published pharmacological data available on the *in vitro* and *in vivo* animal effects of this family of drugs. One of the many questions that requires further investigation is the ability to discriminate between drugs that are monoamine uptake blockers and those that are substrate-type releasers, since both types of drugs interact with the same transporter proteins to elevate extracellular levels of neurotransmitters. The reason for this need lies in the fact that substrate-type releasing agents might erroneously appear as uptake blockers depending on the assay used [97,98].

The aminoindanes described here have shown promising potential for use as pharmacological tools and many novel structures can be imagined that have not yet been described. While some appear to show selectivity for serotonergic systems, others are explored as dopaminergic substances, for example, as dopamine D_3 receptor ligands. For those who have to deal with frontline exposure to psychoactive substances,

e.g. toxicologists, clinicians or forensic practitioners, the ability to identify these candidates is not a trivial challenge and reference materials (parent drug and probable metabolites) are often absent. The situation is often hampered by the presence of isomeric entities and differentiation between them can be very challenging, particularly under routine laboratory conditions that involve a high turnover rate of specimens submitted for analysis. It seems quite likely that not all emerging substances show psychoactive properties. However, it is quite possible that they are produced in order to circumvent legislation and control. The past few years have seen a rapid growth in the number and availability of novel psychoactive substances that are not regulated under international or national drug control legislation. In many cases, such substances do not diffuse beyond small groups, perhaps often due to the fact that their effects are not acceptable to users. This may also include the potential for serious acute toxicity and require a rapid response by stakeholders in order to protect individual health. Those substances that do diffuse more widely can pose significant public and social harm. In response to the emergence of novel psychoactive substances, countries are exploring a range of policy measures that are designed to protect public health and reduce social harms [e.g. 99,100]. Evaluation of these measures will allow intended and unintended consequences to be identified so that these can help inform and shape the policy responses to this highly dynamic phenomenon.

REFERENCES

[1] Sumnall HR, Evans-Brown M, McVeigh J. Social, policy, and public health perspectives on new psychoactive substances. Drug Test Anal 2011;3:515–23.

[2] Sainsbury PD, Kicman AT, Archer RP, King LA, Braithwaite RA. Aminoindanes – the next wave of 'legal highs'? Drug Test Anal 2011;3:479–82.

[3] Kenner J, Mathews AM. 2-Hydrindamine. J Chem Soc Trans 1914;105:745–8.

[4] Nichols DE, Brewster WK, Johnson MP, Oberlender R, Riggs RM. Nonneurotoxic tetralin and indan analogues

of 3,4-(methylenedioxy)amphetamine (MDA). J Med Chem 1990;33:703–10.

[5] Quinn JF, Razzano DA, Golden KC, Gregg BT. 1,4-cyclohexadiene with Pd/C as a rapid, safe transfer hydrogenation system with microwave heating. Tetrahedron Lett 2008;49:6137–40.

[6] Johnson MP, Frescas SP, Oberlender R, Nichols DE. Synthesis and pharmacological examination of 1-(3-methoxy-4-methylphenyl)-2-aminopropane and 5-methoxy-6-methyl-2-aminoindan: similarities to 3,4-(methylenedioxy)methamphetamine (MDMA). J Med Chem 1991;34:1662–8.

[7] Koo J. Studies in polyphosphoric acid cyclizations. J Am Chem Soc 1953;75:1891–5.

[8] Nichols DE, Barfknecht CF, Long JP, Standridge RT, Howell HG, Partyka RA, et al. Potential psychotomimetics. 2. Rigid analogs of 2,5-dimethoxy-4-methylphenylisopropylamine (DOM, STP). J Med Chem 1974;17:161–6.

[9] Coutts RT, Malicky JL. The synthesis of analogs of the hallucinogen 1-(2,5-dimethoxy-4-methylphenyl)-2-aminopropane (DOM). II. Some ring-methoxylated 1-amino-and 2-aminoindanes. Can J Chem 1974;52:381–9.

[10] Cannon JG, Dushin RG, Long JP, Ilhan M, Jones ND, Swartzendruber JK. Synthesis and dopaminergic activity of (R)- and (S)-4-hydroxy-2-(di-n-propylamino) indan. J Med Chem 1985;28:515–8.

[11] Cannon JG, Pease JP, Hamer RL, Ilhan M, Bhatnagar RK, Long JP. Resorcinol congeners of dopamine derived from benzocycloheptene and indan. J Med Chem 1984;27:186–9.

[12] Paleo MR, Dominguez D, Castedo L. Synthesis of 4-aryl-2-benzazepine-1,5-diones by photocyclization of N-(2-arylethyl)phthalimides. Tetrahedron 1994;50:3627–38.

[13] Cushman M, Dikshit DK. Formation of the 5-benzo[d]naphtho[2,3-b]pyran system during an attempted benzophenanthridine synthesis. J Org Chem 1980;45:5064–7.

[14] Dargan P, Archer J, Hudson S, Rintoul-Hoad S, Wood D. Nightclub urinals – a novel way of knowing what drugs are being used in nightclubs. Abstract Book – Club Health Prague, The seventh international conference on Nightlife, Substance use and related health Issues, 12–14 December 2011, Prague; 2011.

[15] Peters FT, Meyer MR. In vitro approaches to studying the metabolism of new psychoactive compounds. Drug Test Anal 2011;3:483–95.

[16] Casale JF, Hays PA. Characterisation of the 'methylenedioxy-2-aminoindans'. Microgram J 2011;8:43–52.

[17] Casale JF, Hays PA. The characterisation of 4- and 5-Iodo-2-aminoindan. Microgram J 2012;9:18–26.

[18] Malmusi L, Dukat M, Young R, et al. 1,2,3,4-tetrahydroisoquinoline and related analogs of the phenylalkylamine designer drug MDMA. Med Chem Res 1996;6:412–26.

[19] Oberlender R, Nichols DE. (+)-N-methyl-1-(1,3-benzodioxol-5-yl)-2-butanamine as a discriminative stimulus in studies of 3,4-methylenedioxy-methamphetamine-like behavioral activity. J Pharmacol Exp Ther 1990;255:1098–106.

[20] Oberlender R, Nichols DE. Structural variation and (+)-amphetamine-like discriminative stimulus properties. Pharmacol Biochem Behav 1991;38:581–6.

[21] Johnson MP, Huang X, Nichols DE. Serotonin neurotoxicity in rats after combined treatment with a dopaminergic agent followed by a nonneurotoxic 3,4-methylenedioxymethamphetamine (MDMA) analogue. Pharmacol Biochem Behav 1991;40:915–22.

[22] Sprague JE, Johnson MP, Schmidt CJ, Nichols DE. Studies on the mechanism of p-chloroamphetamine neurotoxicity. Biochem Pharmacol 1996;52:1271–7.

[23] Johnson MP, Conarty PF, Nichols DE. [^{3}H]Monoamine releasing and uptake inhibition properties of 3,4-methylenedioxymethamphetamine and p-chloroamphetamine analogues. Eur J Pharmacol 1991;200:9–16.

[24] Nichols DE, Johnson MP, Oberlender R. 5-Iodo-2-aminoindan, a nonneurotoxic analogue of p-iodoamphetamine. Pharmacol Biochem Behav 1991;38:135–9.

[25] Rudnick G, Wall SC. Non-neurotoxic amphetamine derivatives release serotonin through serotonin transporters. Mol Pharmacol 1993;43:271–6.

[26] Scorza C, Silveira R, Nichols DE, Reyes-Parada M. Effects of 5-HT-releasing agents on the extracelullar hippocampal 5-HT of rats. Implications for the development of novel antidepressants with a short onset of action. Neuropharmacology 1999;38:1055–61.

[27] Walline CC, Nichols DE, Carroll FI, Barker EL. Comparative molecular field analysis using selectivity fields reveals residues in the third transmembrane helix of the serotonin transporter associated with substrate and antagonist recognition. J Pharmacol Exp Ther 2008;325:791–800.

[28] Johnson MP, Nichols DE. Combined administration of a non-neurotoxic 3,4-methylenedioxymethamphetamine analogue with amphetamine produces serotonin neurotoxicity in rats. Neuropharmacology 1991;30:819–22.

[29] Huang XM, Nichols DE. 5-HT$_2$ receptor-mediated potentiation of dopamine synthesis and central serotonergic deficits. Eur J Pharmacol 1993;238:291–6.

[30] Huang XM, Marona-Lewicka D, Nichols DE. p-Methylthioamphetamine is a potent new non-neurotoxic serotonin-releasing agent. Eur J Pharmacol 1992;229:31–8.

[31] Monte AP, Maronalewicka D, Cozzi NV, Nichols DE. Synthesis and pharmacological examination of benzofuran, indan, and tetralin analogues of 3,4-(methylenedioxy)amphetamine. J Med Chem 1993;36:3700–6.

[32] Marona-Lewicka D, Rhee GS, Sprague JE, Nichols DE. Psychostimulant-like effects of p-fluoroamphetamine in the rat. Eur J Pharmacol 1995;287:105–13.

[33] Li Q, Murakami I, Stall S, et al. Neuroendocrine pharmacology of three serotonin releasers: 1-(1,3-benzodioxol-5-yl)-2-(methylamino)butane (MBDB), 5-methoxy-6-methyl-2-aminoindan (MMAI) and p-methylthioamphetamine (MTA). J Pharmacol Exp Ther 1996;279:1261–7.

[34] Marona-Lewicka D, Nichols DE. The effect of selective serotonin releasing agents in the chronic mild stress model of depression in rats. Stress 1997;22:91–100.

[35] Marona-Lewicka D, Nichols DE. Behavioral effects of the highly selective serotonin releasing agent 5-methoxy-6-methyl-2-aminoindan. Eur J Pharmacol 1994;258:1–13.

[36] Callaway CW, Wing LL, Nichols DE, Geyer MA. Suppression of behavioral activity by norfenfluramine and related drugs in rats is not mediated by serotonin release. Psychopharmacology 1993;111:169–78.

[37] Smith RL, Gresch PJ, Barrett RJ, Sanders-Bush E. Stimulus generalisation by fenfluramine in a quipazine-ketanserin drug discrimination is not dependent on indirect serotonin release. Pharmacol Biochem Behav 2002;72:77–85.

[38] Marona-Lewicka D, Rhee GS, Sprague JE, Nichols DE. Reinforcing effects of certain serotonin-releasing amphetamine derivatives. Pharmacol Biochem Behav 1996;53:99–105.

[39] Marona-Lewicka D, Nichols DE. Drug discrimination studies of the interoceptive cues produced by selective serotonin uptake inhibitors and selective serotonin releasing agents. Psychopharmacology 1998;138:67–75.

[40] Cozzi NV, Frescas S, Marona-Lewicka D, Huang XM, Nichols DE. Indan analogs of fenfluramine and norfenfluramine have reduced neurotoxic potential. Pharmacol Biochem Behav 1998;59:709–15.

[41] Solomons E, Sam J. 2-Aminoindans of pharmacological interest. J Med Chem 1973;16:1330–3.

[42] Levin N, Graham BE, Kolloff HG. Physiologically active indanamines. J Org Chem 1944;9:380–91.

[43] Witkin LB, Spitaletta P, Galdi F, Heubner CF, Okeefe E, Plummer AJ. Pharmacology of 2-amino-indane hydrochloride (Su-8629): a potent non-narcotic analgesic. J Pharmacol Exp Ther 1961;133:400–8.

[44] Horn AS, Snyder SH. Steric requirements for catecholamine uptake by rat brain synaptosomes: studies with rigid analogs of amphetamine. J Pharmacol Exp Ther 1972;180:523.

[45] Mrongovius RI, Bolt AG, Hellyer RO. Comparison of the anorectic and motor activity effects of some aminoindanes, 2-aminotetralin and amphetamine in the rat. Clin Exp Pharmacol Physiol 1978;5:635–40.

[46] Glennon RA, Young R, Hauck AE, McKenney JD. Structure-activity studies on amphetamine analogs using drug discrimination methodology. Pharmacol Biochem Behav 1984;21:895–901.

[47] Bar-Am O, Weinreb O, Amit T, Youdim MBH. The neuroprotective mechanism of 1-(R)-aminoindan, the major metabolite of the anti-parkinsonian drug rasagiline. J Neurochem 2010;112:1131–7.

[48] Irwin S. Comprehensive observational assessment: Ia. A systematic, quantitative procedure for assessing the behavioral and physiologic state of the mouse. Psychopharmacologia 1968;13:222–57.

[49] Ma SX, Long JP, Flynn JR, Leonard PA, Cannon JG. Dopaminergic structure-activity relationships of 2-aminoindans and cardiovascular action and dopaminergic activity of 4-hydroxy, 5-methyl, 2-di-n-propylaminoindan (RD-211). J Pharmacol Exp Ther 1991;256:751–6.

[50] Cannon JG, Furlano DC, Dushin RG, Chang YA, Baird SR, Soliman LN, et al. Assessment of a potential dopaminergic prodrug moiety in several ring systems. J Med Chem 1986;29:2016–20.

[51] Cannon JG, Perez JA, Bhatnagar RK, Long JP, Sharabi FM. Conformationally restricted congeners of dopamine derived from 2-aminoindan. J Med Chem 1982;25:1442–6.

[52] Arnerić SP, Roetker A, Long JP. Potent anorexic-like effects of RDS-127 (2-di-n-propylamino-4,7-dimethoxyindane) in the rat: a comparison with other dopamine-receptor agonists. Neuropharmacology 1982;21:885–90.

[53] Shulgin AT, Shulgin A. PIHKAL:a chemical love story. Berkeley, CA: Transform Press; 1991.

[54] Shulgin AT, Manning T, Daley PF. The Shulgin Index Psychedelic phenethylamines and related compounds, 1. Berkeley, CA: Transform Press; 2011.

[55] Oberlender RA, Kothari PJ, Nichols DE, Zabik JE. Substituent branching in phenethylamine-type hallucinogens: a comparison of 1-[2,5-dimethoxy-4-(2-butyl) phenyl]-2-aminopropane and 1-[2,5-dimethoxy-4-(2-methylpropyl)phenyl]-2-aminopropane. J Med Chem 1984;27:788–92.

[56] Leach B. A new drug called MDAI is being advertised across the internet as a replacement for 'miaow miaow'. Available: <www.telegraph.co.uk/news/uknews/law-and-order/7602664/New-drug-to-replace-mephedrone-as-legal-high.html>; 2010 [accessed 18.04.10].

[57] Townsend M. New drug set to replace banned mephedrone as a 'legal high'. Available: <www.guardian.co.uk/society/2010/apr/18/drug-replace-ban-mephedrone>; 2010 [accessed 28.06.10].

[58] Leach B. 'Woof woof' is new miaow miaow. Available: <www.telegraph.co.uk/news/uknews/crime/7858157/Woof-woof-is-new-miaow-miaow.html>; 2010 [accessed 28.06.10].

[59] Crick A, Morton E. Woof woof is the new meow meow. Available: <www.thesun.co.uk/sol/homepage/news/3032064/Woof-woof-is-the-new-meow-meow.html>; 2010 [accessed 28.06.10].

[60] Rogers EM. Diffusion of innovations, 5th ed. New York: Free Press; 2003.

[61] Measham F, Wood DM, Dargan PI, Moore K. The rise in legal highs: prevalence and patterns in the use of illegal drugs and first- and second-generation 'legal highs' in South London gay dance clubs. J Subst Use 2011;16:263–72.

[62] Anonymous. The Mixmag drug survey, pp. Available at: http://issuu.com/mixmagfashion/docs/drugsurvey/6>; 2011 [accessed April 2012].

[63] Kelleher C, Christie R, Lalor K, Fox J, Bowden M, O'Donnell, C. An overview of new psychoactive substances and the outlets supplying them. National Advisory Committee on Drugs, 1–176. Available: <http://arrow.dit.ie/cgi/viewcontent.cgi?article=1023,context=cserrep>; 2011. [accessed August 2012].

[64] Denzin NK, Lincoln YS, editors. The SAGE handbook of qualitative research (4th ed.). Thousand Oaks: Sage Publications; 2011.

[65] EMCDDA. Early-warning system on new psychoactive substances. Operating guidelines, European monitoring centre for drugs and drug addiction. Office for official publications of the European communities; 2007.

[66] Mounteney J. Methods for providing an earlier warning of emerging drug trends. PhD thesis. University of Bergen, Norway; 2009.

[67] Wood DM, Dargan PI. Understanding how data triangulation identifies acute toxicity of novel psychoactive drugs. J Med Toxicol 2012;8:300–3.

[68] EMCDDA. 2-Aminoindane. European Database on New Drugs (EDND). EMCDDA; 2012.

[69] EMCDDA. 5-IAI. European Database on New Drugs (EDND). EMCDDA; 2012.

[70] EMCDDA. MDAI. European Database on New Drugs (EDND). EMCDDA; 2012.

[71] EMCDDA, Europol. EMCDDA–Europol 2006 annual report on the implementation of Council Decision 2005/387/JHA. EMCDDA; 2007.

[72] EMCDDA, Europol. EMCDDA–Europol 2010 annual report on the implementation of Council Decision 2005/387/JHA. EMCDDA; 2011.

[73] EMCDDA, Europol. EMCDDA–Europol 2011 annual report on the implementation of Council Decision 2005/387/JHA. EMCDDA; 2012.

[74] Wood DM, Davies S, Calapis A, Ramsey J, Dargan PI. Novel drugs-novel branding. QJM 2012;105:1125–6.

[75] Brandt SD, Sumnall HR, Measham F, Cole J. Analyses of second generation 'legal highs' in the UK: initial findings. Drug Test Anal 2010;2:377–82.

[76] Ramsey J, Dargan PI, Smyllie M, et al. Buying 'legal' recreational drugs does not mean that you are not breaking the law. QJM 2010;103:777–83.

[77] Baron M, Elie M, Elie L. An analysis of legal highs-do they contain what it says on the tin?. Drug Test Anal 2011;3:576–81.

[78] Elie L, Baron M, Croxton R, Elie M. Microcrystalline identification of selected designer drugs. Forensic Sci Int 2012;214:182–8.

[79] Home Office. Annual report on the home office forensic early warning system (FEWS). A system to identify new psychoactive substances in the UK, Home Office; 2012.

[80] EMCDDA. Online sales of new psychoactive substances/"legal highs": summary of results from the 2011 multilingual snapshots. EMCDDA; 2011.

[81] Anonymous. 5-Iodo-2-aminoindane (5-IAI) – possible full-fledged MDMA substitute. Available: <www.bluelight.ru/vb/threads/496012-5-Iodo-2-aminoindane-(5-IAI)-Possible-full-fledged-MDMA-substitute>; 2012. [accessed 20.05.12].

[82] Anonymous. The big, dandy 5-IAI thread (part 1). Available: <www.bluelight.ru/vb/threads/502846-The-Big-amp-Dandy-5-IAI-Thread-(Part-1)>; 2012. [accessed 20.05.12].

[83] Anonymous. Big, dandy 5-IAI thread v. soli screwed up the last one. Available: <www.bluelight.ru/vb/threads/524194-Big-amp-Dandy-5-IAI-Thread-v.-soli-screwed-up-the-last-one>; 2012. [accessed 20.05.12].

[84] Anonymous. Erowid experience vaults/ 2-Aminoindan (also 2-AI, 2-Indanamine). <www.erowid.org/experiences/subs/exp_2Aminoindan.shtml>; 2012 [accessed 20.05.12].

[85] Anonymous. Erowid experience vaults/ MDAI (also 5,6-Methylenedioxy-2-aminoindan). <www.erowid.org/experiences/subs/exp_MDAI.shtml>; 2012 [accessed 20.05.12].

[86] NPIS. National poisons information service. Annual report 2010/2011. Health Protection Agency 2011.

[87] George NC, James DA, Thomas S. Exposure to MDAI: a case report. Clin Toxicol 2011;49:214–5.

[88] Council Decision 2005/387/JHA. Off J Eur Union 2005;48:32–7.

[89] EMCDDA. Action on new drugs. Available: <www.emcdda.europa.eu/activities/action-on-new-drugs>; 2012 [accessed 20.05.12].

[90] Fortson R. Misuse of drugs and drug trafficking offences Offences, confiscation and money laundering, 6th ed. London: Sweet, Maxwell; 2011.

[91] Anonymous. Convention on Psychotropic Substances. United Nations; 1971.

[92] Misuse of Drugs Act 1971. Available: <www.legislation.gov.uk/ukpga/1971/38/contents>; 1971 [accessed 20.05.12].

[93] Anonymous. 167/1998 Sb. O návykových látkách a o změně některých dalších zákonů. Available: <http://portal.gov.cz/app/zakony/zakonPar.jsp?idBiblio=46725,nr=167~2F1998,rpp=15#local-content>; 1998 [accessed 20.05.12].

[94] Verordnung des EDI über die Verzeichnisse der Betäubungsmittel, psychotropen Stoffe, Vorläuferstoffe und Hilfschemikalien; 2011.

[95] Commission staff working paper on the assessment of the functioning of Council Decision 2005/387/JHA on the information exchange, risk assessment and control of new psychoactive substances. European Commission; 2011.

[96] Report from the commission on the assessment of the functioning of Council Decision 2005/387/JHA on the information exchange, risk assessment and Control of new psychoactive substances. European Commission; 2011.

[97] Rothman RB, Baumann MH, Dersch CM, et al. Amphetamine-type central nervous system stimulants release norepinephrine more potently than they release dopamine and serotonin. Synapse 2001;39:32–41.

[98] Rothman RB, Baumann MH. Monoamine transporters and psychostimulant drugs. Eur J Pharmacol 2003;479:23–40.

[99] Hughes B, Winstock AR. Controlling new drugs under marketing regulations. Addiction 2012;107: 1894–9.

[100] Wilkins C, Sheridan J, Adams P, Russell B, Ram S, Newcombe D. The new psychoactive substances regime in New Zealand: A different approach to regulation. J. Psychopharmacol, in press, doi: 10.1177/0269881113491441 2013.

This page intentionally left blank

Ketamine

Qi Li[*], Wai Man Chan[†], John A. Rudd[†], Chun Mei Wang[†],
Phoebe Y.H. Lam[**], Maria Sen Mun Wai[†], David M. Wood[‡],
Paul I. Dargan[‡] and David T. Yew[†]

[*]Department of Psychiatry, The University of Hong Kong, Hong Kong
[†]School of Biomedical Sciences, The Chinese University of Hong Kong, Hong Kong
[**]Department of Oncology, Cancer Research UK/Medical Research, Council Gray Institute for
Radiation Oncology and Biology, University of Oxford, Oxford, UK
[‡]Clinical Toxicology, Guy's and St Thomas' NHS Foundation Trust and King's Health Partners,
London, UK

INTRODUCTION

Ketamine is a controlled substance in most countries around the world due to its narcotic and psychotropic properties. It was first introduced in the 1960s as an anaesthetic agent [1], and subsequently been used widely in both veterinary and human medicine; however, more recently it has gained wide recognition as a recreational drug. It is a 'dissociative' narcotic, where the user develops multi-modal hallucinations, floating sensations, paranoia and dissociation [1,2]. Nightmares, reduction or loss of motor activity, changes in sexual, musical and other sensory perceptions as well as flash-backs, tolerance, dependence and withdrawal symptoms are frequently reported [1–4].

Ketamine is a chemical known as 2-chlorophenyl-2-methylamine-cyclohexanone and

is structurally related to phencyclidine [1]. Its antagonistic action on glutamate receptors may account for the analgesic and dissociative effects of ketamine;[2] these effects are discussed in more detail later in the chapter. Despite its simple chemical structure, ketamine users can develop tolerance and dependence [5]. Craving and tolerance are high although physiological withdrawal is relatively mild [6]. In the early years of the recreational use of ketamine, the literature suggested that the adverse effects of the drug did not persist and therefore that chronic use did not lead to chronic problems, with perhaps the exception cognitive impairment leading to increased risk of death from physical trauma/accidents [7]. This has subsequently been shown not to be the case and the perception may have been based on the early literature where the reports were mainly centered on the acute effects of ketamine. In the

Novel Psychoactive Substances.
DOI: http://dx.doi.org/10.1016/B978-0-12-415816-0.00012-2

last 10 years, we have conducted studies on the many aspects of long-term ketamine toxicity, the results of which are discussed below. Recreational ketamine use is often in combination with alcohol [8], and so we will also discuss the effects of the combination of ketamine and alcohol.

PHARMACOLOGY AND ANIMAL MODELS OF KETAMINE TOXICITY

Effects on the Central Nervous System (CNS)

Neural Circuits of Ketamine Abuse

Ketamine at sub-anaesthetic doses (0.5–1 mg/kg) induces a state of 'dissociation' (including a perception of a distortion of space, time and body image) and feelings ranging from euphoria to detachment, an experience described by users as a mind or spiritual exploration [9–11]. Dissociative anaesthesia is a result of reduced activation in the thalamocortical structures and increased activity in the limbic system and hippocampus [12].

In recent years, there has been an accumulation of evidence indicating that ketamine is a non-competitive antagonist of the *N*-methyl-*D*-aspartate (NMDA) receptor which induces effects resembling both positive and negative symptoms of schizophrenia [13–15]. Current anatomical and pharmacological data on schizophrenia suggest that there are alterations in many neural circuits and several neurotransmitter systems, including those that involve dopamine, glutamate and gamma-aminobutyric acid (GABA) [16]. This observation, coupled with the pathological activation of limbic thalamocortical circuits that mediate psychosis produced by NMDA-receptor antagonists (e.g. phencyclidine (PCP), ketamine, dextromethorphan, and dizocilpine (MK801)), has led to the development of an NMDA hypofunctioning model of schizophrenia [17].

Single doses (0.5 mg/kg) of ketamine can induce a transient psychotic state in healthy volunteers characterised by perceptual aberrations, delusion-like ideas, thought disorder, blunt affect, emotional withdrawal and cognitive dysfunction [10]. Previous studies have observed that the administration of ketamine is associated with impairment of cognitive function [10,18–20]. Positron Emission Tomography (PET) studies demonstrate changes in the prefrontal cortex (PFC), medial frontal and inferior frontal cortices with acute ketamine administration during the acute psychotic state [21,22], which suggests that the frontal cortex may be involved in mediating NMDA receptor-induced psychosis.

In recent studies, by combining functional magnetic resonance imaging (fMRI) with cognitive tasks, researchers were able to examine the effects of ketamine on cognitive performance and brain activation concurrently using a paradigm known to be associated with both behavioral and functional neural abnormalities [23–25]. A number of studies have used ketamine to investigate the effects of NMDA antagonists on brain regions engaged by various cognitive fMRI paradigms, including face emotion processing [26], working memory [27], memory recall [28], verbal fluency [29], and learning [30]. Using this approach, ketamine has been found to have altered the brain networks serving cognitive process rather than affecting performance. Pharmaco-MRI (phMRI), which has been used to investigate the direct effects of drugs on the brain, have confirmed that the short-term administration of NMDA channel blockers causes increased glutamate release in frontal cortex (Brodmann area 8, BA8), anterior thalamus, superior temporal gyrus (BA22), posterior cingulated gyrus (BA24), and parahippocampal gyrus, with decreases in medial orbitofrontal cortex (OFC, BA11) and temporal pole (BA38) and shown in Figure 12.1a [31]. Consistent with the hypothesis that hyper-dopaminergic function underlies psychotic symptoms, ketamine alters the firing rate of mesocortical and mesolimbic dopaminergic neurons to increase concentrations of extracellular dopamine in the striatum and

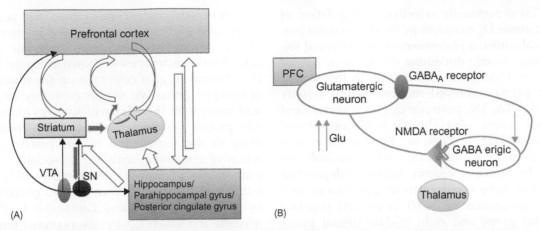

FIGURE 12.1 Schematic overview of neural circuits thought to be relevant in the short-term and long-term ketamine administration; illustration of reported ketamine impact on these regions and circuits in human and animal studies (a), and ketamine block NMDA receptors on the GABAergic interneurons in the cortical-thalamic neural circuit (b). a) Glutamatergic projection from the prefrontal cortex, in conjunction with those from the limbic system (hippocampus, posterior cingulated gyrus and parahippocampal gyrus) connect to thalamus and thereby completes a functional part of the cortical–limbic–thalamic–cortical pathways. GABAergic interneurons in the striatum project to the substantial nigra (SN) which in turn modulate dopamine release through ventral tegmental area (VTA)/SN–cortical–limbic neural circuit; also project to thalamus, which subsequently influence cortical glutamate release. Arrows represent ⇨Glutamate; ➡GABA; →Dopamine. b) NMDA receptor antagonist – ketamine – blocks NMDA receptors on the GABAergic neurons of the thalamus. This decreases GABA-mediated inhibition of the thalamic projection neurons to the prefrontal cortex (PFC). 'Disinhibition' of GABAergic interneurons cause overexcitation of the PFC with the increased expression of glutamate.

prefrontal cortex [32,33]. Such a hypermetabolic pattern is also seen in schizophrenia patients with positive symptoms, in the frontal cortex and thalamus, as well as in the striatum and the temporal cortex [34]. It is likely, therefore, that changes in transmitter patterns after the acute administration of ketamine are similar to those occurring in patients with an acute psychotic state. With regard to the mechanism underlying this effect, there is evidence that glutamate modulates striatal dopamine at the level of the dopaminergic cell bodies of the ventral tegmental area (VTA) and substantial nigra (SN), and also at the level of the glutamatergic afferents of the prefrontal cortex which in turn modulate concentrations of dopamine in the VTA, SN, or in the striatum by providing excitatory input to the GABA interneurons [35,36].

It is also thought that ketamine has damaging actions in the brain by blocking excitatory glutamate receptors on GABAergic inhibitory interneurons in the reticular nucleus of thalamus, and dis-inhibiting excitatory pathways [37,38]. GABAergic reticular nucleus neurons are located adjacent to the thalamus and exert inhibitory control over thalamic projection neurons (see Fig. 12.1b). Olney suggested that NMDA antagonists blocked excitation of GABAergic interneurons, resulting in removal of GABA restraint of cholinergic, serotonergic and glutamatergic afferents to the posterior cingulated cortex [37]. This would induce a triple excitotoxic effect on the posterior cingulated pyramidal cells, accounting for the focal neurodegeneration.

The increasing trend towards chronic ketamine use has inadvertently provided an opportunity to explore drug-induced morphological, functional and neurochemial changes in the central nervous system. Chronic ketamine users exhibit a regionally selective up-regulation of Dopamine D_1 receptors in the dorsalateral prefrontal cortex, a phenomenon also observed following chronic dopamine depletion in animal studies. One recent fMRI study in the adolescent primate showed reduced neural activity in the VTA, SN, posterior cingulate cortex, and visual cortex after chronic ketamine administration [23]. In contrast, hyperfunction was observed in the striatum and entorhinal cortex. In humans, chronic ketamine-dependent patients have a reduction in gray matter volume (measured by MRI) in the left superior frontal gyrus and right middle frontal gyrus [39]. Meanwhile, there is a dose-dependent abnormality of white matter in the bilateral frontal and left temporoparietal regions in chronic ketamine users (measured using diffusion tensor imaging; DTI) [40]. Indeed, a number of brain imaging studies support the hypothesis of cortical 'hypofrontality': namely reduced baseline activity of several regions of the frontal cortex [41], which mirrors core neurobiological deficits seen in schizophrenia; hypofrontality, altered markers of GABAergic interneuron activity and deficits in executive function [42]. As such, chronic ketamine animal models may have unexpected translational value in our attempts to understand the neurobiological mechanisms underlying hypofrontality and to assist in identifying and validating novel drug targets that may restore PFC deficits in schizophrenia or psychosis. An example of the mechanism is depicted in Figure 12.1.

Ketamine Misuse and Neurotransmitters

GLUTAMATE

The major excitatory synaptic transmission in the mammalian CNS is mediated by *L*-glutamate, coming from either glucose, via the Krebs cycle, or glutamine, which is synthesised by glia cells and taken up by the neurons [43]. In common with other neurotransmitters, glutamate is stored in synaptic vesicles and released by calcium-dependent exocytosis; the action of released glutamate is terminated by carrier-mediated reuptake into nerve terminals and neighbouring astrocytes as shown in Figure 12.2 [43]. Moreover, glial cells take up the majority of synaptic glutamate via an excitatory amino acid transporter (EAATs), which serves to maintain proper neuronal functioning by contributing to the termination of the postsynaptic action of glutamate, reducing potentially toxic extracellular glutamate concentrations [44]. In astrocytes, glutamate is converted to glutamine by glutamine synthetase. Glutamine is then released and taken up by the neuronal terminals where it is reconverted to glutamate and to GABA by glutamic acid decarboxylase [43]. Glutamate itself acts via ligand-gated ion channels: ionotropic glutamate receptors (iGluRs, included NMDA, α-amino-3-hydroxy-5-methyl-4-isoxazole propionic acid (AMPA) and kainate) and metabotropic receptors [44]. Indeed, NMDA receptors functioning is also modulated by *glycine*, but the binding site for *glycine* distinct from glutamate-binding site and both have to be occupied for the channel to open.

NMDA RECEPTORS

Seven genes encode for the NMDA receptors (*NR1, NR2A-D* and *NR3A-3B*) and they are located throughout the brain where they are implicated in learning, memory, and experience-dependent forms of synaptic plasticity, such as long-term potentiation (LTP), with *NR2A* and *NR2B* receptors being particularly important in this regard [45]. Like PCP, ketamine blocks the NMDA receptor non-competitively resulting in a 'use-dependent' blockade [46]. The binding site is located within the NMDA receptor and partially overlaps with the magnesium binding site [47]. Thus, ketamine blocks the open channel and reduces channel mean open time. Both (*S* and *R*) stereoisomers of ketamine act via the same binding sites but with different affinities and potencies, resulting in different physiological effects.

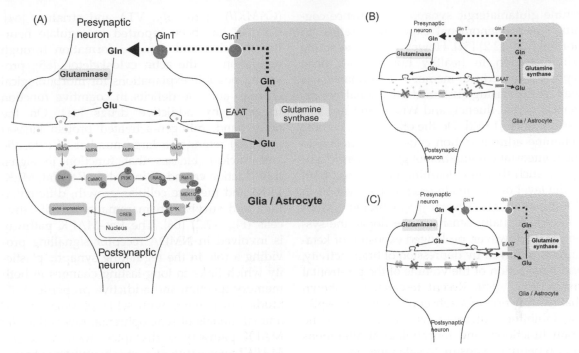

FIGURE 12.2 **The glutamatergic neurotransmission system.** a) NMDA receptor model for the ketamine abuse: Ras-MAPK signalling pathway in the postsynaptic glutamatergic neurons. b) Acute ketamine: acute ketamine administration blocks NMDA receptors, thus impairing the specification of prior expectancies (how much stimulation to expect). Presynaptic glutamate release (via effect on glia/astrocyte glutamate reuptake) increases and, as such, AMPA receptors are excessively and inappropriately stimulated. c) Chronic ketamine: with chronic ketamine exposure, there is a compensatory increase in the number and function of NMDA receptors, but reduction of glutamate release. Besides, glia glutamate reuptake remains impaired. Glu: glutamate; Gln: glutamine; NMDA: *N*-methyl-*D*-aspartate; AMPA: α-amino-3-hydroxy-5-methyl-4-isoxazole propionic acid; EAAT: excitatory amino acid transporter; GlnT: glutamine transporter; CaMKII: calmodulin protein kinase II; PI3K: phosphatidylinositol 3-kinase; MEK1/2: mitogen-activated kinase1/2; ERK: extracellular signal-regulated kinase; CREB: cyclic adenosine monophosphate response element binding protein. 'p' represents phosphorylation.

The glutamate hypothesis of schizophrenia is centered on a deficiency in activity of glutamate at the glutamate synapse, especially in the prefrontal cortex [48,49]. In many brain areas, dopamine inhibits glutamate release, or glutamate excites neurons that dopamine inhibits [49]. Therefore, increased levels of dopamine produce similar effects to decreased levels of glutamate. Taking this further, drugs that block glutamate receptors may increase dopamine effects. The dopamine hypothesis of schizophrenia has been more heavily investigated and is more widely accepted than the glutamate

hypothesis. Nevertheless, there is enough evidence to suggest hypo-functioning NMDA receptors are also involved and correction of neurodegenerative consequences in the brain may lead to pharmacotherapy in schizophrenia or psychosis [16,50].

It is expected that the acute administration of ketamine would enhance presynaptic glutamate release (see above), leading to a rapid change in the individual's subjective experience; thereafter, a decline of effects is seen as the drug is metabolised and excreted [51]. Homeostatic mechanisms involving glial cells retune glutamatergic synapses, and hence cognition and phenomenal experiences return to normal (Fig. 12.2) [51]. However, acute ketamine administration in healthy humans has been reported to reduce memory performance, as well as decreasing sustained attentional performance, verbal fluency, and Wisconsin Card sort performance [19,52]. On the other hand, chronic ketamine administration causes dysfunction to the homeostatic regulation of glutamatergic synapses, such that new learning takes place; both at the level of glutamate receptors, as well as in the dopaminergic system [53]. Chronic NMDA blockade actually sensitises the dopamine system [54]. However, prolonged exposure of ketamine is followed by depression of brain activity, and a reduction of the volume of the pre-frontal cortex (PFC) [55]. Recent research has shown that dysfunction of a subset of cortical fast-spiking inhibitory interneurons might explain the brain functional and morphological alterations after repetitive exposure to ketamine [56].

Animal studies have opened a window to explore the underlying molecular mechanism of ketamine dependence. Following early NMDA receptor inhibition, both working and reference memory in the Morris water maze are impaired in adult rats [57]. In both monkey and rodent experiments, acute ketamine administration impairs executive control and induces a deficit of sensory motor function [58,59]. Genetic models of NMDA receptor hypofunction show constructive validity and also some phenomenological validity. NR1 hypomorphic (NR1−/−) mice showed an increase of locomotor activity, deficits in the prepulse inhibition, social withdrawal, and impairment in spatial memory tests; furthermore, NR2A subunit knockout mice showed a similar phenotype to NR1 (−/−) [60,61]. The genetic animal models point to the direct role of NMDA receptors in the behavioural phenotypes which mimic the clinical symptoms of schizophrenia [62,63].

Acute administration of ketamine induces expression of calmodulin protein kinase II (CAMKII) in the SN, VTA and striatum [64]. CAMKII has been reported to regulate neurite extension and synapse formation through regulation of the actin cytoskeleton [65], providing possible explanations for morphological changes and the deficits in cognitive function triggered by addictive drugs [41]. On the other hand, mitogen-activated protein kinases (MAPKs) refers to a large number of cytosolic and nuclear kinases that function in signal transduction cascades (including Ras/Raf, MEK, ERK) and mediate cellular growth, differentiation, and survival in mammalian proliferative cells (Fig. 12.2) [66]. The Ras-MAPK pathway is involved in NMDA receptor signalling providing a role in the regulating synaptic plasticity which links to long-lasting changes in both memory function and addictive properties [67]. Studies in patients with schizophrenia, and in animal models of schizophrenia, reveal that the MAPK pathway is disrupted with decreased MAPK1 gene expression in schizophrenia brains [68–70]. However, the molecular mechanism of deficits of cognitive function in ketamine abuse has not been widely explored.

GAMMA-AMINOBUTYRIC ACID (GABA)

GABA is the main inhibitory transmitter in the brain. About 20% of CNS neurons are GABAergic and most have short interneurons, but long GABAergic tracts run to the cerebellum and striatum [71]. GABA is formed from glutamate by the action of glutamic acid decarboxylase (GAD), an enzyme found only in GABA-synthesising neurons in the brain [72]. GABA acts on two distinct type of receptor. The $GABA_A$ receptor is a ligand-gated channel, while the $GABA_B$ receptor is G-protein-coupled. $GABA_A$ receptors are generally located postsynaptically and mediate fast postsynaptic inhibition with the channel being selectively permeable to chloride. Thus, chloride is usually negative to the neuron resting potential and increasing chloride permeability hyperpolarises the cell to reduce excitability [72]. $GABA_B$

receptors are located pre- and post-synaptically, and they are typical G-protein-coupled receptors. GABA$_B$ receptors exert their effects by inhibiting voltage-gated calcium channels (thus reducing transmitter release) and by opening potassium channels (thus reducing postsynaptic excitability) resulting from inhibition of adenylate cyclase [72].

The acute administration of ketamine produces the dis-inhibition of glutamatergic activity, and leads to in increased excitatory transmission in the PFC of rodents and primates [73]. The consequence of this is a loss of the 'GABAergic phenotype', leading to the suggestion that dysfunction of these fast-spiking inhibitory interneurons might be a core feature of schizophrenia [74]. In fact, there is a specific decrease in *GAD67* (the main isoform synthesising GABA) and *NR2A* in a subpopulation of GABAergic interneurons in schizophrenic postmortem tissue [75]. However, one recent proton Magnetic Resonance Spectroscopy (^{1}H-MRS) study in healthy volunteers did not find that GABA levels were affected by ketamine in the occipital cortex [76]. It is possible that an acute effect of ketamine on GABA neuron activity can occur differently in different brain regions. After long-term exposure to ketamine, an increased excitatory neurotransmission is followed by a depression of brain activity through unknown mechanisms. Repeated administration of ketamine changed the gene expression level of GABA$_A$ receptors in the frontal cortex through up-regulation of NMDA receptors and over-stimulation of glutamatergic system [77]. A recent microarray study demonstrated an up-regulation of the alpha 5 subunit (*Gabra5*) of the GABA$_A$ receptors in the PFC in chronically ketamine treated mice, with an impairment of reference memory [78]. This may not be unexpected, since it is known that GABA$_A$ receptors play a role in memory and synaptic plasticity [79]. Thus, enhancement of the action of GABA impairs memory processing, while conversely, compounds that reduce the action of GABA

enhance memory, especially the acquisition process [80].

On the other hand, serum levels of brain-derived neurotrophic factor (BDNF) are elevated in chronic ketamine users and in animals that have had a prolonged ketamine exposure [81,82]. Elevation of BDNF induced a rapid increase of GABA$_A$ receptors through activation of TrkB receptor tyrosin kinases [83]. Thus, the derangements of GABA$_A$ receptors observed in the ketamine-abuse models may be regulated by BDNF/TrkB/GABA$_A$ signalling pathways.

DOPAMINE

Dopamine is particularly important in relation to neuropharmacology, because it is involved in several common disorders of brain function, notably Parkinson's disease, schizophrenia, and drug dependence [84–87]. There are two main metabolism products of dopamine: dihydroxyphenylacetic acid (DOPAC) and homovanillic acid (HVA) [88]. The brain content of HVA is often used as an index of dopamine turnover. Dopamine receptors belong to the family of G-protein-coupled transmembrane receptors and their signal transduction mechanisms are linked via adenylyl cyclase to the control of potassium and calcium channels. Two types of dopamine receptor, D$_1$ and D$_2$, are originally distinguished on pharmacological and biochemical grounds. The original D$_1$ receptor family includes D$_1$ and D$_5$ receptors, while the D$_2$ receptor family consists of D$_2$, D$_3$, and D$_4$ receptors. D$_4$ receptors are mainly expressed in the cortex and limbic system, but they are still a focus of interest because of their possible relationship to the mechanism of schizophrenia and drug dependence [85,88–90]. A key component in dopamine signal transduction pathways is the protein DARPP-32 (32-kDa dopamine- and cAMP-regulated phosphoprotein) [91]. When cAMP is increased through activation of D$_1$ receptors and protein kinase A (PKA), DARPP-32 is phosphorylated [91]. Activation of D$_2$

receptors opposes the effect of D_1 receptor activation [91].

The modulation of striatal dopamine by glutamate occurs both at the level of the dopaminergic cell bodies of the VTA and SN, and at the level of glutamatergic afferents in the PFC to, in turn, modulate dopamine in the VTA, SN, or in the striatum by providing excitatory input to GABA interneurons [35]. Following acute treatment with noncompetitive NMDA receptor antagonists, there is a dramatic activation of dopamine transmission in the forebrain, to implicate glutamate in cognitive functioning and dysfunction [32]. In connection with this, impairment of spatial working memory from increased transmission of dopamine in the PFC had been reported [92]. The cognitive and behavioural effects of ketamine can be attenuated by treatment with an AMPA/kainite receptor antagonist, and with a group II mGluR agonist [93]. In addition, the acute administration of ketamine induces the expression of Homer 1a (the immediate early form of Homer, which regulates the metabotropic glutamate receptor GluR1/5) in the ventral striatum and nucleus accumbens, CAMKII and a transporter for dopamine (DAT) in the SN and VTA [64]. The expression of the above-mentioned genes suggests interconnection of NMDA receptors with the dopaminergic system and with mGluR1/5. Indeed, co-activation of D_1 and NMDA receptors is required for certain neuroplastic changes in the nucleus accumbens [94]. Furthermore, dopamine and NMDA receptors function cooperatively at the transcription level on striatal neurons in the CPu [95]. However, dopamine release by NMDA antagonists appears significantly more robust and reproducible in the frontal cortex than in the striatum [96]. D_1 receptor stimulation led to PKA-dependent phosphorylation of NR1; activation of dopamine receptors and PKA also activated ERK through DARPP-32 [97]. As mentioned above, NMDA receptors activate ERK signaling by regulating Ca^{2+} influx and the Ras-GTP/Raf-1 cascade. These findings indicate that the ERK signalling pathway may serve as a convergent intracellular point for mediating neuronal responses to stimulation by dopamine, glutamate, and potentially other drugs of abuse. Finally, D_1 and D_3 receptors play opposite roles in regulating glutamate-induced NR1, CaMKII, and CREB phosphorylation and down-stream signaling events. D_1 receptor stimulation activates PKA, NR1, CAMKII and CREB [98], while D_3 receptor stimulation inhibited them [99]. But how D_1 and D_3 receptors regulate activation of MAPK signalling pathways following acute or chronic ketamine abuse remains a subject of investigation.

It is interesting that repeated ketamine administration induces an increase in basal and resting levels of extracellular dopamine, but not DOPAC or HVA, in the PFC. However, a single-dose of ketamine following a repeated ketamine exposure fails to elevate extracellular dopamine levels, despite an observed increased in extracellular DOPAC and HVA [100]. It is difficult to construct a hypothesis on the mechanisms of how dopamine release is modulated from the above study. On the other hand, there is evidence indicating that D_1 receptor availability increases in chronic ketamine users, which is compatible with the hypothesis that chronic NMDA antagonism leads to D_1 receptor up-regulation via reduced dopamine levels in the pre-frontal cortex [101].

A recent fMRI study in primates with chronic ketamine administration also demonstrated the dysfunction of PFC, with concurrent reduction of tyrosine hydroxylase (TH, rate-limiting enzyme in the synthesis of dopamine) levels also being observed in this brain area [23]. It appears that the effects of repeated ketamine administration on the expression of dopaminergic receptors, and the effects of ketamine in modulating evoked DA release may provide more fruitful avenues for exploration as we try to dissect new treatments for those with an apparent addiction.

SEROTONIN/5-HYDROXYTRYPTAMINE (5-HT)

NMDA receptor blockade increases glutamate and 5-HT release in the medial prefrontal cortex (mPFC) of rats [102]. It is thought that high extracellular levels of 5-HT in the mPFC can be potentially indicative of negative/cognitive symptom of schizophrenia, which can be better treated with atypical antipsychotic drugs [103]. Dopamine and glutamate have been the two neurotransmitters of greatest interest with regard to the aetiology and pathophysiology of ketamine use and schizophrenia/psychosis. However, the role of serotonin in these areas has opened up other possibilities. 5-HT_{2A} antagonism, 5-HT_{1A} agonism, and 5-HT_6 and 5-HT_7 antagonism attenuate the acute effects of NMDA receptor antagonists to increase locomotor activity and impairment of cognitive function. Indeed, clozapine, a potent 5-HT_{2A} receptor blocker, has been shown to block ketamine-induced psychosis [104].

Selective serotonin reuptake inhibitors (SSRIs) and serotonin and norepinephrine reuptake inhibitors (SNRIs) which increase the synaptic availability of monoamines have been used to treat depression for more than 50 years. But the delayed therapeutic effects of most antidepressants suggest that neural adaptation (e.g., BDNF-TrkB receptor signalling pathway), rather than the elevation in synaptic monoamine levels, may be responsible for their therapeutic effects. However, recent clinical research reported that ketamine had rapid antidepressant effects in the patients with Major Depressive Disorder (MDD) [105]. On the contrary, there was increased propensity to psychotomimetic effects and tolerance to ketamine's antidepressant effects after repeated treatment [106]. An animal study has shown that the fast-acting behavioural anti-depressive properties of ketamine depend on the rapid synthesis of BDNF. Deactivation of eukaryotic elongation factor 2 (eEF2, also called CaMKIII) kinase induced by ketamine caused reduced eEF2 phosphorylation and the de-suppression of translation of BDNF [107]. Therefore, the use of eEF2 kinase inhibitors poses as a potential novel MDD treatment with a rapid onset of action. Greater understanding of the molecular mechanism of the interaction between ketamine and serotonin may yield a novel approach in developing new safe and efficient antidepressants.

ACETYLCHOLINE

Acute ketamine administration increases cortical cholinergic function [108]. Acetylcholine has a significant role in cognition and perception. Neuronal acetylcholine receptors (AChRs), in addition to mediating fast synaptic transmission, modulate fast synaptic transmission mediated by the major excitatory and inhibitory neurotransmitters glutamate and GABA respectively [109–112]. Thus, it is possible that a dysfunction of AChRs modulation of glutamatergic and GABAergic synaptic transmission in brain areas associated with processing of learning and memory, such as the hippocampus or the frontal cortex, contributes to the cognitive impairment seen in ketamine abusers. Ketamine itself affects M_2 and M_3 muscarinic receptors (mAChRs), which may contribute to amnesia, mydriasis and bronchodilatation. Evidence from epidemiologic and behavioural studies has indicated that neuronal nicotinic receptors (nAChRs) are likely to be involved in the pathogenesis of schizophrenia, Alzheimer's disease and Parkinson's disease [113]. Formal learning theories showed that increasing cortical acetylcholine engages new explanatory learning and tuned attention to stimuli with unexpected and unpredictable consequences [114]. Cortical acetylcholine might engage the dysfunction of attention after an acute ketamine administration, but chronic ketamine administration did not sensitise cortical cholinergic function [108].

NITRIC OXIDE (NO)

NO is a soluble, short-lived and freely diffusible gas in the nervous system. It is produced mainly by neuronal nitric oxide synthase (nNOS) to raise intracellular calcium [115]. After exposure to NMDAR antagonists *in vitro* or *in vivo*, oxidative stress occurs with a rapid increase of reactive oxygen species (ROS) and non-radical molecules [116,117]. Exposure of mice to ketamine induced a persistent increase in superoxide in the PFC, hippocampus and thalamus, which suggested that increased nicotinamide adenine dinucleotide phosphate (NADPH) oxidase activity was present throughout the brain [118]. NADPH oxidase-derived ROS is necessary for normal cellular function, but excessive oxidative stress can contribute to pathological disease. NADPH oxidase is involved in neuronal signaling, memory and central cardiovascular homeostasis, while overproduction of ROS contributing to neurotoxicity, neurodegeneration and cardiovascular disease. It also seems that ketamine-induced loss of fast-spiking interneurons is mediated by NADPH oxidase [56].

Decreasing superoxide production prevented the effects of ketamine on inhibitory interneurons in the PFC [56]. Dysfunction of GABA-mediated systems with reduced expression of GAD67 and NR2A in schizophrenia has been consistently observed in post-mortem tissue [119]. Therefore, NADPH oxidase may contribute to oxidative mechanisms involved not only in the psychotomimetic effects of ketamine, but also in schizophrenia or other processes involving increased oxidative stress in the brain [120,121]. Further understanding of the mechanisms underlying activation of oxidative stress in the brain may provide new way for drug discovery aimed at the treatment of psychosis and cognitive deficits. The non-selective NOS inhibitor N-nitrol-arginine methylester (L-NAME) antagonises ketamine-induced analgesia through the reduction of NO oxidation products induced by ketamine. Thereby, L-NAME could actually reverse recognition and spatial memory deficits induced by subanaesthetic doses of ketamine administration [122,123]. The effects of L-NAME on cognitive function might act through cyclic GMP (cGMP) or cGMP-dependent protein kinase

G (PKG) signal pathways which has previously been suggested as the main transduction pathway of NO in the CNS [124].

Neuroprotection

Ketamine produces considerably less euphoria and sensory distortion than phencyclidine and is thus better tolerated as an anaesthetic [125]. NMDA receptor antagonists have been investigated for many years as therapeutic agents for the treatment of neurological disorders such as stroke, epilepsy, pain and Parkinson's disease [126,127]. Ketamine also appears to have some anti-inflammatory effects. This might be related to a reduction in glutamate-induced neurotoxicity which results from the blockade of the NMDA receptor. There is evidence that neuronal damage leads to an increase in NMDA receptor density resulting in an increased neuronal glutamate input. This leads to activation and up-regulation of further NMDA receptors and finally induces cell death. Ketamine may reduce the effects of this vicious cycle and decrease neuronal death [128]. But the main drawback of ketamine, despite the safety associated with a lack of overall depressant activity, is that the hallucinations and sometimes delirium and irrational behavior are common during recovery. This limits the usefulness of ketamine, but the effects are said to be less marked in children and, therefore, ketamine is still often used as an anaesthetic agent in paediatrics. It is also commonly used as a field anaesthetic because of its cardiovascular safety profile.

Recent evidence indicates that ketamine produced rapidly acting, long-lasting antidepressant responses in depressed patients depending on the rapid synthesis of BDNF [107]. At rest, ketamine deactivates eukaryotic elongation factor 2 (*eEF2* or *CaMKIII*) kinase, resulting in reduced eEF2 phosphorylation and de-suppression of translation of BDNF. BDNF was linked to the action of traditional antidepressants, since its levels were increased by several antidepressants [129]. Therefore, ketamine may be employed as an appropriate neuropharmacological model for developing novel antidepressants.

Neurotoxicity

The enthusiasm for NMDA blockade to decrease excitoxic neuronal death was tempered by concerns raised by Ikonmidou in Science in 1999 [130]. Increased neuro-apoptosis occurs in late fetal or early neonatal life of rats following injections of ketamine for nine hours (h). This finding was confirmed following investigations that repeated doses of ketamine increased neurodegeneration in many brain regions, whereas single doses of ketamine did not [131]. An increase in cell death also occurred when rat forebrain cell cultures were exposed to ketamine for 12h [132]. Developmental differences between rodents and humans prompted studies in the primate. Frontal cortical cells from 3-year-old rhesus monkeys incubated with ketamine for 24h had apoptosis and neuronal death. In vivo primate experiments reported that increased brain cell death followed 24h of ketamine exposure in gestational day-122 fetuses and postnatal day-5 infant monkeys [133]. Our group in Hong Kong have done studies focusing on the long-term neurotoxic effects of ketamine on the CNS in rodents and primates. We observed an increase of hyperphosphorylated tau protein in the prefrontal and entorhinal cortices of mice and monkeys [134]. Tau is a microtubule-associated protein which can be abnormally phosphorylated in Alzheimer's disease. 15% of hyperphosphorylated tau positive cells co-labelled by terminal dUTP nick end labelling (TUNEL), a marker for apoptotic cells in the CNS, indicated a possible relationship between hyperphosphorylated tau formation and apoptosis following ketamine. However, one should be cautious to translate the all findings of neurotoxicity of ketamine to humans. Should it ultimately be demonstrated that neuroapoptosis does occur with ketamine at clinically relevant doses or after long-term treatment, it will be imperative to develop either alternative safer agents for medical use, or effective mechanisms for blocking those neurotoxic effects.

Chronic long-term treatment may be more relevant to ketamine misuse now appearing worldwide and a series of studies on the chronic effects and long-term research using animals are mandatory. In our laboratory, we have embarked on this direction for the past two to three years, with emphasis on 1) fMRI studies, 2) behaviour tests and 3) morphological, pathological and cytochemical studies. We have so far uncovered several important major findings. Firstly, cell death in the nervous system in response to ketamine was mainly apoptotic [135]. Secondly, apoptotic cell death might appear to occur in some areas more severely than others, e.g. the sensory cortex, hippocampus, PFC and the striatum [8,135]. Thirdly, in addition to apoptosis, hyperphosphorylation of tau protein, a feature of Alzheimer disease, was observed in the neurons and nerve fibers (Layer I) of the brains of chronic ketamine treated monkey and mice [134]. Fourthly, fMRI reported differential areas of hyper- and hypo-regulation [23,136]. These up- and down-regulated areas included the prefrontal cortex, cerebellum, striatum and the midbrain [23,136]. Finally, $GABA_A$ receptor changes were apparent in some of these reactions [78].

Some studies have suggested that ketamine may have a role therapeutically for a range of psychiatric conditions. For example, acute ketamine treatment was useful for the modulation of depression associated with dysfunction of the glutamatergic system. In chronic stress, NMDA affected the respiratory chain in cells and caused the adrenal gland to increase in weight while NMDA antagonists reversed this [137]. Mood changes associated with NMDA receptors were also alleviated by ketamine [138].

Effects on the Urogenital System

There have been a number of studies investigating the effects ketamine on the kidney and

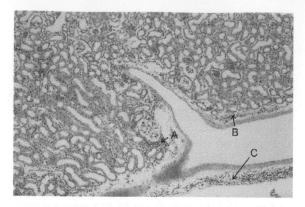

FIGURE 12.3 Stage 1 of urinary bladder toxicity after three months of ketamine treatment. Note mononuclear cells around glomerulus (A), tubules (B) and the pelvis of kidney (C). H & E staining ×100.

bladder; only one has been on the impact of acute ketamine administration. Chen et al. used low dose of ketamine (50 mg/kg) in rats and reported a slight decrease in the blood pressure in terms of minutes, associated with decrease in renal function (glomerular filtration rate and effective renal plasma flow) for two days [139]. The rest of this section will summarise the results of studies investigating the impact of chronic ketamine use on the bladder and kidneys.

Effects on the Bladder

Chronic and long-term ketamine treatment appears to affect muscles of the bladder and the bladder neck (including sphincter muscle) in mice [140]. The optimal toxicity of ketamine usually has a latency of 3–6 months of chronic ketamine treatment. In the urinary bladder, it was very clear it could be separated into distinct stages. The first stage (Fig. 12.3) is associated with simple functional disturbance resulting in stasis, infiltration of mononuclear cells into the bladder and kidney and an increased risk of infection [141]. This can be followed by a second stage where there is degeneration of nerves to

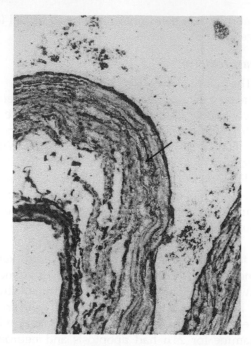

FIGURE 12.4 Degeneration of nerve axons (arrow) to the urinary bladder after four months of ketamine treatment. Silver staining ×400.

the neuromuscular junction (Figs 12.4 and 12.5) and muscle thinning. By the end of the stage 3, the bladder wall muscles show distinct degeneration and the degenerative spaces are occupied by fibrous tissue (Fig. 12.6). This happens not only to the body of the urinary bladder, but also to the neck of the bladder, including the sphincter itself. The consequence of this pathology is the clinical picture of a small and rather rigid bladder such that the patient has to void much more frequently. In addition to these changes, vascular congestion in the lamina propria is seen at stage 2–stage 3 and ketamine intake together with alcohol increases this effect.

Anaesthetics have been long known to cause acute bladder distention after surgery [142]. This was particularly evident when spinal anaesthesia was performed with lidocaine and its analogues [143,144]. Urethane and analogues also act by reducing inhibition of the

FIGURE 12.5 Neuromuscular junctions (arrow) were few in the urinary bladder after four months of treatment. Cholinesterase immunohistochemistry ×400.

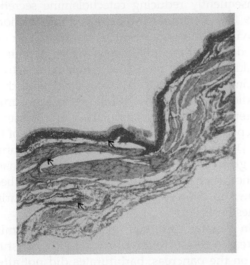

FIGURE 12.6 Increased collagen fibers (arrows) formed in the urinary bladder wall of the ketamine animals of four months. ×400.

visceral nociceptive transmission on the bladder [145]. Other anaesthetics such as TAK 802, an acetycholineterase inhibitor, also act directly on the contraction of urinary bladder muscles. Ketamine used along with diazepam in short duration experiments has been found to interfere with the reflex micturition in rats [146]. Glutamate has also been found to affect urinary

bladder function [147–149]. Sites for glutamate receptors inside the bladder are influenced from receptors at other sites [148,149], including the spinal cord region where nerve fibres control the external sphincter [150]. TAK 802 acted directly on the detrusor muscle [151], our long-term ketamine addicts report consistent pain in the bladder and they passed small amounts of urine every 15–30 minutes. On further clinical investigation, they had small bladders. In our models using rodents, we reported three main phases of bladder changes after chronic and sub-anaesthetic administration of ketamine in mice at 30 mg/kg. The dosage was determined on consideration that the anaesthetic dosage is 150 mg/kg for rodents [152], and the metabolic rate of the rodents (mice in this case) was at least 10 times more rapid than in humans (e.g. heart rates). In addition, our experiments showed that at below 60 mg/kg, there was no abnormal behaviour when mice were placed into a water tank and allowed to swim. However, at 60 mg/kg the limbs, especially the hind limbs, showed reduced movement and were less coordinated [152].

Effects on the Kidney

After between one and three months of ketamine treatment, the majority of mice exhibit a phase one reaction with mononuclear cell infiltration to the kidney (Fig. 12.3); this was similar to phase one damage to the bladder. In the kidney (Fig. 12.3), focal collections of mononuclear cells were seen around the glomeruli and tubules, and even in the calyces and pelvis of the kidney. In the bladder, the collection of mononuclear cells was sub-epithelial, usually in the lamina propria. There was also a minor degree of vascular congestion at this stage, but vigorous inflammation was not evident. We therefore designated this as a ketamine-induced interstitial nephritis, with interstitial cystitis.

Stages 2–3 in the kidney were less distinct, compared with the bladder, and were not easy

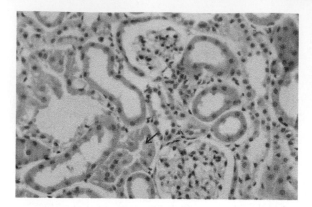

FIGURE 12.7 Hydropic degeneration of the kidney proximal tubule (arrow) after four months of ketamine treatment. ×400.

to separate distinctly. After the first stage of mononuclear cell infiltration, degeneration (mainly hydropic) of the proximal and distal tubules began. Tubular hydropic degeneration and lymphatic infiltration diagram representations were seen (Figs 12.3 and 12.7) [153,154].

In the later part of the third month all the way to sixth month, glomerular atrophy, along with glomerular hypertrophy, was observed [153]. PAS staining confirmed the thickening of basement membrane in many glomeruli. Protein casts were also seen in the distal tubules [153]. Concomitant examination on the chemical pathology of the urine showed that the majority of urine samples of these mice contained protein and glucose. The higher protein levels probably resulted from deterioration of the filtering ability of the glomeruli of the kidneys, confirming the pathological picture of glomerulonephritis. At the same time, the urinary system was not the only part in the pelvic region that was affected by chronic long-term ketamine treatment, as a decrease of 20% of motile sperms was registered after three months of chronic ketamine treatment in our mouse model [153].

In clinical studies, lower urinary tract symptoms (LUTS) were observed in most ketamine users which were related to the dosage and the frequency of ketamine usage [155–158]. Cystoscopy showed that some of the patients had varying degrees of urinary epithelial inflammation [155], whilst ultrasound demonstrated small bladder volume and wall thickening [157].

Effects on the Intestine, Pancreas and Adrenal Gland

It is remarkable that there is limited information on the effects of ketamine on important internal organs such as the intestine, pancreas and the adrenal glands. In cultured adrenal medullary cells, ketamine inhibited Na^+ influx by blocking channels in nicotinic receptors and consequently reducing catecholamine secretion [159,160]. Ketamine also increased the functional activity of the noradrenaline transporter in cultured adrenal medullary cells, which potentially could negatively modulate sympathetic activity during anaesthesia [161]. Consistent with these studies, ketamine also inhibited catecholamine secretion evoked by stimulation of cholinergic receptors in the perfused adrenal medulla of the rat; [162] similar effects are also seen on the adrenal gland of the lizard [163]. In fact, other anaesthetics such as pentobarbital have also been shown to depress catecholamine release during anaesthesia [164].

In isolated pituitary gland cells, ketamine increased the secretion of ACTH [165]. In studies on the pancreas, barbiturates did not affect insulin secretion [166]. Ketamine-xylazine combination decreased blood flow to pancreatic islets [167]. In a short-term anaesthetic study, Reyes et al. reported inhibition of insulin secretion and a glycogenolytic response in the liver after ketamine anaesthesia [168]. In a 24-h study on the rabbit, Illera et al. demonstrated that ketamine combined with diazepam or xylazine led to increase in blood glucose [169]. There have been no long-term studies on the effect of ketamine on either the pancreas and adrenal glands in an intact model.

Most of the studies of the effects of ketamine on the gastrointestinal system have focused on the intestines. Indeed, many of the past studies cited that ketamine was beneficial rather than harmful. However, most of the studies involved only acute ketamine administration. A number of the studies dealt with experimental ischaemia and reperfusion in rodents [170–172]. In these studies, intestinal ischaemia and reperfusion were studied together, usually following 30–45 minutes of ischaemia and an hour of reperfusion. Ketamine was used at doses between 10 mg/kg and 50 mg/kg, depending on the site of the injection. These studies revealed an increase and improvement in the motility of the intestine following ischaemic injury. The ischaemia-induced increases of AST (aspirate aminotransferase, lactic acid dehydrogenase (LDH) and tumour necrosis factor-alpha (TNF-α) returned to normal in animals treated with ketamine and there were also no changes in intestinal morphology, leucocyte infiltration, or serum factors like serum intracellular adhesion molecule 1 [171,172]. There are other reports that ketamine not affect sphincter control [173,174]. Whilst it might be true that ketamine protected the intestine via suppressing NF-Kappa B and other pro-inflammatory cytokines [175], studies to date have only investigated the acute effects of ketamine.

However, the effect of chronic long-term effects of ketamine on internal organs is more complicated. Preliminary studies in our laboratory indicated minimal morphological changes in the intestine, which was normal in appearance (Fig. 12.8). There was neither obvious apoptosis nor necrosis even when ketamine had been used chronically with alcohol. There was, however, a decrease in the degree of proliferation of the mucosa as evident by our immunohistochemical investigations of proliferative nuclear antigens (unpublished data). In the adrenal medulla, there was a very significant decrease of both immunoreactive tyroxine hydroxylase and dopamine β-hydroxylase,

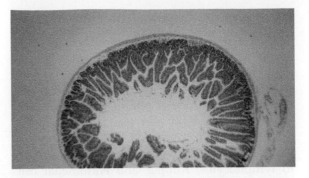

FIGURE 12.8 Normal appearance of the duodenum after six months of ketamine treatment. ×50.

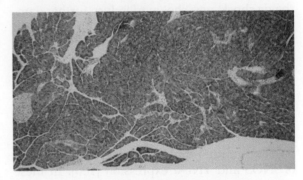

FIGURE 12.9 Significant mononuclear cells infiltration in the pancreas after six months of ketamine treatment. ×100.

indicating a decrease in production of dopamine analogues. In the pancreas, the proliferation (regeneration perhaps) of both acinar cells and cells in the islets of Langerhans were down regulated after ketamine or combined ketamine-alcohol long-term intoxication. In this same organ, we observed an increase in necrotic cells and a decrease in the number of large size islets of Langerhans without an increase in apoptosis (unpublished data). In addition, many of the organs in the body of the ketamine-treated animals showed lymphocytic infiltration, pointing to an increase vulnerability to infection (Fig. 12.9). There have been no human studies on these organs and the pathophysiological changes seen after chronic ketamine use.

Effects on the Liver and Immune System

Ketamine was first introduced clinically as a routine anaesthetic agent in the 1990s: one study showed that out of 295 procedures, there were nine cases of 'severe reactions' [176]. Like the kidney, the liver is one of most vulnerable sites for drug toxicity. Ten years ago, a study by Thompson et al. concluded that ketamine, particularly combined with xylazine, causes injury to lymphocytes, Kupffer and endothelial cells in the liver [177]. As an anaesthetic, ketamine in combination with diazepam and ketamine with xylazine, when used in patients with haemorrhagic shock, was shown to induce further damage in the kidney and the liver [178]. Further, in an early review concerning ketamine-induced anaesthesia in children, Krüger suggested that only low doses of ketamine could be used, but should not be combined with other CNS agents because this combination could induce further liver and/or kidney toxicity [179].

Apart from the liver, ketamine could act on the immune system as well via inhibition of iNOS and TNF-α expression [180]. In these studies, the function of phagocytosis of the Kupffer cells in the liver was affected [181]. The potential of ketamine to affect the immune system has been suspected for at least 20 years, following observations that it increased IL-6 concentrations in the body, leading to activation of the inflammatory cascade and subsequent organ dysfunction [182]. Increased accumulation of CD4 reactive products in other organs of the chronic ketamine-treated mice, such as the brain (Fig. 12.10), also supported that ketamine could indeed modulate the immune system.

The chronic long-term effect of ketamine on the liver has been rarely studied. Our own research indicated that fatty regeneration of liver cells starts to occur six weeks after chronic ketamine treatment in mice [159]. By 16 weeks, fibrosis appeared in the periphery, and by 28 weeks the fibrosis (or cirrhosis) extended into the parenchyma of the liver [159]. Liver cirrhosis

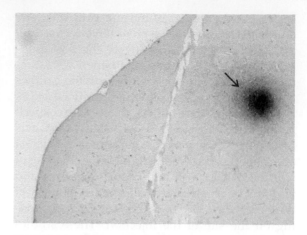

FIGURE 12.10 CD4 positive products (arrow) along with microglia in brains of mice treated with ketamine for four months. ×50.

was usually accompanied by an increase in the number of proliferative cells [159], and was observed initially after 16 weeks of chronic ketamine treatment. This increase of proliferative cells probably indicated an attempt of the organ to regenerate. Glutamate oxaloacetic traminase (GOT) was also seen, confirming liver damage, with a 28.7% increase (138.55 ± 19.341 Karmen units versus 107.645 ± 2.722 Karmen units) after 28 weeks [159]. Like the ketamine interaction with cocaine, liver damage was elevated to a higher extent after just four weeks of combined ketamine and alcohol treatment at the end of the 28 weeks. There was a further 37.6% increase (190.635 ± 21.230 Karmen units versus 138.55 ± 19.341 Karmen units) of the GOT enzymes along with the extensive liver cirrhosis [159].

Both existing ketamine enantiomers (S)-ketamine and (R)-ketamine are presented to the liver for metabolism to norketamine, dehydro-norketamine and hydroxylated norketaimne [183,184]. The metabolism and elimination of the metabolites of ketamine is via the microsomal cytochrome P450 system and there is probably induction of CYP3A4 and CYP2B6 isoenzymes [185,186]. When ketamine is combined with other abusive agents, e.g. cocaine and alcohol, the damage to the liver is synergistic compared to ketamine alone [159,187]. In the case of cocaine, the interaction of ketamine and cocaine led to a decrease of the activity of hepatic glutathione peroxidase and catalase [187]. Ketamine also has pharmacodynamic effects on the liver via NMDA receptors, and can cause the same excitotoxic changes as in the CNS [188].

Effects on the Heart

In rabbits, the combination of ketamine with xylazine (9.4–38.9 mg) suppressed tachyarrhythmias at a high dose of >399 mg/kg [189]. As an anaesthetic, ketamine is often hailed as a good agent for short-term surgery in both humans and animals, because the patients recover rapidly and in addition, particularly in large animal surgery, ketamine is less depressive on the heart [190,191]. In children with congenital heart malfunction and in the elderly, both of whom are more vulnerable to cardiovascular compromise, ketamine has been shown to be safe [192,193]. Ketamine is also frequently employed as analgesic in minor surgery in hospital settings [194].

However, there are some reports where ketamine has been associated with unwanted effects. For example, Spotoft et al. [195] and Jacobson and Hartsfield [196] noted that ketamine produced vasoconstriction and respiratory rate depression associated with an increase in heart rate. The increase in heart rate was accompanied by an increase in blood pressure and heart rate (blood pressure increased by 30%, heart rate by 15%) [197,198]. These were seemingly sporadic reactions that happened only during the operation and it is not likely that this degree of tachycardia/hypertension will be a risk with acute ketamine recreational use (except perhaps when ketamine is taken with a sympathomimetic agent). However, with the chronic misuse of the drug in the community, these effects should not be overlooked.

In the course of our chronic ketamine administration studies, we observed that few mice

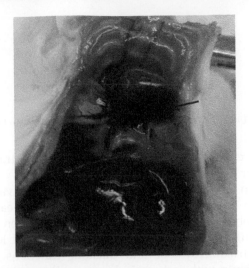

FIGURE 12.11   A dilated heart (arrow) of a ketamine-treated mouse of six months

(30 per 300 animals) had hearts that were visibly dilated by volume (Fig. 12.11). Detailed studies indicated cardiac cell hypertrophy followed by lytic and coagulative changes in their cardiac muscles following four months of chronic treatment [199]. By six months of treatment, the pathological picture was even more marked, and lysis was observed with cell shrinkage or hypertrophy, and many animals died as a result of coagulative necrosis [199]. ECG recordings showed biphasic QRS peaks following three months of treatment. After six months of treatment, ST depression was observed [199]. This suggests that ketamine could induce features of ischaemic heart disease, further strengthened by elevated serum troponin I and creatine kinase levels in these rodent studies of chronic ketamine administration [199]. These changes were not seen in age-matched controls at four and six months after chronic treatment [199]. Troponin I and creatine kinase are present in cardiac muscles where they are necessary for normal contractile activity, Troponin I binds calcium to cause conformational changes so that myosin can attach onto the myofibrillar filament [200–204].

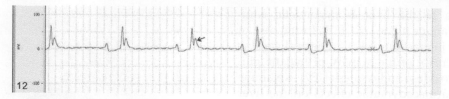

FIGURE 12.12   ST inversion (arrow) in a six-month ketamine-treated mouse depicting an abnormal ECG.

An example of an ischaemic heart ECG trace after ketamine is shown in Figure 12.12.

In the human, there are also data available on the effects of ketamine on the heart: Jakobsen et al. [205]. reported that even a small dose of 0.5mg/kg ketamine reduced left ventricular systolic and diastolic functions in patients with ischaemic heart, whilst a single dose of 1mg/kg ketamine did not affect oxygenation after cardiopulmonary bypass [206]. Using isolated atrial myocytes, ketamine was shown at a low dose of 1.8 μmol/L to inhibit influx of calcium [207]. In other human studies, volunteers were infused with S-(+)-ketamine for two hours to a dosage of 320ng/ml [208]. At such time, an increase of 40–50% cardiac output was recorded when compared with normal. Hudetz et al. studied patients that had intravenous bolus of 0.5mg/kg ketamine to treat postoperative delirium and minimal cardiovascular effects were seen related to ketamine [209].

PHYSICAL AND CHEMICAL DESCRIPTION

Ketamine exists as two optical isomers – S-(+) and (R)-(−)-2-(2-chlorophenyl)-2-(methylamino)cyclohexanone. As discussed earlier in this chapter, the S-(+) isomer is more potent that the R(−) isomer. However, they have similar pharmacokinetics.

Ketamine is available medically in liquid form; however, it is generally sold to users in powder form or as tablets [210]. It is typically used by nasal insufflation, but some users also inject ketamine [210]. Significant first-pass metabolism reduces bioavailability following oral administration [210,211]. Ketamine has an alpha half-life of 2–4 minutes and beta-half-life of 8–16 minutes [212].

Epidemiology

Ketamine was first introduced as an anaesthetic agent in the 1960s: the first reported recreational use of ketamine dates back to 1967 [210]. Initially, recreational use was greater amongst healthcare professionals and those who had easy access to ketamine. However, in the 1980s and 1990s, ketamine started to become more widely used, particularly in the US 'rave scene', leading on to more international use. In terms of population prevalence, the patterns and prevalence of use is only mentioned in descriptive terms in the United Nations Office on Drugs and Crime annual report, and there is limited specific data on the prevalence of ketamine use at a general population level worldwide [213].

Data is available on the prevalence of ketamine use for a few countries around the world. In the United Kingdom (UK), the annual British Crime Survey has included information on the life-time, last year and last month use of ketamine in those aged 15–59 years old since the 2006/7 survey [214]. The life-time prevalence of use has gradually increased form 1.3% in 2006/7 to 2.5% in 2011/12, whereas the annual survey reported last year and last month use rates have remained relatively stable between 0.3–0.6% and 0.1–0.3%, respectively. Similar to other recreational drugs, the prevalence of use is higher in younger individuals and annual survey reported last year and last month use in those aged 15–24 years old is 0.8–2.1% and 0.3–0.9%, respectively.

Data has been available in Australia on the prevalence of ketamine use since 2001 through the Australian Institute of Health and Welfare (AIHW) 'National Drug Strategy Household Surveys'. In 2010, lifetime and last year ketamine use amongst those aged over 14 years was comparable to the British Crime Survey (lifetime and last year prevalence rates of use of 1.0–1.4% and 0.3–0.4% respectively) [215].

In the US the annual National Survey on Drug Use and Health does not routinely report on the prevalence of use of ketamine. However, in the 2006 survey, data was collected on the use of specific hallucinogens, which included ketamine [216]. They reported that an estimated 23 million people had tried ketamine in their lifetime, and that the prevalence of use in the last year of those aged over 12 years old was 0.1%.

Although there is a generally accepted trend of increasing use of ketamine in certain areas of Asia, in particular Hong Kong, China, Indonesia and Malaysia, where ketamine is often reported to be the recreational drug of choice, there is limited national data on the actual prevalence rates of use [214,217]. In Hong Kong, data on the prevalence of recreational drugs is collected through the Central Registry of Drug Abuse (CRDA), and information is available freely on line [218]. There has been increasing use of ketamine in those aged under 21 from 36.9% of 'young drug users' in 2000 to 84.3% of 'young drug users' in 2009. It appears that more recently there has been a slight decline in ketamine use as this fell to 80.1% and 70.4% in 2010 and 2011 respectively. The prevalence of use in those aged over 21 years appears to be lower but more stable (23.1%, 23.9% and 23.3% in 2009, 2010 and 2011, respectively).

In addition to this national population level data, there are numerous sub-population surveys which have reported prevalence of use amongst groups such as students and 'clubbers'.

In the 2009/10 and 2010/11 annual MixMag survey in the UK, the life-time reported use rates of ketamine amongst respondents was 67.8% and 62%, respectively [219]. More recent use was reported as last month use (32.4%) in the 2009/10 survey and last year use (41.2%) in the 2010/11 survey. The survey was expanded to the Global Drugs Survey in the 2011/2012 survey and the prevalence of life-time use amongst UK respondents was 47.8%. There were significant differences in last year use depending on the location of respondents, with last year rates of 24.5% in all UK respondents and 40% in UK 'regular clubbers' compared to only 5.5% in all US respondents [220].

In addition to these Internet based surveys, there have been studies in the night-time economy or club/rave environment that have attempted to collect data on the prevalence of use of ketamine. In a survey of 315 individuals attending 'gay friendly' nightclubs in South East London, UK, self-reported use or intended use on the night of ketamine was 12.5% of those surveyed [221]. A similar structured interview survey of 186 individuals attending a rave in Ontario, Canada reported a use rate of ketamine at the last rave they attended of 8.6% [222].

In one survey of 1285 ketamine users in the UK, 31% reported using 0.125 g or less during a typical session, 35% reported using 2.5–5 g, 34% reported use of 1 g or more and 5.2% reported using more than 3 g per session [223].

ACUTE TOXICITY

The numerous animal models describing the pharmacology and mechanisms of acute toxicity of ketamine have been summarised above. We will now summarise the literature on the acute toxicity of ketamine in human reports.

Ketamine is rarely associated with severe, life-threatening clinical features and therapeutically ketamine is generally considered to have a wide safety-margin. Even at anaesthetic doses sufficient to cause unconsciousness, ketamine generally spares airway reflexes and is not associated with significant haemodynamic instability [212]. These properties mean that recreational users do not generally develop significant compromising clinical features [224]. The main pattern of toxicity is related to the hallucinogenic effects of ketamine and if other significant clinical features are present, the clinician should consider whether these are related to the use of other drugs alongside ketamine.

The most common features of toxicity reported after ketamine user are neuropsychiatric and include hallucinations, agitation, aggression and paranoia [224–226]. Hallucinations can be unpleasant and some users refer to ketamine-associated hallucinations and dissociative features as 'falling into a K-hole' [224,226]. As summarised above, ketamine can increase heart rate and blood pressure and there are reports of moderate sinus tachycardia in patients with ketamine toxicity, with occasional reports of palpitations and chest pain [227]. Generally these cardiovascular effects are self-limiting and not associated with significant risk; however, there is the potential in individuals with underlying cardiovascular disease and/or when ketamine is used with a sympathomimetic agent that these effects may be of more significance.

One of the main acute risks associated with ketamine use is the potential for physical harm and trauma [210,228]. This occurs because users have a decreased awareness of their physical surroundings with a reduced perception of pain, hallucinations, paranoia and other effects which can include poor coordination, temporary paralysis and inability to speak [229]. In one longitudinal study of 30 ketamine users, two died during a 12-month follow up – one from hypothermia and one from drowning in a bath [230]. In another study of 90 ketamine users, 14% reported that they had been involved in an accident after taking ketamine and 83% reported that they knew someone who had [231]. In Hong Kong, between 1996 and 2000,

ketamine was involved in 9% of fatal drugs- and alcohol-related road traffic accidents in Hong Kong [232].

Ketamine-Related Fatalities

As noted earlier, ketamine is rarely associated with life-threatening clinical features and, other than deaths due to physical injury, death related to lone-ketamine use is rare. When ketamine is reported in drug-related fatalities, it is usually found together with other drugs that are more likely to have contributed to death [224,233]. In one UK study assessing 23 deaths over the period 1993–2006 in which ketamine was detected, in only four cases was ketamine the only drug detected [233]. In another study in the USA, of 15 hospital deaths in which ketamine was detected in post-mortem toxicology screening, deaths were most likely to result from other drugs (n = 12) or trauma (n = 2) [234].

CHRONIC TOXICITY

The animal studies and models of ketamine-related acute toxicity have been reviewed in detail earlier in this chapter. We will now summarise human reports of ketamine-related chronic toxicity.

Bladder and Renal Toxicity

Ketamine can cause a wide range of effects within the bladder and renal tract. Ketamine-related bladder toxicity was first reported in Canada in 2007 [235]. In this series of nine long-term users, findings on CT were of bladder wall thickening and a small bladder capacity and cystoscopy demonstrated ulcerative cystitis. Microscopic findings on biopsies taken during cystoscopy included denuded bladder mucosa with ulceration and inflammation with vascular granulation. There have been numerous reports since this time, principally from the UK and

Hong Kong/Taiwan, describing reports or series of patients with ketamine-related bladder problems [236–243]. One recent review pooled data from three case series comprising a total of 93 patients [224]. Almost three-quarters (74.0%) had a reduced bladder volume; 54% had hydronephrosis. This is likely to be a secondary phenomenon and due to factors such as vesci-ureteric reflux, peri-ureteric thickening or ureteric obstruction by gelatinous debris [224,239–242]. Cystitis was seen in all 42 individuals who had a cystoscopy, with abnormal histology in 94.1% of biopsies [224]. Histological changes reported on the bladder biopsies included ulceration and inflammation with eosinphilic infiltration. In some reports it is noted that the histological appearances are similar to carcinoma *in situ* (CIS); however, unlike CIS, ketamine-related cystitis is negative for the CK20 protein [238].

There have also been reports from the palliative care setting of patients developing lower urinary tract symptoms related to the use of ketamine [237,244]. In one of these reports, urinary symptoms were documented to worsen with the use and re-introduction of ketamine and settle with repeated discontinuation [237].

In a recent study of 1285 ketamine users in the UK, 340 (26.6%) reported urinary symptoms – common symptoms included urinary frequency (17.4%), lower abdominal pain (11.3%) and dysuria (8.1%); incontinence was reported by 3.3% and haematuria by 1.5% [223]. These symptoms were significantly associated with higher dose and more frequent ketamine use.

The most important aspect of management of ketamine-related bladder pathology is ketamine cessation. In the recent UK study exploring ketamine-related urinary symptoms, of the 251 of those surveyed who reported on the temporal nature of their urinary symptoms, 51% reported improvement in urinary symptoms when they stopped ketamine [223].

In those who have continued symptoms on ketamine cessation associated with significant bladder pathology surgical intervention including augmentation cysteroplasty or cystectomy may be required [241–243]. There has been one small study in six patients with ketamine-related bladder pathology which suggests that intravesical hyaluronic acid may be beneficial [242].

Gastrointestinal Toxicity

Regular ketamine use can be associated with abdominal pain in up to one-third of users, this is often referred to by users as 'K cramps' [226,231]. There have been reports of some users taking higher doses of ketamine to try and manage this abdominal pain resulting in worsening of their symptoms [226]. The cause of the abdominal pain associated with ketamine is not known but in one retrospective series of 37 ketamine users, abdominal pain was associated with *Helicobacter pylori*-negative gastritis [245].

There have also been increasing reports of abnormal liver function associated with ketamine use (both recreational misuse and therapeutic use), although not all of these reports have included abdominal pain as a prominent symptom [245–249]. In a number of these reports, choledochal cysts were found [245,248,249]. It appears that generally the common bile duct dilatation and liver function tests improve with ketamine cessation [245,248,249]. The cause of these ketamine-related liver effects is poorly understood; however, data from an *in vitro* study in human hepatoma G2 cells has suggested that S(+) ketamine is directly hepatotoxic [250].

Neurological and Neuropsychiatric Chronic Toxicity

As summarised at the beginning of this chapter, there is extensive literature describing animal models which have investigated the neurological effects of ketamine. We will now summarise reports from ketamine users and human studies of the neurological and neuropsychological/neuropsychiatric effects of ketamine.

Psychosis and Depression

As discussed above, ketamine is used in animal models of schizophrenia. A number of volunteer studies have also shown that ketamine administration can result in schizophrenia-like positive and negative symptoms [251,252]. Studies in patients with schizophrenia have shown that ketamine can induce psychotic symptoms, in a similar pattern to patients presenting with schizophrenic symptoms, both in patients treated with antipsychotics and those stabilised off antipsychotic therapy [13,253,254]. In one study comparing 20 volunteers who reported acute recreational ketamine use with 19 polydrug using volunteers with no reported ketamine use, the ketamine users had higher scores for schizotypy at baseline and on three-day follow-up [255]. In a longitudinal study of 150 ketamine users, there was a ketamine dose-dependent effect seen with greater delusional and dissociative symptoms in frequent ketamine users that persisted at one-year follow up [230]. There have, however, been no studies investigating whether schizophrenia is more common in frequent ketamine users.

In the longitudinal study of 150 chronic ketamine users discussed above, increased depressions score (Beck Depression Inventory), but not clinical depression, was associated with frequent (daily) current and previous ketamine use [230]. Depression scores were not increased in less frequent (less than three times per week) ketamine users [230]. There have been some pilot studies which have investigated the use of ketamine as an antidepressant in patients with depression and of intravenous ketamine in patients with resistant depression [256–258]. The results of these preliminary studies are not conclusive, but suggest that ketamine may have a role in this setting.

Neurological and Cognitive Effects

A number of studies have shown that acute ketamine use results in impairment of both working and episodic memory; it appears that these effects are greater in men than in women [259–261]. Long-term ketamine use is also associated with deficits of both long- and short-term memory [259,262]. Many of these studies involved polydrug users and so it is difficult to be certain that all of the effects seen were due solely to ketamine; however, in a longitudinal study the pattern of ketamine use correlated with memory impairment, with more frequent users more likely to have deficits in short- and long-term memory and pattern recognition tasks [230].

More recently, imaging studies in ketamine users have been published which demonstrate structural changes in chronic ketamine users. In one study, white-matter volume was assessed using diffusion tensor MRI in 41 individuals with chronic ketamine use and 44 drug-free volunteers [263]. White matter changes were seen in the frontal lobes bilaterally and the left temporoparietal cortex; the frontal white matter changes correlated with estimated ketamine dose. In another report from the same group, voxel based morphometry in conjunction with statistical parametric MRI mapping was reported in 41 individuals with chronic ketamine use and 44 drug-free volunteers to assess grey matter volume [264]. Decreased grey-matter volume was seen in the left superior frontal gyrus and right middle frontal gyrus in the ketamine users and there was a correlation between the duration/dose of ketamine used and these changes.

DEPENDENCE AND ABUSE LIABILITY

A number of animal models have shown that ketamine has dependence potential – studies in rats show self-administration of ketamine and also that ketamine will substitute for alcohol in drug discrimination studies; ketamine self-administration is also seen in studies in pigeons and rhesus monkeys [265–268].

Ketamine tolerance has been shown in animal models and also in therapeutic use of ketamine as an anaesthetic [269–272]. Human users also report tolerance with increasing doses required to achieve the same effect, in one study there was a six-fold increase from first ever dose to current dose in chronic ketamine users [226,273].

There are reports of ketamine dependence in human ketamine users [274–276], but few large scale studies to be able to determine the incidence of ketamine dependence. In one study of 90 ketamine users, 57% of frequent users noted that they were concerned about ketamine addiction and many reported that they continued to use ketamine until their supplies were exhausted [232]. In another questionnaire based study, 17% (218 of 1285) users fulfilled DSM IV criteria for dependence [223]. In this study, dependence was associated with both the dose used per session and the frequency of ketamine use.

Ketamine is generally associated with psychological rather than physical dependence [224]. There are some reports of physical symptoms on ketamine cessation that include 'shaking', sweating and palpitations; however, the main problem seen in individuals stopping ketamine is craving [275].

REFERENCES

[1] Arditti J, Spadari M, de Haro L, Brun A, Bourdon JH, Valli M. Ketamine – dreams and realities. Acta Clin Belg Suppl 2002;1:31–3.

[2] Wolff K, Winstock AR. Ketamine: from medicine to misuse. CNS Drugs 2006;20:199–218.

[3] Pal HR, Berry N, Kumar R, Ray R. Ketamine dependence. Anaesth Intensive Care 2002;30:382–4.

[4] Lim DK. Ketamine associated psychedelic effects and dependence. Singapore Med J 2003;44:31–4.

[5] Pouget P, Wattiez N, Rivaud-Péchoux S, Gaymard B. Rapid development of tolerance to sub-anaesthetic dose of ketamine: an oculomotor study in macaque monkeys. Psychopharmacology (Berl.) 2010;209:313–8.

[6] Jansen KL, Darracot-Cankovic R. The nonmedical use of ketamine, part two: a review of problem use and dependence. J Psychoactive Drugs 2001;33:151–8.

[7] Degenhardt L, Copeland J, Dillon P. Recent trends in the use of 'club drugs': an Australian review. Subst Use Misuse 2005;40:1241–56.

[8] Wai MS, Luan P, Jiang Y, et al. Long term ketamine and ketamine plus alcohol toxicity – what can we learn from animal models? Mini Rev Med Chem 2012 Apr 17.

[9] Hansen G, Jensen SB, Chandresh L, Hilden T. The psychotropic effect of ketamine. J Psychoactive Drugs 1988;20:419–25.

[10] Krystal JH, Karper LP, Seibyl JP, et al. Subanesthetic effects of the noncompetitive NMDA antagonist, ketamine, in humans. Psychotomimetic, perceptual, cognitive, and neuroendocrine responses. Arch Gen Psychiatry 1994;51:199–214.

[11] Jansen KL. A review of the nonmedical use of ketamine: use, users and consequences. J Psychoactive Drugs 2000;32:419–33.

[12] Domino EF, Chodoff P, Corssen G. Pharmacologic effects of Ci-581, a new dissociative anesthetic, in man. Clin Pharmacol Ther 1965;6:279–91.

[13] Olney JW, Newcomer JW, Farber NB. NMDA receptor hypofunction model of schizophrenia. J Psychiatr Res 1999;33:523–33.

[14] Ebert A, Haussleiter IS, Juckel G, Brune M, Roser P. Impaired facial emotion recognition in a ketamine model of psychosis. Psychiatry Res 2012. Available: <http://dx.doi.org/10.1016/j.psychres.2012.06.034> [accessed 07.03.13].

[15] Zuo D, Bzdega T, Olszewski RT, Moffett JR, Neale JH. Effects of N-acetylaspartylglutamate (NAAG) peptidase inhibition on release of glutamate and dopamine in prefrontal cortex and nucleus accumbens in phencyclidine model of schizophrenia. J Biol Chem 2012;287:21773–21782.

[16] Olney JW, Farber NB. Glutamate receptor dysfunction and schizophrenia. Arch Gen Psychiatry 1995;52:998–1007.

[17] Sharp FR, Hendren RL. Psychosis: atypical limbic epilepsy versus limbic hyperexcitability with onset at puberty? Epilepsy Behav 2007;10:515–20.

[18] Ghoneim MM, Hinrichs JV, Mewaldt SP, Petersen RC. Ketamine: behavioural effects of subanesthetic doses. J Clin Psychopharmacol 1985;5:70–7.

[19] Malhotra AK, Pinals DA, Weingartner H, et al. NMDA receptor function and human cognition: the effects of ketamine in healthy volunteers. Neuropsychopharmacol 1996;14:301–7.

[20] Adler CM, Malhotra AK, Elman I, et al. Comparison of ketamine-induced thought disorder in healthy volunteers and thought disorder in schizophrenia. Am J Psychiatry 1999;156:1646–9.

[21] Breier A, Malhotra AK, Pinals DA, Weisenfeld NI, Pickar D. Association of ketamine-induced psychosis

with focal activation of the prefrontal cortex in healthy volunteers. Am J Psychiatry 1997;154:805–11.

[22] Holcomb HH, Lahti AC, Medoff DR, Weiler M, Tamminga CA. Sequential regional cerebral blood flow brain scans using PET with H2(15)O demonstrate ketamine actions in CNS dynamically. Neuropsychopharmacol 2001;25:165–72.

[23] Yu H, Li Q, Wang D, et al. Mapping the central effects of chronic ketamine administration in an adolescent primate model by functional magnetic resonance imaging (fMRI). Neurotoxicology 2012;33:70–7.

[24] Musso F, Brinkmeyer J, Ecker D, et al. Ketamine effects on brain function – simultaneous fMRI/EEG during a visual oddball task. Neuroimage 2011;58: 508–25.

[25] Liao Y, Tang J, Fornito A, et al. Alterations in regional homogeneity of resting-state brain activity in ketamine addicts. Neurosci Lett 2012;522:36–40.

[26] Abel KM, Allin MP, Kucharska-Pietura K, et al. Ketamine and fMRI BOLD signal: distinguishing between effects mediated by change in blood flow versus change in cognitive state. Hum Brain Mapp 2003;18:135–45.

[27] Honey RA, Honey GD, O'Loughlin C, et al. Acute ketamine administration alters the brain responses to executive demands in a verbal working memory task: an FMRI study. Neuropsychopharmacol 2004;29: 1203–14.

[28] Northoff G, Richter A, Bermpohl F, et al. NMDA hypofunction in the posterior cingulate as a model for schizophrenia: an exploratory ketamine administration study in fMRI. Schizophr Res 2005;72:235–48.

[29] Fu CH, Abel KM, Allin MP, et al. Effects of ketamine on prefrontal and striatal regions in an overt verbal fluency task: a functional magnetic resonance imaging study. Psychopharmacology (Berl.) 2005;183: 92–102.

[30] Corlett PR, Honey GD, Aitken MRF, et al. Frontal responses during learning predict vulnerability to the psychotogenic effects of ketamine – linking cognition, brain activity, and psychosis. Arch Gen Psychiatry 2006;63:611–21.

[31] Deakin JF, Lees J, McKie S, Hallak JE, Williams SR, Dursun SM. Glutamate and the neural basis of the subjective effects of ketamine: a pharmaco-magnetic resonance imaging study. Arch Gen Psychiatry 2008;65:154–64.

[32] Verma A, Moghaddam B. NMDA receptor antagonists impair prefrontal cortex function as assessed via spatial delayed alternation performance in rats: modulation by dopamine. J Neurosci 1996;16:373–9.

[33] Smith GS, Schloesser R, Brodie JD, et al. Glutamate modulation of dopamine measured in vivo with positron emission tomography (PET) and 11C-raclopride

in normal human subjects. Neuropsychopharmacol 1998;18:18–25.

[34] Soyka M, Koch W, Moller HJ, Ruther T, Tatsch K. Hypermetabolic pattern in frontal cortex and other brain regions in unmedicated schizophrenia patients. Results from a FDG-PET study. Eur Arch Psychiatry Clin Neurosci 2005;255:308–12.

[35] Sesack SR, Pickel VM. Prefrontal cortical efferents in the rat synapse on unlabeled neuronal targets of catecholamine terminals in the nucleus accumbens septi and on dopamine neurons in the ventral tegmental area. J Comp Neurol 1992;320:145–60.

[36] Chambers RA, Taylor JR, Potenza MN. Developmental neurocircuitry of motivation in adolescence: a critical period of addiction vulnerability. Am J Psychiatry 2003;160:1041–52.

[37] Olney JW, Labruyere J, Wang G, Wozniak DF, Price MT, Sesma MA. NMDA antagonist neurotoxicity: mechanism and prevention. Science 1991;254:1515–8.

[38] Sharp FR, Tomitaka M, Bernaudin M, Tomitaka S. Psychosis: pathological activation of limbic thalamocortical circuits by psychomimetics and schizophrenia? Trends Neurosci 2001;24:330–4.

[39] Liao Y, Tang J, Corlett PR, et al. Reduced dorsal prefrontal gray matter after chronic ketamine use. Biol Psychiatry 2010;69:42–8.

[40] Liao Y, Tang J, Ma M, et al. Frontal white matter abnormalities following chronic ketamine use: a diffusion tensor imaging study. Brain 2010;133:2115–22.

[41] Nestler EJ. Is there a common molecular pathway for addiction? Nat Neurosci 2005;8:1445–9.

[42] Corlett PR, Honey GD, Krystal JH, Fletcher PC. Glutamatergic model psychoses: prediction error, learning, and inference. Neuropsychopharmacol 2011;36:294–315.

[43] Albrecht J, Sidoryk-Wegrzynowicz M, Zielinska M, Aschner M. Roles of glutamine in neurotransmission. Neuron Glia Biol 2010;6:263–76.

[44] Niciu MJ, Kelmendi B, Sanacora G. Overview of glutamatergic neurotransmission in the nervous system. Pharmacol Biochem Behav 2012;100:656–64.

[45] Nicoll RA, Malenka RC. Expression mechanisms underlying NMDA receptor-dependent long-term potentiation. Ann N Y Acad Sci 1999;868:515–25.

[46] Bubenikova-Valesova V, Horacek J, Vrajova M, Hoschl C. Models of schizophrenia in humans and animals based on inhibition of NMDA receptors. Neurosci Biobehav Rev 2008;32:1014–23.

[47] Thomson AM, West DC, Lodge D. An N-methylaspartate receptor-mediated synapse in rat cerebral cortex: a site of action of ketamine? Nature 1985;313:479–81.

[48] Rolls ET. Glutamate, obsessive-compulsive disorder, schizophrenia, and the stability of cortical attractor

neuronal networks. Pharmacol Biochem Behav 2012;100:736–51.

[49] Sendt KV, Giaroli G, Tracy DK. Beyond dopamine: glutamate as a target for future antipsychotics. ISRN Pharmacol 2012;2012:427267.

[50] Corlett PR, Honey GD, Krystal JH, Fletcher PC. Glutamatergic model psychoses: prediction error, learning, and inference. Neuropsychopharmacol 2010;36:294–315.

[51] Field JR, Walker AG, Conn PJ. Targeting glutamate synapses in schizophrenia. Trends Mol Med 2011;17:689–98.

[52] Harris JA, Biersner RJ, Edwards D, Bailey LW. Attention, learning, and personality during ketamine emergence: a pilot study. Anesth Analg 1975;54:169–72.

[53] Corlett PR, Honey GD, Krystal J,H, Fletcher PC. Glutamatergic model psychoses: prediction error, learning, and inference. Neuropsychopharmacol 2011;36:294–315.

[54] Jentsch JD, Roth RH. The neuropsychopharmacology of phencyclidine: from NMDA receptor hypofunction to the dopamine hypothesis of schizophrenia. Neuropsychopharmacol 1999;20:201–25.

[55] Liao Y, Tang J, Corlett PR, et al. Reduced dorsal prefrontal gray matter after chronic ketamine use. Biol Psychiatry 2011;69:42–8.

[56] Behrens MM, Ali SS, Dao DN, et al. Ketamine-induced loss of phenotype of fast-spiking interneurons is mediated by NADPH-oxidase. Science 2007;318:1645–7.

[57] Mickley GA, Kenmuir CL, McMullen CA, et al. Long-term age-dependent behavioural changes following a single episode of fetal N-methyl-D-Aspartate (NMDA) receptor blockade. BMC Pharmacol 2004;4:28.

[58] Stoet G, Snyder LH. Effects of the NMDA antagonist ketamine on task-switching performance: evidence for specific impairments of executive control. Neuropsychopharmacol 2006;31:1675–81.

[59] Linn GS, O'Keeffe RT, Lifshitz K, Schroeder C, Javitt DC. Behavioural effects of orally administered glycine in socially housed monkeys chronically treated with phencyclidine. Psychopharmacology (Berl) 2007;192:27–38.

[60] Duncan GE, Moy SS, Perez A, et al. Deficits in sensorimotor gating and tests of social behaviour in a genetic model of reduced NMDA receptor function. Behav Brain Res 2004;153:507–19.

[61] Miyamoto Y, Yamada K, Nagai T, et al. Behavioural adaptations to addictive drugs in mice lacking the NMDA receptor epsilon1 subunit. Eur J Neurosci 2004;19:151–8.

[62] Sakimura K, Kutsuwada T, Ito I, et al. Reduced hippocampal LTP and spatial learning in mice lacking NMDA receptor epsilon 1 subunit. Nature 1995;373:151–5.

[63] Enomoto T, Noda Y, Nabeshima T. Phencyclidine and genetic animal models of schizophrenia developed in relation to the glutamate hypothesis. Methods Find Exp Clin Pharmacol 2007;29:291–301.

[64] Iasevoli F, Polese D, Ambesi-Impiombato A, Muscettola G, de Bartolomeis A. Ketamine-related expression of glutamatergic postsynaptic density genes: possible implications in psychosis. Neurosci Lett 2007;416:1–5.

[65] Fink CC, Bayer KU, Myers JW, Ferrell Jr. JE, Schulman H, Meyer T. Selective regulation of neurite extension and synapse formation by the beta but not the alpha isoform of CaMKII. Neuron 2003;39:283–97.

[66] Peyssonnaux C, Eychene A. The Raf/MEK/ERK pathway: new concepts of activation. Biol Cell 2001;93:53–62.

[67] Wang JQ, Fibuch EE, Mao L. Regulation of mitogen-activated protein kinases by glutamate receptors. J Neurochem 2007;100:1–11.

[68] Kyosseva SV, Elbein AD, Griffin WS, Mrak RE, Lyon M, Karson CN. Mitogen-activated protein kinases in schizophrenia. Biol Psychiatry 1999;46:689–96.

[69] Arion D, Unger T, Lewis DA, Levitt P, Mirnics K. Molecular evidence for increased expression of genes related to immune and chaperone function in the prefrontal cortex in schizophrenia. Biol Psychiatry 2007;62:711–21.

[70] Deng MY, Lam S, Meyer U, et al. Frontal-subcortical protein expression following prenatal exposure to maternal inflammation. PLoS One 2011;6:e16638.

[71] Yeganeh-Doost P, Gruber O, Falkai P, Schmitt A. The role of the cerebellum in schizophrenia: from cognition to molecular pathways. Clinics (Sao Paulo) 2011;66(Suppl. 1):71–7.

[72] Gonzalez-Burgos G, Fish KN, Lewis DA. GABA neuron alterations, cortical circuit dysfunction and cognitive deficits in schizophrenia. Neural Plast 2011 723184.

[73] Coyle JT. Glutamate and schizophrenia: beyond the dopamine hypothesis. Cell Mol Neurobiol 2006;26:365–84.

[74] Lewis DA, Gonzalez-Burgos G. Pathophysiologically based treatment interventions in schizophrenia. Nat Med 2006;12:1016–22.

[75] Woo TU, Walsh JP, Benes FM. Density of glutamic acid decarboxylase 67 messenger RNA-containing neurons that express the N-methyl-d-aspartate receptor subunit NR2A in the anterior cingulate cortex in schizophrenia and bipolar disorder. Arch Gen Psychiatry 2004;61:649–57.

[76] Stone JM, Dietrich C, Edden R, et al. Ketamine effects on brain GABA and glutamate levels with 1H-MRS: relationship to ketamine-induced psychopathology. Mol Psychiatry 2012. doi:10.1038/mp.2011.171.

[77] Shi Q, Guo L, Patterson TA, et al. Gene expression profiling in the developing rat brain exposed to ketamine. Neuroscience 2010;166:852–63.

[78] Tan S, Rudd JA, Yew DT. Gene expression changes in GABA(A) receptors and cognition following chronic ketamine administration in mice. PLoS One 2011;6:e21328.

[79] Jones EG. GABAergic neurons and their role in cortical plasticity in primates. Cereb Cortex 1993;3:361–72.

[80] Chapouthier G, Venault P. GABA-A receptor complex and memory processes. Curr Top Med Chem 2002;2:841–51.

[81] Ricci V, Martinotti G, Gelfo F, et al. Chronic ketamine use increases serum levels of brain-derived neurotrophic factor. Psychopharmacology (Berl) 2011;215:143–8.

[82] Ibla JC, Hayashi H, Bajic D, Soriano SG. Prolonged exposure to ketamine increases brain derived neurotrophic factor levels in developing rat brains. Curr Drug Saf 2009;4:11–16.

[83] Mizoguchi Y, Kanematsu T, Hirata M, Nabekura J. A rapid increase in the total number of cell surface functional GABAA receptors induced by brain-derived neurotrophic factor in rat visual cortex. J Biol Chem 2003;278:44097–44102.

[84] Li Q, Lu G, Antonio GE, et al. The usefulness of the spontaneously hypertensive rat to model attention-deficit/hyperactivity disorder (ADHD) may be explained by the differential expression of dopamine-related genes in the brain. Neurochem Int 2007;50:848–57.

[85] Dagher A, Robbins TW. Personality, addiction, dopamine: insights from Parkinson's disease. Neuron 2009;61:502–10.

[86] Okai D, Samuel M, Askey-Jones S, David AS, Brown RG. Impulse control disorders and dopamine dysregulation in Parkinson's disease: a broader conceptual framework. Eur J Neurol 2011;18:1379–83.

[87] Beaulieu JM. A role for Akt and glycogen synthase kinase-3 as integrators of dopamine and serotonin neurotransmission in mental health. J Psychiatry Neurosci 2012;37:7–16.

[88] Li QYD. A review of the dopamine system in the animal models of attention-deficit hyperactivity disorder Stuart M, Gordon A, editors. Nova; 2009. p. 167–88.

[89] Zhang J, Xiong B, Zhen X, Zhang A. Dopamine D1 receptor ligands: where are we now and where are we going. Med Res Rev 2009;29:272–94.

[90] Heinz A, Schlagenhauf F. Dopaminergic dysfunction in schizophrenia: salience attribution revisited. Schizophr Bull 2010;36:472–85.

[91] Frank MJ, Fossella JA. Neurogenetics and pharmacology of learning, motivation, and cognition. Neuropsychopharmacol 2011;36:133–52.

[92] Jentsch JD, Andrusiak E, Tran A, Bowers Jr. MB, Roth RH. Delta 9-tetrahydrocannabinol increases prefrontal cortical catecholaminergic utilization and impairs spatial working memory in the rat: blockade of dopaminergic effects with HA966. Neuropsychopharmacol 1997;16:426–32.

[93] Moghaddam B, Adams BW. Reversal of phencyclidine effects by a group II metabotropic glutamate receptor agonist in rats. Science 1998;281:1349–52.

[94] Smith-Roe SL, Kelley AE. Coincident activation of NMDA and dopamine D1 receptors within the nucleus accumbens core is required for appetitive instrumental learning. J Neurosci 2000;20:7737–42.

[95] Berretta S, Robertson HA, Graybiel AM. Dopamine and glutamate agonists stimulate neuron-specific expression of Fos-like protein in the striatum. J Neurophysiol 1992;68:767–77.

[96] Rabiner EA. Imaging of striatal dopamine release elicited with NMDA antagonists: is there anything there to be seen?. J Psychopharmacol 2007;21:253–8.

[97] Valjent E, Pascoli V, Svenningsson P, et al. Regulation of a protein phosphatase cascade allows convergent dopamine and glutamate signals to activate ERK in the striatum. Proc Natl Acad Sci USA 2005;102:491–6.

[98] Dudman JT, Eaton ME, Rajadhyaksha A, et al. Dopamine D1 receptors mediate CREB phosphorylation via phosphorylation of the NMDA receptor at Ser897-NR1. J Neurochem 2003;87:922–34.

[99] Jiao H, Zhang L, Gao F, Lou D, Zhang J, Xu M. Dopamine D(1) and D(3) receptors oppositely regulate NMDA- and cocaine-induced MAPK signaling via NMDA receptor phosphorylation. J Neurochem 2007;103:840–8.

[100] Lindefors N, Barati S, O'Connor WT. Differential effects of single and repeated ketamine administration on dopamine, serotonin and GABA transmission in rat medial prefrontal cortex. Brain Res 1997;759:205–12.

[101] Narendran R, Frankle WG, Keefe R, et al. Altered prefrontal dopaminergic function in chronic recreational ketamine users. Am J Psychiatry 2005;162:2352–9.

[102] Lopez-Gil X, Babot Z, Amargos-Bosch M, Sunol C, Artigas F, Adell A. Clozapine and haloperidol differently suppress the MK-801-increased glutamatergic and serotonergic transmission in the medial prefrontal cortex of the rat. Neuropsychopharmacol 2007;32:2087–97.

[103] Lopez-Gil X, Artigas F, Adell A. Unraveling monoamine receptors involved in the action of typical and atypical antipsychotics on glutamatergic and serotonergic transmission in prefrontal cortex. Curr Pharm Des 2010;16:502–15.

[104] Malhotra AK, Adler CM, Kennison SD, Elman I, Pickar D, Breier A. Clozapine blunts N-methyl-D-aspartate antagonist-induced psychosis: a study with ketamine. Biol Psychiatry 1997;42:664–8.

[105] Zarate Jr. CA, Singh JB, Carlson PJ, et al. A randomized trial of an N-methyl-D-aspartate antagonist in treatment-resistant major depression. Arch Gen Psychiatry 2006;63:856–64.

[106] Machado-Vieira R, Salvadore G, Luckenbaugh DA, Manji HK, Zarate Jr. CA. Rapid onset of antidepressant action: a new paradigm in the research and treatment of major depressive disorder. J Clin Psychiatry 2008;69:946–58.

[107] Autry AE, Adachi M, Nosyreva E, et al. NMDA receptor blockade at rest triggers rapid behavioural antidepressant responses. Nature 2011;475:91–5.

[108] Nelson CL, Burk JA, Bruno JP, Sarter M. Effects of acute and repeated systemic administration of ketamine on prefrontal acetylcholine release and sustained attention performance in rats. Psychopharmacology (Berl.) 2002;161:168–79.

[109] Radcliffe KA, Dani JA. Nicotinic stimulation produces multiple forms of increased glutamatergic synaptic transmission. J Neurosci 1998;18:7075–83.

[110] Ji D, Dani JA. Inhibition and disinhibition of pyramidal neurons by activation of nicotinic receptors on hippocampal interneurons. J Neurophysiol 2000;83:2682–90.

[111] Alkondon M, Albuquerque EX. Nicotinic acetylcholine receptor alpha7 and alpha4beta2 subtypes differentially control GABAergic input to CA1 neurons in rat hippocampus. J Neurophysiol 2001;86:3043–55.

[112] Hatton GI, Yang QZ. Synaptic potentials mediated by alpha 7 nicotinic acetylcholine receptors in supraoptic nucleus. J Neurosci 2002;22:29–37.

[113] Fratiglioni L, Wang HX. Smoking and Parkinson's and Alzheimer's disease: review of the epidemiological studies. Behav Brain Res 2000;113:117–20.

[114] Sarter M, Parikh V. Choline transporters, cholinergic transmission and cognition. Nat Rev Neurosci 2005;6:48–56.

[115] Horn TF, Wolf G, Duffy S, Weiss S, Keilhoff G, MacVicar BA. Nitric oxide promotes intracellular calcium release from mitochondria in striatal neurons. FASEB J 2002;16:1611–22.

[116] Xia S, Cai ZY, Thio LL, Kim-Han JS, Dugan LL, Covey DF, et al. The estrogen receptor is not essential for all estrogen neuroprotection: new evidence from a new analog. Neurobiol Dis 2002;9:282–93.

[117] Zuo DY, Wu YL, Yao WX, Cao Y, Wu CF, Tanaka M. Effect of MK-801 and ketamine on hydroxyl radical generation in the posterior cingulate and retrosplenial cortex of free-moving mice, as determined by in vivo microdialysis. Pharmacol Biochem Behav 2007;86:1–7.

[118] Behrens MM, Ali SS, Dao DN, Lucero J, Shekhtman G, Quick KL, et al. Ketamine-induced loss of phenotype of fast-spiking interneurons is mediated by NADPH-oxidase. Science 2007;318:1645–7.

[119] Woo TU, Kim AM, Viscidi E. Disease-specific alterations in glutamatergic neurotransmission on inhibitory interneurons in the prefrontal cortex in schizophrenia. Brain Res 2008;1218:267–77.

[120] Prabakaran S, Swatton JE, Ryan MM, et al. Mitochondrial dysfunction in schizophrenia: evidence for compromised brain metabolism and oxidative stress. Mol Psychiatry 2004;9:684–97, 643.

[121] Prabakaran S, Wengenroth M, Lockstone HE, Lilley K, Leweke FM, Bahn S. 2-D DIGE analysis of liver and red blood cells provides further evidence for oxidative stress in schizophrenia. J Proteome Res 2007;6:141–9.

[122] Wang C, Sadovova N, Patterson TA, et al. Protective effects of 7-nitroindazole on ketamine-induced neurotoxicity in rat forebrain culture. Neurotoxicology 2008;29:613–20.

[123] Boultadakis A, Georgiadou G, Pitsikas N. Effects of the nitric oxide synthase inhibitor L-NAME on different memory components as assessed in the object recognition task in the rat. Behav Brain Res 2009;207:208–14.

[124] Klahr S. Can L-arginine manipulation reduce renal disease? Semin Nephrol 1999;19:304–9.

[125] Boctor SY, Ferguson SA. Altered adult locomotor activity in rats from phencyclidine treatment on postnatal days 7, 9 and 11, but not repeated ketamine treatment on postnatal day 7. Neurotoxicology 2010;31:42–54.

[126] Marcoux FW, Goodrich JE, Dominick MA. Ketamine prevents ischemic neuronal injury. Brain Res 1988;452:329–35.

[127] Himmelseher S, Durieux ME. Revising a dogma: ketamine for patients with neurological injury?. Anesth Analg 2005;101:524–34.

[128] Himmelseher S, Durieux ME. Ketamine for perioperative pain management. Anesthesiology 2005;102:211–20.

[129] Adachi M, Barrot M, Autry AE, Theobald D, Monteggia LM. Selective loss of brain-derived neurotrophic factor in the dentate gyrus attenuates antidepressant efficacy. Biol Psychiatry 2008;63:642–9.

[130] Ikonomidou C, Bosch F, Miksa M, et al. Blockade of NMDA receptors and apoptotic neurodegeneration in the developing brain. Science 1999;283:70–4.

[131] Hayashi H, Dikkes P, Soriano SG. Repeated administration of ketamine may lead to neuronal degeneration in the developing rat brain. Paediatr Anaesth 2002;12:770–4.

[132] Wang C, Sadovova N, Fu X, et al. The role of the N-methyl-D-aspartate receptor in ketamine-induced apoptosis in rat forebrain culture. Neuroscience 2005;132:967–77.

[133] Slikker Jr. W, Zou X, Hotchkiss CE, et al. Ketamine-induced neuronal cell death in the perinatal rhesus monkey. Toxicol Sci 2007;98:145–58.

[134] Yeung LY, Wai MS, Fan M, et al. Hyperphosphorylated tau in the brains of mice and monkeys with long-term administration of ketamine. Toxicol Lett 2010;193:189–93.

[135] Sun L, Lam WP, Wong YW, et al. Permanent deficits in brain functions caused by long-term ketamine treatment in mice. Hum Exp Toxicol 2011;30: 1287–96.

[136] Chan WM, Xu J, Fan M, et al. Downregulation in the human and mice cerebella after ketamine versus ketamine plus ethanol treatment. Microsc Res Tech 2011 doi:10.1002/jemt.21052.

[137] Garcia LS, Comim CM, Valvassori SS, et al. Ketamine treatment reverses behavioural and physiological alterations induced by chronic mild stress in rats. Prog Neuropsychopharmacol Biol Psychiatry 2009;33:450–5.

[138] Rezin GT, Gonçalves CL, Daufenbach JF, et al. Acute administration of ketamine reverses the inhibition of mitochondrial respiratory chain induced by chronic mild stress. Brain Res Bull 2009;79:418–21.

[139] Chen CF, Chapman BJ, Munday KA. The effect of althesin, ketamine or pentothal on renal function in saline loaded rats. Clin Exp Pharmacol Physiol 1985;12:99–105.

[140] Tan S, Chan WM, Wai MS, et al. Ketamine effects on the urogenital system – changes in the urinary bladder and sperm motility. Microsc Res Tech 2011;74:1192–8.

[141] Yeung LY, Rudd JA, Lam WP, Mak YT, Yew DT. Mice are prone to kidney pathology after prolonged ketamine addiction. Toxicol Lett 2009;191:275–8.

[142] Darrah DM, Griebling TL, Silverstein JH. Postoperative urinary retention. Anesthesiol Clin 2009;27:465–84.

[143] Kamphuis ET, Kuipers PW, van Venrooij GE, Kalkman CJ. The effects of spinal anesthesia with lidocaine and sufentanil on lower urinary tract functions. Anesth Analg 2008;107:2073–8.

[144] Kreutziger J, Frankenberger B, Luger TJ, Richard S, Zbinden S. Urinary retention after spinal anaesthesia with hyperbaric prilocaine 2% in an ambulatory setting. Br J Anaesth 2010;104:582–6.

[145] Blatt LK, Lashinger ES, Laping NJ, Su X. Evaluation of pressor and visceromotor reflex responses to bladder distension in urethane anesthetized rats. Neurourol Urodyn 2009;28:442–6.

[146] Matsuura S, Downie JW. Effect of anesthetics on reflex micturition in the chronic cannula-implanted rat. Neurourol Urodyn 2000;19:87–99.

[147] Yoshiyama M, Roppolo JR, Thor KB, de Groat WC. Effects of LY274614, a competitive NMDA receptor antagonist, on the micturition reflex in the urethane-anaesthetized rat. Br J Pharmacol 1993;110:77–86.

[148] Yoshiyama M, Roppolo JR, De Groat WC. Alteration by urethane of glutamatergic control of micturition. Eur J Pharmacol 1994;264:417–25.

[149] Tanaka H, Kakizaki H, Shibata T, Ameda K, Koyanagi T. Effects of a selective metabotropic glutamate receptor agonist on the micturition reflex pathway in urethane-anesthetized rats. Neurourol Urodyn 2003;22:611–6.

[150] Yoshiyama M, de Groat WC. Role of spinal metabotropic glutamate receptors in regulation of lower urinary tract function in the decerebrate unanesthetized rat. Neurosci Lett 2007;420:18–22.

[151] Nagabukuro H, Okanishi S, Doi T. Effects of TAK-802, a novel acetylcholinesterase inhibitor, and various cholinomimetics on the urodynamic characteristics in anesthetized guinea pigs. Eur J Pharmacol 2004;494:225–32.

[152] Tan S, Chan WM, Wai MS, et al. Ketamine effects on the urogenital system – changes in the urinary bladder and sperm motility. Microsc Res Tech 2011;74:1192–8.

[153] Wai MS, Chan WM, Zhang AQ, Wu Y, Yew DT. Long-term ketamine and ketamine plus alcohol treatments produced damages in liver and kidney. Hum Exp Toxicol 2011. doi:10.1177/0960327112436404.

[154] Yeung LY, Rudd JA, Lam WP, Mak YT, Yew DT. Mice are prone to kidney pathology after prolonged ketamine addiction. Toxicol Lett 2009;191:275–8.

[155] Chu PS, Ma WK, Wong SC, et al. The destruction of the lower urinary tract by ketamine abuse: a new syndrome?. BJU Int 2008;102:1616–22.

[156] Mak SK, Chan MT, Bower WF, et al. Lower urinary tract changes in young adults using ketamine. Urol 2011;186:610–4.

[157] Mason K, Cottrell AM, Corrigan AG, Gillatt DA, Mitchelmore AE. Ketamine-associated lower urinary tract destruction: a new radiological challenge. Clin Radiol 2010;65:795–800.

[158] Winstock AR, Mitcheson L, Gillatt DA, Cottrell AM. The prevalence and natural history of urinary symptoms among recreational ketamine users. BJU Int 2012. doi:10.1111/j.1464-410X.2012.11028.x.

[159] Purifoy JA, Holz RW. The effects of ketamine, phencyclidine and lidocaine on catecholamine secretion from cultured bovine adrenal chromaffin cells. Life Sci 1984;35:1851–7.

[160] Takara H, Wada A, Arita M, Sumikawa K, Izumi F. Ketamine inhibits 45Ca influx and catecholamine secretion by inhibiting 22Na influx in cultured bovine adrenal medullary cells. Eur J Pharmacol 1986;125: 217–24.

[161] Hara K, Minami K, Ueno S, et al. Up-regulation of noradrenaline transporter in response to prolonged exposure to ketamine. Naunyn Schmiedebergs Arch Pharmacol 2002;365:406–12.

[162] Ko YY, Jeong YH, Lim DY. Influence of ketamine on catecholamine secretion in the perfused rat adrenal medulla. Korean J Physiol Pharmacol 2008;12:101–9.

[163] Varano L, Laforgia V, De Vivo P. Effects of ketamine on the catecholamine containing cells of the adrenal gland of the lizard Podarcis s. sicula Raf. Res Commun Chem Pathol Pharmacol 1985;48:145–8.

[164] Lim DY, Kang TJ, Hong SP, et al. Influence of pentobarbital-Na on stimulation-evoked catecholamine secretion in the perfused rat adrenal gland. Korean J Intern Med 1997;12:163–75.

[165] Kudo M, Kudo T, Matsuki A. Pituitary-adrenal response to ketamine. Masui 1991;40:1107–12.

[166] Bailey CJ, Atkins TW, Matty AJ. Blood glucose and plasma insulin levels during prolonged pentobarbitone anaesthesia in the rat. Endocrinol Exp 1975;9:177–85.

[167] Hindlycke M, Jansson L. Glucose tolerance and pancreatic islet blood flow in rats after intraperitoneal administration of different anesthetic drugs. Ups J Med Sci 1992;97:27–35.

[168] Reyes Toso CF, Linares LM, Rodríguez RR. Blood sugar concentrations during ketamine or pentobarbitone anesthesia in rats with or without alpha and beta adrenergic blockade. Medicina (B Aires) 1995;55:311–6.

[169] Illera JC, González Gil A, Silván G, Illera M. The effects of different anaesthetic treatments on the adreno-cortical functions and glucose levels in NZW rabbits. J Physiol Biochem 2000;56:329–36.

[170] Cámara-Lemarroy CR, Guzmán-de la Garza FJ, Alarcón-Galván G, Cordero-Pérez P, Fernández-Garza NE. The effects of NMDA receptor antagonists over intestinal ischemia/reperfusion injury in rats. Eur J Pharmacol 2009;621:78–85.

[171] Guzmán-de la Garza FJ, Cámara-Lemarroy CR, Ballesteros-Elizondo RG, Alarcón-Galván G, Cordero-Pérez P, Fernández-Garza NE. Ketamine and the myenteric plexus in intestinal ischemia/reperfusion injury. Dig Dis Sci 2010;55:1878–85.

[172] Guzmán-De La Garza FJ, Cámara-Lemarroy CR, Ballesteros-Elizondo RG, Alarcón-Galván G, Cordero-Pérez P, Fernández-Garza NE. Ketamine reduces intestinal injury and inflammatory cell infiltration after ischemia/reperfusion in rats. Surg Today 2010;40:1055–62.

[173] Dalal PG, Taylor D, Somerville N, Seth N. Adverse events and behavioural reactions related to ketamine based anesthesia for anorectal manometry in children. Paediatr Anaesth 2008;18:260–7.

[174] Varadarajulu S, Tamhane A, Wilcox CM. Prospective evaluation of adjunctive ketamine on sphincter of Oddi motility in humans. J Gastroenterol Hepatol 2008;23:e405–9.

[175] Sun J, Wang XD, Liu H, Xu JG. Ketamine suppresses intestinal NF-kappa B activation and

proinflammatory cytokine in endotoxic rats. World J Gastroenterol 2004;10:1028–31.

[176] Slonim AD, Ognibene FP. Sedation for pediatric procedures, using ketamine and midazolam, in a primarily adult intensive care unit: a retrospective evaluation. Crit Care Med 1998;26:1900–4.

[177] Thompson JS, Brown SA, Khurdayan V, Zeynalzadedan A, Sullivan PG, Scheff SW. Early effects of tribromoethanol, ketamine/xylazine, pentobarbitol, and isoflurane anesthesia on hepatic and lymphoid tissue in ICR mice. Comp Med 2002;52:63–7.

[178] Bahrami S, Benisch C, Zifko C, Jafarmadar M, Schöchl H, Redl H. Xylazine-/diazepam-ketamine and isoflurane differentially affect hemodynamics and organ injury under hemorrhagic/traumatic shock and resuscitation in rats. Shock 2011;35:573–8.

[179] Krüger AD. Current aspects of using ketamine in childhood. Anaesthesiol Reanim 1998;23:64–71.

[180] Helmer KS, Cui Y, Chang L, Dewan A, Mercer DW. Effects of ketamine/xylazine on expression of tumor necrosis factor-alpha, inducible nitric oxide synthase, and cyclo-oxygenase-2 in rat gastric mucosa during endotoxemia. Shock 2003;20:63–9.

[181] Takahashi T, Kinoshita M, Shono S, et al. The effect of ketamine anesthesia on the immune function of mice with postoperative septicemia. Anesth Analg 2010;111:1051–8.

[182] Roytblat L, Talmor D, Rachinsky M, et al. Ketamine attenuates the interleukin-6 response after cardiopulmonary bypass. Anesth Analg 1998;87:266–71.

[183] Schmitz A, Portier CJ, Thormann W, Theurillat R, Mevissen M. Stereoselective biotransformation of ketamine in equine liver and lung microsomes. J Vet Pharmacol Ther 2008;31:446–55.

[184] Schmitz A, Thormann W, Moessner L, Theurillat R, Helmja K, Mevissen M. Enantioselective CE analysis of hepatic ketamine metabolism in different species in vitro. Electrophoresis 2010;31:1506–16.

[185] Lupp A, Kerst S, Karge E, Quack G, Klinger W. Investigation on possible antioxidative properties of the NMDA-receptor antagonists ketamine, memantine, and amantadine in comparison to nicanartine in vitro. Exp Toxicol Pathol 1998;50:501–6.

[186] Noppers I, Olofsen E, Niesters M, et al. Effect of rifampicin on S-ketamine and S-norketamine plasma concentrations in healthy volunteers after intravenous S-ketamine administration. Anesthesiology 2011;114:1435–45.

[187] Rofael HZ. Effect of ketamine pretreatment on cocaine-mediated hepatotoxicity in rats. Toxicol Lett 2004;152:213–22.

[188] Sato Y, Kobayashi E, Hakamata Y, et al. Chronopharmacological studies of ketamine in

normal and NMDA epsilon1 receptor knockout mice. Br J Anaesth 2004;92:859–64.

[189] Inaba H, Hayami N, Ajiki K, et al. Deep anesthesia suppresses ventricular tachyarrhythmias in rabbit model of the acquired long QT syndrome. Circ J 2010;75:89–93.

[190] Fontes-Sousa AP, Moura C, Carneiro CS, Teixeira-Pinto A, Areias JC, Leite-Moreira AF. Echocardiographic evaluation including tissue Doppler imaging in New Zealand white rabbits sedated with ketamine and midazolam. Vet J 2008;181:326–31.

[191] Bettschart-Wolfensberger R, Bowen IM, Freeman SL, Weller R, Clarke KW. Medetomidine-ketamine anaesthesia induction followed by medetomidine-propofol in ponies: infusion rates and cardiopulmonary side effects. Equine Vet J 2003;35:308–13.

[192] Akin A, Esmaoglu A, Guler G, Demircioglu R, Narin N, Boyaci A. Propofol and propofol-ketamine in pediatric patients undergoing cardiac catheterization. Pediatr Cardiol 2005;26:553–7.

[193] Sungur Ulke Z, Kartal U, Orhan Sungur M, Camci E, Tugrul M. Comparison of sevoflurane and ketamine for anesthetic induction in children with congenital heart disease. Paediatr Anaesth 2008;18:715–21.

[194] Reboso Morales JA, González Miranda F. Ketamine. Rev Esp Anestesiol Reanim 1999;46:111–22.

[195] Spotoft H, Korshin JD, Sørensen MB, Skovsted P. The cardiovascular effects of ketamine used for induction of anaesthesia in patients with valvular heart disease. Can Anaesth Soc J 1979;26:463–7.

[196] Jacobson JD, Hartsfield SM. Cardiorespiratory effects of intravenous bolus administration and infusion of ketamine-midazolam in dogs. Am J Vet Res 1993;54:1710–4.

[197] Schultetus RR, Paulus DA, Spohr GL. Haemodynamic effects of ketamine and thiopentone during anaesthetic induction for caesarean section. Can Anaesth Soc J 1985;32:592–6.

[198] Bone L, Battles AH, Goldfarb RD, Lombard CW, Moreland AF. Electrocardiographic values from clinically normal, anesthetized ferrets (Mustela putorius furo). Am J Vet Res 1988;49:1884–7.

[299] Chan WM, Xu J, Fan M, et al. Downregulation in the human and mice cerebella after ketamine versus ketamine plus ethanol treatment. Microsc Res Tech 2012;75:258–64.

[200] Harada K, Morimoto S. Inherited cardiomyopathies as a troponin disease. Jpn J Physiol 2004;54:307–18.

[201] Parmacek MS, Solaro RJ. Biology of the troponin complex in cardiac myocytes. Prog Cardiovasc Dis 2004;47:159–76.

[202] Parvatiyar MS, Pinto JR, Dweck D, Potter JD. Cardiac troponin mutations and restrictive cardiomyopathy. J Biomed Biotechnol 2010;2010:350706.

[203] Aarones M, Gullestad L, Aakhus S, et al. Prognostic value of cardiac troponin T in patients with moderate to severe heart failure scheduled for cardiac resynchronization therapy. Am Heart J 2011;161:1031–7.

[204] Kociol RD, Pang PS, Gheorghiade M, Fonarow GC, O'Connor CM, Felker GM. Troponin elevation in heart failure prevalence, mechanisms, and clinical implications. J Am Coll Cardiol 2010;56:1071–8.

[205] Jakobsen CJ, Torp P, Vester AE, Folkersen L, Thougaard A, Sloth E. Ketamine reduce left ventricular systolic and diastolic function in patients with ischaemic heart disease. Acta Anaesthesiol Scand 2010;54:1137–44.

[206] Parthasarathi G, Raman SP, Sinha PK, Singha SK, Karunakaran J. Ketamine has no effect on oxygenation indices following elective coronary artery bypass grafting under cardiopulmonary bypass. Ann Card Anaesth 2011;14:13–18.

[207] Deng CY, Yu XY, Kuang SJ, et al. Electrophysiological effects of ketamine on human atrial myocytes at therapeutically relevant concentrations. Clin Exp Pharmacol Physiol 2008;35:1465–70.

[208] Sigtermans M, Dahan A, Mooren R, et al. S(+)-ketamine effect on experimental pain and cardiac output: a population pharmacokinetic-pharmacodynamic modeling study in healthy volunteers. Anesthesiology 2009;111:892–903.

[209] Hudetz JA, Patterson KM, Iqbal Z, et al. Ketamine attenuates delirium after cardiac surgery with cardiopulmonary bypass. J Cardiothorac Vasc Anesth 2009;23:651–7.

[210] Jansen KL. A review of the non-medical use of ketamine: use, users and consequences. J Psychoactive Drugs 2000;32:419–33.

[211] Clements JA, Nimmo WS, Grant IS. Bioavailability, pharmacokinetics, and analgesic activity of ketamine in humans. J Pharm Sci 1982;71:539–42.

[212] Sinner. B, Garf. BM. Ketamine. Handb Exp Pharmacol 2008;182:313–33.

[213] United Nations Office on Drugs and Crime. World drug report 2012. Available: <http://www.unodc.org/documents/data-and-analysis/WDR2012/WDR_2012_web_small.pdf> [accessed 10.02.13].

[214] Drug Misuse Declared: findings from the 2011/12 Crime Survey for England and Wales, 2nd ed. Available: <http://www.homeoffice.gov.uk/publications/science-research-statistics/research-statistics/crime-research/drugs-misuse-dec-1112/> [accessed 10.02.13].

[215] NDSHS AIHW survey 2010. Available: <http://www.aihw.gov.au/WorkArea/DownloadAsset.aspx?id=10737421314&libID=10737421314> [accessed 10.02.13].

[216] NSDUH survey 2006. Available:<http://www.samhsa.gov/data/2k8/hallucinogens/hallucinogens.pdf> [accessed 10.02.13].

[217] Global Smart Update 2011. Available: <https://www.unodc.org/documents/scientific/Global_Smart_Update-5.pdf [accessed 10.02.11].

[218] Central Registry and Drug Abuse (Hong Kong). Available: <http://www.nd.gov.hk/en/drugstatistics.htm> [accessed 10.02.13].

[219] Dick D, Torrance C. Drugs Survey. MixMag 2010;225: 44–53.

[220] Winstock A. Global drug survey. MixMag 2012;251: 68–73.

[221] Wood DM, Measham F, Dargan. PI. 'Our favourite drug': prevalence of use and preference for mephedrone in the London night time economy one year after control. J Subst Use 2012;17:91–7.

[222] Barrett. SP, Gross. SR, Garand. I, Pihl. RO. Patterns of simultaneous polysubstance use in Canadian rave attendees. Subst Use Misuse 2005;40:1525–37.

[223] Winstock AR, Mitchseon L, Gillatt D, Cottrell AM. The prevalence and natural history of urinary symptoms among recreational ketamine users. BJU Int 2012;110:1762–6.

[224] Kalsi SS, Wood DM, Dargan PI. The epidemiology and patterns of acute and chronic toxicity associated with recreational ketamine use. Emerg Health Threats J 2011;4:1–9.

[225] Morgan CJA, Curran HV. Ketamine use: a review. Addiction 2011;107:27–38.

[226] Erowid: Ketamine. Available: <http://www.erowid.org/chemicals/ketamine/ketamine.shtml> [accessed 10.02.13].

[227] Weiner AL, Vieira L, McKay CA, Bayer MJ. Ketamine abusers presenting to the emergency department. J Emerg Med 2000;16:246–51.

[228] Stewart CE. Ketamine as a street drug. Emerg Med Serv 2001;30:2–4.

[229] Dillon P, Copelan J, Kansen K. Patterns of use and harms associated with non-medical ketamine use. Drug Alcohol Depend 2003;69:23–8.

[230] Morgan CJ, Meutzelfeldt L, Curran HV. Consequences of chronic ketamine self-administration upon neurocognitive function and psychological well being: a 1 year longitudinal study. Addiction 2010;105:121–33.

[231] Muetzelfeldt L, Kamboj SK, Rees H, Taylor J, Morgan CJ, Curran HV. Journey through the K-hole: phenomenological aspects of ketamine use. Druh Alcohol Depend 2005;95:219–29.

[232] Cheng JY, Chan DT, Mok VK. An epidemiological study on alcohol/drugs related fatal traffic crash cases of deceased drivers in Hong Kong between 196 and 2000. Forensic Sci Int 2005;153:196–201.

[233] Schifano F, Corkery J, Oyefenso A, Tonia T, Ghodse AH. Trapped in the K-hole: overview of deaths associated with ketamine misuse in the UK (1993-2006). J Clin Psychopharmacol 2008;28:114–6.

[234] Gill JR, Stajic M. Ketamine in non-hospital and hospital deaths in New York City. J Forensic Sci 2000;45:655–68.

[235] Shahani R, Streutker C, Dickson B, Stewart RJ. Ketamine-associated ulcerative cystitis: a new clinical entity. Urology 2007;69:810–2.

[236] Cottrell A, Warren K, Ayres R, Weinstock P, Kumar V, Gillatt D. The destruction of the lower urinary tract by ketamine abuse: a new syndrome? BJU Int 2008;102:1178–9.

[237] Gregoire MC, MacLellan DL, Finley GA. A pediatric case of ketamine-associated cystitis. Urology 2008;71:1232–3.

[238] Oxley JD, Cottrell AM, Adams S, Gillatt D. Ketamine cystitis as a mimic of carcinoma in situ. Histopathology 2009;55:705–8.

[239] Chu PS, Ma WK, Wong SC, Chu RW, Cheng CH, Wong S, et al. The destruction of the lower urinary tract by ketamine abuse: a new syndrome? BJU Int 2008;102:1616–22.

[240] Cottrell AM, Gillatt D. Consider ketamine misuse in patients with urinary symptoms. Practitioner 2008;252:5.

[241] Mason K, Cotrell AM, Corrigan AG, Gillatt DA, Mitchelmore AE. Ketamine-associated lower urinary tract destruction: a new radiological challenge. Clin Radiol 2010;65:795–800.

[242] Tsai TH, Cha TL, Lin CM, et al. Ketamine associated bladder dysfunction. Int J Urol 2009;55:705–8.

[243] Chu PS, Kwok KF, Lam KM, et al. Street ketamine associated bladder dysfunction: a report of ten cases. Hong Kong Med J 2007;13:311–3.

[244] Storr TM, Quibell R. Can ketamine prescribed for pain cause damage to the urinary tract? Palliat Med 2009;23:670–2.

[245] Poon TL, Wong KF, Chan MY, et al. Upper gastrointestinal problems in inhalational ketamine abusers. J Dig Dis 2010;11:106–10.

[246] Selby NM, Anderson J, Bungay P, Chesterton LJ, Kolhe NV. Obstructive nephropathy and kidney injury associated with ketamine abuse case report. Nephrol Dial Transplant Plus 2008;1:310–2.

[247] Kimura F, Hashimoto Y, Shimodate Y, Hashimoto H, Ishihara H, Matsuki A. Clinical study on total intravenous anesthesia with droperidol, fentanyl and ketamine. Hepatic and renal functions following prolonged surgical operation of over 10 hours in Japanese. Masui 1991;40:1371–5.

[248] Ng SH, Lee HK, Chan YC, Lau FL. Dilated common bile ducts in ketamine abusers. Hong Kong Med J 2009;15:157.

[249] Wong SW, Lee KF, Wong J, Ng WW, Cheung YS, Lai PB. Dilated common bile ducts mimicking choledochal cysts in ketamine abusers. Hong Kong Med J 2009;15:53–6.

[250] Lee ST, Wu TT, Yu PY, Chen RM. Apoptotic insults to human HepG2 cells induced by S-(+)-ketamine occurs through activation of a Bax-mitochondria-caspase protease pathway. Br J Anaesth 2009;102:80–9.

[251] Krystal JH, Karper LP, Seibyl JP, et al. Subanesthetic effects of the noncompetitive NMDA antagonist, ketamine, in humans. Psychotomimetic, perceptual, cognitive, and neuroendocrine responses. Arch Gen Psychiatry 1994;51:199–214.

[252] Driesen NR, McCarthy G, Bhagwagar Z, et al. Relationship of resting brain hyperconnectivity and schizophrenia-like symptoms produced by the NMDA receptor antagonist ketamine in humans. Mol Psychiatry 2013. Epub ahead of print.

[253] Lahti AC, Koffel B, LaPorte D, Tamminga CA. Subanesthetic doses of ketamine stimulate psychosis in schizophrenia. Neuropsychopharmacol 1995;13:9–19.

[254] Malhotra AK, Pinals DA, Adler CM, et al. Ketamine-induced exacerbation of psychotic symptoms and cognitive impairment in neuroleptic-free schizophrenics. Neuropsychopharmacol 1997;17:141–50.

[255] Curran HV, Morgan C. Cognitive, dissociative and psychotogenic effects of ketamine in recreational users on the night of drug use and 3 days later. Addiction 2000;95:575–90.

[256] Berman RM, Cappiello A, Anand A, et al. Antidepressant effects of ketamine in depressed patients. Biol Psychiatry 2000;47:351–4.

[257] Zarate Jr CA, Singh JB, Carlson PJ, et al. A randomized trial of an N-methyl-D-aspartate antagonist in treatment resistant major depression. Arch Gen Psychiatry 2006;63:856–64.

[258] aan het Rot M, Collins KA, Murrough JW, et al. Safety and efficacy of repeated-dose intravenous ketamine for treatment resistant depression. Biol Psychiatry 2009;67:139–45.

[259] Morgan CJ, Curran HV. Acute and chronic effects of ketamine upon human memory: a review. Psychopharmacology (Berl) 2006;188:408–24.

[260] Morgan CJ, Rossell SL, Pepper F, et al. Semantic priming after ketamine acutely in healthy volunteers and following chronic self-administration in substance users. Biol Psychiatry 2006;59:265–72.

[261] Morgan CJ, Perry EB, Cho HS, Krystal JH, D'Souza DC. Greater vulnerability to the amnestic effects of ketamine in males. Psychopharmacology (Berl) 2006;187:405–14.

[262] Narendran R, Frankle WG, Keefe R, Gil R, Martinez D, Slifstein M, et al. Altered prefrontal dopaminergic function in chronic recreational ketamine users. Am J Psychiatry 2005;162:2352–9.

[263] Liao Y, Tang J, Ma M, et al. Frontal white matter abnormalities following chronic ketamine use: a diffusion tensor imaging study. Brain 2010;133:2115–22.

[264] Liao Y, Tang J, Corlett PR, et al. Reduced dorsal prefrontal gray matter after chronic ketamine use. Biol Psychiatry 2011;69:42–8.

[265] Winger G, Hursh SR, Casey KL, Woods JH. Relative reinforcing strength of three N-methyl-D-aspartate antagonists with different onsets of action. J Pharmacol Exp Ther 2002;301:690–7.

[266] Meyer PJ, Phillips TJ. Behavioural sensitization to ethanol does not result in cross-sensitization to NMDA receptor antagonists. Psychopharmacology (Berl.) 2007;195:103–15.

[267] Lu Y, France CP, Woods JH. Tolerance to the cataleptic effect of the N-methyl-D-aspartate (NMDA) receptor antagonists in pigeons: cross-tolerance between PCP-like compounds and competitive NMDA antagonists. J Pharmacol Exp Ther 1992;263:499–504.

[268] Moreton JE, Meisch RA, Stark L, Thompson T. Ketamine self-administration by the rhesus monkey. J Pharmacol Exp Ther 1997;203:303–9.

[269] Cumming JF. The development of an acute tolerance to ketamine. Anesth Analg 1976;55:788–91.

[270] Bree MM, Feller I, Corssen G. Safety and tolerance of repeated anesthesia with CI 581 (ketamine) in monkeys. Anesth Analg 1967;46:596–600.

[271] Byer DE, Gould Jr. AB. Development of tolerance to ketamine in an infant undergoing repeated anesthesia. Anesthesiology 1981;54:255–6.

[272] Livingston A, Waterman AE. The development of tolerance to ketamine in rats and the significance of hepatic metabolism. Br J Pharmacol In Vitro 1978;64:63–9.

[273] Morgan CJ, Rees H, Curran HV. Attentional bias to incentive stimuli in frequent ketamine users. Psychol Med 2008;38:1331–40.

[274] Jansen KL, Darracot-Cankovic R. The nonmedical use of ketamine, part two: a review of problem use and dependence. J Psychoactive Drugs 2001;33:151–8.

[275] Pal HR, Berry N, Kumar R, Ray R. Ketamine dependence. Anaesth Intensive Care 2002;30:382–4.

[276] Hurt PH, Ritchie EC. A case of ketamine dependence. Am J Psychiatry 1992;151:779.

Synthetic Cannabinoid Receptor Agonists

Volker Auwärter[*], *Paul I. Dargan*[†] *and David M. Wood*[†]

[*]Institute of Forensic Medicine, Forensic Toxicology, University Medical Center Freiburg,
Freiburg, Germany
[†]Clinical Toxicology, Guy's and St Thomas' NHS Foundation Trust and King's Health Partners,
London, UK

INTRODUCTION

In 1965, Mechoulam and Gaoni [1] described the synthesis of Δ9-tetrahydrocannabinol (THC) and soon thereafter the first cannabinoid analogues were synthesised; initially these had a chemical structure very similar to THC. After the discovery of the endocannabinoid system and the cannabinoid receptors type 1 (CB_1) and type 2 (CB_2) in the early 1990s [2,3], research activities in this field increased tremendously, as it became possible to identify completely new structural types such as the aminoalkylindoles as potent cannabinoid receptor ligands. Since then, there has been constantly growing scientific interest in these compounds and hundreds of substances with high or medium affinity to one or both of the cannabinoid receptors have been synthesised.

Cannabinoid receptor agonists and antagonists can potentially be used for treating a multitude of diseases and disorders. Applications considered for CB_1 agonists include the following diseases and effects (overviews: [4–6]):

- multiple sclerosis (muscle relaxation)
- pain (analgesia)
- cancer (anti-emetic and appetite-stimulating effect during chemotherapy) and
- glaucoma (lowering intraocular pressure).

Type 1 cannabinoid receptor antagonists have been successfully used in the treatment of obesity, although they can have adverse psychiatric side-effects. For example rimonabant (Acomplia®), a diarylpyrazole, was withdrawn from the market soon after its introduction because of an unfavourable benefit-risk assessment (depression and an increased rate of suicide had been reported as adverse effects) [7].

For several years, cannabinoid research has concentrated on selective CB_2 agonists as they may be of therapeutic use (e.g. analgesia) without having any major central side-effects mediated via the CB_1 receptors [6].

Novel Psychoactive Substances.
DOI: http://dx.doi.org/10.1016/B978-0-12-415816-0.00013-4

317

Up to 2008 only a few experimental drug users had access to synthetic cannabinoid receptor agonists and they were not used as recreational drugs on a larger scale; although there is evidence that they have been available on the Internet since at least 2005/6 [8]. The situation changed fundamentally when the first products consisting of herbal mixtures and synthetic cannabinoids became increasingly available and were sold as a natural substitute for cannabis [9,10]. The first products of this type were marketed in Europe under the name 'Spice', which has become the generic term for a meanwhile uncountable number of products and brands; the term 'K2' is often used in North America for the same products/compounds. Often, there is limited or no information on the active ingredients in the packaging of these products and the contents can vary greatly, both in the actual synthetic cannabinoid receptor agonists present and their quantity [11–15]. In the meantime, many synthetic cannabinoid receptor agonists have been classified in a number of countries around the world (see Chapter 1 for more information), under either narcotics or medicines acts, but it is very difficult to control their sale via the Internet and these products continue to be widely available [16–18].

Synthetic cannabinoid receptor agonists are mostly sold as either incense or smoking mixtures. These are composed of plant materials treated with synthetic cannabinoids, e.g. by spraying with a suspension/solution containing the active ingredient(s) and subsequent drying [19]. There are also sellers marketing the active ingredients as 'research chemicals' under their chemical name as pure substances. These are often crystalline powders of relatively high purity [20]. Most of them are ordered online and sent by post (often across international borders) [20,21].

In the last four years, the European Monitoring Centre for Drugs and Drug Addiction (EMCDDA) reported record numbers of new psychoactive substances (NPS) each year; 24 substances in 2009, 41 substances in 2010, 49 substances in 2011 and 73 substances in 2012. Synthetic cannabinoid receptor agonists have been the most common substance amongst these, amounting for approximately half of all of the NPS reported. The number of branded or 'end products' is even higher as meanwhile a multitude of smaller sellers have entered this market (the amount of active ingredients required is small and does not appear to significantly influence the price); prices are currently generally in the region of 20–30 Euros (approximately UK£ 17–26 or US$ 26–39) per unit product (3 grams) [8]. This highly dynamic development is potentially due to the pressure exercised on the market by successive steps to ban the substances. Figure 13.1 shows the results of blood samples analyzed in the Institute of Forensic Medicine Freiburg, Germany, between May 2010 and March 2012. It is evident that manufacturers and sellers are already reacting to the recommendations of the German Expert Committee for Narcotics to reclassify a substance under the Narcotics Act (BtMG) many months before it is actually reclassified. It is clearly seen that the market gradually shifts from substances that are about to be banned to 'legal' alternatives. Similar observations have been made in Japan, Australia and other countries [17,22,23].

When monitoring the products sold on the market within the EU project 'SPICE and synthetic cannabinoids' it became apparent that active ingredients gradually became more potent, and 'exotic' substituents such as adamantyl or tetramethylcyclopropyl groups were increasingly introduced to products available on the market. This development is alarming as it may potentially further increase the risk of highly toxic compounds entering the market. In other countries similar observations were made [24,25]. In the UK, shortly after implementation of a classification using a generic approach covering a wide range of chemical structure types and many of their derivatives, new substances emerged that were not covered by the legislation. However, classified substances usually do not instantly disappear from the market [25].

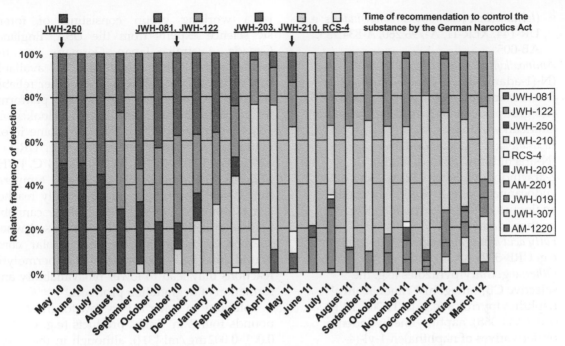

FIGURE 13.1 Relative frequency of selected synthetic cannabinoids detected in serum samples (n = 734) between May 2010 and March 2012. The arrows mark the time the Expert Committee for Narcotics recommended to classify the substance under the Narcotics Act. (For color version of this figure, the reader is referred to the online version of this book).

PHARMACOLOGY

Physical and Chemical Description

The term 'synthetic cannabinoids' covers all synthetic substances binding to one of the two known cannabinoid receptors (CB_1 or CB_2) and their structural/chemical analogues. Thus synthetic cannabinoids include compounds with completely different chemical structures and can be classified as follows: [26–28]

- *Classical cannabinoids*: THC and other plant-derived cannabinoids as well as substances with a related chemical structure such as HU-210, AM-411, AM-906, O-1184
- *Non-classical cannabinoids*: Cyclohexylphenoles or 3-arylcyclohexanoles such as CP-47,497-C8, CP-55,940 or CP-55,244

- *Aminoalkylindoles* (AAI), which can be sub-classified e.g. as follows:
  - *naphthoylindoles*: e.g. JWH-007, JWH-015, JWH-018, JWH-019, JWH-022, JWH-073, JWH-081, JWH-122, JWH-200, JWH 210, JWH-213, JWH-387, JWH-398, AM-1220, AM-2201, MAM-2201, EAM-2201, AM-2232, WIN-55,212
  - *phenylacetylindoles*: e.g. JWH-203, JWH-250, JWH-251, RCS-8
  - *benzoylindoles*: e.g. WIN-48,098, RCS-4, AM-694, AM-1241, AM-2233
  - *adamantoylindoles*: e.g. AB-001 (1-pentyl-3-(1-adamantoyl)indole), AM-1248
  - *(N-adamantyl-carboxamide)-indoles*: e.g. APICA (N-(1-adamantyl)-1-pentyl-1H-indole-3-carboxamide), STS-135 (1-adamantyl)-1-(5-fluoropentyl)-1H-indole-3-carboxamide)

- *(tetramethylcyclopropylcarbonyl)indoles*: e.g. UR-144, XLR-11, A-796,260, A-834,735, AB-005
- *Aminoalkylindazoles*: e.g. APINACA (N-(1-adamantyl)-1-pentyl-1H-indazole-3-carboxamide), APINACA-5F; further sub-groups in analogy to the group of the aminoalkylindoles
- *Eicosanoides*: e.g. anandamide (N-arachidonoylethanolamide), 2-arachidonoylglycerol (2-AG) and other endocannabinoids as well as synthetic analogues of these compounds like methanadamide
- *Fatty acid amide hydrolase (FAAH) inhibitors*: e.g. URB-597
- *Other*: e.g. diarylpyrazoles (e.g. the selective CB_1-antagonist rimonabant), naphthoylpyrroles (e.g. JWH-307, JWH-030, JWH-368), naphthylmethylindenes, or derivatives of naphthalen-1-yl-(4-pentyloxynaphthalen-1-yl)methanone (CRA13 or CB-13).

Figure 13.2 gives an overview of the basic structures of these substance classes with some examples for each class.

Most of the compounds have been identified in ready-to-smoke products, but some have occurred only as pure substances in powder form. Pure substances are typically offered as 'research chemicals' and are often available in larger quantities (up to kg scale) [29].

In pure form, most of the compounds are white crystalline powders. Melting points are not available in the original literature. However, some data is available in data sheets available at www.swgdrug.org. These range from 55°C to 145°C (5-fluoro-AKB48: 55°C, AKB48: 64°C, UR-144: 68°C, XLR-11: 73°C, 5-chloro-UR-144: 79°C, JWH-019: 90°C, 5-bromo-UR-144: 93°C, AM-2201: 94°C, JWH-073: 99.8°C, A796,260: 135°C, STS-135: 139°C, AM-1248: 143°C, AM2233: 145°C). The 'Scientific Working Group for the Analysis of Seized Drugs'(SWGDRUG)

is a working group consisting of foren- sic science experts from the USA, England, Canada, Australia, Japan, Germany and the Netherlands. Boiling points are not available in the scientific literature or from other reliable sources. However, Internet trading platforms for 'research chemicals' such as www.lookchem. com or www.guidechem.com give some infor- mation in substance data sheets (e.g. UR-144: 426°C, JWH-203: 498°C, JWH-073: 524°C, JWH-122: 557°C, JWH-200: 602°C). As the tempera- ture inside a burning cigarette usually reaches a maximum of 800–900°C [30], these cannabi- noids are likely to be readily vapourised when smoked, although for the more polar com- pounds like JWH-200 or AM-1220, thermolytic reactions may lead to lower bioavailability and formation of degradation products.

Solubility in water is very low for most com- pounds found in 'Spice' products (e.g. CRA13: 0.001–0.002 mg/ml [31]), although in the scien- tific literature a number of water soluble can- nabinoids have been described [32,33]. These may be used to produce preparations for intra- venous application and are of high value for sci- entific research, but are less likely to be suitable for smoking. The rather non-polar compounds usually occurring in 'Spice' products are solu- ble in ethanol or less polar solvents like acetone or ethyl acetate. However, a few compounds carrying polar groups like the 1-methylpiperi- din-2-yl-methyl moiety (AM-1220, AM-2233, AM-1248) or the 2-morpholino-4-ylethyl moiety (WIN 48,098, JWH-200, WIN 55,212-2) have also appeared in 'Spice' products and show similar solubility in more polar solvents.

Typical reactions for synthesis of aminoalky- lindoles – which are the most prevalent com- pounds found in herbal products – include Friedel-Crafts acylation at C3 followed by N-alkylation of a substituted indole or vice versa [34]. Common precursors are 1-alkyl- indoles / 1-alkyl-2-methylindoles (alkyl: butyl, pentyl, hexyl or others) and 1-naphthoyl chlo- rides (substituted at C4 in some cases) [35].

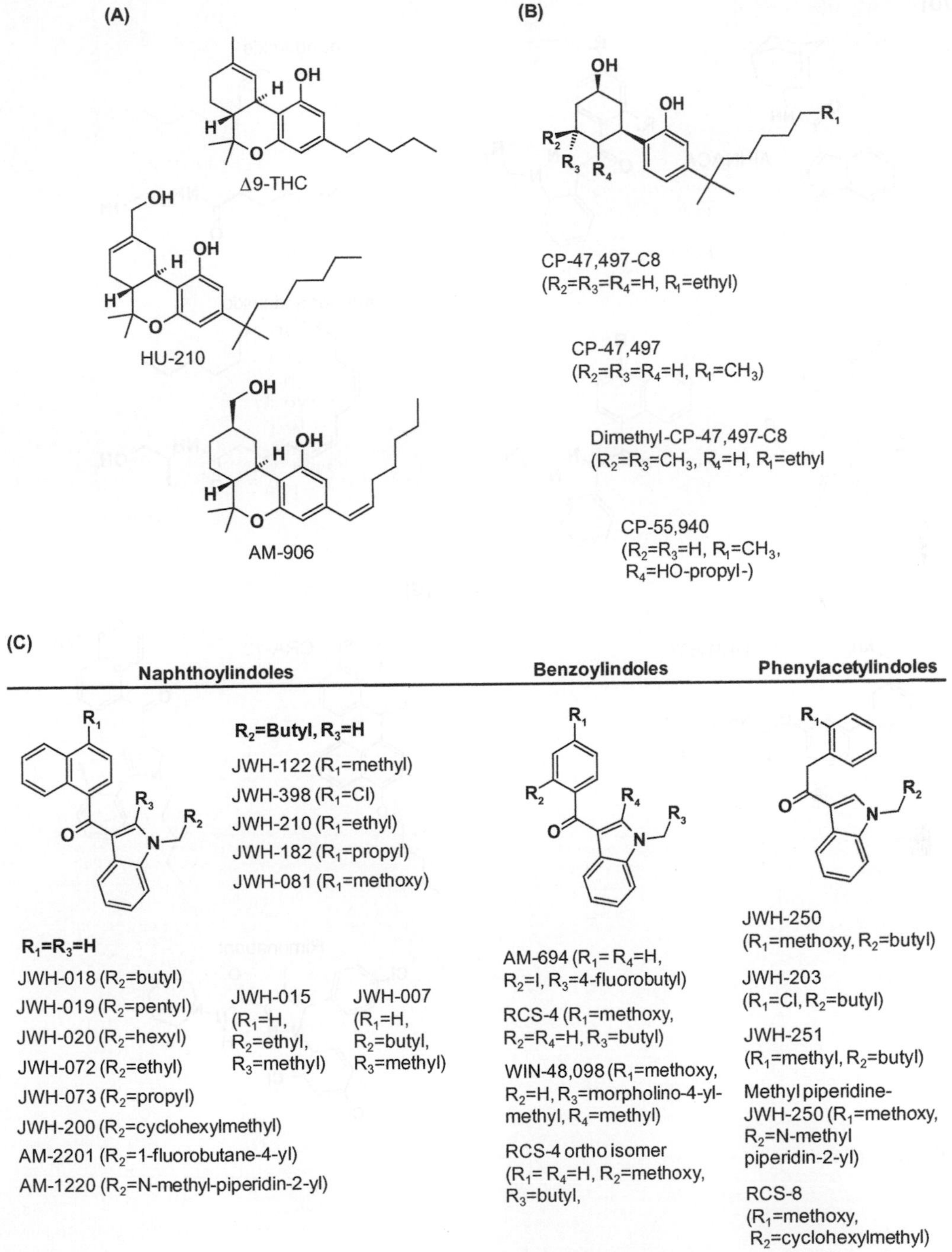

(A)

Δ9-THC

HU-210

AM-906

(B)

CP-47,497-C8
(R₂=R₃=R₄=H, R₁=ethyl)

CP-47,497
(R₂=R₃=R₄=H, R₁=CH₃)

Dimethyl-CP-47,497-C8
(R₂=R₃=CH₃, R₄=H, R₁=ethyl

CP-55,940
(R₂=R₃=H, R₁=CH₃,
R₄=HO-propyl-)

(C)

Naphthoylindoles	Benzoylindoles	Phenylacetylindoles

R₂=Butyl, R₃=H

JWH-122 (R₁=methyl)

JWH-398 (R₁=Cl)

JWH-210 (R₁=ethyl)

JWH-182 (R₁=propyl)

JWH-081 (R₁=methoxy)

R₁=R₃=H

JWH-018 (R₂=butyl)

JWH-019 (R₂=pentyl)

JWH-020 (R₂=hexyl)

JWH-072 (R₂=ethyl)

JWH-073 (R₂=propyl)

JWH-200 (R₂=cyclohexylmethyl)

AM-2201 (R₂=1-fluorobutane-4-yl)

AM-1220 (R₂=N-methyl-piperidin-2-yl)

JWH-015
(R₁=H,
R₂=ethyl,
R₃=methyl)

JWH-007
(R₁=H,
R₂=butyl,
R₃=methyl)

AM-694 (R₁= R₄=H,
R₂=I, R₃=4-fluorobutyl)

RCS-4 (R₁=methoxy,
R₂=R₄=H, R₃=butyl)

WIN-48,098 (R₁=methoxy,
R₂=H, R₃=morpholino-4-yl-
methyl, R₄=methyl)

RCS-4 ortho isomer
(R₁= R₄=H, R₂=methoxy,
R₃=butyl,

JWH-250
(R₁=methoxy, R₂=butyl)

JWH-203
(R₁=Cl, R₂=butyl)

JWH-251
(R₁=methyl, R₂=butyl)

Methyl piperidine-
JWH-250 (R₁=methoxy,
R₂=N-methyl
piperidin-2-yl)

RCS-8
(R₁=methoxy,
R₂=cyclohexylmethyl)

FIGURE 13.2 Basic chemical structures of synthetic cannabinoids. (A) Classical cannabinoids; (B) non-classical cannabinoids; (C) aminoalkylindoles; (D) aminoalkylindazoles; (E) eicosanoids; (F) eicosanoids; (G) others.

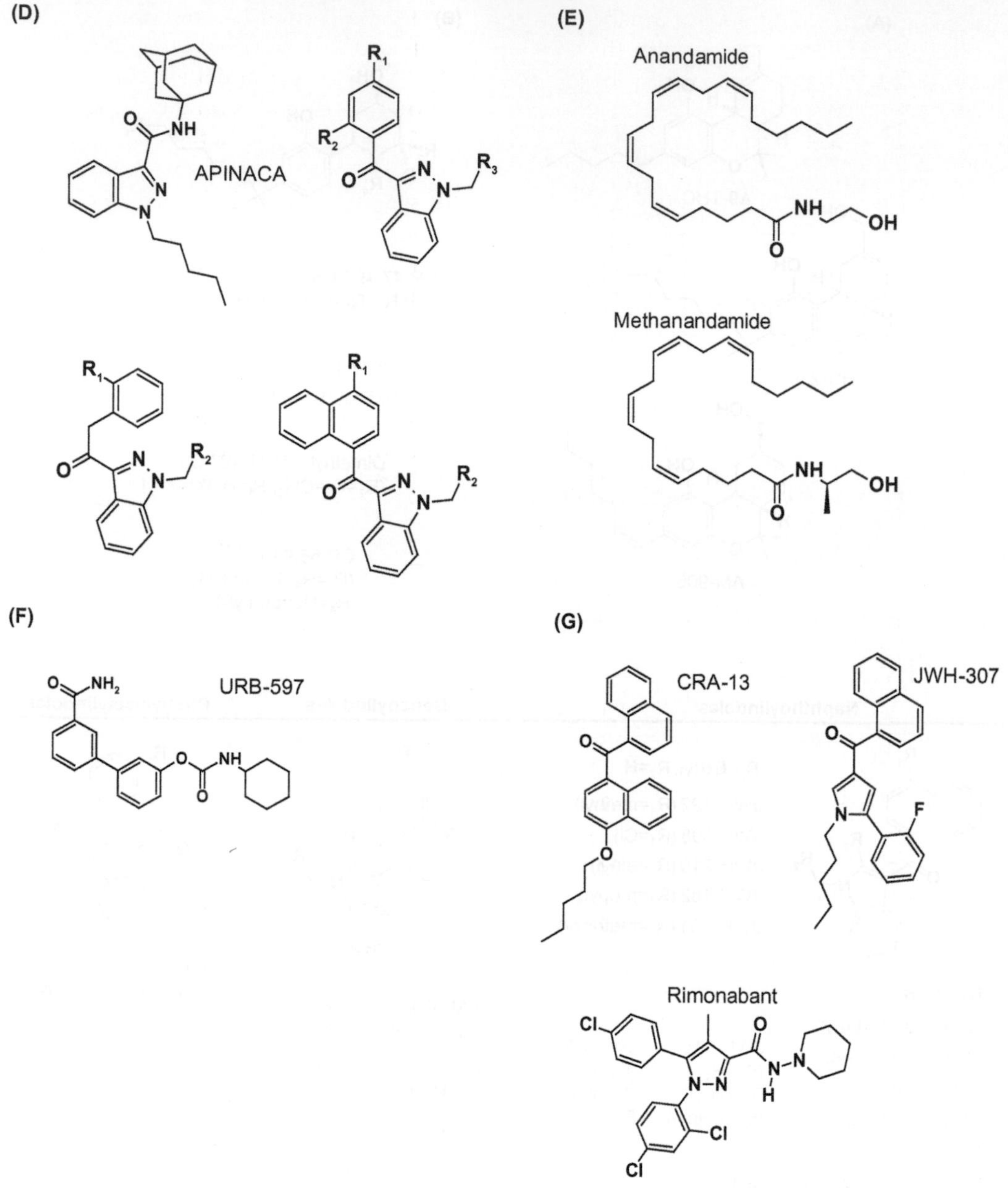

(D)

APINACA

(E)

Anandamide

Methanandamide

(F)

URB-597

(G)

CRA-13

JWH-307

Rimonabant

FIGURE 13.2   (Continued).

Two classes of synthetic cannabinoids deserve a closer look because of the thermolytic reactions which may occur when they are smoked in a joint:

1. 2,2,3,3,-Tetramethylcyclopropane derivatives (e.g. UR-144).

The cyclopropane ring undergoes thermolytic ring opening to produce a ring-opened isomer in the case of UR-144 [36]. This isomer often makes up the major part detected, e.g. in serum samples of UR-144 users (own unpublished data), which at least in part may also be a consequence of the ring-opened species being present in ready-to-smoke mixtures – most probably as an undesired by-product from synthesis. The same reaction can be expected for structurally related cannabinoids carrying this substituent.

Another compound detected in 'Spice' products containing UR-144 is the product of the electrophilic addition of water to the cyclopropyl moiety [37].

Currently no data on the biological activity of either of these compounds is available; however, they may be involved in producing the cannabis-like effects seen after smoking UR-144. This seems likely as the binding affinity of UR-144 itself to the CB_1 receptor suggests low potency.

2. N-fluoroalkyl compounds carrying the fluorine at the terminal C atom (e.g. 5-fluoropentylindoles like AM-2201).

It has been shown that these compounds undergo thermally induced reactions under smoking conditions leading to unsaturated (elimination of hydrofluoric acid) and de-fluorinated compounds (fluorine/hydrogen exchange) [38]. Smoking products containing AM-2201 can therefore lead to artefactual generation of JWH-022 and JWH-018.

Recently, a substance combining the structural elements of the 2,2,3,3,-tetramethylcyclopropyl and 5-fluoropentyl moieties was identified in a 'Spice' product seized in South Korea [39].

Another noteworthy finding is the presence of azepane isomers in preparations containing synthetic cannabinoids carrying an (N-methylpiperidin-2-yl)methyl moiety like AM-1220 or AM-2233, which can be explained as a side product in synthesis of these compounds[15,40]. Again, the relative content of the isomers greatly varies from product to product and it is unknown to which extent the isomers alter either the desired or adverse effects of the smoked drug.

Herbal Components of 'Spice' Products

Although labels on the packages sometimes indicate that the products contain some rare plants like 'Siberian motherwort' (*Leonurus sibiricus*), 'Dwarf skullcap' (*Scutellaria nana*), 'Indian warrior' (*Pedicularis densiflora*) or 'Wild dagga' (*Leonotis leonurus*), which are potentially psychoactive, these have not been detected in these products to date. In addition it seems rather unlikely that they would produce clearly perceptible effects in the amounts usually smoked in a joint. However, in some herbal products sold as 'legal highs', substances like the monoamine oxidase inhibitors, harmine and harmaline [11] or mescaline [41], have been detected in products that were free of synthetic cannabinoids. These are known plant constituents of, for example, *Peganum harmala* and *Banisteriopsis caapi* (harmine, harmaline) or *Lophophora williamsii* and *Trichocereus pachanoi* (mescaline) and may well produce effects when sufficient amounts are smoked. In other products declared to contain 'kratom' (*Mitragyna speciosa*), the synthetic opioid O-desmethyltramadol was found as an active ingredient [11].

However, one of the components often declared as ingredient of 'Spice' products called Damiana (*Turnera diffusa*) has been identified recently in various brands and seems to be one of the major constituents of the majority of products (data from BKA [BKA stands for "Bundeskriminalamt" (German Federal Criminal Police Office).], personal communication, accepted for publication in *Forensic Toxicology* and Ogata et al. [42]). Damiana contains, among others, a number of monoterpene compounds (like pinene and thymol) and arbutin, which may have some pharmacological activity when smoked, but are unlikely to produce significant psychotropic effects. It is noteworthy that Damiana is reportedly used as a substitute for tobacco, particularly for preparing cannabis joints, and the price is rather low (approximately 30–40 EUR (UK£ 26–34 or US$ 39–52) per kg). Only a few other plant species present in 'Spice' products have been described in the literature (e.g. Lamiaceae herbs like *Melissa*, *Mentha* and *Thymus* species [42]) and it seems reasonable that a variety of smokeable herbs available at low prices are likely to be arbitrarily used.

To summarise, although additional effects produced by herbal constituents of the products cannot be excluded, in most cases the synthetic cannabinoid receptor agonists can be regarded as the main pharmacologically active ingredients.

Pharmacokinetics

So far, the pharmacokinetic profiles of synthetic cannabinoid receptor agonists used as active ingredients in 'Spice' products are largely unknown [43], however, there is evidence for oral and inhalational bioavailability derived from clinical cases [44], self experiments [45–47] and clinical studies [31].

After oral doses of 10 mg of AM-694 [46], 26 mg of AB-001 [47] or 5 mg of AM-2201 (own unpublished data) no noticeable effects were seen in humans, whereas the same amount smoked led to typical symptoms of intoxication [9]. After smoking not more than 2 mg of CP-47,497-C8, a comparably potent CB_1 receptor agonist, considerable physical and mild psychotropic effects were obtained. In a clinical trial involving oral administration of varying amounts of the synthetic cannabinoid CRA13 in fasted subjects or after high-fat and high-calorie breakfast (1–80 mg, n = 48 [31]), higher rates of adverse effects including dizziness, headache and nausea were only seen at higher doses (i.e. 40 and 80 mg) and occurred more often after the high-fat/high-calorie breakfast – a phenomenon that was also seen in other lipophilic drugs and may be explained by solubilisation effects and enhancement of intestinal lymphatic transport [48]. Maximum plasma concentrations (C_{max}) occurred 1.5–2 hours after administration and linear pharmacokinetics were observed (i.e. C_{max} and AUC increased with dose). The terminal elimination half-lives ranged, on average, between 21 and 34 hours. Enterohepatic recirculation of the drug was postulated as in a number of subjects a double-peak in the concentration-time profile emerged (second peak around six hours post-dose). Concentrations decreased in a biphasic manner, reflecting distribution of the drug into deeper compartments as it can be expected in the case of highly lipophilic drugs.

In an online anonymous survey, 'Spice' users reported that 'synthetic cannabis' had a shortened time to peak effect (within 5 minutes) compared to 'natural cannabis' (6–10 minutes), a shorter duration (two-thirds feeling 'stoned' for one hour after smoking 'synthetic cannabis' compared to three-quarters feeling the effects of 'natural cannabis' for two hours or longer) [49].

It appears that drug availability when smoking 'Spice' products is comparable to THC availability in a joint. Experiments with smoking apparatus showed that main stream and side stream smoke from a joint consisting of 500 mg tobacco and 500 mg of a herbal mixture (178 mg/g JWH-210, 63 mg/g JWH-122 and 14.5 mg/g JWH-018) contained approximately equal amounts of active compounds (own unpublished data). The amount of substance recovered from main stream and side stream smoke made up approximately 50% of the total substance present in the joint.

The available data suggests that elimination of synthetic cannabinoid receptor agonists follows a biphasic kinetic pattern similar to

THC [50]. After smoking, initially high serum concentrations occur which drop quickly due to primary distribution from the central compartment into other tissues. For JWH-018, Teske et al. reported a decrease to values below 10% of C_{max} within 3 hours [45]. Similar behaviour can be expected for other, structurally related compounds. In contrast, terminal elimination half-lives after continued and high dose misuse have been estimated to be in the range of up to several weeks, pointing towards extensive distribution into deeper compartments [51] and slow redistribution after cessation of use. Again, similar observations can be made regarding serum THC concentrations after chronic cannabis abuse, where terminal elimination half-lives were estimated in a similar range [52–54].

Although many of the synthetic cannabinoid receptor agonists available in 'Spice' products have not been fully characterised regarding their metabolism, there is data for a number of representatives demonstrating extensive oxidative metabolism. In general, among the main phase I metabolites are the monohydroxylated, carboxylated (at the N-alkyl chain) and desalkylated compounds. In naphthoylindoles like JWH-015 [55], JWH-018 [56–58] and AM-2201 [59,60] also dihydrodiol metabolites (formed after epoxidation of the naphthyl moiety), ketones (oxidation of the N-alkyl chain) and dihydroxylated compounds alongside further minor metabolites were detected. In case of the benzoylindoles AM-694 [46] and RCS-4 [61] and the phenylacetylindole JWH-250 [62] similar observations were made, with the exception of dihydrodiols, which were not detected.

The substances fluorinated at the terminal N-alkyl position like AM-694 and AM-2201 undergo *in vitro* enzymatic defluorination/oxidation in addition to the reactions mentioned above [46,59,60].

The metabolic transformation of the 2,2,3,3-tetramethyl cyclopropyl moiety present, for example in UR-144 and XLR-11, has not been systematically investigated. However, for UR-144, mono-hydroxylated metabolites seem to be most abundant *in vitro* and *in vivo* [59]. In these compounds oxidative ring opening may occur in addition to the already reported metabolic reactions, leading to highly reactive unsaturated compounds.

Grigoryev et al. carried out metabolic studies for AB-001, a synthetic cannabinoid receptor agonist carrying an adamantoyl moiety, and found similar patterns of metabolic reactions with monohydroxylation also occurring at the adamantane unit [47].

Hutter et al. investigated urine samples of patients with confirmed consumption of seven different cannabinoids [63] and were able to show that monohydroxylation either at the N-alkyl chain, the naphthoyl moiety or the indole moiety, together with carboxylation are the metabolic reactions leading to the most valuable target analytes in urine for detection of drug consumption using LC-MS/MS techniques.

Chimalakonda et al. incubated AM-2201 and JWH-018 with different cytochrome P450 enzymes and identified CYP2C9 and CYP1A2 as the major isoforms involved in oxidative metabolic reactions [60]. CYP1A2 also plays a major role in metabolism of AM-694, RCS-4 and JWH-122, but in these compounds also CYP2D6 (RCS-4) and CYP3A5 (JWH-122, AM-694) contribute significantly (own unpublished data).

The only available data on phase II metabolism shows that JWH-018 and JWH-073 are conjugated to glucuronic acid effectively via various UDP-glucuronosyltransferase (UGT) enzymes. Due to the fact that most of the metabolites are extensively transformed to ester or ether glucuronides before excretion in bile or urine [64], enzymatic cleavage is a prerequisite for sensitive detection of the aforementioned oxidative metabolites, particularly in urine samples.

Pharmacodynamics

The so-called endocannabinoid system consists of receptor structures and pertinent endogenous substances called endocannabinoids which bind to these receptor structures, as well as subsequent signal transduction mechanisms. So far, CB_1 and CB_2 receptors have been identified in human tissue: [2,3]

- CB_1 receptors are found in high density in the brain and the spinal cord
- CB_2 receptors are primarily located in the spleen and in cells of the immune system.

Both are G protein-coupled receptors, stimulation results in inhibition of adenylate cyclase and reduction of cellular concentrations of cyclic adenosine monophosphate (cAMP) [65,66]. Moreover, interactions with calcium channels have been demonstrated [67,68]. Known endogenous ligands binding to and activating CB_1 and CB_2 receptors include the eicosanoids anandamide (arachidonylethanolamide) and 2-arachidonoylglycerol [69–71]. These lipids derived from arachidonic acid are involved in a large number of control mechanisms in brain metabolism, but also in the peripheral tissues [5].

The endocannabinoid system can interact with various signal transduction paths and receptor families and influence a multitude of physiological processes. It is involved in numerous processes of homoeostasis, among them the regulation of blood pressure, the sleep–wake cycle and the so-called reward system [72]. Potential therapeutic uses include analgesia, spasmolysis, immunosuppression, control of intraocular pressure, and control of appetite, memory and brain development.

The CB_1 receptors responsible for the psychotropic effects of cannabinoids are primarily localised pre-synaptically. Principal effects of their activation are reduced cellular excitability and the reduced release of several neurotransmitters including dopamine, noradrenaline, glutamate and serotonin [73–77]. This inhibition of neurotransmission allows cannabinoid receptor agonists to have profound impact on neuronal communication [78].

In contrast, the CB_2 receptors are responsible for immunomodulatory and, at least partially, also for the analgesic properties of cannabinoids [79]. It has also been demonstrated that palmitoylethanolamide mediates cannabis-like effects without binding to CB_1 or CB_2 receptors [80]. This suggests that there are likely to be further cannabinoid receptors, although these have not been identified to date [81]. Furthermore, synthetic cannabinoid receptor agonists may also interact with non-cannabinoid receptor targets, although such interaction was not described so far and was even excluded for some compounds [82].

Typical effects of CB_1 agonists include: [5]

- sedation
- cognitive dysfunction
- tachycardia
- postural hypotension
- dry mouth
- ataxia
- immunosuppression
- psychotropic effects.

Most synthetic cannabinoid receptor agonists found in 'Spice' products to date show a markedly higher affinity for the CB_1 receptor than THC (Table 13.1 [83–93]) and are full agonists at this site [78,94], unlike THC, which acts as a partial agonist [95]. The resulting overstimulation of the receptors is presumably the reason why individuals who develop acute toxicity associated with use of these products develop not only symptoms typically associated with cannabis intoxication such as tachycardia, sedation, cognitive deficits, psychotic experiences, anxiety, panic attacks and reddened eyes, but often further effects have been observed which are considered rather untypical for cannabis (among them agitation, seizures, hypertension, nausea and intractable vomiting) [44,96–99].

Although no data from clinical studies are available for most of the synthetic cannabinoid receptor agonists used in 'Spice' products, it

TABLE 13.1 Synthetic Cannabinoids and their CB_1-Receptor Binding Affinities [83–93]

Substance	Sum Formula	CB1-Receptor Affinity K_I [nM]
THC	$C_{21}H_{30}O_2$	40.7 ± 1.7
1-(5-Chloropentyl)-3-(2-iodobenzoyl)indole	$C_{20}H_{19}INOCl$	?
4-Methoxyphenyl-(1-butyl-1H-indol-3-yl) methanone	$C_{20}H_{21}NO_2$	?
AB-001 (JWH-018 adamantyl derivative)	$C_{24}H_{31}NO$	?
AM-1220	$C_{26}H_{26}NO_2$	0.75
AM-1220 azepane derivative	$C_{26}H_{26}NO_2$	?
AM-1248	$C_{26}H_{34}N_2O$	11.9
AM-2201	$C_{24}H_{22}FNO$	1.0
AM-2232	$C_{24}H_{20}N_2O$	0.28
AM-2233	$C_{22}H_{23}IN_2O$	2.8
AM-694	$C_{20}H_{19}FINO$	0.08
CP-47,497-C8	$C_{22}H_{36}O$	4.7
CRA-13	$C_{26}H_{24}O_2$	6.1 ± 1.1
HU-210	$C_{25}H_{38}O_3$	0.73 ± 0.11
JWH-007	$C_{25}H_{25}NO$	9.5 ± 4.5
JWH-015	$C_{23}H_{21}NO$	336 ± 36
JWH-018	$C_{24}H_{23}NO$	9 ± 5
JWH-019	$C_{25}H_{25}NO$	9.8 ± 2
JWH-073	$C_{23}H_{21}NO$	8.9 ± 1.8
JWH-073 methyl homologue	$C_{24}H_{23}NO$	?
JWH-081	$C_{25}H_{25}NO_2$	1.2 ± 0.03
JWH-122	$C_{25}H_{25}NO$	0.69 ± 0.05
JWH-122-5-fluoropentyl derivative	$C_{25}H_{24}NOF$	?
JWH-200 (WIN 55,225)	$C_{25}H_{24}N_2O_2$	42 ± 5
JWH-203	$C_{21}H_{22}ClNO$	8.0 ± 0.9
JWH-210	$C_{26}H_{27}NO$	0.46 ± 0.03
JWH-250	$C_{22}H_{25}NO_2$	11 ± 2
JWH-250 1-(2-methylene-N-methyl-piperidyl) derivative	$C_{24}H_{28}N_2O_2$	?
JWH-251	$C_{22}H_{25}NO_2$	29 ± 3
JWH-307	$C_{26}H_{24}FNO$	7.7 ± 1.8
JWH-387	$C_{24}H_{22}BrNO$	1.2 ± 0.1

(Continued)

TABLE 13.1   (Continued)

Substance	Sum Formula	CB1-Receptor Affinity K_I [nM]
JWH-398	$C_{24}H_{22}ClNO$	2.3 ± 0.1
N-(1-adamantyl)-1-pentyl-1H-indazole-3-carboxamide (APINACA)	$C_{22}H_{30}N_3O$	?
N-(1-adamantyl)-1-pentyl-1H-indole-3-carboxamide (APICA)	$C_{23}H_{30}N_2O$	?
RCS-4	$C_{21}H_{23}NO_2$	?
RCS-4 ortho isomer	$C_{21}H_{23}HNO_2$	?
RCS-8	$C_{25}H_{29}NO_2$	?
UR-144	$C_{21}H_{29}NO$	150
WIN 48,098 (pravadoline)	$C_{23}H_{26}N_2O_3$	3155 ± 54

?: K_I not measured/unknown.

can be assumed that they show similar properties to THC [100] or the rather polar compound Org 28611 [32] regarding hysteresis when plotting plasma concentrations against subjectively perceived effects (effects are delayed relative to the plasma concentrations and persist after the concentrations dropped already).

The extensive metabolism and the fact that a number of metabolites were shown to retain binding affinity to the CB_1 receptor leads to a complex picture which has only just begun to be explored. For example, Brents et al. found that hydroxylated metabolites of JWH-018 show *in vitro* and *in vivo* CB_1 receptor affinity and agonistic activity [101]. Chimalakonda et al. added similar findings for the AM-2201 metabolite hydroxylated in position 4 of the 5-fluoropentyl moiety [60]. Additionally, some of the hydroxylated JWH-073 metabolites retain nanomolar affinity to CB_1 receptors, exerting partial agonist activity (with the exception of one of the indole hydroxylated metabolites showing neutral antagonist activity) [102]. Finally, the glucuronidated 5-hydroxypentyl metabolite of JWH-018 has been shown to have CB_1 binding affinity in the upper nanomolar range and neutral antagonist capacity [103].

The onset of action occurs typically within minutes of smoking and the duration of action has been reported to range from one to two hours [49] (most probably referring to aminoalkylindole type cannabinoids) to up to around six hours for the cyclohexylphenol CP-47,497-C8 [9], but it has to be kept in mind that these figures may strongly depend on the specific substance, the dose used and the preparation.

PREVALENCE OF USE

There is data on the prevalence of use of synthetic cannabinoid receptor agonists from both population and sub-population surveys. The majority of the data is from the UK and the USA, with smaller amounts of data available from elsewhere. One thing to note is that the terminology used in these surveys is not consistent, with some referring to these compounds as 'Spice' or 'K2', whereas others refer to them as 'legal weed' or 'synthetic cannabis'. The use of different terminology may mean that respondents interpret the question differently, which may influence their responses.

Population Level Surveys

Collection of data on the use of synthetic cannabinoid receptor agonists at a population level is not undertaken in the preparation of multi-national reports such as the World Drug Report from the United Nations Office on Drugs and Crime (UNODC) and the European Monitoring Centre for Drugs and Drug Addiction (EMCDDA). This is similar to the situation for other NPS as discussed in Chapter 4.

Currently it appears that data on population level prevalence on the use of synthetic cannabinoid receptor agonists are only systematically collected at a national level in England and Wales [104,105]. Data was first collected on the use of 'Spice (and other synthetic cannabinoids)' in England and Wales in the 2010/2011 British Crime Survey, with 0.2% of those aged 16–59 years old reporting use within the last year (0.4% for those aged 16–24 years old and 0.1% for those aged 25–59 years old) [104]. This was significantly lower than the use of cannabis, with 6.8% of those surveyed reporting that they had used cannabis within the last year [104]. The overall last year use of 'Spice (and other synthetic cannabinoids)' had fallen in the 2011/2012 Crime Survey England and Wales to 0.1% of those aged 16–59 years old, whereas last year use of cannabis remained relatively stable (6.9%) [105]. In Germany, the 2009 Epidemiological Survey of Substance Abuse surveyed 8030 individuals aged 18–64 years old through questionnaires, telephone surveys and Internet surveys, and reported the last year use of 'Spice' was 0.8% (the response rate was 50.1%) [106].

Subpopulation Surveys

There is emerging information from a series of subpopulation surveys which appears to suggest that use of synthetic cannabinoid receptor agonists is higher than general population use amongst certain groups, including younger individuals, those continuing in education, those who frequent the night-time economy and those more likely to be subjected to routine occupational drug screening (e.g. the military and athletes) [107–116]. In one study, 5956 urine samples from athletes as part of anti-doping testing regimens in the USA were screened for the presence of two synthetic cannabinoid receptor agonists (JWH-018 and JWH-073) and their metabolites [112]. Parent JWH-018 and JWH-073 were not detected in any of the samples analyzed. However, metabolites of JWH-018 and/or JWH-073 were detected in 4.5% of the samples analyzed (mixed JWH-018/JWH-073 metabolites in 50% of positive samples, JWH-018 metabolites only in 49% and JWH-073 metabolites only in 1%). There have also been similar concerns about use of these products amongst the US military [117].

There is greatest evidence and data reported on the use of synthetic cannabinoid receptor agonists amongst students and young people [110,111,113,116]. A representative survey among pupils aged 15–18 conducted in Germany in 2010 reported that 9% had previously used 'Spice and other smokeable blends', and 2% reported use within the last 30 days [116]. In Florida, USA, 2396 college students were mailed electronically a questionnaire on their use of tobacco, cannabis and 'Spice' (the questionnaire indicated that this was also known as 'K2' or 'legal weed') [111]. Of the 852 (36%) who responded: life-time use of tobacco was 34%, cannabis 36% and spice 8%. The use of 'Spice' was significantly associated with younger age (18–19 years old 10% vs 20+ years old 5%, p = 0.02), being male (female 6% vs male 10%, p < 0.01), less time at college (third year or above in college 4% vs 1st or 2nd year in college 10%, p < 0.01), or previous smoking of tobacco (yes 18% vs no 3%, p < 0.01) and previous cannabis use (yes 21% vs no 1%, p < 0.01). There were no differences in prevalence of use across different racial groups or relationship status.

The Monitoring the Future programme is a longitudinal study programme of secondary school and college students in the USA [114]. The programme surveys approximately 50 000 students in 8th Grade, 10th Grade and 12th Grade from a number of different public and private schools. Additionally, this programme follows up students after graduation by developing two panels of 1200 students randomly selected from each graduating class, with each panel surveyed on a biannual basis. Inclusion of questions regarding the use of 'synthetic marijuana' was included in the 2011 survey for 12th Grade students only; the last year reported use rate was 11.4% [115]. This was higher than college students aged 19–22 years old (8.5%) and young adults aged 19–28 years old (7.4%) [115]. In the 2012 national survey of 4449 students attending 395 schools in the USA, use of synthetic marijuana was included for all three study populations. The last year use prevalence rates were 4.4%, 8.8% and 11.3% for those aged 13–14 years (8th Grade), 15–16 years (10th Grade) and 17–18 years (12th Grade), respectively [118]. In the follow-up study of 'Early and Middle Adulthood' participants (those graduates aged 19–30), overall reported last year use of 'synthetic marijuana' is 6.5%, with the last-use reported use being highest in those who have just graduated (11.7% in those aged 19–20 years old and 10.3% in those aged 21–22 years old), decreasing to 2.0% and 2.1% in those 25–26 years old and 29–30 years old, respectively [119]. Interestingly the rate amongst those aged 27–28 years old was 4.6%.

Questions relating to the use of 'Spice/Magic' were first included in the 2009/2010 annual MixMag survey of UK clubbers and those who associate themselves with the *MixMag* magazine and online website [107]. Lifetime (ever used) and last month use of 'Spice/Magic' were 12.7% and 2% respectively, which was considerably lower than the lifetime (93%) and last month (54.4%) use of cannabis [107]. There was a similar pattern of greater use of cannabis compared to 'Spice/Magic'

in the 2010/2011 MixMag annual survey: 'Spice/Magic' Lifetime use 10.3% compared to 86.5% for cannabis; last year use of 'Spice/Magic' 2.2% compared to 64% for cannabis [108]. The MixMag survey was expanded in the subsequent year to become the Global Drug Survey with 7700 UK respondents and 3300 US respondents [109]. Overall lifetime use of 'synthetic cannabis' in the UK respondents was 14.2% (lifetime use of the US respondents was not reported). Interestingly there was greater last year use of synthetic cannabis amongst US respondents (14%) compared to UK respondents (3.3%), whereas there was no real difference in last year use of cannabis (US: 69.3%; UK 68.2%). Additionally, those people who were classified as 'frequent clubbers' in the UK had slightly higher last year use (5%) than the total UK survey population. In a further analysis of the total dataset, the prevalence of lifetime use of 'synthetic cannabis' amongst the 14 966 participants was 16.8%, and prevalence of use within the last month was 6.5% [49]. A questionnaire survey of 308 attendees at South London 'gay friendly' nightclubs undertaken *in situ* within the nightclub environment, reported life-time, last month and use on the night of the survey/planned use on the night of the survey of synthetic cannabinoid receptor agonists of 9.0%, 2.2% and 0.6%, respectively [120].

Motivation for Consumption

An online survey conducted by the Centre of Drug Research of Goethe University Frankfurt in 2011 shows that apart from curiosity and the wish to get high – motives frequently mentioned also in the context of use of illicit drugs – legal availability was also an important motive for consumption [121]. This applies in particular to regular users of incense mixtures. The lack of rapid tests to detect these substances (e.g. in traffic checks) also plays an important role. Most regular users of synthetic cannabinoid receptor agonists seemed, in this survey, to be experienced drug users who have used not only cannabis but other illicit drugs as well.

Whereas Pabst et al. [106] stated in the 'Epidemiological Survey on Addiction' of 2009 that the prevalence of 'Spice' use in the German adult population was 0.8% for the 18–64-year age group (2.5% in young adults), a survey conducted among pupils aged 15–18 years in the Frankfurt area showed a prevalence of 9% in 2010 (compared to 6% in 2009) [116]. In view of the fact that circumventing drug tests will continue to be an important motive (no reliable rapid tests are to be expected in the foreseeable future and only a small number of specialised laboratories are able to cover the complete spectrum of substances available in the market), it seems unlikely that this trend will significantly decline, at least in the near future.

In an Australian survey of 316 synthetic cannabinoid receptor agonist users, the most common reason for trying these products in 50% of those surveyed was that they were curious to compare the effects to cannabis [122]. Other reasons for using included legal status (39%), easier to obtain than cannabis (23%), produced desirable effects (20%), something different to cannabis (11%), offered by friends (10%), therapeutic effects (9%), avoiding workplace or drug-driver testing (8%) and to aid reduction or cessation of cannabis use (5%). Most (76%) believed that the synthetic cannabinoid receptor agonists were less likely to be detected through workplace testing than cannabis. Similarly, 79% believed that they were less likely to be detected by drugs dogs.

In the more detailed analysis of the 2011/12 Global Drugs Survey, those who reported a preference for 'synthetic cannabis' over 'natural cannabis' were asked which of four reasons was most important in determining this preference [49]. 69 individuals reported a preference for 'synthetic cannabis' – the most important reason was effect (58%), followed by availability (18.9%), not as easily detected in urine screens (14.5%) and cost (8.7%).

ACUTE TOXICITY

Animal Studies

There have been no animal studies investigating the patterns or mechanisms of acute toxicity of the synthetic cannabinoid receptor agonists. However, there has been one study investigating the potential cytotoxic effects of three non-classical synthetic cannabinoid receptor agonists (CP-55,940; CP-47,497; CP-47,497-C8) in the neuroblastoma-glioma hybrid cell line NG108-15 [123]. All three synthetic cannabinoid receptor agonists were associated with concentration dependent cytotoxicity. Pre-incubation with the selective CB_1 antagonist AM-251 resulted in suppression of the cytotoxicity, whereas pre-incubation with the selective CB_2 antagonist AM-630 had no effect on the cytotoxicity seen. In this cell model the synthetic cannabinoid receptor agonists were associated with activation of the caspase cascade, pre-incubation with the irreversible inhibitor of capsase-3, Z-DEVD-4NK, attenuated the cytotoxicity seen. This suggests that the synthetic cannabinoid receptor agonists have the potential to cause cellular cytotoxicity through activation of the CB_1 receptor and that the capsase cascade has an important role of this. However, the clinical significance of this to 'Spice' users is unclear at this time.

Koller et al. evaluated cytotoxic, genotoxic, immunomodulatory, and hormonal activities of four naphthoylindoles (JWH-018, JWH-073, JWH-122 and JWH-210) and the benzoylindole AM-694 in human cell lines and primary cells [124]. Toxic effects induced by all compounds included damage to the cell membranes of buccal (TR146) and breast (MCF-7) derived cells and were only seen at concentrations $\geq 75–100 \mu M$ which is several orders of magnitude higher than the concentrations typically expected in humans after recreational use. In single cell gel electrophoresis assays,

JWH-073 and JWH-122 induced DNA migration in buccal and liver cells (HepG2), while JWH-210 was only active in the latter cell line. In bone marrow cells (U2-OS) all compounds caused anti-estrogenic effects at levels between 2.1 and 23.0 μM. Impact on cytokine release (i.e., on IL-10, IL-6, IL-12/23p40 and TNF-alpha levels) was only seen in pilopolysaccharide (LPS)-stimulated human PBMCs after treatment with JWH-210 and JWH-122 which caused a decrease of TNF-alpha and IL-12/23p40. All toxic effects were observed only at concentrations significantly higher than those expected in users. However, since local concentrations may be higher in certain tissues (e.g. epithelial cells or fat tissue where lipophilic substances may accumulate), further experimental work is required to find out if DNA damage may occur in drug users, particularly after chronic use.

Human Data

The information on the acute toxicity related to the use of synthetic cannabinoid receptor agonists is based on user reports on Internet discussion forums and questionnaire surveys, case reports without analytical confirmation, poisons centre case series and case reports/series with analytical confirmation of the synthetic cannabinoid receptor agonists involved [125]. The limitation of many of these reports are that many include use of more than one synthetic cannabinoid receptor agonist and so it can be difficult to determine the role of an individual synthetic cannabinoid receptor agonist in the pattern of toxicity seen.

In an Internet based anonymous survey of 168 people recruited from e-mail lists and Internet discussion forums containing reference to 'Spice' products, recruitment was undertaken from 13 different countries [126]. Unwanted effects included drowsiness (75%), light-headedness (74%), heart racing (59%), feeling of clumsiness (57%), nervousness or anxiety (54%), paranoia (54%), dizziness (51%), nausea (36%), hallucinations (28%), tinnitus (20%), vomiting (10%). The quantity of 'Spice' product consumed did not differ significantly between those who did and did not have a history of unwanted effects. Additionally, 11% reported that recurrent use of the same product resulted in variable and unpredictable effects, suggesting potentially that there is variability in the content of products that individuals were using.

In a more detailed analysis of the 2011/12 Global Drug Survey, participants were asked to compare their experiences following 'synthetic cannabis' use with 'natural cannabis' across 12 broad positive and negative effect domains [49]. 'Natural cannabis' was reported to be associated with more positive effects than 'synthetic cannabis'. Negative effects more associated with 'natural cannabis' included sedation, impairment of memory and addictiveness; however, 'synthetic cannabis' was associated with more hangover effects, paranoia, 'harmful effects on the lungs' and negative effects when high than 'natural cannabis'.

In a review of case records of 11 individuals aged 15–19 years in an addiction treatment centre who were users of 'Spice' products, commonly reported adverse effects were: irritability (36%), anxiety (27%), anger (9%), memory changes (100%), auditory perceptual changes (9%), visual perceptual changes (45%), paranoid thoughts (35%), palpitations (20%), blackouts (9%) and tremor (9%) [113]. Limited information was given in this paper on the frequency of use of 'Spice' and 91% also reported use of cannabis and alcohol; it is therefore difficult to determine to what degree the reported adverse effects were due to synthetic cannabinoid receptor agonists rather than cannabis and/or alcohol.

The final user survey on adverse effects is a questionnaire survey of 316 Australian synthetic cannabinoid receptor agonist users which used purposeful sampling strategies linked to Internet sites selling synthetic cannabinoid receptor agonists products [122]. In this survey, 68% reported at least one adverse effect during their last use of synthetic cannabinoid receptor agonists (median number of adverse effects 1 (IQR 0–4)). Adverse effects reported included: decreased motor coordination (38%), fast or irregular heart beat (33%), dissociation (22%), dizziness (20%), paranoia (18%), confusion (18%), headache (18%), panic (14%), slurred speech (14%), sweating (14%), nausea and/or vomiting (9%), depression (4%) and psychosis (4%). Only 2.1% felt that their adverse effects were serious enough for them to consider seeking help from friends or a healthcare provider. Males reported significantly more adverse effects than females and younger users aged 18–25 years reported more adverse effects than both those aged 26–35 years and ≥36 years. Use of bongs resulted in significantly more adverse effects than other routes of use. Concurrent alcohol use was associated with more adverse effects, whereas there was no association with concurrent use of synthetic cannabinoid receptor agonists and cannabis.

In a large US poisons centre case series describing self-reported synthetic cannabinoid receptor agonist toxicity cases reported to US poisons centres during a 9-month period in 2010 [97], there were 1898 exposures; 1353 of these were single-agent exposures. The most common clinical effect was tachycardia (n = 510, 37.7%), seizures were reported in 52 patients (3.8%). Data from the UK National Poisons Information Service shows that there were 532 accesses to the online TOXBASE website over the first 6 months of 2012 concerning 'Spice' products; it is not possible to determine how many of these applied to cases of acute toxicity and these are all self-reported/non-analytically confirmed cases [127].

The initial clinical case reports of acute toxicity associated with the use of synthetic cannabinoid receptor agonists described toxicity similar to that seen with cannabis [128,129].

In addition, these and other reports suggest that the use of synthetic cannabinoid receptor agonists in individuals who are susceptible to psychosis may precipitate or worsen underlying psychosis [128–133]. One of these reports describes a 25-year-old male with previous recurrent psychotic episodes related to cannabis managed on mono-therapy with amisulpride [130]. He smoked 3 g of 'Spice' on three separate occasions and developed a worsening psychosis (imperative voices and recurrent paranoid hallucinations). In a further report, a 21-year-old male with underlying attention deficit hyperactivity disorder (ADHD) on methylphenidate consumed 'Spice' and within minutes was reported to have developed blurred vision, an unsteady gait and an acute onset of an 'agonal state' with 'the fear of ignorance of his friends' [131]. This panic attack was accompanied by 'vegetative hyperirritability' lasting more than two hours. The patient was managed with midazolam and overnight observation. One further report describes a 17-year-old female who took a 'single bong hit' of JWH-018 15 minutes prior to arrival in the Emergency Department (ED) [133]. Her friends described that she became 'violent and crazy'. She was restless and anxious on arrival in the ED with heart rate 120/min, BP 135/85 mmHg and normal neurological examination. She was managed with intravenous lorazepam and her symptoms settled.

In more recent reports of acute toxicity related to 'Spice' products, it appears that in addition to 'cannabis-like' effects such as sedation, cognitive deficits, psychosis and anxiety, the synthetic cannabinoid receptor agonists can also cause more significant clinical features. Some examples of more recent case reports of acute toxicity (some of which also include analytical confirmation of the synthetic cannabinoid receptor agonist(s)) include the following.

There have been a number of reports of convulsions related to the use of 'Spice' products, but not all of these cases have included analytical confirmation on biological samples. The first report was of a 21-year-old male who presented to the ED with vomiting and a convulsion one hour after smoking 'Spicy XXX'; no analytical confirmation of either biological samples or the product was undertaken in this case [134]. In another report, a 19-year-old male had a convulsion whilst smoking 'Happy Tiger Incense' and a further convulsion on arrival in the ED [99]. Four synthetic cannabinoid receptor agonists were found on analysis of the product smoked: JWH-018, JWH-081, JWH-250 and AM-2201; however, there was no analysis of biological samples. In another report of convulsions potentially related to synthetic cannabinoid receptor agonists, a 48-year-old man ingested an unknown white powder and within 30 minutes became agitated and had a convulsion [135]. On arrival in the ED he had a heart rate of 106 and blood pressure of 140/88 and temperature of 37.7. Following further seizure activity in the ED, he was treated with IV lorazepam. On the second day of his admission, he developed a supra-ventricular tachycardia requiring electrical cardioversion. Analysis of the powder used confirmed it was JWH-018 and subsequent analysis of the urine detected the major omega carboxyl metabolite of JWH-018 (concentration 200 nM). The most recent report was from our group, reporting a 20-year-old male who presented to the ED following a generalised self-terminating tonic clonic convulsion lasting between two and three minutes after smoking two tokes of a 'Black Mamba' spice product [136]. Toxicological screening of a urine sample taken at presentation detected three mono-hydroxylated metabolites of AM-2201; no other drugs or alcohol were detected other than nicotine and caffeine.

There has been a recent report of 16 cases of acute kidney injury (AKI) in six US states [137]. There was a range of different products used and screening of biological samples was undertaken in six individuals. The synthetic cannabinoid receptor agonists detected were (1-(5-fluoropentyl)-1H-indole-3-yl)(2,2,3,3-tetramethylcyclopropyl)methanone (also known as XLR-11) and/or its metabolite in four individuals and UR-144 in one individual. It is noteworthy that these tetramethylcyclopropyl derivatives may undergo ring-opening reactions when smoked, resulting in reactive unsaturated molecular species. No synthetic cannabinoid receptor agonists were detected in two individuals, one had a sample collected nine days after use and the other had insufficient serum collected two days after use and nothing detected in urine collected four days after use. Fifteen (93.8%) had nausea and vomiting as the chief symptom at presentation and therefore there is the potential that the cause of AKI was pre-renal renal failure secondary to dehydration rather than actual synthetic cannabinoid receptor agonist related renal toxicity. This is substantiated by renal biopsy findings of acute tubular necrosis in six of the eight patients who had a renal biopsy and three of eight people had acute interstitial nephritis on renal biopsy; however, limited information was provided on the extent to which alternative causes of an interstitial nephritis were excluded. In summary, whilst this report highlights the potential for renal toxicity associated with the synthetic cannabinoid receptor agonists, we feel that it is more likely that the AKI seen in these patients was due to dehydration related to vomiting.

The largest series of analytically confirmed synthetic cannabinoid receptor agonist acute toxicity is from Germany [44]. This was a series of 29 patients and the following synthetic cannabinoid receptor agonists were detected: CP-47-497-C8 (one case), JWH-015 (one), JWH-018 (eight), JWH-073 (one), JWH-081 (seven), JWH-122 (eleven), JWH-250 (four), AM-694 (one), JWH-210 (eleven). More than one synthetic cannabinoid receptor agonist was detected in a number of cases; two synthetic cannabinoid receptor agonists were detected in seven patients and three synthetic cannabinoid receptor agonists were detected in five patients.

There appeared to be a change in the synthetic cannabinoid receptor agonists detected over time with detection of JWH-018 in 2008/2009, JWH-081 and JWH-122 frequently seen in 2010, JWH-210 first identified at the end of 2010 and being identified in all but one case in 2011. Generally clinical features resolved within four to 14 hours, although one patient had an acute psychosis that lasted for several days. 'Cannabis-like' changes in perception and hallucinations were common (38%) and agitation was also common (41%). Tachycardia was commonly reported (76%) with heart rate of 90–170 beats per minute and 34% had hypertension (150–200 mmHg systolic, median 160 mmHg; 80–100 mmHg diastolic, median 85 mmHg). Nausea and vomiting occurred in 28%, 21% had shortness of breath, 15% had confusion, 17% had 'unconsciousness' lasting up to an hour, 14% had 'shaking and shivering', 14% had an elevated creatine kinase and 28% of patients had hypokalaemia (serum potassium 3.1–3.4 mmol/L (four patients), 2.8–2.9 mmol/L (three patients), 2.3 mmol/L (one patient). Other significant clinical features that were less common and reported in fewer than three patients (10%) included muscle pain, myoclonus, chest pain and convulsions. There appeared to be a greater frequency of adverse effects reported with the JWH-group synthetic cannabinoid receptor agonists than with CP-47,497-C8.

In conclusion, the reports to date of toxicity associated with synthetic cannabinoid receptor agonists suggest that in addition to cannabis-like effects, additional toxicity associated with these products includes agitation, tachycardia, and hypertension to a greater extent than that seen with cannabis, and also hypokalaemia and convulsions.

Deaths

There have been reports in the popular press and on Internet discussion forums of deaths potentially related to the use of 'Spice' products [138,139]. However, there are no reports in the published literature of deaths that have been confirmed to be related to synthetic cannabinoid receptor agonists.

CHRONIC TOXICITY

There are currently no published data describing studies in either animals or humans which have investigated the potential for chronic toxicity associated with the use of synthetic cannabinoid receptor agonists. However, in single cell gel electrophoresis assays the aminoalkylindoles JWH-073, JWH-122 and JWH-210 were shown to induce DNA migration which may indicate a potential for genotoxicity [124].

DEPENDENCE AND ABUSE POTENTIAL

Animal Data

In a rat discrimination study, where rats were initially trained to discriminate THC from vehicle, synthetic cannabinoid receptor agonists such as CP-55,940 and WIN-55,212-2 were able to substitute for THC [140]. These effects of synthetic cannabinoid receptor agonists could be attenuated by the effects of the CB_1 antagonist SR-141,716A.

In a rhesus monkey model, animals were trained to discriminate THC, JWH-018, JWH-073 from vehicle in drug lever pressing studies [141]. The ED_{50} values were 0.044, 0.013, 0.058 mg/kg and duration of action was four, two and one hour for THC, JWH-018 and JWH-073, respectively. All three compounds could dose-dependently attenuate the rimonabant discriminative stimulus in animals chronically treated with THC. The authors of this study concluded that the synthetic cannabinoid receptor agonists JWH-018 and JWH-073 had similar subjective effects to THC although the shorter duration of action may lead to more frequent use and in turn, there is the potential that this could increase dependence risk.

In another rhesus monkey study, animals were given either three or 14 days of pre-treatment with THC to study its effects on tolerance to three synthetic cannabinoid receptor agonists (CP-55,940; JWH-073; JWH-018) [142]. Three days of pre-treatment with THC did not result in cross tolerance to any of these three synthetic cannabinoid receptor agonist compounds. Fourteen days pre-treatment with THC decreased sensitivity to THC, CP-55,940, JWH-018 and JWH-073 9.2-fold, 3.6-fold, 4.3-fold and 5.6-fold respectively. The differences in sensitivity to pre-treatment with THC may indicate differences in the potential for dependence associated with THC and synthetic cannabinoid receptor agonists and also between different synthetic cannabinoid receptor agonists.

Human Data

In the multi-national Internet survey of 168 'Spice' users, 37% met the DSM IV criteria for abuse and 12% met the dependence criteria [126]. In terms of the dependence criteria, 38% had been unable to cut down or stop 'Spice' use, 36% were describing symptoms of tolerance, 28% were using for longer periods than originally intended and 18% stated that 'Spice' use was interfering with other activities. No individuals reported that they had sought or received treatment in relation to their problematic use of 'Spice'. Withdrawal symptoms following cessation of 'Spice' were described as 'rare', although data not provided in the paper on the percentage of individuals who reported withdrawal effects. Commonly reported effects included headache (15%), anxiety/nervousness (15%), insomnia (14%), anger/irritability (13%), poor concentration (9%), nausea (7%) and depression (6%).

Nabilone is a prescription synthetic cannabinoid receptor agonist that is used in the treatment of chemotherapy-induced vomiting as well as off-label use for chronic pain. The abuse potential of nabilone has been reviewed using published scientific literature, popular press and Internet database reviews [143]. Overall, nabilone was perceived to have more undesirable effects, longer onset of action and to be more expensive than 'natural cannabis'. In addition, it was felt that there was limited evidence of abuse potential for nabilone.

There have been reports of cannabis dependent users who have been able to transition to use of 'Spice' without development of withdrawal symptoms [144]. These reports support the previously described rhesus monkey studies where synthetic cannabinoid receptor agonists can substitute for THC. There have also been anecdotal reports of 'dependent' use of synthetic cannabinoid receptor agonists on Internet discussion forums [145]. The first published report of dependence and withdrawal associated with the use of synthetic cannabinoid receptor agonists was from Germany [146]. This was a 20-year-old male who had been initially using 1 g of 'Spice Gold' per day for eight months, which had increased to 3 g per day in three to four divided doses due to decreasing effects. During a phase of abstinence associated with difficulty in supply he developed withdrawal symptoms of profuse sweating, 'internal unrest', tremor, palpitations, insomnia, headache, nausea, vomiting and diarrhoea. These symptoms resolved on recommencing use of the 'Spice Gold' product. He was therefore eventually admitted for withdrawal treatment and started to develop symptoms of withdrawal on day two of his admission with increasing 'internal unrest'. Symptoms continued to develop such that on day four he had increasing 'internal unrest', strong desire to use 'Spice' products, nightmares, nausea, sweating, tremor, tachycardia, hypertension and headache. He was treated with a combination of zopiclone, promethazine, clonidine and pramipexole. He was discharged on day 18 and was able to discontinue the pramipexole one month later.

In another report, a 23-year-old male with previous high use of cannabis (5–10 g per day) switched to 'Spice Gold', with increasing dose up to 10 g per day [147]. Concerned about his use, self-attempted cessation resulted in the development of severe withdrawal symptoms. Eventually he was admitted to a psychiatry unit for withdrawal. On the day after his admission he had a (18F)fallypride PET scan to measure cerebral D2/3 receptor availability. This demonstrated reduction in cerebral D2/3 receptors compared to healthy controls in the striatum, caudate nucleus, anterior/posterior putamen, thalamus, insular cortex, hippocampus, lateral temporal cortex and amygdala. Following one week of observed abstinence in hospital, the PET scan was repeated and demonstrated that receptor availability had recovered to 'normal values'. This single case report demonstrates that 'Spice Gold' – which during product monitoring in the years 2008–2010 contained CP-47,497-C8 as main active substance, sometimes alongside with JWH-018 (own unpublished data) – is associated with changes in cerebral D2/3 availability, which may explain its dependence potential and associated withdrawal syndromes.

FORENSIC RELEVANCE

Synthetic cannabinoid receptor agonists are important in all fields in which the use and detection of cannabis may be of relevance.

These include:

- road traffic offences
- unexplained deaths
- occupational screening
- assessment of criminal responsibility
- assessment of fitness to drive following drug use and
- abstinence testing in therapeutic facilities.

Offences against narcotics acts or medicines acts may also be of relevance.

It is currently difficult to predict the extent to which the synthetic cannabinoid receptor agonist products may be used in these areas. One significant limiting step is that the complex analytical techniques are often not widely available in all of these areas and keeping the analytical libraries up to date can be a challenge.

As with natural cannabinoids and other substances with high volumes of distribution it is important when interpreting analytical findings to be aware of the problem of accumulation due to redistribution from deep compartments, which may lead to a prolonged elimination time after the beginning of abstinence [148,149]. Evaluation of follow-up checks in psychiatric patients showed that after chronic consumption of incense mixtures positive results for metabolites can be expected in urine samples for many weeks and even up to several months. The same applies to serum samples where the initially often very high concentrations will continuously decline with a terminal elimination half-life of several days up to more than a week [51]. After a single consumption, the substance can usually be detected in blood and urine samples for at least several days. New consumption can therefore be assumed only if the concentrations in consecutive samples show a significant increase or new substances not identified before are detected.

CONCLUSIONS

Products containing synthetic cannabinoid receptor agonists are widely available over the Internet and emerging data show that these products are increasingly used amongst certain sub-populations. There are many dozen synthetic cannabinoid receptor agonists across different chemical classes and new chemicals continue to be described with increasing frequency over the last two to three years.

Although there is limited data available on the pharmacology of the synthetic cannabinoid receptor agonists used in 'Spice' products, the available data suggests that most of them are of higher potency and intrinsic activity at CB_1 receptors than cannabis. It also appears that these substances have a wider and more severe pattern of toxicity than cannabis.

REFERENCES

[1] Mechoulam R, Gaoni. Y. A total synthesis of Dl-delta-1-tetrahydrocannabinol, the active constituent of hashish. J Am Chem Soc 1965;87:3273–5.

[2] Matsuda LA, Lolait SJ, Brownstein MJ, Young AC, Bonner TI, et al. Structure of a cannabinoid receptor and functional expression of the cloned cDNA. Nature 1990;346(6284):561–4.

[3] Munro S, Thomas KL, Abu-Shaar M. Molecular characterization of a peripheral receptor for cannabinoids. Nature 1993;365(6441):61–5.

[4] Croxford JL. Therapeutic potential of cannabinoids in CNS disease. CNS Drugs 2003;17(3):179–202.

[5] Porter AC, Felder CC. The endocannabinoid nervous system: unique opportunities for therapeutic intervention. Pharmacol Ther 2001;90(1):45–60.

[6] Thakur GA, Tichkule R, Bajaj S, Makriyannis A. Latest advances in cannabinoid receptor agonists. Expert Opin Ther Pat 2009;19(12):1647–73.

[7] Christensen R, Kristensen PK, Bartels EM, Bliddal H, Astrup A, et al. Efficacy and safety of the weight-loss drug rimonabant: a meta-analysis of randomised trials. Lancet 2007;370(9600):1706–13.

[8] EMCDDA. Understanding the 'spice' phenomenon. EMCDDA. Thematic Papers; 2009.

[9] Auwärter V, Kristensen PK, Bartels EM, Bliddal H, Astrup A, et al. 'Spice' and other herbal blends: harmless incense or cannabinoid designer drugs? J Mass Spectrom 2009;44(5):832–7.

[10] Uchiyama N, Kikura-Hanajiri R, Kawahara N, Haishima Y, Goda Y, et al. Identification of a cannabinoid analog as a new type of designer drug in a herbal product. Chem Pharm Bull (Tokyo) 2009;57(4):439–41.

[11] Dresen S, Ferreiros N, Pütz M, Westphal F, Zimmermann R, Auwärter V, et al. Monitoring of herbal mixtures potentially containing synthetic cannabinoids as psychoactive compounds. J Mass Spectrom 2010;45(10):1186–94.

[12] Lindigkeit R, Boehme A, Eiserloh I, et al. Spice: a never ending story? Forensic Sci Int 2009;191(1–3):58–63.

[13] Uchiyama N, Kikura-Hanajiri R, Ogata J, Goda Y, et al. Chemical analysis of synthetic cannabinoids as designer drugs in herbal products. Forensic Sci Int 2010;198(1–3):31–8.

[14] Nakajima J, Takahashi M, Seto T, et al. Identification and quantitation of two benzoylindoles AM-694 and (4-methoxyphenyl)(1-pentyl-1H-indol-3-yl) methanone, and three cannabimimetic naphthoylindoles JWH-210, JWH-122, and JWH-019 as adulterants in illegal products obtained via the Internet. Forensic Toxicol 2011;29(2):95–110.

[15] Nakajima J, Takahashi M, Seto T, et al. Analysis of azepane isomers of AM-2233 and AM-1220, and detection of an inhibitor of fatty acid amide hydrolase (3'-(aminocarbonyl)(1,1'-biphenyl)-3-yl)-cyclohexyl-carbamate (URB597) obtained as designer drugs in the Tokyo area. Forensic Toxicol 2013;31:76–85.

[16] Schmidt MM, Sharma A, Schifano F, Feinmann C. 'Legal highs' on the net-Evaluation of UK-based websites, products and product information. Forensic Sci Int 2011;206(1–3):92–7.

[17] Vardakou I, Pistos C, Spiliopoulou C. Spice drugs as a new trend: mode of action, identification and legislation. Toxicol Lett 2010;197(3):157–62.

[18] Rosenbaum CD, Carreiro S, Babu KM. Here today, gone tomorrow ... and back again? A review of herbal marijuana alternatives (K2, Spice), synthetic cathinones (bath salts), kratom, Salvia divinorum, methoxetamine, and piperazines. J Med Toxicol 2012;8(1):15–32.

[19] Piggee C. Investigating a not-so-natural high. Anal Chem 2009;81(9):3205–7.

[20] Ginsburg BC, McMahon LR, Sanchez JJ, Javors MA. Purity of synthetic cannabinoids sold online for recreational use. J Anal Toxicol 2012;36(1):66–8.

[21] Kneisel S, Auwärter V. Analysis of 30 synthetic cannabinoids in serum by liquid chromatography-electrospray ionization tandem mass spectrometry after liquid-liquid extraction. J Mass Spectrom 2012;47(7):825–35.

[22] Kikura-Hanajiri R, Uchiyama N, Kawamura M, Goda Y, et al. Changes in the prevalence of synthetic cannabinoids and cathinone derivatives in Japan until early 2012. Forensic Toxicol 2013;31:44–53.

[23] de Jager AD, Warner JV, Henman M, Ferguson W, Hall A. LC-MS/MS method for the quantitation of metabolites of eight commonly-used synthetic cannabinoids in human urine – An Australian perspective. J Chromatogr B Analyt Technol Biomed Life Sci 2012;897:22–31.

[24] Nakajima J, Takahashi M, Seto T, et al. Identification and quantitation of two new naphthoylindole drugs-of-abuse,(1-(5-hydroxypentyl)-1H-indol-3-yl)(naphthalen-1-yl) methanone (AM-2202) and (1-(4-pentenyl)-1H-indol-3-yl)(naphthalen-1-yl) methanone, with other synthetic cannabinoids in unregulated 'herbal' products circulated in the Tokyo area. Forensic Toxicol 2012;30(1):33–44.

[25] Dargan PI, Hudson S, Ramsey J, et al. The impact of changes in UK classification of the synthetic

cannabinoid receptor agonists in 'Spice'. Int J Drug Policy 2011;22(4):274–7.

[26] ACMD, Consideration of the major cannabinoid agonists. 2009.

[27] Howlett AC, Barth F, Bonner TI, et al. International union of pharmacology. XXVII. Classification of cannabinoid receptors. Pharmacol Rev 2002;54(2): 161–202.

[28] Thakur GA, Nikas S, Makriyannis A. CB1 cannabinoid receptor ligands. Mini Rev Med Chem 2005;5(7):631–40.

[29] UNODC. Synthetic cannabinoids in herbal products. SCITEC/24: <http://www.unodc.org/documents/scientific/Synthetic_Cannabinoids.pdf>; 2011.

[30] Baker RR. Temperature distribution inside a burning cigarette. Nature 1974;247:405–6.

[31] Gardin A, Kucher K, Kiese B, Appel-Dingemanse S. Cannabinoid receptor agonist 13, a novel cannabinoid agonist: first in human pharmacokinetics and safety. Drug Metab Dispos 2009;37(4):827–33.

[32] Zuurman L, Passier PC, de Kam MI, Kleijn HJ, Cohen AF, van Gerven JM. Pharmacodynamic and pharmacokinetic effects of the intravenously administered CB1 receptor agonist Org 28611 in healthy male volunteers. J Psychopharmacol 2009;23(6):633–44.

[33] Martin BR, Wiley JL, Beletskaya I, et al. Pharmacological characterization of novel water-soluble cannabinoids. J Pharmacol Exp Ther 2006;318(3):1230–9.

[34] Bell MR, D'Ambra TE, Kumar V, et al. Antinociceptive (aminoalkyl)indoles. J Med Chem 1991;34(3): 1099–110.

[35] Huffman JW, Zengin G, Wu M-J, et al. Structure-activity relationships for 1-alkyl-3-(1-naphthoyl) indoles at the cannabinoid CB(1) and CB(2) receptors: steric and electronic effects of naphthoyl substituents. New highly selective CB(2) receptor agonists. Bioorg Med Chem 2005;13(1):89–112.

[36] Shevyrin V, Melkozerov V, Nevero A. Identification and analytical properties of new synthetic cannabimimetics bearing 2,2,3,3-tetramethylcyclopropanecarbonyl moiety. Forensic Sci Int 2012;226(1–3):62–73.

[37] Kavanagh P, Grigoryev A, Savchuk S, Mikhura I, Formanovsky A. UR-144 in products sold via the Internet: identification of related compounds and characterization of pyrolysis products. Drug Test Anal 2013 (Epub ahead of print).

[38] Donohue KM, Steiner RR. JWH-018 and JWH-022 as combustion products of AM2201. Microgram J 2012;9(2):52–6.

[39] Choi H, Heo S, Kim E, Hwang BY, Lee C, Lee J. Identification of (1-pentylindol-3-yl)-(2,2,3,3-tetramethylcyclopropyl)methanone and its 5-pentyl fluorinated analog in herbal incense seized for drug trafficking. Forensic Toxicol 2013;31:86–92.

[40] Kneisel S, Bisel P, Brecht V, Broecker S, Müller M, Auwärter V. Identification of the cannabimimetic AM-1220 and its azepane isomer (N-methylazepan-3-yl)-3-(1-naphthoyl) indole in a research chemical and several herbal mixtures. Forensic Toxicol 2012;30:126–34.

[41] Kikuchi H, Uchiyama N, Ogata J, Kikura-Hanajiri R, Goda Y. Chemical constituents and DNA sequence analysis of a psychotropic herbal product. Forensic Toxicol 2010;28(2):77–83.

[42] Ogata J, Uchiyama N, Kikura-Hanajiri R, Goda Y. DNA sequence analyses of blended herbal products including synthetic cannabinoids as designer drugs. Forensic Sci Int 2012 (Epub ahead of print).

[43] Seely KA, Lapoint J, Moran JH, Fattore L. Spice drugs are more than harmless herbal blends: a review of the pharmacology and toxicology of synthetic cannabinoids. Prog Neuropsychopharmacol Biol Psychiatry 2012;39(2):234–43.

[44] Hermanns-Clausen M, Kneisel S, Szabo B, Auwärter V. Acute toxicity due to the confirmed consumption of synthetic cannabinoids: clinical and laboratory findings. Addiction 2013;108(3):534–44.

[45] Teske J, Weller JP, Fieguth A, Rothämel T, Schulz Y, Tröger HD. Sensitive and rapid quantification of the cannabinoid receptor agonist naphthalen-1-yl-(1-pentylindol-3-yl)methanone (JWH-018) in human serum by liquid chromatography-tandem mass spectrometry. J Chromatogr B Analyt Technol Biomed Life Sci 2010;878(27):2659–63.

[46] Grigoryev A, Kavanagh Melnik A. The detection of the urinary metabolites of 1-((5-fluoropentyl)-1H-indol-3-yl)-(2-iodophenyl)methanone (AM-694), a high affinity cannabimimetic, by gas chromatography–mass spectrometry. Drug Test Anal 2013;5(2):110–5.

[47] Grigoryev A, Kavanagh Melnik A. The detection of the urinary metabolites of 3-((adamantan-1-yl)carbonyl)-1-pentylindole (AB-001), a novel cannabimimetic, by gas chromatography-mass spectrometry. Drug Test Anal 2012;4(6):519–24.

[48] Gershkovich Hoffman A. Effect of a high-fat meal on absorption and disposition of lipophilic compounds: the importance of degree of association with triglyceride-rich lipoproteins. Eur J Pharm Sci 2007;32(1): 24–32.

[49] Winstock AR, Barratt MJ. Synthetic cannabis: A comparison of patterns of use and effect profile with natural cannabis in a large global sample. Drug Alcohol Depend 2013 (Epub ahead of print).

[50] Huestis MA. Pharmacokinetics and metabolism of the plant cannabinoids, delta9-tetrahydrocannabinol, cannabidiol and cannabinol. Handb Exp Pharmacol 2005;168:657–90.

[51] Kneisel S, Teske J, Auwärter V. Analysis of synthetic cannabinoids in abstinence control: long drug

detection windows in serum and implications for practitioners. Drug Test Anal 2013.

[52] Toennes SW, Ramaekers JG, Theunissen EL, Moeller MR, Kauert GF. Comparison of cannabinoid pharmacokinetic properties in occasional and heavy users smoking a marijuana or placebo joint. J Anal Toxicol 2008;32(7):470–7.

[53] Skopp G, Potsch L. Cannabinoid concentrations in spot serum samples 24–48 hours after discontinuation of cannabis smoking. J Anal Toxicol 2008;32(2):160–4.

[54] Karschner EL, Schwilke EW, Lowe RH, et al. Do Delta9-tetrahydrocannabinol concentrations indicate recent use in chronic cannabis users? Addiction 2009;104(12):2041–8.

[55] Zhang Q, Ma P, Cole RB, Wang G. Identification of in vitro metabolites of JWH-015, an aminoalkylindole agonist for the peripheral cannabinoid receptor (CB2) by HPLC-MS/MS. Anal Bioanal Chem 2006;386(5):1345–55.

[56] Grigoryev A, Savchuk S, Melnik A, et al. Chromatography-mass spectrometry studies on the metabolism of synthetic cannabinoids JWH-018 and JWH-073, psychoactive components of smoking mixtures. J Chromatogr B Analyt Technol Biomed Life Sci 2011;879(15–16):1126–36.

[57] Wintermeyer A, Möller I, Thevis M, et al. In vitro phase I metabolism of the synthetic cannabimimetic JWH-018. Anal Bioanal Chem 2010;398(5):2141–53.

[58] Sobolevsky T, Prasolov I, Rodchenkov G. Detection of JWH-018 metabolites in smoking mixture post-administration urine. Forensic Sci Int 2010;200(1–3):141–7.

[59] Sobolevsky T, Prasolov I, Rodchenkov G. Detection of urinary metabolites of AM-2201 and UR-144, two novel synthetic cannabinoids. Drug Test Anal 2012 (Epub ahead of print).

[60] Chimalakonda KC, Seely KA, Bratton SM, et al. Cytochrome P450-mediated oxidative metabolism of abused synthetic cannabinoids found in K2/Spice: identification of novel cannabinoid receptor ligands. Drug Metab Dispos 2012;40(11):2174–84.

[61] Kavanagh P, Grigoryev A, Melnik A, Simonov A. The Identification of the Urinary Metabolites of 3-(4-Methoxybenzoyl)-1-Pentylindole (RCS-4), a Novel Cannabimimetic, by Gas Chromatography-Mass Spectrometry. J Anal Toxicol 2012;36(5):303–11.

[62] Grigoryev A, Melnik A, Savchuk S, Simonov A, Rozhanets V. Gas and liquid chromatography-mass spectrometry studies on the metabolism of the synthetic phenylacetylindole cannabimimetic JWH-250, the psychoactive component of smoking mixtures. J Chromatogr B Analyt Technol Biomed Life Sci 2011;879(25):2519–26.

[63] Hutter M, Broecker S, Kneisel S, Auwärter V. Identification of the major urinary metabolites in man of seven synthetic cannabinoids of the aminoalkylindole type present as adulterants in 'herbal mixtures' using LC-MS/MS techniques. J Mass Spectrom 2012;47(1):54–65.

[64] Wohlfarth A, Scheidweiler KB, Chen X, Liu HF, Huestis MA. Qualitative confirmation of 9 synthetic cannabinoids and 20 metabolites in human urine using LC-MS/MS and library search. Anal Chem 2013 (Epub ahead of print).

[65] Howlett AC. Inhibition of neuroblastoma adenylate cyclase by cannabinoid and nantradol compounds. Life Sci 1984;35(17):1803–10.

[66] Howlett AC, Fleming RM. Cannabinoid inhibition of adenylate cyclase. Pharmacology of the response in neuroblastoma cell membranes. Mol Pharmacol 1984;26(3):532–8.

[67] Caulfield M, Brown DA. Cannabinoid receptor agonists inhibit Ca current in NG108-15 neuroblastoma cells via a pertussis toxin-sensitive mechanism. Br J Pharmacol 1992;106(2):231–2.

[68] Mackie K, Hille B. Cannabinoids inhibit N-type calcium channels in neuroblastoma-glioma cells. Proc Natl Acad Sci U S A 1992;89(9):3825–9.

[69] Devane WA, Hanus L, Breuer A, et al. Isolation and structure of a brain constituent that binds to the cannabinoid receptor. Science 1992;258(5090):1946–9.

[70] Johnson DE, Heald SL, Dally RD, Janis RA. Isolation, identification and synthesis of an endogenous arachidonic amide that inhibits calcium channel antagonist 1,4-dihydropyridine binding. Prostaglandins Leukot Essent Fatty Acids 1993;48(6):429–37.

[71] Mechoulam R, Ben-Shabat S, Hanus L. Identification of an endogenous 2-monoglyceride, present in canine gut, that binds to cannabinoid receptors. Biochem Pharmacol 1995;50(1):83–90.

[72] Mascia MS, Obinu MC, Ledent C, et al. Lack of morphine-induced dopamine release in the nucleus accumbens of cannabinoid CB(1) receptor knockout mice. Eur J Pharmacol 1999;383(3):R1–R2.

[73] Ishac EJ, Jiang L, Lake KD, et al. Inhibition of exocytotic noradrenaline release by presynaptic cannabinoid CB1 receptors on peripheral sympathetic nerves. Br J Pharmacol 1996;118(8):2023–8.

[74] Kathmann M. Cannabinoid CB1 receptor-mediated inhibition of NMDA- and kainate-stimulated noradrenaline and dopamine release in the brain. Naunyn Schmiedebergs Arch Pharmacol 1999;359(6):466–70.

[75] Nakazi M, Bauer U, Nickel T, Kathmann M, Schlicker E. Inhibition of serotonin release in the mouse brain via presynaptic cannabinoid CB1 receptors. Naunyn Schmiedebergs Arch Pharmacol 2000;361(1):19–24.

[76] Shen M, Piser TM, Seybold VS, Thayer SA. Cannabinoid receptor agonists inhibit glutamatergic

synaptic transmission in rat hippocampal cultures. J Neurosci 1996;16(14):4322–34.

[77] Szabo B, Müller T, Koch H. Effects of cannabinoids on dopamine release in the corpus striatum and the nucleus accumbens in vitro. J Neurochem 1999;73(3):1084–9.

[78] Atwood BK, Huffman J, Straiker A, Mackie K. JWH018, a common constituent of 'Spice' herbal blends, is a potent and efficacious cannabino id CB receptor agonist. Br J Pharmacol 2010;160(3):585–93.

[79] Felder CC, Joyce KE, Briley EM, et al. Comparison of the pharmacology and signal transduction of the human cannabinoid CB1 and CB2 receptors. Mol Pharmacol 1995;48(3):443–50.

[80] Lambert DM, Di Marzo V. The palmitoylethanolamide and oleamide enigmas: are these two fatty acid amides cannabimimetic? Curr Med Chem 1999;6(8):757–73.

[81] Wiley JL, Martin BR. Cannabinoid pharmacology: implications for additional cannabinoid receptor subtypes. Chem Phys Lipids 2002;121(1–2):57–63.

[82] Dziadulewicz EK, Bevan SJ, Brain CT, et al. Naphthalen-1-yl-(4-pentyloxynaphthalen-1-yl)methanone: a potent, orally bioavailable human CB1/CB2 dual agonist with antihyperalgesic properties and restricted central nervous system penetration. J Med Chem 2007;50(16):3851–6.

[83] Aung MM, Griffin G, Huffman JW, et al. Influence of the N-1 alkyl chain length of cannabimimetic indoles upon CB(1) and CB(2) receptor binding. Drug Alcohol Depend 2000;60(2):133–40.

[84] Compton DR, Rice KC, De Costa BR, et al. Cannabinoid structure-activity relationships: correlation of receptor binding and in vivo activities. J Pharmacol Exp Ther 1993;265(1):218–26.

[85] D'Ambra TE, Estep KG, Bell MR, et al. Conformationally restrained analogues of pravadoline: nanomolar potent, enantioselective, (aminoalkyl) indole agonists of the cannabinoid receptor. J Med Chem 1992;35(1):124–35.

[86] Deng H, Gifford AN, Zvonok AM, et al. Potent cannabinergic indole analogues as radioiodinatable brain imaging agents for the CB1 cannabinoid receptor. J Med Chem 2005;48(20):6386–92.

[87] Frost JM, Dart MJ, Tietje KR. Indol-3-ylcycloalkyl ketones: effects of N1 substituted indole side chain variations on CB(2) cannabinoid receptor activity. J Med Chem 2010;53(1):295–315.

[88] Huffman JW, Mabon R, Wu MJ, et al. 3-Indolyl-1-naphthylmethanes: new cannabimimetic indoles provide evidence for aromatic stacking interactions with the CB(1) cannabinoid receptor. Bioorg Med Chem 2003;11(4):539–49.

[89] Huffman JW, Padgett LW. Recent developments in the medicinal chemistry of cannabimimetic

indoles, pyrroles and indenes. Curr Med Chem 2005;12(12):1395–411.

[90] Huffman JW, Padgett LW, Isherwood ML, Wiley JL, Martin BR. 1-Alkyl-2-aryl-4-(1-naphthoyl)pyrroles: new high affinity ligands for the cannabinoid CB1 and CB2 receptors. Bioorg Med Chem Lett 2006;16(20):5432–5.

[91] Huffman JW, Szklennik PV, Almond A, et al. 1-Pentyl-3-phenylacetylindoles, a new class of cannabimimetic indoles. Bioorg Med Chem Lett 2005;15(18):4110–3.

[92] Makriyannis A, Deng H. Cannabimimetic indole derivatives, W.I. Organization, Editor. 2001.

[93] Showalter VM, Compton DR, Martin BR, Abood ME. Evaluation of binding in a transfected cell line expressing a peripheral cannabinoid receptor (CB2): identification of cannabinoid receptor subtype selective ligands. J Pharmacol Exp Ther 1996;278(3):989–99.

[94] Nakajima J, Takahashi M, Nonaka R, et al. Identification and quantitation of a benzoylindole (2-methoxyphenyl)(1-pentyl-1H-indol-3-yl) methanone and a naphthoylindole 1-(5-fluoropentyl-1H-indol-3-yl)-(naphthalene-1-yl) methanone (AM-2201) found in illegal products obtained via the Internet and their cannabimimetic effects evaluated by in vitro (35S) GTPÎ³S binding assays. Forensic Toxicol 2011;29(2):132–41.

[95] Paronis CA, Nikas SP, Shukla VG, Makriyannis A. Delta(9)-Tetrahydrocannabinol acts as a partial agonist/antagonist in mice. Behav Pharmacol 2012;23(8):802–5.

[96] Forrester MB, Kleinschmidt K, Schwarz E, Young A. Synthetic cannabinoid exposures reported to Texas poison centers. J Addict Dis 2011;30(4):351–8.

[97] Hoyte CO, Jacob J, Monte AA, Al-Jumaan M, Bronstein AC, Heard KJ. A characterization of synthetic cannabinoid exposures reported to the National Poison Data System in 2010. Ann Emerg Med 2012;60(4):435–8.

[98] Pant S, Deshmukh A, Dholaria B, Kaur V, Ramavaram S, Ukor M, et al. Spicy seizure. Am J Med Sci 2012;344(1):67–8.

[99] Schneir AB, Baumbacher T. Convulsions associated with the use of a synthetic cannabinoid product. J Med Toxicol 2012;8(1):62–4.

[100] Cone EJ, Huestis MA. Relating blood concentrations of tetrahydrocannabinol and metabolites to pharmacologic effects and time of marijuana usage. Ther Drug Monit 1993;15(6):527–32.

[101] Brents LK, Reichard EE, Zimmerman SM, et al. Phase I hydroxylated metabolites of the K2 synthetic cannabinoid JWH-018 retain in vitro and in vivo cannabinoid 1 receptor affinity and activity. PLoS One 2011;6(7):e21917.

[102] Brents LK, Gallus-Zawada A, Radominska-Pandya A, et al. Monohydroxylated metabolites of the K2

synthetic cannabinoid JWH-073 retain intermediate to high cannabinoid 1 receptor (CB1R) affinity and exhibit neutral antagonist to partial agonist activity. Biochem Pharmacol 2012;83(7):952–61.

[103] Seely KA, Brents LK, Radominska-Pandya A, et al. A major glucuronidated metabolite of JWH-018 is a neutral antagonist at CB1 receptors. Chem Res Toxicol 2012;25(4):825–7.

[104] Smith K, Flatley J, Drug misuse declared: findings from the 2010/11 British Crime Survey Home Office Statistical Bulletin 2011.

[105] Blunt D. Drug misuse declared: findings from the 2011/12 crime survey for England and Wales 2nd ed. 2012.

[106] Pabst A, Piontek D, Kraus L, Müller S. Substance use and substance use disorders results of the 2009 epidemiological survey of substance abuse. Sucht 2010;56(5):327–36.

[107] Dick D, Torrance C. MixMag drugs survey. MixMag 2010;225:44–53.

[108] Winstock A. The 2011 drugs survey. MixMag 2011;238:50–9.

[109] Winstock A. MixMag/Global drugs survey. MixMag 2012;251:68–73.

[110] UNODC, World drug report 2012. United Nations publication, Sales No. E.12.XI.1, 2012.

[111] Hu X, Primack BA, Barnett TE, Cook RL. College students and use of K2: an emerging drug of abuse in young persons. Subst Abuse Treat Prev Policy 2011;6:16.

[112] Heltsley R, Shelby MK, Crouch DJ, et al. Prevalence of synthetic cannabinoids in U.S. athletes: initial findings. J Anal Toxicol 2012;36(8):588–93.

[113] Castellanos D, Singh S, Thornton G, Avila M, Moreno A. Synthetic cannabinoid use: a case series of adolescents. J Adolesc Health 2011;49(4):347–9.

[114] Monitoring the future: a continuing study of American youth. <Available: http://monitoringthefuture.org/> [accessed: 17.04.13].

[115] Johnston LD, O'Malley PM, Bachman JG, Schulenberg JE. Monitoring the Future national survey results on drug use, 1975–2011: Volume I, Secondary school students. Ann Arbor: Institute for Social Research, The University of Michigan; 2012.

[116] Werse B, Müller O, Schell C, Morgenstern C. Annual report 'MoSyD'. Drug trends in Frankfurt am Main 2010. Goethe-University Frankfurt am Main, Centre for Drug Research. 2011.

[117] Berry-Caban CS, Kleinschmidt PE, Rao DS, Jenkins J. Synthetic cannabinoid and cathinone use among US soldiers. US Army Med Dep J 2012:19–24.

[118] Johnston LD, O'Malley PM, Bachman JG, Schulenberg JE. Monitoring the Future national results on drug use: 2012 overview, key findings on

adolescent drug use. Ann Arbor: Institute for Social Research, The University of Michigan; 2013.

[119] Johnston LD, O'Malley PM, Bachman JG, Schulenberg JE. Monitoring the Future national survey results on drug use, 1975–2011: Volume II, College students and adults ages 19–50. Ann Arbor: Institute for Social Research, The University of Michigan; 2012.

[120] Wood DM, Hunter L, Measham F, Dargan PI. Limited use of novel psychoactive substances in South London nightclubs. QJM 2012;105(10):959–64.

[121] Werse B, Morgenstern C. Final report–online-survey on 'Legal Highs' (German). Goethe-University Frankfurt am Main, Centre for Drug Research. 2011.

[122] Barratt MJ, Cakic V, Lenton S. Patterns of synthetic cannabinoid use in Australia. Drug Alcohol Rev 2013;32(2):141–6.

[123] Tomiyama K, Funada M. Cytotoxicity of synthetic cannabinoids found in "Spice" products: the role of cannabinoid receptors and the caspase cascade in the NG 108-15 cell line. Toxicol Lett 2011;207(1):12–17.

[124] Koller VJ, Zlabinger GJ, Auwärter V, Fuchs S, Knasmueller S. Toxicological profiles of selected synthetic cannabinoids showing high binding affinities to the cannabinoid receptor subtype CB. Arch Toxicol 2013 (Epub ahead of print).

[125] Wood DM, Dargan I. Novel psychoactive substances: how to understand the acute toxicity associated with the use of these substances. Ther Drug Monit 2012;34(4):363–7.

[126] Vandrey R, Dunn KE, Fry JA, Girling ER. A survey study to characterize use of Spice products (synthetic cannabinoids). Drug Alcohol Depend 2012;120(1–3):238–41.

[127] ACMD. Further consideration of the synthetic cannabinoids. 2012.

[128] Every-Palmer S. Warning: legal synthetic cannabinoid-receptor agonists such as JWH-018 may precipitate psychosis in vulnerable individuals. Addiction 2010;105(10):1859–60.

[129] Every-Palmer S. Synthetic cannabinoid JWH-018 and psychosis: an explorative study. Drug Alcohol Depend 2011;117(2–3):152–7.

[130] Müller H, Sperling W, Köhrmann M, Huttner HB, Kornhuber J, Maler JM. The synthetic cannabinoid spice as a trigger for an acute exacerbation of cannabis induced recurrent psychotic episodes. Schizophr Res 2010;118(1–3):309–10.

[131] Müller H, Huttner HB, Köhrmann M, Wielopolski JE, Kornhuber J, Sperling W. Panic attack after spice abuse in a patient with ADHD. Pharmacopsychiatry 2010;43(4):152–3.

[132] Tung CK, Chiang T, Lam M. Acute mental disturbance caused by synthetic cannabinoid: a potential emerging substance of abuse in Hong Kong. East Asian Arch Psychiatry 2012;22(1):31–3.

[133] Vearrier D, Osterhoudt KC. A teenager with agitation: higher than she should have climbed. Pediatr Emerg Care 2010;26(6):462–5.

[134] Simmons J, Cookman L, Kang C, Skinner C. Three cases of "spice" exposure. Clin Toxicol (Phila) 2011;49(5):431–3.

[135] Lapoint J, James LP, Moran CL, Nelson LS, Hoffman RS, Moran JH. Severe toxicity following synthetic cannabinoid ingestion. Clin Toxicol (Phila) 2011;49(8):760–4.

[136] McQuade D, Hudson S, Dargan PI, Wood DM. First European case of convulsions related to analytically confirmed use of the synthetic cannabinoid receptor agonist AM-2201. Eur J Clin Pharmacol 2013;69(3):373–6.

[137] Schwartz MD, et al. Acute kidney injury associated with synthetic cannabinoid use–multiple states, 2012. MMWR 2013;62(6):93–8.

[138] Nelms B. Beware of the Spice of death – After teen dies, cops warn about legal 'synthetic marijuana'. 2012 (cited 2013 March 17th); Available: <http://www.thecitizen.com/articles/03-14-2012/beware-spice-death-%E2%80%94-after-teen-dies-cops-warn-about-legal-%E2%80%98synthetic-marijuana%E2%80%99>.

[139] Synthetic Marijuana Memorials. Available: <http://tothemaximusblog.org/?page_id=560> [accessed 19.03.13].

[140] Tomiyama K, Funada M. Drug discrimination properties and cytotoxicity of the cannabinoid receptor ligands. Nihon Arukoru Yakubutsu Igakkai Zasshi 2012;47(3):135–43.

[141] Ginsburg BC, Schulze DR, Hruba L, McMahon LR. JWH-018 and JWH-073: Delta(9)-tetrahydrocannabinol-like discriminative stimulus effects in monkeys. J Pharmacol Exp Ther 2012;340(1):37–45.

[142] Hruba L, Ginsburg BC, McMahon LR. Apparent inverse relationship between cannabinoid agonist efficacy and tolerance/cross-tolerance produced by Delta(9)-tetrahydrocannabinol treatment in rhesus monkeys. J Pharmacol Exp Ther 2012;342(3):843–9.

[143] Ware MA, St Arnaud-Trempe E. The abuse potential of the synthetic cannabinoid nabilone. Addiction 2010;105(3):494–503.

[144] Gunderson EW, Haughey HM, Ait-Daoud N, Joshi AS, Hart CL. 'Spice' and 'K2' herbal highs: a case series and systematic review of the clinical effects and biopsychosocial implications of synthetic cannabinoid use in humans. Am J Addict 2012;21(4):320–6.

[145] Erowid. Products – spice and synthetic cannabinoids reports. Erowid experience vaults Available: <http://www.erowid.org/experiences/subs/exp_Products_Spice_and_Synthetic_Cannabinoids.shtml#Addiction_&_Habituation> [accessed 17.04.12].

[146] Zimmermann US, Winkelmann PR, Pilhatsch M, et al. Withdrawal phenomena and dependence syndrome after the consumption of 'spice gold'. Dtsch Arztebl Int 2009;106(27):464–7.

[147] Rominger A, Cumming P, Xiong G, et al. Effects of acute detoxification of the herbal blend 'Spice Gold' on dopamine D receptor availability: A (F)fallypride PET study. Eur Neuropsychopharmacol 2013 (Epub ahead of print).

[148] Fraser AD, Worth D. Urinary excretion profiles of 11-nor-9-carboxy-Delta9-tetrahydrocannabinol: a Delta9-THC-COOH to creatinine ratio study #2. Forensic Sci Int 2003;133(1–2):26–31.

[149] Johansson E, Halldin MM. Urinary excretion half-life of delta 1-tetrahydrocannabinol-7-oic acid in heavy marijuana users after smoking. J Anal Toxicol 1989;13(4):218–23.

This page intentionally left blank

Natural Product (Fungal and Herbal) Novel Psychoactive Substances

Simon Gibbons and Warunya Arunotayanun

Department of Pharmaceutical and Biological Chemistry, UCL School of Pharmacy, London, UK

INTRODUCTION

Natural product drugs of abuse are as old as mankind with *Cannabis* (*Cannabis sativa*), Opium (from *Papaver somniferum*) and cocaine (from *Erythroxylon coca*) being ancient examples of materials which were used in pain relief or in religious ceremonies [1]. Natural product novel psychoactive substances (NPS) ('legal highs') are legally-available products from natural sources including plant or fungal materials that may be either extracts or crude plant or fungal material that contain compounds that elicit a psychoactive effect. This is distinct to the synthetic chemicals (single chemical entities; SCE) covered in the other chapters of Section 3 of this book. Both the natural product and the single chemical NPS are used as substitutes for and/or in addition to establish, classical recreational drugs such as cocaine, amphetamines, hallucinogenic materials related to tryptamine and analgesic drugs of the opiate class.

An important United Kingdom (UK)-based Internet market survey undertaken in 2011 [2] showed that 1308 kinds of product were available with an average price of £9.69. Most of these materials were in the form of pills (46.6%)

while 18.1% were single plant materials or plant extracts. The top five plant-related products by frequency were *Salvia* (*Salvia divinorum*), kratom (*Mitragyna speciosa*), Hawaiian baby woodrose seeds (*Argyreia nervosa*), fly agaric (*Amanita muscaria*) and 'Genie' (a smoking mixture containing multiple plant materials and of dubious pharmacognostical identity).

This chapter will focus on the NPS that are natural products, namely extracts of plants or the crude fungal or herbal materials themselves. The term 'legal' will be applied to the UK at the time of writing this chapter as legality varies from country to country and is time dependent as more is understood about the harms associated with these materials and individual countries consider classification of these substances.

NPS products are widely available and easily affordable from the Internet and retailers. This ready availability has resulted in an expansion of use and increasing sales of these materials in Europe and globally. According to the European Monitoring Centre for Drugs and Drug Addiction (EMCDDA) Report in 2011, the number of online sites selling NPS in July 2011 was found to be two- and three-fold greater

Novel Psychoactive Substances.
DOI: http://dx.doi.org/10.1016/B978-0-12-415816-0.00014-6

than in the last six and 18 months, respectively. A third of 631 online shops found in the study were based in the USA, while a fifth were in the UK [3]. This is cause for considerable concern and some countries, such as the USA, have seen an increase in the use of hallucinogenic substances over the past decade [4]. This increasing use coupled with a steady stream of case reports of adverse effects from NPS products shows that this area will be of societal importance for some considerable time to come [5].

Many of the natural product NPS currently in use were studied for their chemistry some time ago, and it should be noted that there is still a requirement for modern analysis of some of these complex natural materials. The paucity of data casts further doubt on the acute safety of these products and their long-term safety profile. Further natural product chemistry is needed on the analysis of the minor chemical components, their pharmacology at various receptor types, their biological activity as a mixture and their interactions with existing controlled drugs of abuse, and even their interactions with conventional medicines needs

further study. This chapter will cover the main fungal and herbal NPS and give an overview of their chemistry, pharmacological action where known and recent toxicological reports.

FUNGAL NPS

Fly Agaric (*Amanita Muscaria*)

Amanita muscaria is a member of the Basidiomycete group of fungi [6] and is the classic toadstool depicted in literature and art with a red or orange cap that is often mottled with white spots. When dry these specimens have an orange/brown colour but the mottled spotting is still clearly visible (Fig. 14.1) and NPS samples are sold as bagged-up whole basidia (caps).

Figure 14.2 shows a NPS product purchased on the Internet and disingenuously labelled 'not for consumption', despite there being no rational reason other than for research purposes for possession of this material.

Within this genus there are a number of poisonous relatives including the panther (*Amanita pantherina*), the death cap (*Amanita phalloides*)

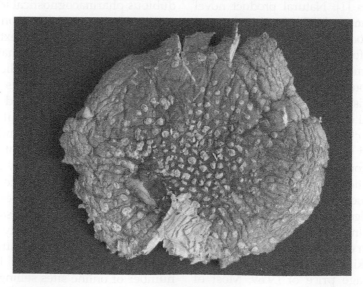

FIGURE 14.1   Cap of *Amanita muscaria* showing the distinctive mottled spots characteristic of this species.

and the delightfully termed destroying angel (*Amanita verna*) [6].

Fly agaric has a long history of use as a sedative material and the main psychoactive compounds within these species are thought to be analogues of the neurotransmitter gamma-aminobutyric acid (GABA) and glutamic acid, notably muscimol and ibotenic acid, respectively [7] (Fig. 14.3).

The natural products muscimol and ibotenic acid are isoxazole alkaloids and possess some structural similarity with GABA and both act at various parts of the GABA receptor. Muscimol is derived from ibotenic acid by decarboxylation [8]. Muscarine is a cholinergic agonist and was thought to contribute to the overall psychoactivity of *A. muscaria*. However, it was reported later that the mushroom contained only trace amount of muscarine so it is unlikely to be responsible for the psychoactive effect [9].

The use of this material is steeped in history and it has been suggested that fly agaric is Soma, the vedic drug consumed by the Indo-Iranians. Poisoning by this species has been described as the '*pantherina–muscaria*' syndrome and is 'atropine like' and comparatively rare. Symptoms can manifest between 30 minutes and two hours and include dizziness, confusion, tiredness, and increased sensitivity to visual and auditory stimuli. The 'atropine-like' effects also include dryness of mouth, pupil dilation and followed by drowsiness with a deep sleep [10,11]. Effects have been compared to alcohol consumption but with hallucinations, incoherent speech, possible seizures, vomiting, transient deep sleep or coma and persistent headache [4]. Treatment of toxicity associated with *A. muscarina* includes gut decontamination and in severe cases the use of benzodiazepines; some authors have suggested that the use of a cholinesterase inhibitor such as

FIGURE 14.2 A legal high purchase of *Amanita muscaria* (Fly Agaric) 'labelled not for consumption'.

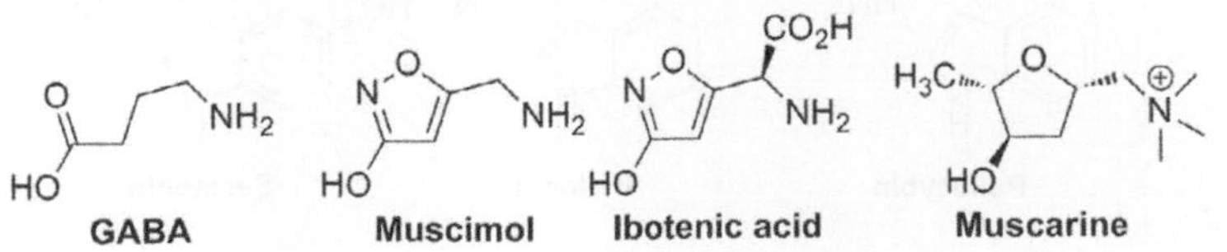

FIGURE 14.3 Structures of Gamma amino butyric acid (GABA), muscimol, ibotenic acid, and muscarine.

physostigmine may need to be considered [12]. In some countries fly agaric is consumed as food stuff but the red skin is removed and the mushroom is soaked or boiled and the resulting water is discarded, and this would presumably have an effect on reducing the concentrations of the isoxazoles, and therefore reduce the risk of psychotropic effects. A few case studies have been reported with one example of a 48- year-old male having consumed fly agaric, vomited and fell asleep 30 minutes after consumption, being found later comatose and having a seizure-like episode. Four hours after admission he was still comatose and was administered activated charcoal and awoke 10 hours after ingestion. At 18 hours his condition deteriorated with paranoid psychosis and visual and auditory hallucinations which persisted for five days. This case report suggested that a delay can occur with the onset of poisoning and that psychosis may last several days [13].

One case report has shown respiratory and cardiovascular depression with ingestion of fly agaric with myoclonus, flushing and mydriasis and treatment included the use of intravenous atropine, diazepam and mechanical/assisted ventilation [14].

Psilocybe and Related Species – 'Magic Mushrooms'

Magic mushrooms of the genus *Psilocybe* are still marketed as 'novel psychoactive substances', notably the spores and mycelia of *Psilocybe cubensis* and *Psilocybe semilanceata*, which are sold with sand as 'specimens for microscopy'. The addition of water allows fruiting of the mushrooms after a few days. Sometimes spores are sold as a suspension in solution in a sterile syringe and when this suspension is added with water to a suitable fruiting matrix, for example a breakfast cereal, the spores will germinate and produce fruiting bodies (the mushroom caps; basidia) [15].

Possession of the whole fresh or dried prepared material of *Psilocybe* is an offence in the UK and 'head shops' and websites circumvent this legislation by marketing the spores in suspension as an aid to microscopy and a microbiological training aid. The main psychoactive components are, however, Class A drugs in the UK.

Specifically, these species contain the phosphate ester psilocybin which is hydrolyzed *in vivo* to psilocin, which is the main psychoactive component of magic mushrooms and shows marked similarity with serotonin (Fig.14.4). Psilocybin is a serotonin agonist and has high affinity for the $5HT_{2A}$ receptors [4].

The symptoms of psilocybin and psilocin intoxication are similar to that of LSD and mescaline, but with a shorter action [16]. The acute toxicity of aqueous extracts of *P. cubensis* has been investigated in rats with an increased rearing behaviour, gnawing and even a gender difference being observed [17]. Psilocybin-like myocardial toxicity has also been investigated [18] and the authors concluded that sub-chronic intoxication may lead to a magnesium imbalance without affecting the concentrations of calcium, sodium, potassium or chloride.

FIGURE 14.4   The structures of the main components of *Psilocybe*, psilocybin and psilocin, show marked similarity with serotonin.

In the 1960s and 1970s collections of the Liberty Cap (*Psilocybe semilanceata*) across various parts of the UK was a popular endeavour; users reported vivid visual hallucinations and auditory disturbances [19]. Case reports for the sensory hallucinogenic effects of magic mushrooms are manifold [20,21] and whilst significant systemic toxicity appears to be rare, the main cause of fatality arises from misadventure due to psychosis [22]. In a Swedish study looking at intoxications by analysis of urine content over a 4-year period covering 103 cases, psilocin was found to be the most frequently observed psychoactive and was found in 54% of cases [23].

An example of this includes resisting arrest and the assault on a police officer [24] where the assailant was intoxicated with psilocin after ingesting 4 mg of magic mushrooms and attempts to subdue the assailant with nine bean bag rounds and multiple Taser usage failed. The man subsequently was shot trying to enter a police car containing a loaded rifle. Analysis of a post-mortem urine sample reported 4200 ng/mL of psilocin. This case demonstrates the psychotic, bizarre and 'seemingly purposeful behavior' [24] that these materials can elicit. However, an important and recent review by Van Amsterdam and co-workers [25] on *Psilocybe* use in the Netherlands has concluded that its use is relatively safe and that only a few and mild adverse effects have been reported. According to the Coordination point Assessment and Monitoring new drugs (CAM), Netherlands, the dependence potential of the mushroom was low as well as chronic toxicity while acute toxicity was moderate. However, the unpredictable nature of panic attacks and flashbacks associated with this material remain a point of concern.

Herbal NPS

Catha edulis (Khat)

Khat (*Catha edulis* Forsk. ex Endl.) is a small tree from the Celastraceae plant family that grows in East Africa and the Southern Arabian Peninsula. The crude drug material has a number of names in various locations such as 'qat' (Yemen), 'tchat' (Ethiopia), 'qaad or jaad' (Somalia) or 'miraa' (Kenya) [26]. The chewing of khat is part of the ethnic culture of Yemeni, Somali and East African societies especially at social gatherings. The fresh leaves of this species are consumed every day by approximately 80–90% of men and 10–60% of the women in East Africa [27], primarily due to its central nervous system (CNS) stimulating and euphoric effects [28].

Khat use has spread with immigration to other parts of the world, such as Europe and America, by East Africans. Users ingest approximately 100–500 g of leaves or stem bark by chewing every one or two hours to ingest the juice. Chewing is the most common form of using khat; other routes of use include drinking as a tea, smoking or nasal insufflation, although these routes are far less common [29,30]. Various natural products of the alkaloid, flavonoid, terpenoid, tannin, and glycoside classes have been characterised but it is the alkaloids of the phenylethylamine group that are the main pharmacological principles [28,29,31]. *S*-cathinone (Fig. 14.5), the

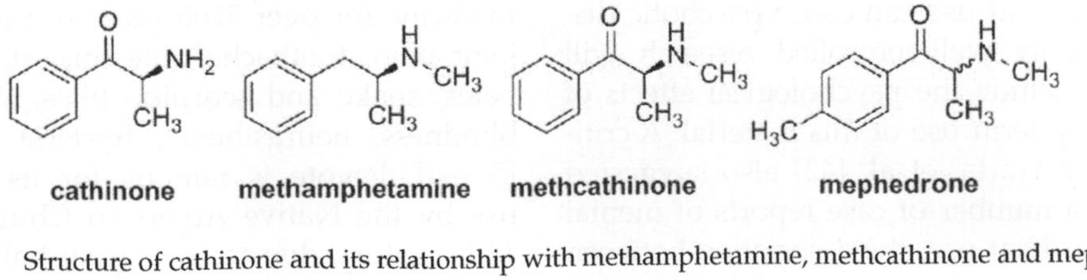

cathinone   methamphetamine   methcathinone   mephedrone

FIGURE 14.5   Structure of cathinone and its relationship with methamphetamine, methcathinone and mephedrone.

main phenylethylamine alkaloid is found to be present at around 78–343 mg/100 g of fresh khat leaves [28,32], and is currently thought to be the major active component responsible for the pharmacological effects [29,32,33]. However, S-cathinone can be easily decomposed by light, heat and human enzymes into cathine (1S,2S(+)-norpseudoephedrine) and 1R,2S–norephedrine [31]. These metabolites of cathinone demonstrate lower psychostimulant effects due to their less lipophilic properties [29].

Khat has often been called 'natural amphetamine' [31] as it has amphetamine-like effects in central and locomotor stimulation via a similar mode of action to amphetamine by inducing dopamine release and inhibiting dopamine reuptake [29,34,35]. The structure of cathinone and its derivatives such as methcathinone are highly similar to amphetamine, but cathinone has half the potency of amphetamine and cathine is 7–10 times less potent [35,36]. During chewing, over 90% of the alkaloids are released from khat material [33] and can be well absorbed through oral mucosa, with absorption efficiencies of 60% and 80% for cathinone and cathine, respectively [37]. In addition to the major phenylethylamine alkaloids, other classes of alkaloids can be also detected, notably merucathinone, pseudomerucathine, merucathine [38] and the highly complex cathedulins [39].

Several reports detailing a wide range of effects on khat chewers have been published and these include hypertension and arrhythmia due to CNS stimulation [40], alertness, insomnia, anxiety, dizziness, anorexia (a typical amphetamine-use condition), impairment of concentration, irritability and even paranoid psychosis [29]. A review by Odenwald [41] has shown that excessive khat use can cause psychotic disorders and only well-controlled research will be able to conclude the psychological effects of short and long-term use of this material. A critical review by Warfa, et al. [42] also suggested that despite a number of case reports of mental disorder from khat use, the association between khat consumption and psychiatric disorder was still ambiguous and well-designed studies are necessary. Chronic use of khat and synthetic cathinone derivatives have been reported to be a risk factor for several diseases, for example acute myocardial infarction [43,44], haemorrhoids [45], acute cerebral infarction [46], duodenal ulcer [47], oesophageal cancer [48], as well as having profound effects on reproductive [49], behavioral and cognitive [36] function. Futhermore, there are a number of acute and chronic liver disease cases associated with khat chewing, especially when taken together with alcohol or other drugs [50,51]. Consumption of khat has also been linked to fatality from impairment of driving ability [52] or serious cardiac disease due to cathinone's activity to inhibit noradrenaline uptake, which may lead to cardiac morbidity as seen in cocaine intoxication [53]. Moreover, several studies reported the addictive potential of khat due to the dopaminergic effects and noradrenaline reuptake inhibition of cathinone [54]. It is notable that khat is still legal in some countries such as the UK and the Netherlands whilst it is banned in a number of countries, including the USA, Canada, France, Denmark, Germany and Ireland [55].

Lophophora williamsii (Peyote)

Lophophora williamsii is a member of the Cactaceae plant family known as peyote or peyotl, and is a well-known psychoactive cactus found in deserts from Central Mexico to Northern Texas [56]. It is available from horticultural specialists in Europe and gardening centres as an attractive house plant. Its traditional use is ancient and it has been widely used in ethnomedicine for over 5700 years to treat influenza, joint pain, toothache, intestinal disorders, diabetes, snake and scorpion bites, skin diseases, blindness, neurasthenia, hysteria and asthma [56–58]. Peyote is famous for its sacramental use by the Native American Church since the 16th century due to the visual hallucination of

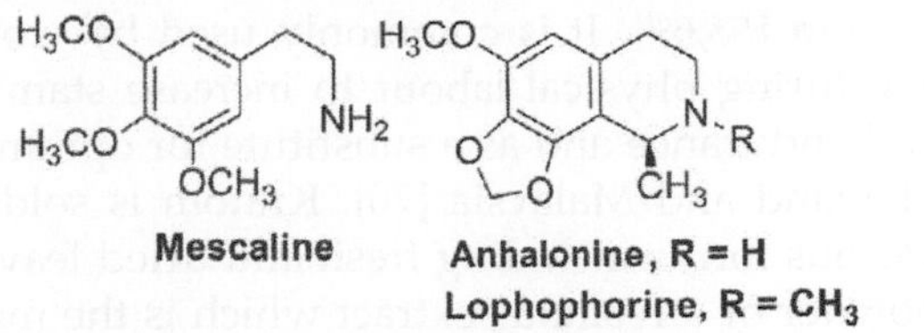

FIGURE 14.6 Mescaline and other components of Peyote, San Pedro and Peruvian Torch cacti.

mescaline, the major phenylethylamine alkaloidal component (Fig. 14.6). Peyote is banned in the USA except for religious purposes whereas in the UK, it is legally sold online and in head shops as fresh or dried plant material. The most common route of administration is by ingestion in forms of fresh or dried buttons, dried powder, capsule or as a tea [59]. Adverse effects experienced by most users are hallucinations, alteration of consciousness and perception, physical reactions such as 'respiratory pressure' and muscle tension can occur and nausea and induction of vomiting (emesis) due to its bitter taste. The hallucinogenic activity and effects in autonomic functions of mescaline are similar to those of LSD, psilocin and psilocybin but with longer onset and duration [16,60]. The principal hallucinogenic constituent is mescaline (3,4,5-trimethoxy-β-phenylethylamine) (Fig. 14.6) and is found up to 6% of dried-button weight [57]. The chemistry of peyote has been extensively studied, and a large spectrum of alkaloids with over 60 different structures from the phenylethylamine and tetrahydroisoquinoline groups, such as anhalonine and lophophorine have been reported (Fig. 14.6). Fresh buttons of peyote can contain up to 8% of total alkaloids [58] and these accumulate with age.

Many studies on L. williamsii highlight its taxonomy and traditional use rather than pharmacological activity. The use of peyote and mescaline is uncommon although the clinical effects are significant and can necessitate medical intervention. However in a series of recent case reports, no life-threatening toxicity was observed.

Only mild to moderate reactions including hallucinations, tachycardia, agitation and mydriasis were reported [61]. The symptoms of mescaline along with many other plant-derived psychoactive compounds have been reviewed. The acute toxicity of mescaline was not as severe as some other natural psychoactive substances mentioned in the review; however, there was a fatal case resulting from peyote consumption [62]. Moreover, there was a report that an inappropriate use of mescaline during pregnancy for religious purpose was associated with fetal abnormality [63]. Studies on the combination of peyote components are needed, particularly the ability of compounds such as hordenine to modulate mescaline activity. The quinoline alkaloids anhanoline and lophophorine are effectively 'masked' phenylethylamines (compounds which have the phenylethylamine moiety within their structure) and further work on their psychoactive properties is also warranted.

Trichocereus Species

Apart from L. williamsii, there are two other popular psychoactive mescaline-containing cacti in the genus Trichocereus, which are Trichocereus pachanoi (Echinopsis pachanoi) and Trichocereus peruvianus. These are known as the 'San Pedro' and 'Peruvian Torch' cacti respectively. These cacti are native to the Andean region of South America and are commonly used in shamanistic treatments by decoction of sliced pieces of the cacti [4,59,64]. Although they are one of the most popular legal highs [2,59], there are very few studies relating to their chemistry, activities and toxicity. Mescaline is a major component and the key constituent contributing to their psychoactive activities. T. pachanoi was found to contain mescaline at approximately 0.12% of fresh plant [64] and from 0.33% to 2.37% by dry weight while the mescaline content of dried T. peruvianus was found to vary from 0% to 0.82% [4]. A recent study quantifying mescaline concentrations in 14 taxa/cultivars of plants in Trichocereus showed that *T. pachanoi* (Matucana) contained the highest mescaline concentration (4.7% of dry weight) whilst *T. peruvianus* contained only 0.24% of mescaline [65,66]. This study can clearly explain why *T. pachanoi* was the selected plant among the *Trichocereus* species in indigenous practice, with users selecting this particular species. Only mescaline has been the main focus in the majority of studies on the San Pedro cactus and much further chemical analysis and biological evaluation remains to be completed.

Mitragyna Speciosa

Mitragyna speciosa Korth. (Rubiaceae) is a psychoactive plant grown in Southeast Asia, especially in Thailand and Malaysia where it is called 'kratom' and 'Biak-Biak', respectively. Growing and buying kratom in the source countries is illegal whilst in several western countries it is freely purchased and appears to be widely used and the most commonly sold legal highs identified in 2011 [3]. Traditionally, Kratom has been used an analgesic and a material to reduce fever [67]. It has potential medical use as an alternative for chronic pain and opioid withdrawal self-therapy due to the opioid agonist activity of the major alkaloids in

kratom [68,69]. It is commonly used by workers during physical labour to increase stamina and endurance and as a substitute for opium in Thailand and Malaysia [70]. Kratom is sold in various forms including fresh and dried leaves, powder or a resinous extract which is the main form of NPS (Fig 14.7) in the UK.

A common route of administration is by chewing the fresh leaves at a dosage of normally 10 to 30 leaves per day. Kratom can be ingested as crushed dried leaves by taking powder, drinking as a tea or by smoking the leaves or the extract [71]. Mitragynine (Fig. 14.8) is the major alkaloid (up to 66% in the extract) in kratom, and is the principle compound responsible for analgesic activity due to its potent opioid agonist property [70,72]. Although mitragynine can act on the mu (μ)- and kappa (κ)-opioid receptors, it is structurally different from morphine and other opioid narcotic pain-killers. Mitragynine and its analogues in kratom are indole alkaloids of the *Corynanthe*-type possessing a monoterpene (iridoid) moiety. The mitragynine concentration in kratom leaves from Malaysia (12%) has been found to be less than the leaves from Thailand (66%) [70]. Several 9-methoxy-*Corynanthe*-type monoterpene indole alkaloids are also present

FIGURE 14.7   (A and B) Legal high samples of kratom (*Mitragyna speciosa*) resin.

as constituents in *M. speciosa* leaves and these include speciogynine (7%), paynantheine (9%), speciociliatine (1%) (Fig. 14.9) [70].

Recently, 7-hydroxymitragynine (Fig. 14.9), a minor constituent (2%) of *M. speciosa*, was isolated and demonstrated potent antinociceptive activity in mice. It is now considered to be a major contributory factor for the analgesic properties of *M. speciosa* due to its selectivity for μ- and κ-opioid receptors. The presence of an hydroxyl group at C-7 increases the potency of 7-hydroxymitragynine to be 13- and 46-fold higher than morphine and mitragynine, respectively [70,73]. This clearly indicates that this is one of the main pharmacological markers of kratom products' quality and potency.

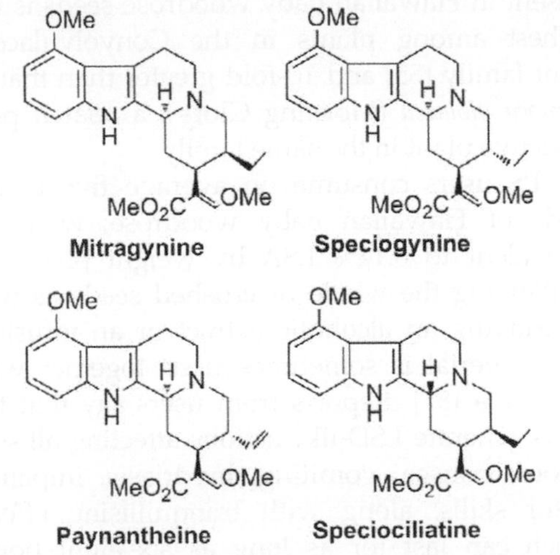

In addition to analgesic activity, mitragynine is also a key component for the anti-inflammatory properties of kratom by suppressing prostaglandin E2 (PGE-2) production in the cyclooxygenase 2 (COX-2) pathway [73]. Whilst kratom is reputed to be a potent analgesic, it has also been shown to demonstrate a wide range of adverse effects. Opioid-like adverse effects have been observed and include constipation, dry mouth and loss of appetite [74]. There have also been reports of patients suffering from intra-hepatic cholestasis after two weeks of kratom use [71] and seizure and coma [75,76] which might result from opioid agonist action of the major components in kratom. Currently, information on the safety of using this material is scarce but there have been studies in mice showing serious conditions after administration, for example, elevated blood pressure and hepatic enzymes after a single dose [74], impaired cognition and behaviour from long-term use [77] and acute lethally hepatotoxic and mild nephrotoxic effects after high dose administration [74]. Kratom extracts and mitragynine have also been shown to possess cytotoxicity to some human cancer cell lines namely SH-SY5Y cells (neuronal cells) [78].

Whilst kratom metabolites could have the potential to be developed as new therapeutic agents, for example for pain and narcotic withdrawal treatment, there are of course possible serious adverse effects of these materials including potential addiction [79]. Despite the increasing use of kratom, reports of severe toxicity in the literature are rare and its adverse effects are not well understood [71]. A report of liver toxicity [80] and a combination of mitragynine and *O*-desmethyltramadol have been published [81] as has a case of seizure and coma recently reported following kratom use [75].

A study looking at 'kratom dependence syndrome' has suggested that as it is a short-acting μ-opioid receptor agonist, therapeutic agents such as dihydrocodeine and lofexidine are effective

FIGURE 14.8 Mitragynine and related indole alkaloids are the main psychoactive constituents of kratom.

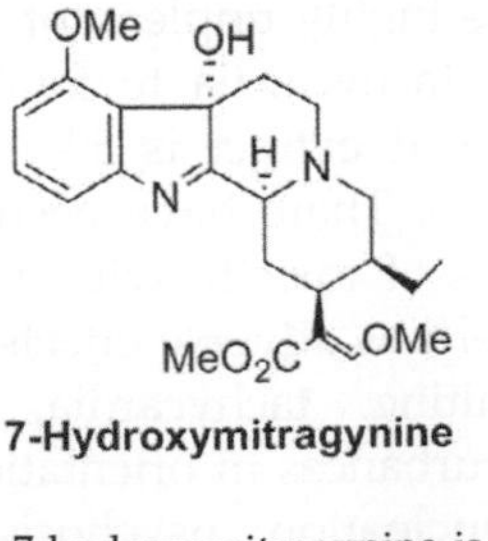

7-Hydroxymitragynine

FIGURE 14.9 7-hydroxymitragynine is a recently characterized psychoactive component of *Mitragyna speciosa*.

in aiding detoxification [82]. Further studies on kratom toxicology and other natural NPS are crucial to understand the harms associated with this material due to their increasing popularity.

Argyreia nervosa (Hawaiian Baby Woodrose)

Argyreia nervosa syn. *Argyreia speciosa*, also known as Hawaiian baby woodrose, elephant creeper and woolly morning glory, is a large climber in the Convolvulaceae plant family and is a relative of the morning glories and bindweeds [83]. In Ayurvedic medicine, every part of the plant including the seed, leaf, bark and root have usage as they possess a broad-range of pharmacological activities such as antimicrobial, antidiarrhoeal, hepatoprotective, anticonvulsant, antioxidant, aphrodisiac, immunomodulatory, analgesic and anti-inflammatory activity [83]. The seeds are the main NPS materials used as a hallucinogen, and have been used traditionally in a number of diseases in India because of their hypotensive, spasmolytic and anti-inflammatory properties while in Hawaii they are used for religious and sacramental purposes [83].

A. nervosa is recognised as a plant containing lysergic acid amide (LSA), also known as ergine (0.04% by weight) (Fig. 14.10) [84], a precursor to lysergic acid diethylamide (LSD, LSD-25), a well-known synthetic hallucinogenic substance and controlled drug of abuse.

However, neurological effects of LSA are similar to those of scopolamine and not to LSD despite the high degree of similarity between both structures (Fig. 14.10). The major components in seeds of *A. nervosa* are alkaloids (0.5–0.9% by weight) [85], mainly the ergoline-type alkaloids including ergine (d-lysergic acid amide, LSA) and isoergine (l-lysergic acid amide, the isomer of LSA). These two natural products are found in the highest percentage at 0.136% and 0.188%, respectively, of total alkaloids along with ergometrine, lysergol, isolysergol and chanoclavine [84,85]. The amount of indole alkaloids

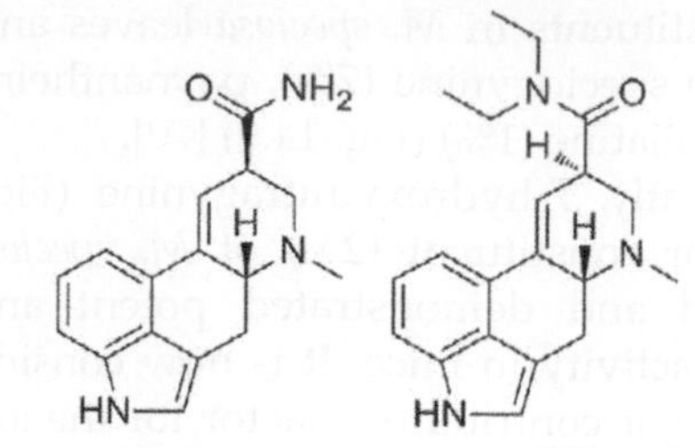

Ergine　　　　**LSD**
(Lysergic acid amide)

FIGURE 14.10　Lysergic acid amide (ergine), a component of Hawaiian baby woodrose is structurally similar to LSD.

present in Hawaiian baby woodrose seeds is the highest among plants in the Convolvulaceae plant family [85] and 10-fold greater than that of *Ipomoea violacea* (Morning Glory), a related psychoactive plant in the same family.

NPS users consume on average five to ten seeds of Hawaiian baby woodrose, which is equivalent to 0.14% LSA by weight [4,86], by swallowing the whole or crushed seeds as well as drinking an alcoholic extract or an infusion. This material is sometimes used together with marijuana [87]. Reports from users say that the seeds generate LSD-like actions affecting all sensations, nausea, vomiting, mydriasis, impaired motor skills, along with tranquillising effects which can last for as long as six–eight hours [86,88]. Hawaiian baby woodrose seeds can often be confused with the seeds of *I. violacea* which are normally dosed at 100–300 seeds (0.02% LSA) [4]. Ingesting more than 12 seeds of *A. nervosa* can cause highly unpleasant effects such as agitation and tachycardia to fatal doses where the LD_{50} of seed extract is 500mg/kg of body weight [4,87,89]. There have been a number of clinical reports of toxicity with reports describing mild to serious adverse effects ranging from nausea, vomiting, tachycardia, hypertension, agitation, disturbances in orientation, visual and auditory hallucination, psychosis and anxiety [86,90]. In one case an individual experienced hallucinations after ingesting the seeds together with smoking *Cannabis* and he was found dead after jumping from a fourth floor [91].

Banisteriopsis caapi and *Psychotria viridis* (Ayahuasca)

Banisteriopsis caapi is a South American hallucinogenic vine in the Malpighiaceae plant family, and is well recognised as a main ingredient of the famous sacred drink called 'ayahuasca' along with the plant *Psychotria viridis* [92,93]. The brew has been traditionally used by ethnic groups for ritual, medicinal and recreational purposes [94,95]. Over the last decade, the use of ayahuasca has spread outside of South America to some religious groups in the USA and European countries as a NPS material [96]. The beverage is usually prepared by boiling or soaking two or more potent psychotropic plants that are native to the Amazon. The most commonly used plants are the stems of the vine *B. caapi* together with an adjuvant plant for instance the leaves of certain Rubiaceae species such as *P. viridis*, or *Diplopterys cabrerana* (Malpighiaceae; syn. *Banisteriopsis rusbyana*) as well as plants in the Solanaceae family such as *Nicotiana* sp., *Datura* sp. and even *Capsicum* sp [92,97,98]. The major components reported in *B. caapi* are β-carboline alkaloid derivatives (0.05–1.95% of dry weight), which mainly include harmine, harmaline and tetrahydroharmine (Fig. 14.11) [92].

The concentration of alkaloids detected in *B. caapi* depends on its origin and part used. For example, the root was found to contain the highest percentage of alkaloids by dry weight compared to other parts of the same plant specimen [97]. McKenna et al., 1984 [94], reported that Peruvian Ayahuasca possessed a high alkaloid content and an average dose 100 mL of the Ayahuasca drink contained 728 mg of total alkaloids consisting of 467 mg of harmine, 160 mg of tetrahydroharmine, 41 mg of harmaline, and 60 mg of dimethyltryptamine (DMT), the active

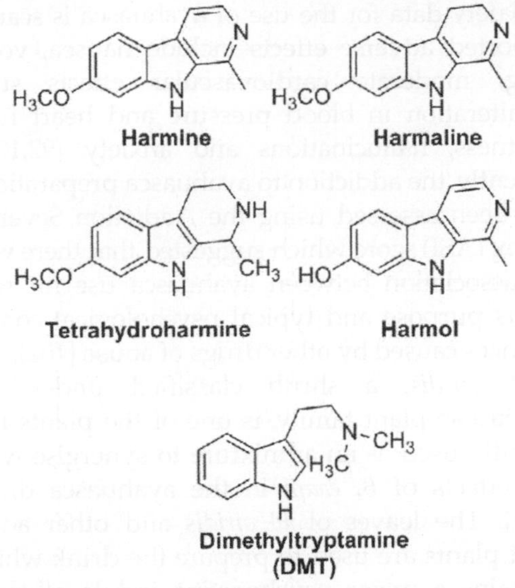

FIGURE 14.11   β-carboline and tryptamine alkaloids of 'Ayahuasca'.

constituent in the admixture plant, which is normally *P. viridis* (Fig. 14.11) [94].

The β-carboline harmaline-type compounds are useful as markers for the identification and standardisation of *B. caapi* samples [98]. Apart from the religious and recreational use of *B. caapi*, it has been shown to have potential for the treatment of neurological disorders. According to Samoylenko et al., 2010, harmine and harmaline demonstrated potent *in vitro* inhibitory activity against monoamine oxidase (MAO)-A and -B enzymes in human brain as well as having the ability to stimulate dopamine release [98,99]. Additionally, proanthocyanidins (−)-epicatechin and (−)-procyanidin that are also present in *B. caapi* showed potent moderate MAO-B inhibitory activities and antioxidant properties which is helpful for the protection of neuronal cell damage from oxidative free radicals [98]. These results support the use of *B. caapi* stem extract for as having potential as a lead for the development of novel therapeutics for Parkinson's disease and other neurodegenerative disorders [100].

Safety data for the use of ayahuasca is scarce. Reported adverse effects include nausea, vomiting, moderate cardiovascular effects such as alteration in blood pressure and heart rate, alertness, hallucinations and anxiety [92,101]. Recently, the addiction to ayahuasca preparations has been assessed using the Addiction Severity Index (ASI) score which suggested that there was no association between ayahuasca use for religious purpose and typical psychological consequences caused by other drugs of abuse [102].

P. viridis, a shrub classified under the Rubiaceae plant family, is one of the plants frequently used as an admixture to synergise with the effects of B. caapi in the ayahuasca drink [103]. The leaves of P. viridis and other adjuvant plants are used to prepare the drink which contains a major psychoactive indole alkaloid N,N-dimethyltryptamine (DMT) (Fig. 14.11), a practically ubiquitous natural product in many species of Leguminosae [94]. The total alkaloid concentrations detected in the leaves of P. viridis ranged from 0.1 to 0.66 % of dry weight [94,97]. DMT is known as a potent hallucinogen, and its structure closely resembles that of serotonin, our endogenous monoamine neurotransmitter and produces similar effects via the various serotonin receptors. Ingesting DMT can cause mood swings and visual, auditory, sensational and perceptual alterations [104]. However, DMT is orally inactive as an hallucinogen since peripheral monoamine oxidase (MAO) can break down DMT before reaching the central nervous system [92,94]. Therefore, it has to be taken together with a plant containing an MAO inhibitor like B. caapi to prevent DMT degradation [92]. The combination of these two plants helps to synergise the psychoactive effect.

It has been found that P. viridis samples in markets contain a wide range of alkaloids and some samples had only minute or undetectable amounts of DMT [105]. DMT is classified as a Schedule I Controlled Substance in the USA and is a Class A controlled drug in the UK [103,106], but it is worrying that many plants containing DMT can be readily bought online or in head shops in different forms without any form of regulation.

Fatal toxicity of ayahuasca preparations have been recorded with subsequent analysis showing the presence of Psychotria and Banisteriopsis natural products [107]. The risks of ingesting plant materials which drastically effect perceptual alterations are obvious especially visual hallucinations. An Internet survey concluded that the online vendors of ayahuasca preparations did not provide any advice or instructions on usage with regard to safety and toxicity and certainly no indication that these materials could interact with other drugs [108].

A study looking at the risks associated with oral use of N,N-dimethyl tryptamine (DMT) and harmaline alkaloids has concluded that their safety margin is comparable to codeine, mescaline or methadone. The risk of sustained psychological disturbances is minimal as the prevalence rate was approximately 1.3 % [109].

Ayahuasca preparations have also been proposed as potential treatments for drug addiction however too few studies have been conducted to substantiate this [110].

Salvia divinorum ('Psychedelic Sage')

Salvia divinorum L., a plant in the Lamiaceae family and native to the Mexican Mazatec, is a prominent NPS product being sold on UK websites in the forms of live plants, dried leaves and extracts in the name of 'Salvia' [2]. S. divinorum is an attractive horticultural plant (Fig. 14.12) and is variously known as Psychedelic sage, Salvia, Diviner's sage, Ska Maria, Ska Pastora, Hojas de Mariais and Hojas de Petora [111].

The plant is exceptionally easy to cultivate and, like culinary sage (Salvia officinalis), it can be cultured with cuttings with compost, not even requiring rooting hormone. Traditionally, the Mazatec used these leaves as a potent hallucinogen which was administered by either chewing, drinking or smoking [112]. Recreational users of

FIGURE 14.12 The flowers of *Salvia divinorum* (Lamiaceae), a popular UK legal high.

this NPS normally ingest an infusion or the fresh leaves of *S. divinorum*, or the material is smoked causing both a rapid onset and a short duration of the hallucinogenic effect [113]. Users report depersonalisation, laughter, weightlessness and self-consciousness disappearing within 30 minutes of usage [114]. There are limited data on the clinical effects of this material and *S. divinorum* may have long-term effects such as déjà vu and a recent review discusses evidence for potential abuse [115].

Salvinorin A, a non-nitrogenous *neo*clerodane diterpene (Fig. 14.13), was determined as the key active substance for the psychoactive activity due to its selective κ-opioid receptor (KOR) agonist properties.

Salvinorin A was found to be the first non-alkaloidal KOR selective drug having a unique structure being different from previous known hallucinogens [116] and was regarded as a potent hallucinogen equivalent to the synthetic lysergic acid diethylamide (LSD; LSD25) and 4-bromo-2,5-dimethoxyphenylisopropylamine (DOB) [112,116]. *S. divinorum* also contains a whole range of other diterpenoid compounds and the chemistry of this species is becoming better delineated [111].

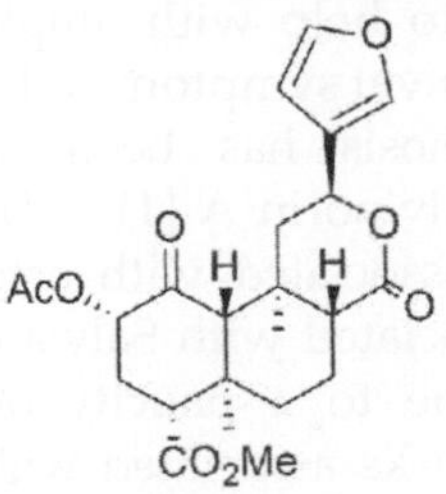

Salvinorin A

FIGURE 14.13 Salvinorin A, a kappa-opioid receptor agonist from the 'Psychedelic sage' *Salvia divinorum*.

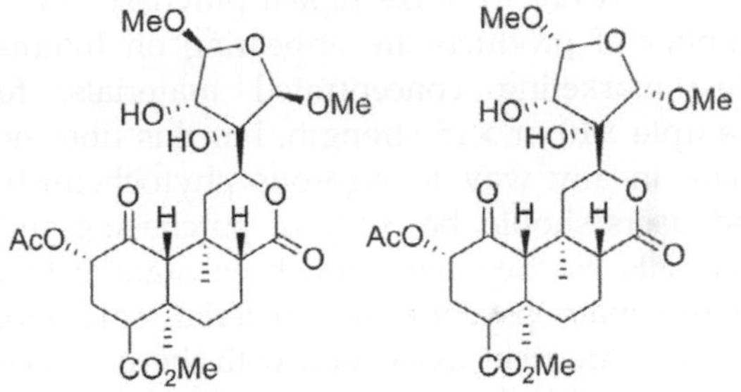

Salvinicin A   **Salvinicin B**

FIGURE 14.14 Salvinicins A and B from *S. divinorum* are partial κ- and μ-opioid receptor agonists.

Only salvinorin A has been demonstrated to have high KOR affinity and the methyl ester and furan ring were found to be essential for its activity. However, there are recent reports concerning the biological activity of two new *neo*cleodane diterpenes, salvinicins A and B (Fig. 14.14) as partial κ- and μ-opioid receptor agonists respectively, but their relevance to the hallucinogenic activity of *S. divinorum* is still unclear [111,117].

Due to its KOR agonist ability, *S. divinorum* is proposed to demonstrate the pharmacological activities relating to the KOR, for example, analgesia, sedation and depressant effects and this has an implied potential utility in the treatment of insomnia, schizophrenia, depression, the hallucinations associated with dementia, for example, Alzheimer's disease, and as a potential aid to help with amphetamine- and opiate-withdrawal symptoms [116].

Toxic psychosis has been observed after ingestion of salvinorin A [118]. Users report an intense high associated with consumption but the harms associated with Salvia use are poorly understood due to a paucity of toxicological data and the risks associated with this are that users may feel that the material is consequently safe due to a lack of adverse reports. The strong psychotic effects of Salvia could put users at risk due to impairment of judgment [119]. However, the plant does not appear to be addictive and users tend not to make repeat purchases [3]. A number of products are appearing on Internet sites marketing concentrated materials, for example ×10 or ×15 strength, but this does not relate in any way to rigorous phytochemistry and users should be wary of purchasing such materials as they have not been standardised on the main psychoactive material, salvinorin A. There are risks associated with these concentrated forms if the extracts turn out to be much higher in salvinorin A concentration. For a concise review of the science including the chemistry, pharmacology and toxicology of *S. divinorum* the reader should consult Prisinzano [120].

CONCLUSIONS

Natural product NPS offer many challenges in terms of analysis of their chemistry and assessment of their pharmacology and toxicology. Firstly in acquiring crude plant or fungal material, the user has little idea of the true botanical or mycological identity of the material, with substitution being a real possibility. This could be further compounded by spraying of the crude drug with a synthetic psychoactive compound as was seen with the Spice smoking mixtures being adulterated with synthetic cannabinoid receptor agonists.

Natural materials are inherently variable in terms of their chemistry with many factors such as weather, soil, geographic location, effects by microbes and herbivory effecting the concentration of natural products within a sample and therefore having the opportunity to drastically change the biological properties of this material. This is further complicated by the existence of chemical races within a single species, where the chemistry may be different from once race to another.

Some herbal NPS are also marketed as extracts, with examples such as kratom and Salvia being popular. The type of extraction methodology used will drastically affect the phytochemical quality of the end product, and if the manufacturers have no idea of the merits of an extraction protocol, they are likely to produce extracts of high variability. Some extracts are marketed as ×5, ×10 or ×15 but this is a meaningless concentration factor as it does not take in to account standardisation of the extract on a psychoactive component, for example salvinorin A or mitragynine.

There is also the temptation that NPS suppliers will adulterate an extract at the point of its manufacture, thereby adding a further level of complexity to the area and potential drug–drug interactions.

Natural product NPS materials are exceptionally complex in terms of their chemistry. This greatly enhances the complexity of their biology with a paucity of data relating to the toxicology of these materials, and even less regarding their interactions with conventional drugs of abuse such as cocaine and other stimulants. The level of complexity, variability, the unknown nature of these samples, coupled with the risks associated with taking psychotic materials, particularly the hallucinogenic tryptamine-containing materials, could offer further risks of ill health by misadventure, with potentially life-threatening consequences.

ACKNOWLEDGMENTS

Dr Wolfgang Schuehly of the Department of Pharmacognosy at the University of Graz and Mr Michael Wasescha (Zurich) are thanked for the beautiful image of *Salvia divinorum*. Ms Sabine Heinrich (UCL School of Pharmacy) is thanked for her help with literature and database searching.

REFERENCES

[1] Musto DF. Opium, cocaine and marijuana in American history. Sci Am 1991;265(1):40–7.

[2] Schmidt MM, Sharma A, Schifano F, Feinmann C. 'Legal highs'on the net – Evaluation of UK-based websites, products and product information. Forensic Sci Int 2011;206(1-3):92–7.

[3] EMCDDA. Online sales of new psychoactive substances/'Legal highs': Summary of results from the 2011 multilingual snapshots. 2011.

[4] Halpern JH. Hallucinogens and dissociative agents naturally growing in the United States. Pharmacol Therapeut 2004;102(2):131–8.

[5] Wood DM, Dargan PI. Novel psychoactive substances: how to understand the acute toxicity associated with the use of these substances. Ther Drug Monit 2012;34(4):363–7.

[6] Bonnet MS, Basson PW. The toxicology of: the destroying angel. Homeopathy 2004;93(4):216–20.

[7] Michelot D, Melendez-Howell LM. Amanita muscaria: chemistry, biology, toxicology, and ethnomycology. Mycol Res 2003;107(2):131–46.

[8] Feeney K. Revisiting Wasson's Soma: exploring the effects of preparation on the chemistry of Amanita muscaria. J Psychoactive Drugs 2010/12/01;42(4):499–506.

[9] Schultes RE. The botanical and chemical distribution of hallucinogens. Annu Rev Plant Physiol 1970:571–98.

[10] Benjamin DR. Mushroom poisoning in infants and children: the Amanita pantherina/muscaria group. Clin Toxicol 1992;30(1):13–22.

[11] Davis DP, William SR. Visual diagnosis in emergency – Amanita muscaria. J Emerg Med 1999;17:739.

[12] Siptak C, Banerji S, Shaw M, Bronstein A. A summer of mushroom poisoning: cluster of 23 human exposures to *Amanita pantherina* and *Amanita muscaria*. Clin Toxicol 2006;44:698.

[13] Brvar M, Možina M, Bunc M. Prolonged psychosis after *Amanita muscaria* ingestion. Wien Klin Wochenschr 2006;118(9):294–7.

[14] Daubert GP, Bora K, Wilson J, Hedge M. Cardiovascular suppression in acute Amanita muscaria overdose. Clin Toxicol 2006;44(5):699.

[15] Lott JP, Marlowe DB, Forman RF. Availability of websites offering to sell psilocybin spores and psilocybin. J Psychoactive Drugs 2009;41(3):305–7.

[16] Wolbach AB, Miner EJ, Isbell H. Comparison of psilocin with psilocybin, mescaline and LSD-25. Psychopharmacology 1962;3(3):219–23.

[17] Kirsten TB, Bernardi MM. Acute toxicity of Psilocybe cubensis (Ear.) Sing., Strophariaceae, aqueous extract in mice. Rev Bras Farmacogn 2010;20:397–402.

[18] Majdanik S, Borowiak K, Brzenzinska M, Machoy-Mokrzynska A. Concentration of selected microelements in blood serum of rats exposed to the action of psilocin and phenylethylamine. Ann Acad Med Stetin 2007;53:153–8.

[19] Antkowiak R, Antkowiak WZ. Alkaloids from Mushroomsroza. In: Arnold B, editor. The alkaloids: chemistry and pharmacology. Waltham, MA: Academic Press; 1991. p. 189–340.

[20] Satora L, Goszcz H, Ciszowski K. Poisonings resulting from the ingestion of magic mushrooms in Krakow. Przegl Lek 2005;62:394–6.

[21] Olsen E, Knudsen L. Mushroom poisoning in the Faeroe Islands. General aspects of mushroom poisoning in the Faeroe Islands following a case of deliberate poisoning with *Psilocybe semilanceata*. Ugeskr Laeger 1983;145(15):1154–5.

[22] Marciniak B, Ferenc T, Kusowska J, Ciećwierz J, Kowalczyk E. Poisoning with selected mushrooms with neurotropic and hallucinogenic effect. Med Pr 2010;61(5):583–95.

[23] Björnstad K, Hultén P, Beck O, Helander A. Bioanalytical and clinical evaluation of 103 suspected cases of intoxications with psychoactive plant materials. Clin Toxicol 2009;47(6):566–72.

[24] French LK, Burton BT. Liberty and death. Clin Toxicol 2010;48:631.

[25] Amsterdam JV, Opperhuizen A, Brink WVD. Harm potential of magic mushroom use: a review. Regul Toxicol Pharmacol 2011;59(3):423–9.

[26] Krikorian AD. Kat and its use: an historical perspective. J Ethnopharmacol 1984;12(2):115–78.

[27] Odenwald M, Klein A, Warfa N. Introduction to the special issue: the changing use and misuse of khat (Catha edulis) – Tradition, trade and tragedy. J Ethnopharmacol 2010;132(3):537–9.

[28] Dhaifalah I, Šantavý J. Khat habit and its health effect. A natural amphetamine. Biomed Pap 2004;148:11–15.

[29] Feyissa AM, Kelly JP. A review of the neuropharmacological properties of khat. Prog Neuro-Psychoph 2008;32(5):1147–66.

[30] Measham F, Moore K, Newcombe R, Welch Z. Tweaking, bombing, dabbing and stockpiling: the emergence of mephedrone and the perversity of prohibition. Drug Alcohol Today 2010;10(1):14–21.

[31] Kelly JP. Cathinone derivatives: a review of their chemistry, pharmacology and toxicology. Drug Test Anal 2011;3(7-8):439–53.

[32] Szendrei K. The chemistry of khat. B Narcotics 1980;32(3):5–35.

[33] Toennes SW, Kauert GF. Excretion and detection of cathinone, cathine and phenylpropanolamine in urine after khat chewing. Clin Chem 2002;48(10):1715–9.

[34] Patel NB. Mechanism of action of cathinone: The active ingredient of khat (Catha edulis). E Afr Med J 2000;77(6):329–32.

[35] Zelger JL, Schorno HX, Carnili EA. Behavioural effects of cathinone, an amine obtained from Catha edulis Forsk.: comparisons with amphetamine, norpseudoephedrine, apomorphine and nomifensine. B Narcotics 1980;32:67–81.

[36] Hoffman R, Al'Absi M. Khat use and neurobehavioral functions: suggestions for future studies. J Ethnopharmacol 2010;132(3):554–63.

[37] Toennes SW, Harder S, Schramm M, Niess C, Kauert GF. Pharmacokinetics of cathinone, cathine and norephedrine after the chewing of khat leaves. Brit J Clin Pharmacol 2003;56(1):125–30.

[38] Brenneisen R, Geisshüsler S. Phenylpentenylamines from Catha edulis. J Nat Prod 1987;50(6):1188–9.

[39] Kite GC, Ismail M, Simmonds MSJ, Houghton PJ. Use of doubly protonated molecules. Rapid Commun Mass Sp 2003;17(4):1553–64.

[40] Brenneisen R, Fisch HU, Koelbing U, Geisshüsler S, Kalix P. Amphetamine-like effects in humans of the khat alkaloid cathinone. Brit J Clin Pharmacol 1990;30(6):825–8.

[41] Odenwald M. Chronic khat use and psychotic disorders: a review of the literature and future prospects. J Addict Res Pros 2007;53(1):9–22.

[42] Warfa N, Klein A, Bhui K, Leavey G, Craig T, Alfred Stansfeld S. Khat use and mental illness: a critical review. Soc Sci Med 2007;65(2):309–18.

[43] Al-Motarreb A, Baker K, Broadley KJ. Khat: pharmacological and medical aspects and its social use in Yemen. Phytother Res 2002;16(5):403–13.

[44] Al-Motarreb A, Briancon S, Al-Jaber N, et al. Khat chewing is a risk factor for acute myocardial infarction: a case-control study. Brit J Clin Pharmacol 2005;59(5):574–81.

[45] Al-Hadrani AM. Khat induced hemorrhoidal disease in Yemen. Saudi Med J 2000;21(5):475–7.

[46] Hadi M, Mujlli XB, Zhang L. The effect of khat (Catha edulis) on acute cerebral infarction. Neuroscience 2005;10(3):219–22.

[47] Rajaa YA, Noman TA, Warafi AKMA, Mashraki NAA, Yosof AMAA. Khat chewing is a risk factor of duodenal ulcer. Saudi Med J 2000;21(9):887–8.

[48] Balint EE, Falkay G, Balint GA. Khat-a controversial plant. Wien Klin Wochenschr 2009;121(19):604–14.

[49] Mwenda JM, Arimi MM, Kyama MC, Langat DK. Effect of khat (Catha edulis) consumption on reproductive functions: review. E Afr Med J 2003;80(6):318–23.

[50] Coton T, Simon F, Oliver M, Kraemer P. Hepatotoxicity of khat chewing. Liver Int 2011;31(3):434.

[51] Chapman MH, Kajihara M, Borges G, et al. Severe, acute liver injury and khat leaves. New Engl J Med 2010;362(17):1642–4.

[52] Toennes SW, Kauert GF. Driving under the influence of khat – alkaloid concentrations and observations in forensic cases. Forensic Sci Int 2004;140(1):85–90.

[53] Cleary L, Docherty JR. Actions of amphetamine derivatives and cathinone at the noradrenaline transporter. Eur J Pharmacol 2003;476(1-2):31–4.

[54] Manghi RA, Broers B, Khan R, Benguettat D, Khazaal Y, Zullino DF. Khat use: lifestyle or addiction? J Psychoactive Drugs 2009;41(1):1–10.

[55] Al-Motarreb A, Al-Habori M, Broadley KJ. Khat chewing, cardiovascular diseases and other internal medical problems: the current situation and directions for future research. J Ethnopharmacol 2010;132(3):540–8.

[56] Gottlieb A. Peyote and other psychoactive cacti. Berkeley, CA: Ronin Publishing; 1977.

[57] Rodriguez DJD, Angulo-Sanchez JL, Hernandez-Castillo FD, Mahendra R, Maria Cecilia C. An overview of the antimicrobial properties of Mexican medicinal plants Advances in phytomedicine. Amsterdam: Elsevier; 2006. p. 325-77.

[58] Bruhn JG, De Smet PAGM, El-Seedi HR, Beck O. Mescaline use for 5700 years. Lancet 2002;359(9320):1866.

[59] Halpern JH, Sewell RA. Hallucinogenic botanicals of America: a growing need for focused drug education and research. Life Sci 2005;78(5):519–26.

[60] Wolbach AB, Isbell H, Miner EJ. Cross tolerance between mescaline and LSD-25 with a comparison of the mescaline and LSD reactions. Psychopharmacology 1962;3(1):1–14.

[61] Carstairs SD, Cantrell FL. Peyote and mescaline exposures: a 12-year review of a statewide poison center database. Clin Toxicol 2010;48(4):350–3.

[62] Beyer J, Drummer OH, Maurer HH. Analysis of toxic alkaloids in body samples. Forensic Sci Int 2009;185:1–9.

[63] Gilmore HT. Peyote use during pregnancy. S Dak J Med 2001;54:27–9.

[64] Barre WL. Peyotl and mescaline. J Psychedel Drug 1979;11(1-2):33–9.

[65] Ogunbodede OO. Alkaloid content in relation to ethnobotanical use of trichocereus pachanoi and related taxa. Texas: Sul Ross State University; 2009.

[66] Ogunbodede O, McCombs D, Trout K, Daley P, Terry M. New mescaline concentrations from 14 taxa/cultivars of Echinopsis spp. (Cactaceae) ('San Pedro') and their relevance to shamanic practice. J Ethnopharmacol 2010;131(2):356–62.

[67] Burkill LH, Haniff M. Malay village medicine. The garden's bulletin straits settlement 1930;6:165–207.

[68] Vicknasingam B, Narayanan S, Beng GT, Mansor SM. The informal use of ketum (Mitragyna speciosa) for opioid withdrawal in the northern states of peninsular Malaysia and implications for drug substitution therapy. Int J Drug Policy 2010;21(4):283–8.

[69] Boyer EW, Babu KM, Adkins JE, McCurdy CR, Halpern JH. Self-treatment of opioid withdrawal using kratom (Mitragynia speciosa korth). Addiction 2008;103(6):1048–50.

[70] Takayama H. Chemistry and Pharmacology of analgesic indole alkaloids from the rubiaceous plant, Mitragyna speciosa. Chem Pharm Bull 2004;52(8):916–28.

[71] Kapp F, Maurer H, Auwärter V, Winkelmann M, Hermanns-Clausen M. Intrahepatic cholestasis following abuse of powdered kratom (Mitragyna speciosa). J Med Toxicol 2011;7(3):227–31.

[72] Watanabe K, Yano S, Horie S, Yamamoto LT. Inhibitory effect of mitragynine, an alkaloid with analgesic effect from Thai medicinal plant Mitragyna speciosa, on electrically stimulated contraction of isolated guinea-pig ileum through the opioid receptor. Life Sci 1997;60(12):933–42.

[73] Utar Z, Majid MIA, Adenan MI, Jamil MFA, Lan TM. Mitragynine inhibits the COX-2 mRNA expression and prostaglandin E2 production induced by lipopolysaccharide in RAW264.7 macrophage cells. J Ethnopharmacol 2011;136(1):75–82.

[74] Harizal SN, Mansor SM, Hasnan J, Tharakan JKJ, Abdullah J. Acute toxicity study of the standardized methanolic extract of Mitragyna speciosa Korth in rodent. J Ethnopharmacol 2010;131(2):404–9.

[75] Nelsen J, Lapoint J, Hodgman M, Aldous K. Seizure and coma following kratom (Mitragynina speciosa Korth) exposure. J Med Toxicol 2010;6(4):424–6.

[76] Roche KM, Hart K, Sangalli B, Lefberg J, Bayer M. Kratom: a case of a legal high. Clin Toxicol 2008;46(7):598.

[77] Apryani E, Taufik Hidayat M, Moklas MAA, Fakurazi S, Farah Idayu N. Effects of mitragynine from Mitragyna speciosa Korth leaves on working memory. J Ethnopharmacol 2010;129(3):357–60.

[78] Saidin NA, Randall T, Takayama H, Holmes E, Gooderham NJ. Malaysian Kratom, a phyto-pharmaceutical of abuse: studies on the mechanism of its cytotoxicity. Toxicology 2008;253:19–20.

[79] Babu KM, McCurdy CR, Boyer EW. Opioid receptors and legal highs: salvia divinorum and Kratom. Clin Toxicol 2008;46(2):146–52.

[80] Kupferschmidt H. Toxic hepatitis after Kratom (Mitragyna sp.) consumption. Clin Toxicol 2011;49:532.

[81] Kronstrand R, Roman M, Thelander G, Eriksson A. Unintentional fatal intoxications with mitragynine and O-d from the herbal blend Krypton. J Anal Toxicol 2011;35(4):242–7.

[82] McWhirter L, Morris S. A case report of inpatient detoxification after kratom (Mitragyna speciosa) dependence. Eur Addict Res 2010;16(4):229–31.

[83] Modi AJ, Khadabadi SS, Farooqui IA, Deore SL. Agyreia speciosa Linn.F : phytochemistry, pharmacognosy and pharmacological studies. Int J Pharm Sci Rev Res 2010;2(2):14–21.

[84] Miller MD. Isolation and identification of lysergic acid amide and isolysergic acid amide as the principle ergoline alkaloids in Argyreia nervosa, a tropical wood rose. J AOAC 1970;53(1):123–8.

[85] Chao J-M, Marderosian AHD. Ergoline alkaloidal constituents of Hawaiian baby woodrose, Argyreia nervosa (Burm.f.) bojer. J Pharm Sci 2006;62(4):588–91.

[86] Al-Assmar SE. The seeds of the Hawaiian baby woodrose are a powerful hallucinogen. Arch Int Med 1999;159(17):2090.

[87] Mandarin F. Vines of the serpent: a morning glory ethnobotanical. Available: <www.goa-shoom.net> [accessed 12.08.11].

[88] Björnstad K. Mass spectrometric investigation of intoxications with plant-derived psychoactive substances. Stockholm: Karolinska Institute; 2009.

[89] Joseph A, Mathew S, Skaria BP, Sheeja EC. Medicinal uses and biological activities of Argyreia speciosa Sweet (Hawaiian baby woodrose) – An overview. Indian J Nat Prod Resour 2011;2(3):286–91.

[90] Gopel C, Maras A, Schmidt MH. Hawaiian baby rose wood: case report of an Agyreia nervosa induced toxic psychosis. Psychiat Prax 2003;30(4):223–4.

[91] Klinke HB, Muller IB, Steffenrud S, Dahl-Sørensen R. Two cases of lysergamide intoxication by ingestion of seeds from Hawaiian Baby Woodrose. Forensic Sci Int 2010;197(1-3):e1–e5.

[92] McKenna DJ, Callaway JC, Grob CS. The scientific investigation of Ayahuasca: a review of past and current research. Heffter Rev Psyched Res 1998;1:65–76.

[93] Freedland CS, Mansbach RS. Behavioral profile of constituents in ayahuasca, an Amazonian psychoactive plant mixture. Drug Alcohol Depend 1999;54(3):183–94.

[94] McKenna DJ, Towers GHN, Abbott F. Monoamine oxidase inhibitors in South American hallucinogenic plants: tryptamine and β-carboline constituents of Ayahuasca. J Ethnopharmacol 1984;10(2):195–223.

[95] Bennett BC. Hallucinogenic plants of the Shuar and related indigenous groups in Amazonian Ecuador and Peru. Brittonia 1992;44(4):483–93.

[96] Moura S, Carvalho FG, de Oliveira CDR, Pinto E, Yonamine M. qNMR: an applicable method for the determination of dimethyltryptamine in ayahuasca, a psychoactive plant preparation. Phytochem Lett 2010;3(2):79–83.

[97] Rivier L, Lindgren J-E. 'Ayahuasca' the South American hallucinogenic drink: an ethnobotanical and chemical investigation. Econ Bot 1972;26(2):101–29.

[98] Samoylenko V, Rahman MM, Tekwani BL, et al. Banisteriopsis caapi, a unique combination of MAO inhibitory and antioxidative constituents for the activities relevant to neurodegenerative disorders and Parkinson's disease. J Ethnopharmacol 2010;127(2):357–67.

[99] Schwarz MJ, Houghton PJ, Rose S, Jenner P, Lees AD. Activities of extract and constituents of Banisteriopsis caapi relevant to parkinsonism. Pharmacol Biochem Be 2003;75(3):627–33.

[100] Houghton PJ, Howes M-J. Natural products and derivatives affecting neurotransmission relevant to Alzheimer's and Parkinson's disease. Neurosignals 2005;14:6–22.

[101] Riba J, Barbanoj MJ. Bringing ayahuasca to the clinical research laboratory. J Psychedel Drug 2005;37(2):219–30.

[102] Fábregas JM, González D, Fondevila S, et al. Assessment of addiction severity among ritual users of ayahuasca. Drug Alcohol Depend 2010;111(3):257–61.

[103] Blackledge RD, Taylor CM. Psychotria viridis – A botanical source of dimethyltryptamine (DMT). Microgram J 2003;1(1-2):18–22.

[104] Cozzi NV, Gopalakrishnan A, Anderson LL, et al. Dimethyltryptamine and other hallucinogenic tryptamines exhibit substrate behavior at the serotonin uptake transporter and the vesicle monoamine transporter. J Neural Transm 2009;116(12):1591–9.

[105] Callaway J, Brito G, Neves E. Phytochemical analyses of Banisteriopsis caapi and Psychotria viridis. J Psychedel Drug 2005;37(2):145–50.

[106] King LA. Forensic chemistry of substance misuses: a guide to drug control. Cambridge: The Royal Society of Chemistry; 2009.

[107] Sklerov J, Levine B, Moore KA, King T, Fowler D. Case report: a fatal intoxication following the ingestion of 5-Methoxy-N,N-Dimethyltryptamine in an Ayahuasca preparation. J Anal Toxicol 2005;29(8):838–41.

[108] Dalgarno P. Buying Ayahuasca and other entheogens online: a word of caution. Addict Res Theory 2008;16(1):1–4.

[109] Gable RS. Risk assessment of ritual use of oral dimethyltryptamine (DMT) and harmala alkaloids. Addiction 2007;102(1):24–34.

[110] Pires APS, Oliveira CDR, Yonamine M. Ayahuasca: a review of pharmacological and toxicological aspects. Rev Ciênc Farm Básica Apl 2010;31(1):15–30.

[111] Grundmann O, Phipps SM, Zadezensky I, Butterweck V. Salvia divinorum and Salvinorin A: an update on pharmacology and analytical methodology. Planta Med 2007;73(1039):46.

[112] Daniel JS. Salvia divinorum and salvinorin A: new pharmacologic findings. J Ethnopharmacol 1994;43(1):53–6.

[113] McClatchey WC, Mahady GB, Bennett BC, Shiels L, Savo V. Ethnobotany as a pharmacological research tool and recent developments in CNS-active natural products from ethnobotanical sources. Pharmacol Therapeut 2009;123(2):239–54.

[114] Singh S. Adolescent salvia substance abuse. Addiction 2007;102(5):823–4.

[115] Schneider RJ, Ardenghi P. Salvia divinorum Epling and Játiva ('ska Maria Pastora') and Salvinorina A: increasing recreational use and abuse potential. Rev Bras Plantas Med 2010;12:358–62.

[116] Roth BL, Baner K, Westkaemper R, et al. Salvinorin A: a potent naturally occurring nonnitrogenous κ opioid selective agonist. Proc Natl Acad Sci 2002;99(18):11934–11939.

[117] Harding WW, Tidgewell K, Schmidt M, et al. Salvinicins A and B, new neoclerodane diterpenes from Salvia divinorum. Org Lett 2005;7(14):3017–20.

[118] Paulzen M, Gründer G. Toxic psychosis after intake of the hallucinogen salvinorin A. J Clin Psychiatry 2008;69(9):1501–2.

[119] Ahern N, Greenberg C. Psychoactive herb use and youth: a closer look at Salvia divinorum. J Psychosoc Nurs Ment Health Serv 2011;49(8):16–19.

[120] Prisinzano TE. Psychopharmacology of the hallucinogenic sage Salvia divinorum. Life Sci 2005;78(5):527–31.

Tryptamines

Shaun L. Greene

Victorian Poisons Information Centre, Melbourne, Australia

TRYPTAMINES

The naturally occurring tryptamine dimethyltryptamine (DMT) has been used for centuries in South America as a psychoactive substance during religious sacraments [1]. Hallucinogenic mushrooms contain the tryptamines psilocybin and psilocin: these are discussed in more detail in Chapter 14, which is on natural products. Bufotenin, an isomer of psilocin, is found in the skin of various species of the toad *Bufo* genus [2]. Serotonin and melatonin are naturally occurring tryptamines derived from the amino acid tryptophan. Lysergic acid diethylamide (LSD) is perhaps the best-known synthetic tryptamine and the most potent known hallucinogen [3]. Recently numerous synthetic tryptamines have emerged as recreational psychoactive substances [4].

PHARMACOLOGY

Physical and Chemical Description

Tryptamine itself is a monoamine alkaloid related to the amino acid tryptophan. Tryptamines are built around an indole ring structure, a fused double ring comprising a pyrole ring and benzene ring, with the addition of a 2-carbon side chain (Fig. 15.1). Additions to the aromatic ring or 2-carbon side chain give rise to a multitude of naturally occurring and synthetic tryptamines. Substitutions at positions six and seven of the indole ring reduce potency and therefore modifications are usually made at positions four and five [4].

Tryptamines can be classified as 'simple tryptamines' or 'ergolines' [3,5]. Simple tryptamines are classified according to substitution at the fourth and fifth positions on the indole ring. Structures of common simple unsubstituted, 4-substituted and 5-substituted tryptamines are illustrated in Figures 15.1–15.3. The synthetic simple tryptamines exist as powders or crystals at room temperature.

LSD is the representative member of the ergolines (named as they were originally synthesised from an ergot fungus). Ergolines have a more complex structure based around an indole system and tetracyclic ring (Fig. 15.4). Ergine is an ergot-type psychoactive alkaloid found in seeds of the morning glory family (*Convolvulaceae*) [6]. Ergine has a similar structure to LSD and is also known as lysergamide or lysergic acid amide [7].

Novel Psychoactive Substances.
DOI: http://dx.doi.org/10.1016/B978-0-12-415816-0.00015-8

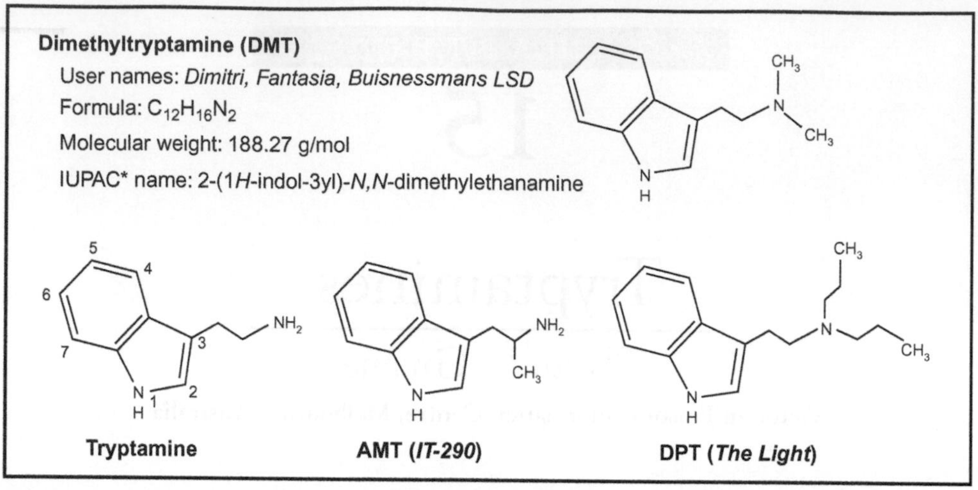

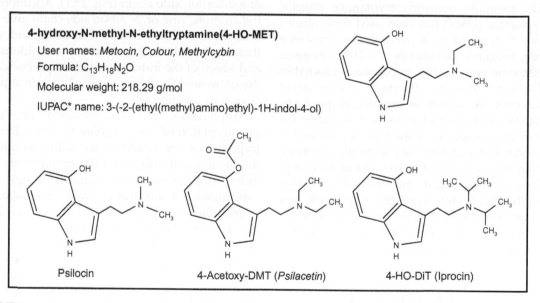

FIGURE 15.1 Chemical structures of simple, un-substituted tryptamines. IUPAC:International Union of Pure and Applied Chemistry.

FIGURE 15.2 Chemical structures of simple, 4-substituted tryptamines. *IUPAC: International Union of Pure and Applied Chemistry.

Mitragynine is an indole alkaloid with a tryptamine-like structure (Fig. 15.4) that acts as a μ-and δ-opioid receptor agonist [8,9]. Mitragynine also shares structural similarity with yohimbine. Extracts of the plant kratom (*Mitragyna speciosa*) containing mitragynine are used as recreational drugs in South-east Asia [10].

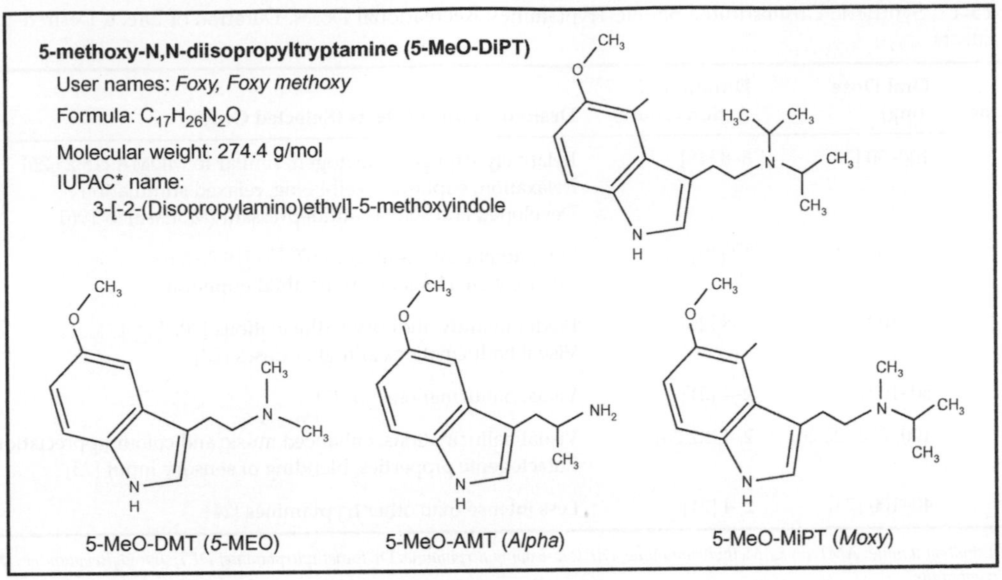

5-methoxy-N,N-diisopropyltryptamine (5-MeO-DiPT)

User names: *Foxy, Foxy methoxy*

Formula: C₁₇H₂₆N₂O

Molecular weight: 274.4 g/mol

IUPAC* name:
   3-[-2-(Disopropylamino)ethyl]-5-methoxyindole

5-MeO-DMT (*5-MEO*)   5-MeO-AMT (*Alpha*)   5-MeO-MiPT (*Moxy*)

FIGURE 15.3   Chemical structures of simple, 5-substituted tryptamines. *IUPAC: International Union of Pure and Applied Chemistry.

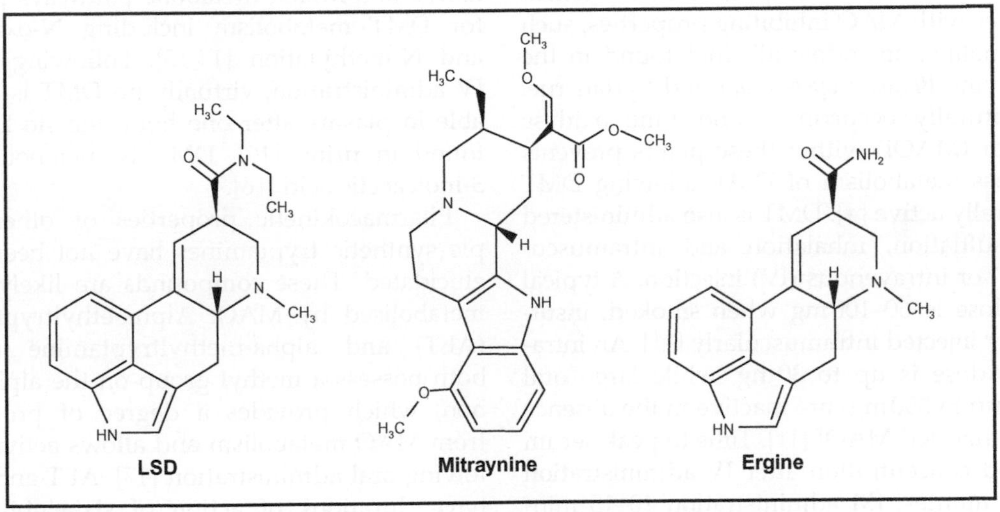

LSD   Mitraynine   Ergine

FIGURE 15.4   Chemical structures of LSD, mitraynine and ergine.

Pharmacokinetics

Pharmacokinetic properties of synthetic simple tryptamines have not been extensively studied; however, information exists for a number of naturally occurring tryptamines. DMT undergoes extensive first pass metabolism via monoamine oxidase (MAO) facilitated degradation, and is therefore not orally active [4].

TABLE 15.1 Synthetic Unsubstituted Simple Tryptamines: Recreational Doses, Duration of Effect, Desired Clinical Effects

TryptaMine	Oral Dose (mg)	Duration of Action (hours)	Desired Clinical Effects (Selected Only)
AET	100–50 [18]	6–8 [18]	Relatively stronger entactogen, similar to MDMA [18,25,26] Relaxation, euphoria, well-being, relaxed stimulation Developed and sold as antidepressant (Monase) in 1961
AMT	20–30 [19]	12 [19]	Sympathetic effects at lower doses [19,25,26] Intense visual hallucinations, mild euphoria
DiPT	25–100 [20]	6–8 [20]	Predominantly auditory hallucinations [20,25,26] Visual hallucinations at higher doses [27]
DET	50–10 [21]	2–4 [21]	Visual hallucinations [21,25]
DPT	100–250 [22,26]	2–4 [22,26]	Visual hallucinations, enhanced music and colour appreciation [17,25] Entactogenic properties, blending of sensory input [23]
DALT	40–100 [24]	2–4 [24]	Less intense than other tryptamines [24]

AET: alpha-Ethyltryptamine; AMT: alpha-Methyltryptamine; DiPT: di-isopropyltrptamine; DET: diethyltrptamine; DPT: dipropyltryptamine; DALT: N,N-diallyltryptamine.

Ayahuascan mixtures containing DMT as the active psychoactive component also contain additives with MAO inhibiting properties, such as harmaline, an indole alkaloid found in the jungle vine *Banisteriopsis caapi* and Syrian rue. The naturally occurring monoamine oxidase inhibitor (MAOI) within these plants prevents first pass metabolism of DMT, allowing DMT to be orally active [4]. DMT is also administered via insufflation, inhalation and intramuscular (IM) or intravenous (IV) injection. A typical DMT dose is 60–100mg when smoked, insufflated or injected intramuscularly [11]. An intravenous dose is up to 30mg, while large oral doses (up to 350mg) are inactive in the absence of a co-ingested MAOI [11]. Time to peak serum or blood concentration after IV administration is two minutes, IM administration 10–15 minutes and ingestion 90–120 minutes; however, there is wide individual variation in time to peak clinical effects [12,13]. DMT's duration of effect is usually less than one hour, but varies with dose [11].

DMT is metabolised by MAO, most likely via oxidative deamination of the side chain [14]. Recent studies also suggest there are alternative non-MAOI metabolic pathways present for DMT metabolism including N-oxidation and N-methylation [14,15]. Following IM or IV administration, virtually no DMT is detectable in plasma after one hour and no DMT is found in urine [16]. DMT is metabolised to 3-inoleacetic acid [16].

Pharmacokinetic properties of other simple synthetic tryptamines have not been fully elucidated. These compounds are likely to be metabolised by MAO. Alpha-ethyltryptamine (AET) and alpha-methyltryptamine (AMT) both possess a methyl group on the alpha carbon, which provides a degree of protection from MAO metabolism and allows activity following oral administration [17]. AET and AMT have durations of action of six–eight hours and 12 hours, respectively, when taken orally [18,19]. Di-isopropyltryptamine (DiPT), diethyltryptamine (DET), dipropyltryptamine (DPT), N,N-diallyltryptamine (DALT) are all orally active and have durations of action of two–four hours (DiPT reported up to eight hours) following oral administration (Table 15.1) [18–24].

Psilocin is a 4-substituted simple tryptamine found within psychedelic mushroom species and has been used as a hallucinogenic substance for thousands of years. Psilocin is metabolised via hepatic glucoronidation [28]. User reports suggest durations of action varying from two–six hours following ingestion of the synthetic 4-substitiuted simple tryptamines including 4-hydroxy-N-methyl-N-ethyltryptamine (4-HO-MET), 4-hydroxy-N,N-diethyltryptamine (4-HO-DET), 4-hydroxy-di-isopropyltryptamine (4-HO- DiPT), 4-hydroxy-N-methyl-N-isopropyltryptamine (4-HO- MiPT), 4-acetoxy-N, N-dimethyltryptamine (4-Acetoxy-DMT) and 4-acetoxy-N,N-diisopropyltryptamine (4-Acetoxy-DiPT) [29–33].

5-substituted simple tryptamines, including 5-methoxy-dimethyltryptamine (5-MeO-DMT), 5-methoxy-alpha-methyltryptamine (5-MeO-AMT), 5-methoxy-diisopropyltryptamine (5-MeO-DiPT), 5-methoxy-N-methyl-N-isopropyltryptamine (5-MeO-MiPT) and N,N-diallyl-5-methoxytryptamine (5-MeO-DALT) are metabolised through 6-hydroxylation, O-methylation or N-dealkylation by hepatic cytochrome P450 enzymes [28]. The metabolites are then conjugated with glucuronide or sulphide [28]. 5-MeO-DMT is metabolised by polymorphic cytochrome P450 2D6 to bufoteine, which is also a pharmacologically active tryptamine. Bufoteine is subsequently inactivated via deamination by monoamine oxidase A (MAO-A) [34]. 5-substituted simple tryptamines are administered orally and have durations of action varying between one and 18 hours (Table 15.1). 5-MeO-DMT has minimal activity when administered orally due to significant MAOI-mediated first pass metabolism, and is therefore smoked, insufflated or injected [34].

Pharmacodynamics

Mechanism of Action

The predominant clinical effect produced by tryptamine exposure is hallucinations, mediated by agonism at $5HT_{1A}$ and $5HT_{2A}$ receptors [3,35,36]. Tryptamines exhibit less selectivity and affinity for $5HT_{2A}$ receptors compared to hallucinogenic phenylethylamines [4]. However, most tryptamines exhibit hallucinogenic properties, rather than entactogenic or stimulant properties. The alpha methylated tryptamines possess an alpha carbon methyl group, and exhibit relatively greater stimulant activity, like the amphetamines with similar structures [17]. Examples include AMT and 5-methoxy-alpha-methyltryptamine (5-MeO-AMT). Other receptors implicated in tryptamine central nervous system (CNS) interaction include vesicular monoamine transporter 2 (VMAT2), sigma-1 receptor, trace-amine-associated receptors (TAAR) and serotonin transporter (SERT) [35,37–39].

DMTs pharmacodynamic properties have been studied in more detail than other tryptamines. DMT binds to $5HT_{1A}$, $5HT_{1B}$, $5\text{-}HT_{1D}$, $5HT_{2A}$, $5HT_{2B}$, $5HT_{2C}$, $5HT_6$, and $5HT_7$ receptors [35,40–42]. Agonist activity has been demonstrated at $5HT_{1A}$, $5HT_{2A}$ and $5HT_{2C}$ receptors [36,40,42]. DMT also has affinity for sigma-1, alpha1-adrenergic, alpha2-adrenergic, SERT, VMAT2, imidazoline-1 and TAAR receptors [38,39]. Like many hallucinogens the primary receptor responsible for DMTs psychedelic effects appears to be $5HT_{2A}$, but there is also indirect evidence of $5HT_{2C}$ contribution; DMTs EC_{50} for the $5HT_{2C}$ receptor is lower than the EC_{50} for the $5HT_{2A}$ receptor [42]. Following administration of typical psychedelic doses of DMT, blood and plasma DMT concentrations are within the range required to produce 50% of the maximal effect (EC_{50}) at $5HT_{2A}$ receptors, suggesting a likely agonist $5HT_{2C}$ effect following recreational DMT doses [12,13,36,40,43].

Other unsubstituted tryptamines including AMT and AET have hallucinogenic and stimulatory effects [24]. AMT is a reuptake inhibitor and releaser of serotonin, noradrenaline and dopamine [4,44]. DPT is unique in producing aural rather than visual hallucinations. Evidence from a rodent study suggests both $5HT_{1A}$ and $5HT_{2A}$

receptors mediate DPTs pharmacological effects, but the exact mechanism behind its unique psychoactive auditory effect remains unknown [43].

The pharmacodynamic properties of the simple 4-substituted tryptamines have not been studied. User reports of exposure to these substances illustrate similar clinical effects to psilocin exposure, suggesting a similar mechanism of action. Psilocin is a partial $5HT_{2A}$ agonist and has stimulatory effects at other serotonin receptors, but has little effect at dopamine or noradrenergic receptors [4,25,45].

Simple 5-substituted tryptamines inhibit monoamine reuptake, but appear to have minimal effect on monoamine release [44]. An *in vitro* study demonstrated 5-MeO-DiPTs ability to act as a selective high affinity inhibitor of SERT, but did not produce serotonin release [27]. Simple 5-substituted tryptamines possess a methoxyl or hydroxyl group at position five of the tryptamine ring, increasing the potency of the molecule compared to its unsubstituted relation [46]. 5-MeO-DiPT is seven times more potent than DMT [47].

Although extracts of plants from the *Convolvulaceae* family containing alkaloids including ergine have been used as psychoactive substances for centuries, little is known about their pharmacodynamic properties. Ergines tryptamine ring structure, structural similarity to LSD and reported psychoactive effects suggest that 5-HT receptor agonism is the predominant pharmacodynamic effect [6,48].

Mitragynine is a μ- and δ-opioid receptor agonist; however, it is structurally similar to the alpha-2 receptor antagonist yohimbine and exhibits yohimbine-like binding to alpha-adrenergic receptors [9,49]. At low doses (10–30 mg) users report predominantly stimulatory effects consistent with a possible yohimbine-like effect [50]. Mitragynine may also activate noradrenergic and serotonergic pathways in the spinal cord, block alpha-2 adrenergic receptors and stimulate $5-HT_{2A}$ receptors [51,52]. Kratom, the plant containing mitragynine, also contains a number of additional alkaloids including

7-hydroxymitragyine that possess potent opioid agonism. Recently it has been suggested that 7-hydroxymitragyine may be the predominant active alkaloid within kratom, rather than mitragynine [53].

In summary the hallucinogenic effect of tryptamines appear to be produced predominantly through agonism at $5HT_{2A}$ receptors, although numerous other receptors including $5HT_{1A}$ and $5HT_{2C}$ are also likely to contribute. A number of 4-substituted tryptamines have greater noradrenergic effect, while potency is increased by substitution at position five on the tryptamine ring. Pharmacodynamic properties of most of the designer synthetic recreational tryptamines have not been studied in detail.

Positive or Desired Clinical Effects

Published substance-user reports on Internet sites, including Erowid (http://www.erowid.org) and Drugs-Forum (http://www.drugs-forum.com), are the primary source of information regarding the positive effects experienced by users following use of these substances. Few published controlled studies exist. Although useful, these sites are subject to reporting bias and inability to accurately confirm exposure.

DMT, the archetypical tryptamine, offers an insight into the psychoactive properties of other tryptamines. In recreational doses DMT produces sudden onset of intense visual hallucinations [11,14,43]. Sympathomimetic effects such as dilated pupils, tachycardia and hypertension are more prominent at lower doses [14,25]. Some users report that the experience is more intense than that following LSD exposure [25].

Unlike DMT, synthetic unsubstituted simple tryptamines are orally active. Smoking and nasal insufflation are also common methods of administration. Positive effects reported by users vary, but visual hallucinations are universal [25,26]. The exception is DiPT, which is unique in producing auditory distortion at lower doses and auditory hallucinations at higher doses [20,26]. AET and AMT have stimulatory effects at lower

TABLE 15.2 Synthetic 4-Substituted Simple Tryptamines: Recreational Doses, Duration of Effect, Desired Clinical Effects

Tryptamine	Oral Dose (mg)	Duration of Action (hours)	Desired Clinical Effects (Selected Only)
4-HO-MET	10–0 [29]	4–6 [29]	Wave-like pattern with distortion of colour/sound/form [26,29] Increased appreciation of music
4-HO-DET	10–5 [30]	4–6 [30]	Hallucinations similar to those produced by LSD [26,30]
4-HO-DiPT	15–0 [31]	2–3 [31]	Rapid onset, very short acting [31] Visual and auditory hallucinations
4-HO-MiPT	12–5 [32]	4–6 [32]	Increased mood, laughing, creative thinking, appetite, sexual interest, music appreciation. Reduced anxiety [32,54]
4-Acetoxy-DMT	10–0 [33]	3–6 [33]	Increased energy, euphoria, colourful visual effects Abstract, associative thought patterns [25,26,33]
4-Acetoxy-DiPT	15–0 [31]	2–3 [31]	Gentle, less intense than other tryptamines [31,54] Enhanced sensuality, libido, increased appreciation of music

4-HO-MET: 4-hydroxy-N-methyl-N-ethyltryptamine; 4-HO_DET: 4-hydroxy-N,N-diethyltryptamine; 4HO-DiPT: 4-hydroxy-di-isopropyltryptamine; 4-HO-MiPT: 4-hydroxy-N-methyl-N-isopropyltryptamine; 4-Acetoxy-DMT: 4-acetoxy-N,N-dimethyltryptamine; 4-Acetoxy-DiPT: 4-acetoxy-N,N-diisopropylryptamine.

doses [18,19,26]. AET is described by users as having psychoactive properties similar to 3,4-methylenedioxymethamphetamine (MDMA) [18,26]. General positive effects reported by users of unsubstituted simple tryptamines include: 'rushing' sensation, both open and closed eye 'pleasant' visual hallucinations, increased mood, energy, libido, concentration and empathogenic qualities [25,26]. Oral doses, duration of action and characteristic positive effects of common synthetic simple unsubstituted tryptamines are summarised in Table 15.1.

Synthetic 4-substituted simple tryptamines are orally active. User reports suggest synthetic 4-substituted simple tryptamines produce clinical effects very similar to those mediated by psilocin [25]. Visual hallucinations predominate, with varying reports of other effects including euphoria, increased libido, increased energy and enhanced thought processes and appreciation of music [25,26,28–33]. General positive effects reported by users of 4-substituted simple tryptamines include: increased laughing, intense visual hallucinations, 'rushing' sensation,

euphoria, increased libido, enhanced tactile sensations, increased concentration and a feeling of warmth and inner peace [25,26]. Oral doses, duration of action and characteristic positive effects of the more common synthetic simple 4-substituted tryptamines are summarised in Table 15.2.

Although 5-substituted simple tryptamines are more potent than their unsubstituted cousins, clinical effects are similar to the unsubstituted molecule [4,46]. Visual hallucinations are reported following exposure to all of the synthetic 5-substituted simple tryptamines. Other positive reported effects include: intense 'rushing' sensation (when smoked), euphoria, increased libido, increased energy, sexually 'interesting' interactions, increased concentration and sociability, a reduction in fear and anxiety, enhanced appreciation of music and food, and facilitation of 'life-changing spiritual experiences' [25,26,34,55,56]. Users of 5-MeO MiPT report marked increase in enjoyment obtained from tactile stimulation, and synaesthetic effects [25,26]. Oral doses, duration of action and positive effects

TABLE 15.3   Synthetic 5-Substituted Simple Tryptamines: Recreational Doses, Duration of Effect, Desired Clinical Effects

Tryptamine	Oral Dose (mg)	Duration Action (hours)	Desired Clinical Effects (Selected Only)
5-MeO-DMT	Smoked 6–20 [57] Insufflated 2–15 [57]	1–2 [57] 1–2 [57]	Rapid onset of fast, intense, incoherent hallucinations Changes in all sensory perceptions, abrupt cessation Powerful 'rushing' sensation, occasional euphoria Profound life changing spiritual experiences [25,26]
5-MeO-AMT	2.5–4.5 [58]	12–18 [58]	Increased mood, energy, sociability, creative thinking, sexual interest, laughing [45]
5-MeO-DiPT	6–12 [59]	4–8 [59]	Emotionally opening, increased mood, libido, euphoria, improved self-confidence, reduced fear/anxiety, enhancement of and appreciation for music [25,26]
5-MeO-MiPT	4–6 [60]	4–6 [60]	Increased tactile stimulation, synaesthetic effects, libido Increased appreciation of food, sound [26,43]
5-MeO-DALT	12–20 [61]	2–4 [61]	Reduced visual hallucinations, but reported changes in perception of space and time [26]

5-MeO-DMT: 5-methoxy-dimethyltrptamine; 5-MeO-AMT: 5-methoxy-alpha-methyltrptamine; 5-MeO-DiPT: 5-methoxy-diisopropyltrptamine; 5-MeO-MiPT: 5-methoxy-N-methyl-N-isopropyltrptamine; 5-MeO-DALT: N,N-diallyl-5-methoxytrptamine.

of the synthetic simple 5-substituted tryptamines are summarised in Table 15.3.

Leaves of the *Mitragyna* species (also known as kratom) containing alkaloids including mitragynine and 7-hydroxymitragynine have been used as an opioid substitute, stimulant and antidiarrhoeal agent in Thailand and other areas of Southeast Asia for centuries [62]. Kratom has been advocated as an adjunct to chronic pain therapy, and facilitator of reducing opioid dependence. Kratom users report mild caffeine like stimulation at low does (2–4g), with subtle calming opioid-like effects at higher doses (>5g). Others report simultaneous sedation and stimulation. Some users report very mild open and closed eye visual effects. Effects are dose dependent and begin 10–20 minutes after chewing the leaves and last two–three hours. Kratom can also be smoked and brewed as a tea [62,63].

Argyreia nervosa is a climbing vine more commonly known as Hawaiian baby woodrose, which belongs to the morning glory family (Convolvulaceae). Seeds of Hawaiian baby woodrose contain psychoactive alkaloids including ergine [6]. Other members of the morning glory family also contain ergine within their seeds. Seeds are ingested with a typical dose being between 25 and 400 seeds depending on the species used and desired effect level. Onset of effect occurs in 20–40 minutes and total duration is 5–8 hours. Positive effects reported by users include: euphoria, increased mood, feeling of insight and engagement, closed and open eye visual hallucinations, increased sensual and aesthetic appreciation, and increased laughter [54].

PREVALENCE OF USE

Compared to other psychoactive substances, rates of recreational use of synthetic tryptamines are low worldwide [64]. This is despite the ready availability of precursors and relative ease of manufacture of these substances [65]. Large population surveys specifically examining prevalence of drug (including synthetic tryptamines) use have only recently been conducted. The Mixmag/Guardian 2012 Global Drugs Survey obtained

TABLE 15.4 Frequency of European Reported Drug Seizures and User Reports for Synthetic Simple Tryptamines Published on the Website Erowid (www.erowid.org)

Tryptamine	EDND Reports [67] (Year/number of reports, country)	2011 (number of reports, country)	Erowid User Reports [25] Prior to 2010 (years/number)	2011–12 (number)
AMT	2001–2009/10, FI, SE, UK, FR, DK	3, BG, NO, HU	2000–2010/227	4
AET	–	0	2004–2008/7	0
DiPT	2005–2008/5, SE, FI	0	2000–2010/60	0
DET	–	0	2000–2008/6	0
DPT	2004–2009/10, UK, SE, FI, NO, DK	0	2000–2010/159	7
DALT	–	0	2000–2011/0	0
4-HO-MET	2007–2009/7, SE, FI	2, BG, NO	2007–2010/7	16
4-HO-DET	2004–2007/3, SE	0	2006–2010/8	0
4-HO-DiPT	2004–2008/4,SE, UK	0	2000–2010/49	0
4-HO-MiPT	2006–2007/2, SE	0	2005–2010/29	1
4-Acetoxy-DMT	2009/1, FI	2, BG, CZ	2006–2010/43	16
4-Acetoxy-DiPT	2005–2008/5, SE, DK	0	2000–2010/46	0
5-MeO-DMT	2000-2009/24, SE, UK, CZ, FI, NO, FR, BG, NO, FI, DK, BE, NL	0	2000–2010/280	7
5-MeO-AMT	2003–2007/8	2, HU, DE	2000–2009/114	2
5-MeO-DiPT	2001–2009/27, DK, SE, UK, GR, FI, FR, NO	0	2000–2010/280	3
5-MeO-MiPT	2004–2009/4, UK, FI	1, NO	2003–2009/31	3
5-MeO-DALT	2007–2010/2, FI, UK	5, DE, UK, SE, BG, BE	2004–2010/22	9

BG: Bulgaria; NO: Norway; HU: Hungary; UK: United Kingdom; SE: Sweden; FI: Finland; DK: Denmark; CZ: Czech Republic; FR: France; DE: Germany; BE: Belgium; NL: Netherlands; GR: Greece.

over 15500 responses, of which 7700 were United Kingdom (UK) respondents 7.7% reported DMT use on at least one occasion, with 3.4% reporting use in the previous 12 months [66]. 6.6% of US respondents (n = 3360) reported DMT in the previous 12 months [66].

The European Information System and Database on New Drugs (EDND) administered by the European Monitoring Centre for Drugs and Drug Addiction (EMCDDA) provides information on seizures and exposures to new drugs within member states [67]. Frequency

and source of seizures for simple tryptamines in European countries are detailed in Table 15.4.

Comparative numbers of published user reports on Internet user sites provide an estimate of tryptamine use compared to mainstream recreational psychoactive substances. The number of user reports for each tryptamine published on the Erowid website (www.erowid. org) is summarised in Table 15.4. The 1375 reports detailing synthetic tryptamine experiences account for 6.2% of total user reports published on this website [25].

Although there are significant limitations in utilising seizure data and user reports to draw conclusions, it is notable that 4-HO-MET, 4-Acetoxy-DMT and 5-MeO-DALT had seizures reported during 2011 and all have a large proportion of their total Erowid user reports occurring in 2011–2012. This may indicate a recent increase in use of these substances.

In 2009 the EDND received two reports of 4-Acetoxy-N-methyl-N-ethyltryptamine (4-Acetoxy-MET, a relatively unknown synthetic tryptamine) seizures in Finland and Cyprus [68]. A seizure of 5-methoxy-N,N-dipropyl-tryptamine (5-MeO-DPT) in Finland was reported for the first time to the EDND in 2010 [69]. One survey of Internet drug-user forums in 2011 identified 4-hydroxy-N-ethyl-N-ethyl-tryptamine (4-HO-AET), a relatively unknown synthetic tryptamine [70]. Very little is known about the pharmacology and clinical effects of these tryptamines.

Kratom is freely available for purchase as an herbal extract from numerous Internet suppliers. At the time of writing kratom was a controlled substance in Thailand, Malaysia and Australia [7]. There are 193 user reports on the Erowid website regarding kratom exposure, 14 of those during 2011 [25]. Between 2007 and 2011 there were 16 kratom seizures reported to the EDND (one in 2011) originating from countries throughout Europe [71].

Seeds containing ergot-type alkaloids including ergine are readily available on the Internet, often marketed as 'ethnobotanicals' touted to enhance memory [7]. There are 276 user reports on the Erowid website regarding Hawaiian baby woodrose exposure, 17 of those during 2010 and one during 2011 [25].

ACUTE TOXICITY

Animal Data

Studies in mice found an LD_{50} of 47 mg/kg and 32 mg/kg for DMT administered intraperitoneally and intravenously, respectively [72]. Other rodent studies found the LD_{50} for psilocin and 5-MeO-DMT to be less than that of DMT (5-MeO-DMT 48–278 mg/kg depending on route of administration) [73,74]. 5-MeO-DMT administered to mice, rats, sheep and monkeys produced ataxia, mydriasis, tremors and seizures [74]. A pattern of clinical toxicity identical to serotonin toxicity in humans has been described in rats administered tryptamines, and appears to mediate via $5HT_{2A}$ receptors [75]. Lessin et al. found an LD50 of 160 mg/kg for AMT injected intraperitoneally in mice, and also demonstrated that co-administration of a MAOI potentiated the clinical effects of tryptamines [17].

An animal study of mitragynine demonstrated stimulant effects at low doses and opioid-like toxic effects at high doses [76].

Human Data

Controlled Studies in Humans

Eleven hallucinogen users administered 0.4 mg/kg of DMT intravenously experienced a 35 mmHg increase in systolic blood pressure (BP), 30 mmHg increase in diastolic BP, heart rate increase of 26 beats per minute (b.p.m.), and an increase in pupil diameter (mean of 3.5 mm 2–5 minutes post exposure) and rectal temperature (mean of +0.13°C at 60 minutes post exposure) [13]. DMT at a dose of 0.48 mg/kg body weight ingested with the naturally occurring MAOIs harmine (dose of 2.4 mg/kg body weight) and harmaline (dose of 0.4 mg/kg body weight) by 15 healthy male volunteers produced a mean increase in heart rate of eight b.p.m. 20 minutes post exposure and a mean increase in systolic BP of 11 mmHg 40 minutes post exposure. Pupillary diameter increased from 3.7±0.2 mm to a maximum of 4.9±0.2 mm 180 minutes post exposure. Oral temperature increased from a mean basal value of 37.1°C to 37.3°C [77]. DMT ingested at doses of 0.5, 0.75 and 1.0 mg/kg body weight by six healthy male volunteers produced a non-significant rise in systolic BP (13.8 mmHg with the 1.0 mg/kg body weight dose) and diastolic BP (10.4 mmHg with the 1.0 mg/kg body weight dose), which occurred at 60–90 minutes post exposure. There were only modest changes in heart rate with the largest dose producing a mean increase of 9.2 b.p.m [78]. Nausea was the most predominant reported adverse effect. Five volunteers described the overall experience as pleasant while one volunteer experienced an 'intensely dysphoric reaction' [78].

Controlled studies of human exposure to naturally occurring tryptamines demonstrate little clinical toxicity using typical recreational doses. Hallucinogen-naïve adults administered 30 mg/kg of psilocybin experienced no adverse physiological effect [79]. An oral dose of 0.85 mg of DMT combined with a MAOI produced only modest increases in heart rate (a mean maximum increase of four b.p.m. occurring 60 minutes post exposure) and BP (maximum mean systolic BP increase of six mmHg occurring 75 minutes post exposure) [80]. High doses (up to 315 µg/kg body weight) of psilocybin in healthy subjects were reported to cause moderate increases in BP (only statistically significant at 60 minutes post exposure correlating with a 15 mmHg rise in systolic BP) with no effect on axillary body temperature [81]. In all of these studies, recreational-level hallucinogenic effects were described by the volunteers.

Three out of four subjects administered Hawaiian baby woodrose seeds, in an experiment designed to test driving ability, experienced adverse effects including tremor, nausea and vomiting. One subject developed a psychotic-like state. All subjects experienced an increase in BP (mean increase in systolic BP of 30 mmHg). Adverse effects lasted for up to nine hours [82].

Published Case Reports of Tryptamine-associated Toxicity

Relatively few published case reports of synthetic simple tryptamine toxicity exist. A 23-year-old Caucasian male presented to an emergency department (ED) with paranoia and sensory hallucinations (formication) following ingestion of a capsule containing 5-MeO-DiPT. He recovered following a 4-hour period of observation. 5-MeO-DiPT was detected in serum and urine [83]. Another report describes a 25-year-old male who arrived in an ED 30 minutes following ingestion of 25 mg of 5-MeO-DiPT. He was agitated, hallucinating, tachycardic, hypertensive and hyperpyrexic. The patient settled with supportive care, but investigations revealed renal impairment (creatinine concentration of 150 µmol/L – normal range 80–124 µmol/L), metabolic acidosis with an anion gap of 44 (10–20 mmol/L) and a serum bicarbonate of 9 mmol/L (22–30 mmol/L) and rhabdomyolysis (peak creatinine kinase 38 855 U/L, myoglobin 13 145 mg/L (0–110 mg/L)). Urine toxicology screen was negative for amphetamines, cannabinoids, cocaine, ethanol and barbiturates; however, the presence of 5-MeO-DiPT was not analytically confirmed. The patient made a full recovery following rehydration and forced alkaline dieresis [84].

A 19-year-old male presented to an ED with hypertension, tachycardia, mydriasis, hallucinations and cataplexy following ingestion of a larger than normal dose of his 5-MeO-DiPT. There was no limb rigidity, but his limbs remained in whatever position they were placed in. Laboratory investigation revealed hyperglycaemia, glycosuria and an increased white cell count. A urine drug screen was positive for cocaine and phencyclidine (the patient subsequently denied taking these); use of 5-MeO-DiPT was not analytically confirmed. He recovered over a number of hours following treatment with lorazepam [85]. DPT toxicity was reported in a 19-year-old female who presented to the ED 90 minutes after ingesting an unknown amount of DPT. A commercial vial labelled DPT was found with the patient. A label indicated the product was 'for research purposes only'. Clinical toxicity included hallucinations, extreme agitation and tachycardia (200 b.p.m.). Agitation resolved following administration of 3 mg of lorazepam, although the route of lorazepam administration is not recorded. She developed rhabdomyolysis (serum creatinine kinase greater than 8000 U/L) requiring treatment with IV fluids and was discharged well 60 hours post admission. The presence of DPT was not analytically confirmed [86]. A 21-year-old male presented to an ED one hour after ingesting 270 mg of AMT, after miscalculating his normal dose by a factor of 10. He had visual hallucinations, tremor and an exaggerated startle reaction. Sympathomimetic effects were present, but not prominent: heart rate 52 b.p.m., BP 183/93 mmHg, respiratory rate 20/minute and temperature 36.4°C. He was only orientated to person and had dilated pupils (10 mm diameter). Visual hallucinations resolved 10 hours post exposure and he was discharged well [87].

A 17-year-old male regular hallucinogen user decided he wanted to extend the length of his relatively short 5-MeO-DMT induced hallucinogenic experiences. After obtaining information via the Internet he purchased some Syrian rue seeds containing the natural MAOI harmaline. He ingested the seeds, smoked 10 mg of 5-Meo-DMT and insufflated a further 15–20 mg. Friends found him collapsed a few hours later, hallucinating and agitated. Mydriasis and marked diaphoresis were noted. In the ED he was tachycardic (heart rate 186 b.p.m.) and hyperpyrexic (40.7°C). He required physical restraint and subsequently settled with 2.5 mg of IV lorazepam. Over the following 24 hours he exhibited autonomic system lability (lowest BP 80/35 mmHg) and rhabdomyolysis (treated with IV fluid therapy) [88]. He made a full recovery; however, this case illustrates exacerbation of tryptamine toxicity with concurrent MAOI exposure. A 37-year-old male presented with agitation and a sympathomimetic toxidrome after ingestion of a mixture of methylone (2-methylamino-l-[3,4–methylenedioxyphenyl]propan-l-one) and 5-MeO-MiPT [89]. In another case a 21-year-old male in

Canada presented to an ED with hallucinations and inability to move his limbs after ingesting a pill called 'Foxy'. He recovered after two hours. 5-MeO-DiPT was identified in urine at a concentration of 1.7 μg/ml [90].

Review of the American Association of Poison Control Centers' Total Exposure Surveillance System (TESS) database during 2002–2003 found 41 exposures to 5-MeO-DiPT resulting in moderate to severe toxicity in 68% of these cases. Effects included hypertension, tachycardia, hallucinations and agitation [83]. The Erowid website warns that 5-MeO-AMT has been sold as LSD in the USA. 5-MeO-AMT has a steep dose response curve compared to LSD and so dosing errors are likely to result in a greater chance of toxicity; one associated death has been reported on the Erowid website [91].

A 2007 literature review supplemented by interviews with users of ayahuasca brews (containing DMT and a naturally occurring β-carboline MAOIs) concluded that human consumption of a mixture of DMT and β-carboline MAOIs posed no greater risk than therapeutic or recreational doses of codeine, mescaline and methadone. However, few controlled human studies with quantified exposure to DMT were included, and a number of deaths associated with ayahuasca brews were identified [73].

Kratom use has been associated with seizures. A 64-year-old male arrived in an ED after a witnessed seizure at home following kratom tea ingestion. On examination he had a Glasgow Coma Scale (GCS) of six and a heart rate of 110 b.p.m. Shortly after he had a further generalised seizure and was intubated. Magnetic resonance imaging (MRI) of the brain was unremarkable. He made a full recovery after a period of supportive care. Mitragynine was detected in a urine sample at a concentration of 167 ng/ml [92]. A 32-year-old male obtained a substance via the Internet after searching for *Mitragyna speciosa*. He was subsequently found unconscious with seizure-like movements and required intubation.

A general toxicology screen was negative. The patient recovered after a period of supportive care and treatment of aspiration pneumonia [93]. A male who used kratom to self-manage opioid withdrawal ingested 100 mg of modafinil and kratom. Twenty minutes later he experienced a generalised seizure lasting five minutes. A general toxicology screen, computerised tomography (CT) and MRI of the brain and serum electrolytes were normal and the patient made an unremarkable recovery Analysis of the plant material the patient was using confirmed its identity as kratom [94].

Two patients developed psychoactive symptoms after ingesting Hawaiian baby woodrose seeds. One recovered, while the other jumped from a height and died [7,95].

Published Case Reports of Tryptamine-associated Deaths

A 29-year-old male died after his male partner injected an aqueous solution of 5-MeO-DiPT rectally with the aim of enhancing sexual pleasure. The patient developed abdominal symptoms and agitation and died in hospital 3.5 hours later. 5-MeO-DiPT was identified in urine (1.67 μg/ml) and blood (0.412 μg/ml). An autopsy revealed polyarteritis nodosa, an ischaemic area of myocardium, pulmonary congestion and pulmonary alveolar haemorrhages. The exact cause of death was not stated [96]. Unconfirmed reports on the Erowid website state that a 100 mg dose was used rectally, while a normal dose would be 10 mg [97].

Several deaths have been attributed to AET; the compound originally was originally marketed in 1961 as an antidepressant (Monase®) but subsequently discontinued because of associated agranulocytosis [98]. A 19-year-old female ingested a glass of beer containing two 'hits' of white powder she had been told was MDMA. Shortly after the ingestion she became confused, vomited and had a cardiac arrest. An autopsy revealed bilateral pulmonary congestion. AET was found in blood (5.6 mg/L) and urine

(80.4 mg/L). No MDMA was identified [99]. Prior to sudden death a young male developed agitation, hyperpyrexia and stimulant-like effects following ingestion of AET. AET was detected in post-mortem blood [100].

In 2005 a 25-year-old male was found dead in the USA after ingesting a herbal hallucinogenic extract. There was no anatomical cause for death found at autopsy. Analysis of post-mortem blood revealed DMT (0.02 mg/L) and 5-MeO-DiPT (1.88 mg/L). Naturally occurring β-carboline MAOIs were also detected. Death was attributed to 'hallucinogenic amine intoxication' [101].

Psychoactive substances are implicated in numerous deaths resulting from altered perception and subsequent dangerous or irrational behaviour while under the influence of the substance. For example, a 26-year-old male was killed after walking onto a motorway and being hit by a lorry. A coroner found his traumatic death had occurred while under the influence of 5-MeO-DALT [6,102]. A male jumped from a building and died after ingesting Hawaiian baby woodrose seeds. Ergine was found in post-mortem blood and urine [7,102].

Nine unintentional deaths in Sweden have been linked to the use of a recreational product named 'Krypton', available for purchase on the Internet. Krypton comprises powdered kratom leaves mixed with the μ-receptor agonist O-desmethyltramadol. Both mitragynine and O-desmethyltramadol were found in post-mortem blood samples. Mitragynine blood concentrations ranged from 0.02–0.18 μg/ml. O-desmethyltramadol concentrations ranged from 0.4–4.3 μg/ml [103,104].

User Reports of Tryptamine Toxicity

Internet drug user fora, including Erowid publish user reports of effects experienced following psychoactive substance exposure. The accuracy of reports is limited by possible user bias, difficulty in ascertaining the identity and amount of the substance exposure and the possibility of co-ingestants. General user reports of adverse or unwanted effects of unsubstituted simple tryptamines include: restlessness, yawning, anxiety, tension, nausea, vomiting, palpitations, muscle pain, bruxism, headache, frightening or distressing hallucinations, overwhelming fear (DMT), abdominal discomfort, nasal irritation when insufflated and respiratory discomfort or distress when inhaled [26]. User reports of adverse or unwanted effects of AMT include nausea and vomiting (particularly common), anxiety, restlessness, muscle tension and palpitations [26]. There have been reports of ataxia, confusion and inner ear discomfort associated with DiPT exposure [105].

General user reports of adverse or unwanted effects for the 4-substituted simple tryptamines include: lethargy, fatigue, anxiety, fear, paranoia, frightening hallucinations, intense overwhelming thoughts or visual disturbances, diaphoresis, flushing, elevated heart rate, muscle pain, confusion and difficulty speaking [26].

General user reports of adverse or unwanted effects for the 5-substituted simple tryptamines include: terror, anxiety, fear, paranoia, frightening hallucinations, intense overwhelming experiences, respiratory discomfort or distress when inhaled, difficulty integrating experiences into normal life, nausea and vomiting at higher doses, headache, fatigue, muscle pain, abdominal discomfort, diarrhoea, minor bruxism [26]. Recreational use of 5-MeO-DiPT seems particularly associated with nausea, abdominal discomfort and diarrhoea [26,106]. User reports appear to indicate that 5-MeO-DALT has fewer side-effects compared to the other 5-substituted simple tryptamines; however, retrograde amnesia is associated with higher doses [107].

User reports of unwanted or adverse effects associated with kratom exposure include: bitter taste, nausea, vomiting, diarrhoea, mild depression following use, nystagmus and tremor [54,108]. User reports of unwanted or adverse effects associated with Hawaiian baby woodrose

seed exposure include: anxiety, nausea, vomiting, abdominal pain, delirium, dizziness, confusion, fear and paranoia [54].

CHRONIC TOXICITY

Animal Data

Examination of the brains of rats seven days after subcutaneous administration of AET revealed reduced 5-hydroxytryptamine (5-HT) and 5-hydroxyindoleacetic acid concentrations in the hippocampus and frontal cortex. In addition there were reduced numbers of 5-HT uptake sites in the frontal cortex. The authors concluded AET produces chronic serotonergic toxicity similar to that associated with MDMA use [109].

Compton et al. studied the effect of 5-MeO-DiPT on the cognitive development of adolescent rats. Performance of rats administered 5 mg/kg of 5-MeO-DiPT and tested 80 days later using spatial and non-spatial memory tasks revealed markedly reduced performance compared to rats administered placebo. The authors concluded that 5-MeO-DiPT acts as a toxin compromising CNS system serotonin systems [110]. The same group subsequently studied a group of adolescent rats administered 5 mg/kg or 20 mg/kg of 5-MeO-DiPT, or placebo. Rats receiving either dose of 5-MeO-DiPT performed comparably to placebo rats on learning-related tasks performed with a fixed goal-location, but significantly worse when there was a change in goal location. The authors concluded that 5-MeO-DiPT rats were unable to adapt their learning and behaviour, potentially signifying damage to forebrain serotonergic systems, similar to those caused by MDMA exposure [111].

Further evidence of the detrimental effects of 5-MeO-DiPT comes from a study in which rats were administered saline, MDMA or 5-MeO-DiPT. Both the 5-MeO-DiPT and MDMA rats demonstrated significant spatial learning deficits compared to the placebo group [112]. Sogawa et al., while studying the *in vitro* effects of 5-MeO-DiPT on monoamine systems within monkey fibroblast-like kidney cells, found that 5-MeO-DiPT was toxic to cells at high concentrations [27]. Work using a culture containing a slice of the mesencephalic area of rat brain demonstrated that administration of 5-MeO-DiPT into the culture markedly decreased the concentration of intracellular serotonin. In addition there was a marked decrease in binding to serotonin receptors by labelled citalopram, suggesting 5-MeO-DiPT has neurotoxic properties [113].

Human Data

There is sparse data examining the chronic toxicological effects of tryptamine use in humans. AET was originally developed as an antidepressant, but was discontinued in the 1960s due to a significant rate of agranulocytosis [98]. A young German male developed temporary intrahepatic cholestasis following regular ingestion of powdered kratom over a two-week period. No other causative agent was identified [114].

DEPENDENCE AND ABUSE POTENTIAL

Animal Data

A small number of animal studies in four different species demonstrate varying degrees of tolerance to DMT, but do not allow any firm conclusions to be drawn regarding tolerance in humans [112,115,116].

Human Data

Internet user reports suggest that DMT has little propensity to induce tolerance and dependence in humans [25,26]. It has been postulated that the extremely intense short-lived duration of the psychoactive experience may explain this. There are a number of human studies supporting user's observations. Strassman et al. administered volunteers four hallucinogenic doses (0.3 mg/kg) of DMT at 30-minute intervals on two consecutive days [117]. There was no psychological tolerance. Biological responses including heart rate, cortisol and prolactin concentrations decreased with repeated doses, although BP changes did not [117].

In a double blind controlled trial nine experienced psychoactive substance users were administered an ayahyasca mixture containing DMT (0.75 mg/kg dose) or placebo, in two doses four hours apart. This was repeated a week later. There was no difference in subjective, neuropsychological, autonomic or immunologic effects following two consecutive doses of DMT. There was, however, a trend toward lower BP and heart rate following the second dose of DMT [117].

There are no published studies examining the dependence and abuse potential of synthetic simple tryptamines, but a number of factors suggest they have low potential for dependence:

1. There is no evidence in humans that significant tolerance develops and therefore escalating doses are unlikely to be required to obtain regular effects.
2. Users report very minimal 'after' or 'comedown' effects, negating the need to re-dose to minimize these.
3. These substances have not been associated with dependence as evidenced by user reports and the published medical literature.

Addiction to kratom has been reported in South-east Asia in older publications, but is not a recognised issue with recreational use of kratom in the developed world [118,119]. There are reports of a withdrawal syndrome similar to opioid withdrawal: yawning, irritability, rhinorrhea, myalgia and diarrhoea [57].

CONCLUSIONS

Tryptamines are a diverse group of compounds which cause hallucinations. They are $5HT_{2A}$ and $5HT_{1A}$ agonists and are metabolised by a number of pathways including monoamine oxidase, limiting the oral bioavailability of many compounds. Dimethyltryptamine (DMT) is a naturally occurring tryptamine, whose psychoactive properties have been used in religious ceremonies for centuries and is also used as a recreational drug in the UK and USA. Psilocin is a 4-substituted simple tryptamine found within psychedelic mushroom species and it is used as a recreational hallucinogen. Recreational synthetic tryptamines including 5-methoxy-diisopropyl-tryptamine (5-MeO-DiPT, Foxy methoxy) and newer compounds (AET, 4-HO-MET, 4-Acetoxy-DMT and 5-Meo-DMT) have been associated with agitation, tachyarrhythmias, hyper-pyrexia and death. Ergoline-type tryptamines include mitragynine, available as herbal supplement and associated with seizures, and ergine, an alkaloid found in seeds of the morning glory family with a similar structure to LSD. Animal models illustrate serotonergic neurotoxicity with chronic tryptamine use, although there is no human data to be able to determine whether this also occurs in humans.

REFERENCES

[1] McKenna DJ. Clinical investigations of the therapeutic potential of ayahuasca: rational and regulatory challenges. Pharmacol Therapeut 2004;102:111–29.
[2] Lyttle T, Goldstein D, Gartz J. Bufo. Toads and bufotenine: fact and fiction surrounding an alleged psychedelic. J Psychoactive Drugs 1996;28:267–90.
[3] Fantegrossi WE, Murnane AC, Reissig CJ. The behavioural pharmacology of hallucinogens. Biochem Pharmacol 2008;75:17–33.
[4] Hill SL, Thomas SHL. Clinical toxicology of newer recreational drugs. Clin Toxicol 2011;49:705–19.
[5] Nichols DE. Hallucinogens. Pharmacol Therapeut 2004;101:131–81.
[6] Hylin JW, Watson DP. Ergoline alkaloids in tropical wood roses. Science 1965;148:499–500.

[7] Gibbons S. 'Legal highs' – novel and emerging psychoactive drugs: a chemical overview for the toxicologist. Clin Toxicol 2012;50:15–24.
[8] Lee CM, Trager WF, Beckett AH. Corynantheidine-type alkaloids.II. Absolute configuration of mitragynine, speciociliatine, mitraciliatine and speciogynine. Tetrahedron 1967;23:375–85.
[9] Thongpradichote S, Matsumoto K, Tohda M, et al. Identification of opioid receptor subtypes in antinociceptive actions of supraspinally-administered mitragynine in mice. Life Sci 1998;62:1371–8.
[10] Beckett AH, Shellard EJ, Phillipson JD, Lee CM. Alkaloids from Mitragyna speciosa (Korth.). J Pharm Pharmacol 1965;17:753–5.
[11] Shulgin A, Shulgin A. TiHKAL. The continuation: 6. Erowid. Available: <http://www.erowid.org/library/books_online/tihkal/tihkal06.shtml> [accessed 11.03.13].
[12] Kaplan J, Mandel LR, Stillman R, et al. Blood and urine levels of N,N-dimethyltryptamine following administration of psychoactive dosages to human subjects. Psychopharmacologia 1974;38:239–45.
[13] Strassman RJ, Qualls CR. Dose-response study of N,N-dimethyltryptamine in humans. Neuroendocrine, autonomic, and cardiovascular effects. Arch Gen Psych 1994;51:85–97.
[14] Sitaram BR, Lockett L, Talomsin R, Blackman GL, McLeod WR. In vivo metabolism of 5-methoxy-N, N-dimethyltryptamine and N,N-dimethyltryptamine in the rat. Biochem Pharmacol 1997;36:1509–12.
[15] Riba J, McIlhenny EH, Valle M, Bouso JC, Barker SA. Metabolism and disposition of N,N-dimethyltryptamine and harmala alkaloids after oral administration of ayahuasca. Drug Test Analysis 2012;4:610–6.
[16] Sz á ra St Dimethyltryptamine: its metabolism in man; the relation of its psychotic effect to the serotonin metabolism. Cell Mol Life Sci 1956;12:441–2.
[17] Lessin AW, Long RF, Parkes MW. Central stimulant actions of α-alkyl substituted tryptamine in mice. Brit J Pharmacol 1965;24:49–67.
[18] Shulgin A, Shulgin A. TiHKAL. The continuation: 11. Erowid. Available: <http://www.erowid.org/library/books_online/tihkal/tihkal11.shtml> [accessed 29.01.12].
[19] Shulgin A, Shulgin A. TiHKAL. The continuation: 48. Erowid. Available:<http://www.erowid.org/library/books_online/tihkal/tihkal48.shtml> [accessed 29.01.12]
[20] Shulgin A, Shulgin A. TiHKAL. The continuation: 4. Erowid. Available: <http://www.erowid.org/library/books_online/tihkal/tihkal04.shtml> [accessed 29.01.12].
[21] Shulgin A, Shulgin A. TiHKAL. The continuation: 3. Erowid. Available: <http://www.erowid.org/library/books_online/tihkal/tihkal03.shtml> accessed 29.01.12].
[22] Shulgin A, Shulgin A. TiHKAL. The continuation: 9. Erowid. Available: <http://www.erowid.org/library/books_online/tihkal/tihkal09.shtml> [accessed 29.01.12].

[23] Erowid. DTP dosage. Available: <http://www.erowid.org/chemicals/dpt/dpt_dose.shtml> [accessed 29.01.12].

[24] BlueLight. Available: <http://www.bluelight.ru/vb/threads/146249-The-Big-amp-Dandy-5-MeO-DALT-Thread?p=2104770#post2104770> [accessed 29.01.12].

[25] Erowid. Erowid Experience Vaults. Available: <http://www.erowid.org/experiences/exp_front.shtml> [accessed 29.01.12].

[26] Erowid. Erowid Psychoactive Vaults. Available: <http://www.erowid.org/chemicals/> [accessed 29.01.12].

[27] Sogawa C, Sogawa N, Tagawa J, et al. 5-Methoxy-N,N-diisopropyltryptamine (Foxy), a selective and high affinity inhibitor of serotonin transporter. Toxicol Lett 2007;170:75–82.

[28] Kamata T, Katagi M, Tsuchihashi H. Metabolism and toxicological analyses of hallucinogenic tryptamine analogues being abused in Japan. Forensic Toxicol 2010;28:1–8.

[29] Shulgin A, Shulgin A. TiHKAL. The continuation: 21. Erowid. Available: <http://www.erowid.org/library/books_online/tihkal/tihkal21.shtml> [accessed 29.01.12].

[30] Shulgin A, Shulgin A. TiHKAL. The continuation: 16. Erowid. Available: <http://www.erowid.org/library/books_online/tihkal/tihkal16.shtml> [accessed 29.01.12].

[31] Shulgin A, Shulgin A. TiHKAL. The continuation: 17. Erowid. Available: <http://www.erowid.org/library/books_online/tihkal/tihkal17.shtml> [accessed 29.01.12].

[32] Shulgin A, Shulgin A. TiHKAL. The continuation: 22. Erowid. Available: <http://www.erowid.org/library/books_online/tihkal/tihkal22.shtml> [accessed 29.01.12].

[33] Shulgin A, Shulgin A. TiHKAL. The continuation: 18. Erowid. Available: <http://www.erowid.org/library/books_online/tihkal/tihkal18.shtml> [accessed 29.01.12].

[34] Shen HW, Jiang XL, Winter JC, Yu AM. Psychedelic 5-methoxy-N,N-dimethyltryptamine: metabolism, pharmacokinetics, drug interactions, and pharmacological actions. Curr Drug Metab 2010;11(8):659–66.

[35] Pierce PA, Peroutka SJ. Hallucinogenic drug interactions with neurotransmitter receptor binding sites in human cortex. Psychopharmacology 1989;97:118–22.

[36] Ray TS. Psychedelics and the human receptorome. PLOS ONE 2010;5:3.

[37] Cozzi NV, Gopalakrishnan A, Anderson LL, et al. Dimethyltryptamine and other hallucinogenic tryptamines exhibit substrate behavior at the serotonin uptake transporter and the vesicle monoamine transporter. J Neural Transm 2009;116:1591–9.

[38] Fontanilla D, Johannessen M, Hajipour AR, Cozzi NV, Jackson MB, Ruoho AE. The hallucinogen N,N-dimethyltryptamine (DMT) is an endogenous sigma-1 receptor regulator. Science 2009;323:934–7.

[39] Su TP, Hayashi T, Vaupel DB. When the endogenous hallucinogenic trace amine N,N-dimethyltryptamine meets the sigma-1 receptor. Sci Signal 2009;2:12.

[40] Keiser MJ, Setola V, Irwin JJ, et al. Predicting new molecular targets for known drugs. Nature 2009;462:175–81.

[41] Deliganis AV, Pierce PA, Peroutka SJ. Differential interactions of dimethyltryptamine (DMT) with 5-HT1A and 5-HT2 receptors. Biochem Pharmacol 1991;41:1739–44.

[42] Smith RL, Canton H, Barrett RJ, Sanders-Bush E. Agonist properties of N,N-dimethyltryptamine at serotonin 5-HT2A and 5-HT2C receptors. Pharmacol Biochem Behav 1998;61:323–30.

[43] Fantegrossi WE, Reissig CJ, Katz EB, Yarosh HL, Rice KC, Winterb JC. Hallucinogen-like effects of N,N-dipropyltryptamine (DPT): possible mediation by serotonin 5-HT1A and 5-HT2A receptors in rodents. Pharmacol Biochem Behav 2008;88:358–65.

[44] Nagai F, Nonaka R, Satoh K, Kamimura H. The effects of non-medically used psychoactive drugs on monoamine neurotransmission in rat brain. Eur J Pharmacol 2007;559:132–7.

[45] Peden NR, Macaulay KEC, Bisset AF, Crooks J, Pelosi AJ. Clinical toxicology of 'magic mushroom' ingestion. PMJ 1981;57:543–5.

[46] Rogawski MA, Aghajanian GK. Serotonin autoreceptors on dorsal raphe neurons: structure-activity relationships of tryptamine analogs. J Neurosci 1981;1:1148–54.

[47] Jacob III P, Shulgin AT. Structure-activity relationships of the classic hallucinogens and their analogs. NIDA Res Monogr 1994;146:74–91.

[48] Shulgin A, Shulgin A. TiHKAL. The continuation: 26. Erowid. Available: <http://www.erowid.org/library/books_online/tihkal/tihkal26.shtml> [accessed 11.03.13].

[49] Wikipedia. Available: <http://en.wikipedia.org/wiki/Mitragynine> [accessed 11.03.13].

[50] Kratom.Net. Available: <http://www.kratom.net/content.php?46-Mitragynine> [accessed 11.03.13].

[51] Matsumoto K, Suchitra T, Murakami Y, et al. Central antinociceptive effects of mitragynine in mice: contribution of descending noradrenergic and serotonergic-systems. Eur J Pharmacol 1996;317:75–81.

[52] Matsumoto K, Yamamoto LT, Watanabe K, et al. Inhibitory effect of mitragynine, analgesic alkaloid from Thai herbal medicine, on neurogenic contraction of the vas deferens. Life Sci 2005;78:187–94.

[53] Babu KM, McCurdy CR, Boyer EW. Opioid receptors and legal highs: salvia divinorum and Kratom. Clin Toxicol 2008;46:146–52.

[54] Erowid. Psychoactive effects. The vaults of Erowid. Available: <http://www.erowid.org/psychoactives/effects/effects.shtml> [accessed 29.01.12].

[55] Shulgin A, Shulgin A. Tryptamines I have known and loved: the chemistry continues. Erowid.Available: <http://www.erowid.org/library/books_online/tihkal/> [accessed 29.01.12].

[56] Ott J. Pharmepena-Psychonautics: human intranasal, sublingual and oral pharmacology of 5-methoxy-N,N-dimethyltryptamine. J Psychoactive Drugs 2001;33:403–7.

[57] Erowid. Available: <http://www.erowid.org/library/books_online/tihkal/tihkal38.shtml> [accessed 29.01.12].

[58] Shulgin A, Shulgin A. TiHKAL. The continuation: 5. Erowid. Available: http://www.erowid.org/library/books_online/tihkal/tihkal05.shtml [accessed 29.01.12].

[59] Shulgin A, Shulgin A. TiHKAL. The continuation: 37. Erowid. Available: <http://www.erowid.org/library/books_online/tihkal/tihkal37.shtml> [accessed 29.01.12].

[60] Shulgin A, Shulgin A. TiHKAL. The continuation: 40. Erowid. Available: <http://www.erowid.org/library/books_online/tihkal/tihkal40.shtml> [accessed 29.01.12].

[61] Shulgin A, Shulgin A. 5-MeO-DALT. The vaults of Erowid. Erowid. Available: <http://www.erowid.org/chemicals/5meo_dalt/5meo_dalt_info1.shtml> [accessed 29.01.12].

[62] Adkins JE, Boyer EW, McCurdy CR. Mitragyna speciosa, a psychoactive tree from Southeast Asia with opioid activity. Curr Top Med Chem 2011;11:1165–75.

[63] Erowid. Kratom: basics. The vaults of Erowid. Available: <http://www.erowid.org/plants/kratom/kratom_basics.shtml> [accessed 29.01.12].

[64] SandersB Lankenau SE, Bloom JJ, Hathazi D. 'Research chemicals': tryptamine and phenylamine use among high-risk youth. Subst Use Misuse 2008;43:389–402.

[65] Collins M. Some new psychoactive substances: precursor chemicals and synthesis-driven end-products. Drug Test Anal 2011;3:404–16.

[66] 2012 Global Drug Survey. Available: <http://globaldrugsurvey.com/run-my-survey/2012-global-drug-survey> [accessed 11.01.13].

[67] European information system and database on new drugs. Available: <http://ednd.emcdda.europa.eu/html.cfm/index7246EN.html> [accessed 29.01.12].

[68] European information system and database on new drugs. Available: <http://ednd.emcdda.europa.eu/html.cfm/index7246EN.html?SUB_ID=96> [accessed 29.01.12].

[69] European information system and database on new drugs. Available: <http://ednd.emcdda.europa.eu/html.cfm?nNodeid=7246&SUB_ID=142> [accessed 29.01.12].

[70] Kelleher C, Christie R, Lalor K, Fox J, O'Donnell C. An overview of new psychoactive substances and the outlets supplying them. : National Advisory Committee on Drugs (Ireland); 2011.

[71] European information system and database on new drugs. Available: http://ednd.emcdda.europa.eu/html.cfm/index7246EN.html?SUB_ID=72 [accessed 29.01.12].

[72] ChemIDplus Advanced. N,N-dimethyltryptamine or serotonin. Bethesda, MD:National Library of Medicine, Specialized Information Services; 2005. Available: http://chem.sis.nim.nih.gov/chemidplus/jsp/common/Toxicity.jsp [accessed 29.01.12].

[73] Gable RS. Risk assessment of ritual use of oral dimethyltryptamine (DMT) and harmala alkaloids. Addiction 2007;102:24–34.

[74] Gillin JC, Tinklenberg J, Stoff D, Stillman R, Shortlidge JS, Wyatt RJ. 5-methoxy- N,N-dimethyltryptamine: behavioral and toxicological effects in animals. Biol Psychiatry 1976;11:355–8.

[75] Van Oekelen D, Megens A, Meert T, Luyten WH, Leysen JE. Role of 5-HT(2) receptors in the tryptamine-induced 5-HT syndrome in rats. Behav Pharmacol 2002;13(4):313–8.

[76] Macko E, Weisbach JA, Douglas B. Some observations on the pharmacology of mitragynine. Arch Int Pharmacodyn Ther 1972;198:145–61.

[77] Callaway JC, McKenna DJ, Grob CS, et al. Pharmacokinetics of Hosca alkaloids in healthy humans. J Ethnopharmacol 1999;65:243–56.

[78] Riba J, Rodriguez-Fornells A, Urbano G, et al. Subjective effects and tolerability of the South American psychoactive beverage ayahuasca in healthy volunteers. Psychopharmacology (Berl) 2001;154:85–95.

[79] Griffiths R, Richards W, Johnson M, McCann U, Jesse R. Mystical-type experiences occasioned by psilocybin mediate the attribution of personal meaning and spiritual significance 14 months later. J Psychopharmacol 2008;22:621–32.

[80] Riba J, Valle M, Urbano G, Yritia M, Morte A, Barbanoj MJ. Human pharmacology of ayahuasca: subjective and cardiovascular effects, monoamine metabolite excretion, and pharmacokinetics. J Pharmacol Exp Ther 2003;306:73–83.

[81] Hasler F, Grimberg U, Benz MA, Huber T, Vollenweider FX. Acute psychological and physiological effects of psilocybin in healthy humans: a double-blind, placebo-controlled dose-effect study. Psychopharmacology (Berl.) 2004;172:145–56.

[82] Kremer C, Paulke A, Wunder C, Toennes SW. Variable adverse effects in subjects after ingestion of equal doses of Argyreia nervosa seeds. Forensic Sci Int 2012;214:6–8.

[83] Wilson JM, McGeorge F, Smolinske S, Meatherall RA. Foxy Intoxication. Forensic Sci Int 2005;148:31–6.

[84] Alatrash G, Majhail NS, Pile JC. Rhabdomyolysis after ingestion of 'foxy', a hallucinogenic tryptamine derivative. Mayo Clin Proc 2006;81:550–1.

[85] Smolinske S, Rastogi R, Schenkel S. Foxy methoxy: a new drug of abuse. J Toxicol Clin Toxicol 2003;41:641.

[86] Dailey RM, Nelson LD, Scaglione JM. Tachycardia and rhabdomyolysis after intentional ingestion of N,N-dipropyltryptamine. J Toxicol Clin Toxicol 2003;1:742.

[87] Holstege CP, Baer AB, Kirk MA. Prolonged hallucinations followi mng ingestion of alpha-methyltryptamine. J Toxicol Clin Toxicol 2003;41:746.

[88] Brush DE, Bird SB, Boyer EW. Monoamine oxidase inhibitor poisoning resulting from internet misinformation on illicit substances. Clin Toxicol 2004;42:191–5.

[89] Shimizu E, Watanabe H, Kojima T, et al. Combined intoxication with methylone and 5-MeO-MIPT. Prog Neuropsychopharmacol Biol Psychiatry 2007; 31(1):288–91.

[90] Meatherall R, Sharma P. Foxy, a designer tryptamine hallucinogen. J Anal Toxicol 2003;27:313–7.

[91] Erowid. Reported LSD-related death was not LSD. The vaults of Erowid. Available: <http://www.erowid.org/chemicals/lsd/lsd_media2.shtml>. [accessed 29.01.12].

[92] Nelsen JL, Lapoint J, Hodgman MJ, Aldous KM. Seizure and coma following Kratom (Mitragynina speciosa Korth) exposure. J Med Toxicol 2010;6(4):424–6.

[93] Roche KM, Hart K, Sangalli B, Lefberg J, Bayer M. Kratom: a case of a legal high. Clin Tox 2008;46:598.

[94] Boyer EW, Babu KM, Adkins JE, McCurdy CR, Halpern JH. Self-treatment of opioid withdrawal using Kratom (Mitragynia speciosa korth). Addiction 2008;103:1048–50.

[95] Klinke HB, Müller IB, Steffenrud S, Dahl-S Sørensen R. Two cases of lysergamide intoxication by ingestion of seeds from Hawaiian Baby Woodrose. Forensic Sci Int 2010;197:1–3.

[96] Tanaka E, Kamata T, Katagi M, Tsuchihashi H, Honda K. A fatal poisoning with 5-methoxy-N,N-diisopropyltryptamine, Foxy. Forensic Sci Int 2006;163(1–2):152–4.

[97] Erowid. Japanese Death Associated with 5-MeO-DIPT. The vaults of Erowid. Available: <http://www.erowid.org/chemicals/5meo_dipt/5meo_dipt_media1.shtml> [accessed 29.01.12].

[98] Butin JW. Agranulocytosis following Monase therapy. J Kans Med Soc 1962;63:338–40.

[99] Morano RA, Spies C, Walker FB, Plank SM. Fatal intoxication involving etryptamine. J Forensic Sci 1993;38:721–5.

[100] Daldrup T, Heller C, Matthiesen U, Honus S, Bresges A, Haarhoff K. Etryptamine, a new designer drug with a fatal effect. Z Rechtsmed 1986;97:61–8.

[101] Sklerov J, Levin B, Moore KA, King T, Fowler F. A fatal intoxication following the ingestion of 5 methoxy-N,N- dimethyltryptamine in an ayahuasca preparation. J Anal Toxicol 2005;29:838–41.

[102] Cambridge news. Available: <http://www.cambridge-news.co.uk/Home/Familys-vow-over-legal-high-drugs-danger.htm> [accessed 29.01.12].

[103] Bäckstrom BG, Classon G, Löwenhielm P, Thelander G. Krypton-new, deadly Internet drug. Since October 2009 have nine young persons died in Sweden. (Swedish). Lakartidningen 2010;107:3196–7.

[104] Kronstrand R, Roman M, Thelander G, Eriksson A. Unintentional fatal intoxications with mitragynine and O-desmethyltramadol from the herbal blend Krypton. J Anal Toxicol 2011;35:242–7.

[105] Wikipedia. Available: <http://en.wikipedia.org/wiki/Diisopropyltryptamine> [accessed 29.01.12].

[106] Wikipedia. Available: <http://en.wikipedia.org/wiki/5-MeO-DIPT> [accessed 29.01.12].

[107] Wikipedia. Available: <http://en.wikipedia.org/wiki/5-MeO-DALT> [accessed 29.01.12].

[108] Grewal K. Observation on the pharmacology of mitragynine. J Pharmacol Exp Ther 1932;46:251–71.

[109] Huang XM, Johnson MP, Nichols DE. Reduction in brain serotonin markers by alpha-ethyltryptamine (Monase). Eur J Pharmacol 1991;200:187–90.

[110] Compton DM, Selinger MC, Testa EK, Larkins KD. An examination of the effects of 5-Methoxy-n, n-di(ISO)propyltryptamine hydrochloride (Foxy) on cognitive development in rats. Psychol Rep 2006;98:651–61.

[111] Compton DM, Dietrich KL, Selinger MC, Testa EK. 5-methoxy-N,N-di(iso)propyltryptamine hydrochloride (Foxy)-induced cognitive deficits in rat after exposure in adolescence. Physiol Behav 2011;103:203–9.

[112] Kovacic B, Domino EF. Tolerance and limited cross-tolerance to the effects of N,N-dimethyltryptamine (DMT)and lysergic acid diethylamide-25 (LSD) on food-rewarded bar pressing in the rat. J Pharmacol Exp Ther 1976;197:495–502.

[113] Nakagawa T, Kaneko S. Neuropsychotoxicity of abused drugs: molecular and neural mechanisms of neuropsychotoxicity induced by methamphetamine, 3,4-methylenedioxymethamphetamine (ecstasy), and 5-methoxy-N,N-diisopropyltryptamine (foxy). J Pharmacol Sci 2008;106:2–8.

[114] Kapp FG, Maurer HH, Auwärter V, Winkelmann M, Hermanns-Clausen M. Intrahepatic cholestasis following abuse of powdered Kratom (Mitragyna speciosa). J Med Toxicol 2011;7:227–31.

[115] Cole JM, Pieper WA. The effects of N,Ndimethyltryptamine on operant behaviour in squirrel monkeys. Psychopharmacology 1973;29:107–13.

[116] Gillin JC, Cannon E, Magyar R, Schwartz M, Wyatt RJ. Failure of N,N-dimethyltryptamine to evoke tolerance in cats. Biol Psychiatry 1973;7:213–20.

[117] Strassman RJ, Qualls CR, Berg LM. Differential tolerance to biological and subjective effects of four closely spaced doses of N,Ndimethyltryptamine in humans. Biol Psychiatry 1996;39(9):784–95.

[118] Thuan LC. Addiction to Mitragyna Speciosa. Proceeding of the Alumni Association, Malaya 1957;10:322–4.

[119] Suwanlert S. A study of kratom eaters in Thailand. Bull Narc 1975;27:21–7.

This page intentionally left blank

16

Benzofurans and Benzodifurans

Shaun L. Greene

Victorian Poisons Information Centre, Melbourne, Australia

BENZOFURANS

The benzofurans are molecules combining a benzene ring and one or more attached heterocyclic furan rings, as illustrated in Figure 16.1 [1]. The benzofuran class of substances are members of the amphetamine and phenylethylamine classes. Benzofurans containing one furan ring that have been implicated as psychoactive recreational drugs include 6-(2-aminopropyl)benzofuran (6-APB) and 5-(2-aminopropyl)benzofuran (5-APB) [2,3]. The structure and chemical properties of these compounds are illustrated in Figure 16.1.

The benzodifurans are a ring-substituted sub-group of the alpha-methylated phenylethylamines (a group which includes amphetamine and methamphetamine). Presence of the difuran ring structure increases $5HT_{2A}$ receptor interaction, making benzodifurans highly potent synthetic hallucinogens [4]. The benzodifurans as a group are known as 'FLY' drugs because their molecular structure resembles an insect in shape [5]. Although numerous compounds belong to this class, there is only significant published human exposure information for bromo-dragonFLY (1-(8-bromobenzo[1,2-b;4,5-b']difuran-4-yl)-2-aminopropane) and 2C-B-FLY

(1-(8-Bromo-2,3,6,7-tetrahydrobenzol[1,2-b:4,5-b']difuran-4-yl)-2-aminoethane).

PHARMACOLOGY

Physical and Chemical Description

5-APB (5-(2-aminopropyl)benzofuran) and 6-APB (6-(2-aminopropyl)benzofuran) are benzofuran analogues of MDA (3,4-methylenedioxyamphetamine) originally synthesised in 1993 by investigators at Purdue University who were examining the role of the MDA dioxle ring structure in interacting with serotonergic neurons [6]. Substitution of MDA's methylenedioxy ring system with the isometric benzofuran ring derivatives produced 5-APB and 6-APB (Fig. 16.1). The formula for 5-APB and 6-APB is $C_{11}H_{13}NO$ yielding a molecular weight of 175.23 g/L. 5-APB and 6-APB exist as crystalline solids at room temperature.

In 1996–1997 Monte et al. reported the synthesis of dihydrobenzofuran analogues of various phenylalkylamine and mescaline derivatives [7]. A tetrahydrobenzodifuran analogue from this work was subsequently utilised in the manufacture of bromo-dragonFLY at Purdue

Novel Psychoactive Substances.
DOI: http://dx.doi.org/10.1016/B978-0-12-415816-0.00016-X

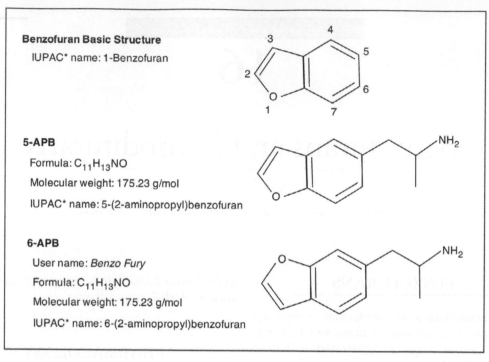

Benzofuran Basic Structure

IUPAC* name: 1-Benzofuran

5-APB

Formula: $C_{11}H_{13}NO$

Molecular weight: 175.23 g/mol

IUPAC* name: 5-(2-aminopropyl)benzofuran

6-APB

User name: *Benzo Fury*

Formula: $C_{11}H_{13}NO$

Molecular weight: 175.23 g/mol

IUPAC* name: 6-(2-aminopropyl)benzofuran

FIGURE 16.1 Common single furan ring benzofurans; structure and chemical properties. *IUPAC: International Union of Pure and Applied Chemistry.

University in 1998 by the same group, with the aim of synthesising a research tool to investigate serotonin receptor activity [8]. Bromo-dragonFLY (1-(8-bromobenzo[1,2-b;4,5-b'] difuran-4-yl)-2-aminopropane), also known as bromo-benzodifuranul-isopropylamine, possess a bromine substituted phenyl ring of the underlying phenylethylamine structure, held between two difuran rings. A methyl group is present on the alpha carbon. Bromo-dragonFLY has a molecular formula of $C_{13}H_{12}BrNO_2$ equating to a weight of 294.15 g/mol. The hydrochloride salt decomposes at 240°C. Bromo-dragonFLY was named after its structural resemblance to a dragonFLY. Bromo-dragonFLY is also known amongst users as 'DOBFLY', 'spamfly', 'placid', 'ABDF' and 'DOB-DragonFLY' [9,10].

2C-B-FLY (1-(8-Bromo-2,3,6,7-tetrahydro-benzol[1,2-b:4,5-b']difuran-4-yl)-2-aminoethane) was first synthesised in 1996 [8]. 2C-B-FLY shares a similar structure to bromo-dragonFLY, but possesses saturated rather than aromatic difuran rings, and lacks a methyl group at the alpha carbon. 2C-B-FLY has a molecular formula of $C_{12}H_{14}BrNO_2$ yielding a molecular weight of 284.15 g/mol. The hydrochloride salt decomposes at 310°C. Benzodifurans are not known to occur naturally. The chemical structures of the common benzodifurans are illustrated in Figure 16.2.

Pharmacokinetics

The pharmacokinetic properties of this group of drugs have not been studied. User reports for 5-APB and 6-APB suggest a relatively slow onset of effect over 30–120 minutes with a psychoactive effect lasting three–four hours and a

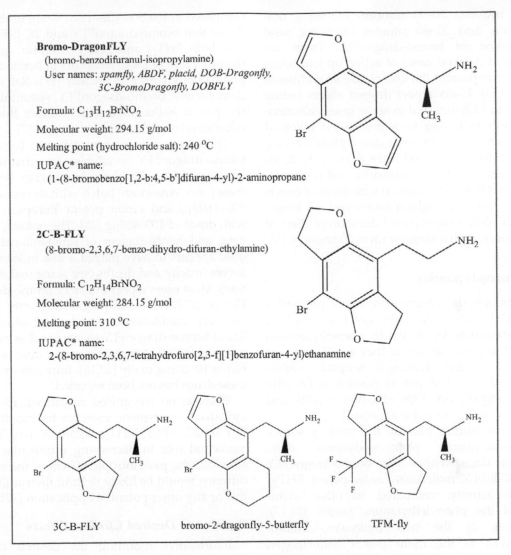

Bromo-DragonFLY
(bromo-benzodifuranul-isopropylamine)
User names: *spamfly, ABDF, placid, DOB-Dragonfly,*
3C-BromoDragonfly, DOBFLY

Formula: $C_{13}H_{12}BrNO_2$

Molecular weight: 294.15 g/mol

Melting point (hydrochloride salt): 240 °C

IUPAC* name:
(1-(8-bromobenzo[1,2-b:4,5-b']difuran-4-yl)-2-aminopropane

2C-B-FLY
(8-bromo-2,3,6,7-benzo-dihydro-difuran-ethylamine)

Formula: $C_{12}H_{14}BrNO_2$

Molecular weight: 284.15 g/mol

Melting point: 310 °C

IUPAC* name:
2-(8-bromo-2,3,6,7-tetrahydrofuro[2,3-f][1]benzofuran-4-yl)ethanamine

3C-B-FLY

bromo-2-dragonfly-5-butterfly

TFM-fly

FIGURE 16.2 Common benzodifurans; structure and chemical properties. *IUPAC: International Union of Pure and Applied Chemistry.

come down phase of up to 24 hours [3,10–13]. Routes of administration include nasal insufflation of powder, direct ingestion of powder or powdered dissolved in a liquid , ingestion of powder wrapped in moist tissue paper ('parachuting') and rectal administration [3,10–12]. Insufflation is reported to result in a more rapid onset of effects, but shorter duration of action [11]. Reported oral doses vary from 30 mg to 130 mg [3,10–12].

There are no published studies examining the pharmacokinetics of Bromo-dragonFLY or 2C-B-FLY; however, some information can be obtained from user reports. Onset of effects appears to occur between 20–90 minutes following oral ingestion, and 30–60 minutes following nasal insufflation of bromo-dragonFLY. There are reports of delayed onset of action up to 6 hours which may prompt the user to take an additional dose [9,13]. Users report desired effects lasting from 6 to 12 hours (but in some cases hallucinogenic effects lasting two–three days) followed by an 'after-effect' or come down phase varying between four hours and three days [9,13]. 2C-B-FLY users report predominantly oral use (occasional rectal administration) with onset of effects varying between eight minutes and 2.5 hours. Most 2C-B-FLY users report duration of effect of 6–10 hours, with a range of up to 20 hours [9,13].

Pharmacodynamics

Although the pharmacodynamic properties of 5-APB and 6-APB have not been studied, their structural similarity to phenethylamines has led to postulation that they are likely to act as catecholamine releasing or re-uptake inhibiting agents [6,14]. A patent granted to Eli Lilly and Company in 2006 classifies 5-APB and 6-APB as $5HT_{2C}$ receptor agonists [15].

Animal studies provide the primary source of benzodifuran pharmacodynamic data. Difuran rings within the Bromo-dragonFLY and 2C-B-FLY molecules confer potent $5HT_{2A}$ agonist activity, compared to other members of the phenylethylamine family [16,17]. However, of the two molecules, Bromo-dragonFLY is the more potent hallucinogen [18]. The presence of a methyl group on the alpha carbon offers protection from monoamine oxidase mediated degradation, increasing the duration of action of bromo-dragonFLY [19]. Additionally the alpha carbon methyl group results in greater $5HT_{2A}$ agonism and hence hallucinogenic effects [4]. Functional studies of synthesised dihydrobenzofuran analogues of hallucinogenic amphetamines (including dimethoxybromoamphetamine, DOB) demonstrated increased R-isomer potency

compared to the S-isomer [17]. There is also evidence that bromo-dragonFLY and 2C-B-FLY are also both $5HT_{2B}$ and $5HT_{2C}$ receptor agonists; the exact role of these receptors in benzodifuran pharmacodynamics is unclear [17]. Rat studies demonstrate bromo-dragonFLY substitutes for the potent hallucinogenic LSD more than any other structurally similar compound [7].

There appear to be two different batches of bromo-dragonFLY available for distribution to users (although there is overlap between these). An 'American' batch with doses varying 500–1600 µg and a more potent 'European' batch with doses of 100–800 µg [20]. The potency of the drug and hence danger of unintentional overdose appears to have played a role in leading to severe toxicity and deaths (see acute toxicity section). Most users report intake of bromo-dragon-FLY or 2C-B-FLY as impregnated blotter paper; however, insufflation of powder and addition of liquid Bromo-dragonFLY to sugar cubes are also reported [13,21]. A typical reported dose of 2C-B-FLY is 10–20 mg orally [13,18]. Intravenous use of these drugs has not been reported.

There is no recognised medicinal, research, industrial or cosmetic uses for bromo-dragon-FLY or 2C-B-FLY. 5-HT_{2A} agonists may have a medicinal role in decreasing intraocular pressure, but the psychotropic effects of the benzodifurans would be likely to limit their utility for this or any other potential application [22].

Positive or Desired Clinical Effects

Information regarding the desired clinical effects of 5-APB and 6-APB is limited to on-line user report forums. Limited user reports indicate that positive effects of 5-APB include increased empathy, variable euphoria, visual disturbances, appreciation for music and dancing and more general 'stimulation' as opposed to 6-APB [12,23–25]. Reported positive effects of 6-APB include increased tactile and visual stimulation, mild euphoria, and appreciation for music, visual hallucinations and increase in mood, feelings of peace, love and self-acceptance [10,11,13,14].

There are some reports from users that concurrent use of 5-APB and 6-APB increases positive effects, especially the degree of euphoria. Some users describe the combination of 5-APB and 6-APB as being similar to MDMA, but more intense [26,27].

No formal studies examining the behavioural or psychological effects of the benzodifurans exist; however, information can be obtained via user reports and published case reports of bromo-dragonFLY and 2C-B-FLY toxicity. Following bromo-dragonFLY exposure, hallucinogenic effects predominate and include altered perception of time and space, increased energy, increased associative thinking, ego softening and various open and closed eye visual changes including 'shimmering lights', 'visual morphing with rainbow trails', 'high resolution colourful visuals', transformation of an object's size and shape and then rapid dissipation into other forms. Some users also report an increased appreciation of music, increased social interaction and conversation with others, and a perception of improved cognition. Some users comment on the slow onset of action of bromo-dragonFLY and residual effects (including hallucinations) lasting for days after use [13,20].

2C-B-FLY users report euphoria, enhanced interpersonal communication, increased mood, closed and open eye visual changes including brightening of colours and visual hallucinations, feelings of insight, mental and physical stimulation, increased tactile stimulation and changes in the perception of time [13,18].

Stimulant effects similar to amphetamines are also reported following bromo-dragonFLY and 2C-B-FLY exposure and include dilated pupils, muscle twitching, restlessness, tachycardia and changes in body temperature [18,21].

PREVALENCE OF USE

Compared to other hallucinogenic substances the benzofurans appear not to be widely used in any part of the world; however, there are few coordinated surveys in Europe, North America or Australasia to determine accurate prevalence of use. The 2012 Mixmag/Guardian Global Drugs Survey documented that 3.2% of United Kingdom (UK) respondents indicated lifetime use of Benzo Fury (2.4% in the previous 12 months) [28]. Less than 0.5% of US respondents reported Benzo Fury use in the previous 12 months [28]. The survey did not report any 5-APB, bromo-dragonFLY or 2C-B-FLY use [28].

Internet user reports of 5-APB and 6-APB date from late 2010 [13]. Identification of bromo-dragonFLY as an emerging drug of abuse came in 2008 with reports of use in Finland, Italy, Norway and Belgium, identified by the Psychonaut Web Mapping Group [29]. Isolated reports of recreational use date from 2001 [13]. Indirect measures of the prevalence of bromo-dragonFLY's use include online user reports, drug seizures by law enforcement agencies and published reports of clinical toxicity or death (see acute toxicity). Compared to MDMA (1294 published experiences 2005–2012), bromo-dragonFLY and 2C-B-FLY had 38 and 16 published user reports on the Erowid website, respectively [13]. Seizures by drug enforcement agencies of bromo-dragonFLY occurred in every year in Sweden 2006–2009, and in sporadic years during the same period in Finland, Poland, Norway, Denmark and the UK [30]. Between 2007 and 2009 there were three seizures of 2C-B-FLY in Finland, and one each in Denmark and the UK [31]. The European Monitoring Centre for Drugs and Drug Addiction (EMCDDA) Early Warning System (EWS) did not detect any benzodifurans amongst 41 newly reported new psychoactive substances available in Europe in 2010 [32].

ACUTE TOXICITY

Animal Data

There are no animal studies investigating the acute toxicity of benzofurans.

Human Data

No formal studies examining human benzofuran toxicity exist. Information is only available via user reports and published case reports of 5-APB, 6-APB, bromo-dragonFLY and 2C-B-FLY toxicity.

User Reports of Benzofuran and Benzodifurans Toxicity

User reports published on the Erowid website and other drug user fora websites provide some information on the unwanted or adverse effects of benzofurans. The accuracy of this information is limited by the possibility of bias in user reports, the difficulty in correctly ascertaining the actual identity and amount of the chemical exposure and the possibility of co-ingestants. An example of the limitations of user reports is illustrated by a study published in 2011 in which analysis of tablets marketed via the Internet as 'Benzo Fury' revealed caffeine in a mix of BZP and 3-TFMPP with no benzofurans or benzodifurans [33].

There are very few user reports of adverse effects of 5-APB. Adverse effects reported in those that are available include nausea, anxiety, increased heart rate, dizziness, sleep disturbance and muscle pain [12,34]. 6-APB users report nausea, vomiting, diarrhoea, headache, bruxism, headache and excessive stimulatory effects including increased heart rate, blood pressure and temperature [10,11,13]. Users also report adverse effects during a relatively long (12–24 hour) 'come-down' period including excess stimulation preventing sleep, muscle aches, muscle twitches, slow cognition and gastrointestinal upset [10,13].

Reported adverse effects of bromo-dragonFLY and 2C-B-FLY are typical of excess $5HT_{2A}$ activity and sympathomimetic stimulant activity. These include paranoia, fear, panic, frightening/troublesome hallucinations, confusion, palpitations, diaphoresis, headache, insomnia, nausea, diarrhoea, temperature disturbance, muscle twitching, dry mouth and difficulty concentrating or communicating [13,20]. The published user experiences do not enable an accurate assessment of the prevalence of these individual adverse effects following bromo-dragonFLY or 2C-B-FLY exposure.

The Swedish Poisons Centre received 22 enquiries regarding self-reported bromo-dragonFLY toxicity in 2006–2007, and then a further 32 in 2008. Common features of toxicity were: agitation, anxiety, tachycardia, visual hallucinations and mydriasis [35,36]. The UK National Poisons Information Service received two enquiries regarding self-reported bromo-dragonFLY exposure in 2008. Features of toxicity reported in these cases included agitation and tachycardia [37].

There are no published cases of toxicity presenting to healthcare services, or of deaths associated with 5-APB or 6-APB.

Published Case Reports of Bromo-dragonFLY Toxicity

An 18-year-old male developed hallucinations and agitation following ingestion of bromo-dragonFLY and nasal insufflation of a 'white powder'. Following two witnessed self-terminating generalised seizures he was transferred to the emergency department (ED). On arrival he was unconscious (GCS 3/15), hypertensive (BP 182/94mmHg) and tachycardic (HR 124b.p.m.). He was sedated, paralyzed and intubated and ventilated for 24 hours. His recovery was uneventful aside from a secondary complication of aspiration pneumonia, which resolved with intravenous antibiotics. Toxicological analysis of serum and urine taken at the time of presentation, using gas chromatography/ mass spectrometry (GC-MS) and liquid chromatography with tandem mass spectrometry (LC-MS/MS), identified bromo-dragonFLY (serum concentration 0.95ng/ml). Additionally ketamine and cannabis were detected, but were judged to be in quantities insufficient to be of clinical significance [38].

Two Swedish males ingested an unknown amount of bromo-dragonFLY reportedly obtained via an Internet supplier. Both men collapsed and were found 17 hours later. One was deceased (see below – Bromo-dragonFLY deaths) and the other (35 years of age) was confused. He was transported to hospital where he had a self-limiting generalised seizure and was noted to have severe peripheral limb vasoconstriction, which worsened during subsequent days. Despite treatment with vasodilators (glyceryl-trinitrate, calcium channel antagonists, ACE inhibitors, nitroprusside and prostacyclin analogues) and a sympathomimetic block he required amputation of the distal phalanges of his left hand. He also developed rhabdomyolysis (creatinine kinase 107000 IU/L) and acute kidney injury (creatinine 400 μmol/L) requiring renal replacement therapy. Analysis of a urine sample obtained at the time of hospital presentation revealed the presence of bromo-dragonFLY, but no analysis was completed to exclude the presence of other recreational drugs or vasoactive compounds [35,36].

Personne et al. reported the case of a 20-year-old Swedish male who ingested five or six blotters each impregnated with 0.5 mg of bromo-dragonFLY. He experienced a period of hallucinations and two days later developed peripheral limb pain. He did not present until six days after use of the bromo-dragonFLY and was found to have cyanotic, pulseless painful distal extremities. He was treated successfully with nifedipine and intravenous glyceryl-trinitrate and required four days of therapy before vasoconstriction resolved [35]. No information is provided in this report regarding analysis of urine or serum for bromo-dragonFLY or other recreational drugs.

An 18-year-old male in Denmark was admitted to hospital with an acute psychotic state following reported ingestion of 2 ml of bromo-dragonFLY. He was reported to be tachycardic, hypertensive, hyperpyrexial, agitated and hallucinating and required treatment with 'large' (unspecified) doses of benzodiazepines. He self-discharged from hospital after four days,

apparently well. No information is provided in this report regarding analysis of urine or serum for bromo-dragonFLY [39].

Published Case Reports of 2C-B-FLY Toxicity

The Erowid website reports a case of toxicity in a 23-year-old male in Spain in 2009, related to 2C-B-FLY exposure. Following licking what he presumed was 2C-B-FLY powder on his finger that he had ordered from an Internet company, this individual developed unpleasant hallucinations, temperature disequilibrium, diaphoresis, palpitations, confusion, weakness, agitation and hypertension requiring treatment with benzodiazepines in the local ED. Erowid reports that subsequent analysis of the powder revealed bromo-dragonFLY with 'synthetic impurities'; no information is available to confirm whether or not analysis of biological samples was undertaken [43].

Published Case Reports of Bromo-dragonFLY Deaths

There have been a number of deaths attributed to bromo-dragonFLY in North America and Europe. In 2008 an 18-year-old female and her boyfriend ingested 1 ml of a liquid they believed to be a hallucinogen at around 2200 hours. They both experienced the onset of drug effects at 2400 hours and shortly afterwards fell asleep. The boyfriend woke at 0500 hours to find his girlfriend dead. Subsequent autopsy revealed lung and brain oedema, splenic enlargement, mucosal irritation of the stomach and renal ischaemic changes. Bromo-dragonFLY was identified in post-mortem femoral blood, urine and vitreous humour using LC-MS/MS analysis. No other drugs or pharmaceuticals were detected in the analysis of these samples. Analysis of liquid from within the bottle containing the ingested 'hallucinogen' confirmed the sole presence of bromo-dragonFLY [40].

There is limited information available regarding a Swedish male who died after ingesting bromo-dragonFLY with a friend (the cases described in the acute toxicity section). There is no post-mortem information available or information on the analysis of biological samples [35,36]. The Erowid website also lists the death of an 18-year-old male in Norway following use of gamma-hydroxybutyrate and bromo-dragonFLY [41].

In May of 2011, a group of young adults in Oklahoma ingested what they thought to be 2C-B-FLY. A 22-year-old female died shortly after taking the drug, eight others developed toxicity requiring hospital treatment. One required critical care treatment and another (a male) subsequently died. Local newspapers reported that analysis of the ingested substance revealed bromo-dragonFLY rather than 2C-B-FLY. Because of the likely differences in potency and the differences in dose reported by users, ingestion of a recreational dose of misidentified 2C-B-FLY that was actually bromo-dragonFLY would potentially result in a ×100 bromo-dragonFLY overdose [42].

Published Case Reports of 2C-B-FLY Deaths

There are no published reports in the medical literature of deaths caused by 2C-B-FLY exposure. There are two reports of unconfirmed 2C-B-FLY-related deaths on Erowid, both possibly related to mistaken exposure to bromo-dragonFLY, rather than 2C-B-FLY. One report is of two males and a female in California who ingested what was thought to be 2C-B-FLY in 2009. The 18-year-old male became unwell two hours post ingestion, and at 3.5 hours post ingestion was said to have had a change in conscious state, possible seizure and cardiac arrest. The other two individuals required supportive care in hospital prior to complete recovery [43]. Erowid reports that the same company supplied the substance as that which resulted in toxicity in the 23-year-old male in Spain (see 2C-B-FLY toxicity section). It is reported that this was manufactured in China, labelled as 'b1' and subsequently analyzed and found to be bromo-dragonFLY [43].

CHRONIC TOXICITY

Animal Data

There are no published animal reports examining the chronic toxicological effects of benzofuran exposure.

Human Data

One individual required amputation of the digits of one hand for 'acute limb ischaemia' following Bromo-dragonFLY use (see acute toxicity section) [35,36]. There are no other reports of chronic long-term physical health effects relating to 5-APB, 6-APB, bromo-dragonFLY or 2C-B-FLY exposure.

DEPENDENCE AND ABUSE POTENTIAL

Animal Data

There are no published animal studies investigating the dependence/abuse potential of benzofurans or benzodifurans.

Human Data

There are no published reports of dependence to benzofurans or benzodifurans. User reports appear to suggest single rather than recurrent use of bromo-dragonFLY possibly due to prolonged come down/after effects following exposure [13].

REFERENCES

[1] Mustafa A. The chemistry of heterocyclic compounds, benzofurans Chemistry of heterocyclic compounds: a series of monographs, Vol 29. New York: Wiley-Interscience; 1974. p 1–2.
[2] Legal Highs Forum. Available: <http://www.legal-highsforum.com/showthread.php?12738-5apb> [accessed 11.09.12].

[3] Erowid Experience Vaults. Available: <http://www.erowid.org/experiences/subs/exp_6APB.shtml> [accessed 11.09.12].

[4] Acuna-Castillo C, Villalobos C, Moya PR, Saez P, Cassels BK, Huidobro-Toro JP. Differences in potency and efficacy of a series of phenylisopropylamine/phenethylamine pairs at 5-HT2a and 5-HT2 receptors. Brit J Pharmacol 2002;136:510–9.

[5] Hill SL, Thomas SHL. Clinical toxicology of newer recreational drugs. Clin Toxicol 2011;49:705–19.

[6] Monte AP, Marona-Lewicka D, Cozzi NV, Nichols DE. Synthesis and pharmacological examination of benzofuran, indan, and tetralin analogues of 3,4-methylenedioxyamphetamine. J Med Chem 1993;36:3700–6.

[7] Monte AP, Marona-Lewicka D, Parker MA, Wainscott DB, Nelson DL, Nichols DE. Dihydrobenzofuran analogues of hallucinogens. 3. Models of 4-substituted (2,5-dimethoxyphenyl) alkylamine derivatives with rigidified methoxy groups. J Med Chem 1996;39:2953–61.

[8] Parker MA, Marona-Lewicka D, Lucaites VL, Nelson DL, Nichols DE. A novel (benzodifuranyl)aminoalkane with extremely potent activity at the 5-HT2A receptor. J Med Chem 1998;41:5148–9.

[9] Erowid. Available: <http://www.erowid.org/chemicals/> [accessed 29.01.12].

[10] Blue Light Forum. Available: <http://www.bluelight.ru/vb/threads/525471-6-APB-report-from-an-icklenoob> [accessed 11.09.12].

[11] Drugs Forum. Available at <http://www.drugs-forum.com/forum/showwiki.php?title=6-APB> [accessed 11.09.12].

[12] Drugs Forum. Available: <http://www.drugs-forum.com/forum/showthread.php?t=154505> [accessed 11.09.12].

[13] Erowid. Erowid Experience Vaults. Available: <http://www.erowid.org/experiences/exp_front.shtml> [accessed 29.01.12].

[14] Wikipedia APB. Available: <http://en.wikipedia.org/wiki/6-APB> [accessed 11.09.12].

[15] Briner K, Burkhart JP, Burkholder TM, et al. Aminoalkylbenzofurans as serotonin (5-HT(2c)) agonists. 7045545 (US patent). Published 2000-01-19, issued 2006-16-03.

[16] Parker MA. Studies of the perceptiotropic phenethylamines: determinants of affinity for the 5-HT2 receptor. A Thesis submitted to the faculty of Prude University 1998. Available: <http://bitnest.ca/external.php?id=%251C%2B95%2522%250D%2519%2518%2505%250C%250Dtz%257D%2500%2501> [accessed 29.01.12].

[17] Chambers JJ, Kurrasch-Orbaugh DM, Parker MA, Nichols DE. Enantiospecific synthesis and pharmacological evaluation of a series of super-potent, conformationally restricted 5-HT(2A/2C) receptor agonists. J Med Chem 2001;44:1003–10.

[18] Erowid. Available: <http://www.erowid.org/chemicals/2cb_fly/2cb_fly.shtml> [accessed 29.01.12].

[19] Nichols DE. Medicinal chemistry and structure activity relationships. In: Cho AK, Segal DS, editors. Amphetamine and its analogs. Psychopharmacology, toxicology and abuse. San Diego, CA: Academic Press; 1994. p. 3–33.

[20] Drugs Forum Bromo-dragonfly. Available: <http://www.drugs-forum.com/forum/showwiki.php?title=Bromo-dragonfly> [accessed 29.01.12].

[21] Erowid. The Vaults of Erowid. Available at <http://www.erowid.org/chemicals/bromo_dragonfly/bromo_dragonfly.shtml> [accessed 29.01.12].

[22] May JA, McLaughlin MA, Sharif NA, Hellberg MR, Dean TR. Evaluation of the ocular hypotensive response to serotonin 5-HT1A and 5-HT2A receptor ligands in conscious ocular hypertensive cynomolgus monkey. J Pharmacol Exp Ther 2003;306:301–9. [accessed 29.01.12].

[23] Herbal Highs. Available at <http://forum.herbal-highs.com/showthread.php?tid=6708&page=6> [accessed 11.09.12].

[24] Blue Light Forum. Available: <http://www.bluelight.ru/vb/threads/603896-5-APB-a-functional-and-fun-alternative-to-MDMA> [accessed 11.09.12].

[25] Drugs Forum. Available; <http://www.drugs-forum.com/forum/showthread.php?t=168765> [accessed 11.09.12].

[26] Blue Light Forum. Available: <http://www.bluelight.ru/vb/threads/622605-My-5-6-APB-experiences-after-months-and-5-5g-s> [accessed 11.09.12].

[27] Drugs Forum. Available: <http://www.drugs-forum.com/forum/showthread.php?t=178526> [accessed 11.09.12].

[28] 2012 Global Drug Survey. Available: <http://globaldrugsurvey.com/run-my-survey/2012-global-drug-survey>.

[29] Psychonaut Web Mapping Project. Alert on new recreational drugs on the web; building up a European-wide Web scanning system. Final Report 2010 Available: <www.psychonautproject.eu/reports/Psychonaut_Project_Executive_Summary.pdf> [accessed 29.09.12].

[30] European information system and database on new drugs. Available: <http://ednd.emcdda.europa.eu/html.cfm/index7246EN.html?SUB_ID=36> [accessed 29.01.12].

[31] European information system and database on new drugs. Available: <http://ednd.emcdda.europa.eu/html.cfm/index7246EN.html?SUB_ID=41> [accessed 29.01.12].

[32] Europol 2010 Annual Report on the implementation of Council Decision 2005/387/JHA. Available: <http://www.emcdda.europa.eu/publications/implementation-reports/2010> [accessed 29.01.12].

[33] Baron M, Elie M, Elie L. An analysis of legal highs: do they contain what it says on the tin?. Drug Test Anal 2011;3:576–81.

[34] Blue Light Forum. Available: <http://www.bluelight.ru/vb/threads/635761-Adverse-Reaction-to-5-APB-Please-Read> [accessed 11.09.12].

[35] Personne M, Hulten P. Bromo-dragonfly: a life threatening designer drug. Clin Tox 2008;46:379–80.

[36] Thorlacius K, Borna C, Personne M. Bromo-dragon fly-life-threatening drug. Can cause tissue necrosis as demonstrated by the first described case. Lakartidningen 2008;105:199–200.

[37] Dargan P, Wood D. Technical profile of Bromo-DragonFLY. EMCDDA Publication. Available: <http://ednd.emcdda.europa.eu/assets/upload/showfile?filename=BDF_Tech_Prof_EMCDDA_Mar_2010.pdf>; 2010 [accessed 29.01.12].

[38] Wood DM, Looker JJ, Shaikh L, et al. Delayed onset of seizures and toxicity associated with recreational use of bromo-dragonfly. J Med Toxicol 2009;5:226–9.

[39] Nielsen VT, Høgberg LC, Behrens JK. Bromo-dragonfly poisoning of 18-year-old male. Ugeskr Laeger 2010;172:1461–2.

[40] Andreasen MF, Telving R, Birkler RI, Schumacher B, Johannsen M. A fatal poisoning involving Bromo-Dragonfly. Forensic Sci Int 2009;183:91–6.

[41] Erowid. Reported GHB-related death may not have been GHB: Bromo-Dragonfly is confirmed by toxicology. Available: <http://www.erowid.org/chemicals/bromo_dragonfly/bromo_dragonfly_death2.shtml> [accessed 29.01.12].

[42] News on 6. Available: <http://www.newson6.com/story/14641463/second-victim-dies-after-taking-designer-drug-in-konawa> [accessed 29.01.12].

[43] Erowid. Information on reported deaths related to 2C-B-Fly: misidentified substance is most likely bromo-dragonfly. Available: <http://www.erowid.org/chemicals/2cb_fly/2cb_fly_death1.shtml> [accessed 29.01.12].

17

Miscellaneous Compounds

Shaun L. Greene

Victorian Poisons Information Centre, Melbourne, Australia

METHOXETAMINE

Methoxetamine is an arylcyclohexylamine analogue of ketamine and has been available recently as a novel psychoactive substance (NPS); it is often marketed as 'bladder friendly' and devoid of the urological side-effects associated with ketamine [1,2]. Limited published user reports and case reports of human toxicity demonstrate that in addition to the dissociative and hallucinogenic effects of ketamine, acute methoxetamine toxicity can be associated with additional effects including stimulant features and cerebellar toxicity.

Pharmacology

Physical and Chemical Description

Methoxetamine (2-(3-methoxyphenyl)-2-(ethylamino)cyclohexanone) was synthesised by a United Kingdom (UK) chemist as an alternative to ketamine [3]. Methoxetamine is a member of the arylcyclohexylamine class and has a similar structure to ketamine (2-(2-chlorophenyl)-2-(methylamino)cyclohexanone), phencyclidine (1-(1-phenylcyclohexyl) piperidine or PCP) and 3-MeO-PCP (Fig. 17.1). The 2-chloro group on the phenyl ring of the ketamine molecule is replaced with a 3-methoxy group and the N-methyl group on the amine ring is replaced with an N-ethyl group [4].

Although not substantiated or published in the mainstream scientific literature, chemists who manufacture methoxetamine suggest substitution of the N-methyl group using an N-ethyl group provides greater potency, and substitution of the 2-chloro group with a 3-methoxy group reduces the analgesic and anaesthetic properties associated with ketamine [3].

A number of other closely related analogues have been found in the UK and elsewhere in Europe during 2012 – including 3-methoxyPCP, 4-methoxyPCP, 2-methoxy-ketamine and N-ethyl-nor-ketamine (NEK) [5].

Synthesis of methoxetamine can be achieved through straightforward chemical processes which also allow synthesis of other analogues of methoxetamine and ketamine [5].

Pharmacokinetics

There are no published formal animal or human studies examining the pharmacokinetic properties of methoxetamine. Users report rapid onset of action (10–20 minutes) following nasal insufflation, rectal or intravenous/intramuscular administration. The sublingual route

Novel Psychoactive Substances.
DOI: http://dx.doi.org/10.1016/B978-0-12-415816-0.00017-1

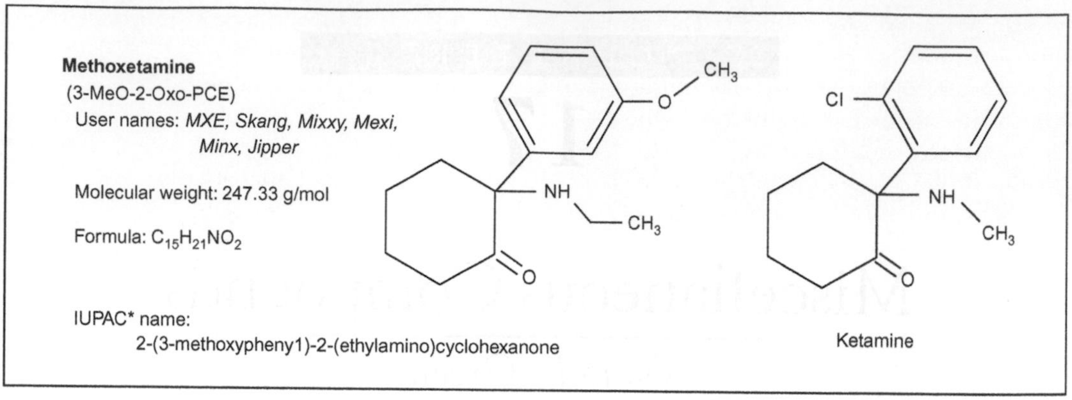

Methoxetamine

(3-MeO-2-Oxo-PCE)

  User names: *MXE, Skang, Mixxy, Mexi,*
                   Minx, Jipper

Molecular weight: 247.33 g/mol

Formula: $C_{15}H_{21}NO_2$

IUPAC* name:
    2-(3-methoxypheny1)-2-(ethylamino)cyclohexanone

Ketamine

FIGURE 17.1   Chemical structure, physical properties and user names for Methoxetamine. *(IUPAC – International Union of Pure and Applied Chemistry.)

is reported as less effective, with oral least effective. Peak effects are reported to occur between 1–3 hours after exposure with a further duration of between three and six hours during which after-effects are experienced. There are a few reports of effects lasting 24 hours; the half-life for clinical effects is three hours [5,6].

The metabolism of methoxetamine has not been formally studied, but its structural similarity to ketamine may mean it is similarly metabolised: extensive hepatic first pass metabolism by CYP2B6, CYP34A and CYP2C9 isoenzymes. The major pathway in ketamine metabolism is N-demethylation to the active metabolite norketamine [7]. Extensive first pass metabolism of methoxetamine may account for its limited effect when administered orally, although currently there is no data which can substantiate this. Preliminary data from analysis of three urine samples from patients with analytically confirmed acute methoxetamine toxicity has suggested that in addition to metabolites expected based on the metabolism of ketamine, additional Phase I metabolites of methoxetamine are produced [8]. The expected metabolites based on the metabolism of ketamine are the N-desethyl (nor), dehydro, dehydro-nor, hydroxy, hydroxy-nor,

hydroxy-dehydro and hydroxy-nor-dehydro metabolites. Greatest responses were seen for the N-desethyl (nor) and hydroxy-nor metablites (response relative to methoxetamine of 38.3 and 13.3%). The additional metabolites seen included O-desmethyl, dihydro-nor, O-desmethyl-hydroxy-nor and O-desmethyl-dehydro metabolites.

Pharmacodynamics

There have been no formal published animal or human studies examining the mechanism of action of methoxetamine. However it is postulated to be a N-methyl D-aspartate (NMDA) receptor antagonist and a dopamine reuptake inhibitor [9,10]. Initial *in vitro* binding studies carried out by the US National Institutes of Mental Health Psychoactive Drug Screening Program (PDSP) by Brian Roth at the University of North Carolina have been published within a UK Advisory Council on the Misuse of Drugs (ACMD) report on methoxetamine [5]. These studies show that methoxetamine is, like ketamine, an NMDA (*N*-methyl-D-aspartate) receptor antagonist but that it has a higher affinity for the NMDA receptor [5]. These studies also show that methoxetamine (unlike ketamine) has affinity for the serotonin transporter (SERT). The affinity of methoxetamine for SERT was similar to its affinity for the NMDA receptor. The differences in the pharmacological activity of methoxetamine compared to ketamine may at least in part explain the differences seen in the patterns of acute toxicity of methoxetamine as discussed in the later sections of this chapter.

POSITIVE OR DESIRED CLINICAL EFFECTS

No formal studies examining the behavioural or psychological effects of methoxetamine exist; however information can be obtained from user reports and published case reports of methoxetamine toxicity. Some users report dissociative effects similar to ketamine, but often lasting for a longer period (up to 24 hours). Others report a dissociative experience, but with a significantly greater loss of cognition and a degree of disorientation compared to ketamine. Other positive reported effects include euphoria, a sense of calm and serenity, increased mood and vivid recall of past memories and dreams. More neutral effects include a loss of space and time perception, open and closed eye visual hallucinations, distortion or loss of sensory perception, feeling of analgesia or numbness and a perception of being 'out of body' [5,11].

Prevalence of Use

Recreational methoxetamine use is a recent and relatively rare occurrence. Published Internet reports of methoxetamine use first appeared in 2011 [6,11]. The Erowid website documents 41 user reports over an 11-month period [6]. A 2011 UK-wide drug trends survey of drug users, enforcement agencies and drug action teams published in *Druglink* (Nov/Dec issue) described the emerging use of methoxetamine as an alternative to ketamine in the UK [12]. The 2012 Mixmag/Guardian Global Drugs Survey found that 4.9% of UK respondents had reported previous lifetime use of methoxetamine (4.2% with the previous 12 months) [13].

Only 1.5% of respondents in the USA reported previous methoxetamine use [13]. The reasons given by users for taking methoxetamine in this survey were: 73% felt it was easier to get hold of than ketamine, 20% felt it was better value for money than ketamine, 18% believed it caused less damage to the liver or kidneys than ketamine and 20% were either curious to try methoxetamine or were mis-sold what they thought was ketamine [13].

A survey of individuals attending gay-friendly night clubs in South East London in 2011 found that 65.8% of 313 survey respondents reported life-time use of a 'legal high'. Life-time use of methoxetamine was reported by 6.4% of respondents, while 1.9% had used methoxetamine in the previous month and 1.6% on the night of the survey [14].

The European Monitoring Centre for Drugs and Drug Addiction (EMCDDA) 2010 report lists methoxetamine as a new psychoactive substance reported for the first time in Europe (UK, November 2010) [1]. The EDND database reports six seizures of methoxetamine (five in 2011, one in 2010), all in in different European countries [15].

There are published reports of methoxetamine use and toxicity in the USA and UK in 2011 and 2012; these are described in more detail below [10,16–18].

Acute Toxicity

Animal Data

There are no animal studies investigating the acute toxicity of methoxetamine.

Human Data

No formal studies examining human methoxetamine toxicity exist. Information is only available through user reports and published case reports of methoxetamine toxicity. User reports of adverse or wanted effects of methoxetamine are similar to those reported with ketamine and include diaphoresis, confusion, disorientation, nasal discomfort on insufflation, loss of consciousness and associated amnesia, depersonalisation, pain at intramuscular injection site, vertigo, nausea/vomiting, frightening hallucinations or sensory perceptions and severe dissociation [5,11].

PUBLISHED CASE REPORTS OF METHOXETAMINE TOXICITY

In August 2011 the first case report of possible methoxetamine toxicity was published. A 32-year-old male hallucinogenic drug user presented to an emergency department (ED) following intramuscular injection of methoxetamine. He was agitated with a heart rate of 105 b.p.m. and a blood pressure (BP) of 140/95 mmHg. He was afebrile and had no focal neurological findings; however, he was disorientated to time and minimally responsive, appearing to be in a dissociative state. Rotatory nystagmus was present. Electrolytes and acid base status were normal. Ethanol concentration was less than 10 mg/dL. Analysis of serum or urine for methoxetamine was not undertaken and so it is not possible to be certain that the toxicity seen was due to methoxetamine. The patient made a full recovery after an eight-hour period of observation [16].

Wood et al. published a case series of three male patients aged between 28 and 42 years who presented to an ED in London, UK with a clinical state consistent with sympathomimetic toxicity following reported use of methoxetamine [10]; further clinical detail on these cases is provided below. Serum and urine toxicology screening confirmed the presence of methoxetamine in all three patients. All three were hypertensive and tachycardic, two had dilated pupils and one was hyperpyrexial.

A 29-year-old was found 'catatonic' by his mother and presented with mild confusion, tremor, visual hallucinations, heart rate 121 b.p.m., BP 201/104 mmHg and dilated pupils. He was treated with 5 mg oral diazepam

and recovered after overnight observation. He admitted to dissolving 200 mg of 'methoxetamine powder' he had purchased from an Internet supplier in water and ingesting it. Methoxetamine (concentration 0.09 mg/L), diphenhydramine and venlafaxine were detected in his serum, no sympathomimetic drugs were detected on a comprehensive toxicology screen [10]. The second patient was a 28-year-old who presented after being found collapsed in a nightclub bathroom. He was drowsy (GCS 10/15), confused, intermittently agitated and had a heart rate of 113 b.p.m., BP 198/78 mmHg, temperature of 36.9°C and dilated pupils. A bag of white powder labelled 'methoxetamine' was found with him and there was white powder around his nostrils. He was treated with 5 mg of intramuscular midazolam and recovered within three hours. He reported purchasing methoxetamine from a high street 'head shop'. Methoxetamine was detected in serum at a concentration of 0.2 mg/L, no other drugs were present [10]. A 42-year-old presented after being found collapsed in the street. He had a heart rate of 135 b.p.m., BP of 187/83 mmHg, GCS of 6/15 and temperature of 38.2°C. He required a nasopharyngeal airway and recovered over the following two hours, although he required diazepam 5 mg for agitation during this period. He admitted snorting 0.75 g of 'benzofury' and 0.5 g of 'methoxetamine' that he had purchased from an Internet supplier. Methoxetamine was detected in serum at a concentration of 0.12 mg/L and in addition 5-APB and 6-APB were also detected [10].

In a further report, a 19-year-old male presented to an ED 30 minutes after injecting an unknown quantity of methoxetamine. He was extremely agitated, confused and ataxic. Examination revealed a BP of 168/77 mmHg, heart rate 134 b.p.m., temperature 37.6°C, dilated pupils and the presence of nystagmus. Midazolam (10 mg IV) and diazepam (10 mg IV) were required to control agitation. Clinical toxicity resolved after a period of supportive care. He had been admitted to the ICU of the same hospital two days previously after self-administering 3,4-methylenedioxymethemphetamine (MDMA) intravenously and developing repeated vomiting, diarrhoea, tachycardia and diaphoresis. Toxicity had resolved following treatment with midazolam and a period of supportive care. Methoxetamine was detected in the patient's serum in a sample taken five hours after the time of reported administration. MDMA was detected in three serum samples taken between 24 hours and 3.5 days post MDMA exposure [17]. It is therefore not possible to determine how whether all of the features seen in this case were due to methoxetamine or whether MDMA could have played a role.

Three further cases of analytically confirmed methoxetamine exposure associated with reversible cerebellar toxicity were reported in the UK in 2012 [18]. A 19-year-old male insufflated a white powder he believed to be ketamine. He subsequently felt drowsy and uncoordinated and fell. On presentation he was disorientated, drowsy and had evidence of cerebellar toxicity (nystagmus, slurred speech and coarse dysdiadochokinesis). Heart rate was 107 b.p.m., temperature 36.7°C and blood pressure 194/110 mmHg. ECG and routine biochemistry including, creatinine kinase, were normal at the time of presentation. He received no specific care and cerebellar toxicity resolved over 48 hours. Serum creatinine kinase peaked at 2271 IU/L 13 hours after exposure [18]. A 17-year-old male felt drunk and uncoordinated before falling over and losing consciousness 20 minutes after nasal insufflation of 'MXE'. He was found collapsed and transported to the ED. At the time of presentation there was minimal response to painful stimuli, heart rate was 72 b.p.m., BP 148/104 mmHg and temperature of 34.5°C. He was passively warmed and administered intravenous fluid. His condition improved and BP and temperature had normalised. Four hours after treatment had started he had severe truncal ataxia, dysarthria, horizontal nystagmus and dysdiadochokinesis. There was no myoclonus and reflexes were normal. He had recovered completely 16 hours after hospital presentation. Renal function and electrolytes were normal. Creatinine kinase was 255 IU/L 16 hours post exposure [18]. An 18-year-old male nasally insufflated 'MXE'. Within 30 minutes he developed severe limb incoordination and imbalance. He reported feeling drunk and thought he would lose consciousness. On ED presentation he was drowsy, responsive to voice with slurred speech, heart rate 67 b.p.m., BP of 151/112 mmHg, horizontal nystagmus, impaired limb coordination and truncal ataxia. His blood pressure returned to normal three hours later and cerebellar toxicity resolved during a 16-hour period of observation. He received no specific intervention [18].

PUBLISHED CASE REPORTS OF METHOXETAMINE DEATHS

The website Drugs-Forum.com reports the death of an individual in Sweden following injection of methoxetamine 100 mg and MDAI (5,6-Methylenedioxy-2-aminoindane) 400 mg [19]. No further details regarding this event have been published. There are no other reported deaths associated with methoxetamine exposure.

Chronic Toxicity

Animal Data

There has only been one animal study investigating chronic methoxetamine toxicity, limited to an investigation of renal and bladder toxicity [20]. This was a study in mice using a model validated to investigate the chronic lower urinary tract pathology associated with chronic ketamine exposure. Two-month-old Institute of Cancer Research (ICR) mice were administered either 30 mg/kg of methoxetamine per day (five mice) or saline control (seven mice) by intraperitoneal injection for three months. There was hydropic degeneration in both the proximal and distal convoluted tubules of the kidney and inflammatory cell infiltration of the kidneys in all of the mice given methoxetamine. In addition, there was glomerular atrophy in three of the methoxetamine-treated mice. All of the methoxetamine-treated mice had mononuclear cell infiltration in the submucosal layer and in the muscle layer of bladder.

Human Data

There are no published human reports examining the chronic toxicological effects of methoxetamine exposure. The UK chemist who first manufactured methoxetamine hypothesised that inclusion of the N-ethyl group on the amine ring would prevent the urinary-tract problems associated with chronic ketamine use [3] but this hypothesis has not been formally studied and there is currently no published human evidence to support it; furthermore the animal studies discussed above suggest that this is not likely.

Dependence and Abuse Potential

Animal Data

There are no published animal or *in vitro* studies investigating the dependence/abuse potential of methoxetamine.

Human Data

There are no published reports of dependence to methoxetamine. User reports appear to suggest that many feel a desire to administer more methoxetamine after initial exposure. It is possible this may lead to repeated use and dependence [5].

DIMETHYLAMYLAMINE (DMAA)

Dimethylamylamine (DMAA), also known as methylhexanamine, was patented in 1944 by the pharmaceutical company Eli Lilly under the trade name Forthane. It was intended for use as a nasal decongestant and treatment for hypertrophied or hyperplastic oral tissues

[21]. DMAA was introduced and advertised as a geranium extract dietary supplement called Geranamine® in the USA in 2005 [22]. Geranium oil was reported to contain less than 0.7% DMAA in a study conducted in 1996; however, this finding has not been able to be replicated in subsequent studies [23]. The first report of DMAA on an Internet drug-user forum appeared in 2006 [19]. Following legislation restricting the use of piperazines including 1-benzylpiperazine (BZP) in New Zealand in March 2008, DMAA emerged as a new 'legal-high' and more recently has been identified in head shops in a number of European countries [24,25]. In 2009 the World Anti-Doping Agency added DMAA to its list of prohibited substances following documented use in athletes from numerous sporting codes [26].

Pharmacology

Physical and Chemical Description

DMAA is an aliphatic amine with structural similarity to monoamines including phenethylamine derivatives such as amphetamine (Fig. 17.2) [27,28]. DMAA is soluble in water [28].

Pharmacokinetics

There are no formal animal or human studies examining the pharmacokinetic properties of DMAA. User reports provide some information. The preferred route of administration of DMAA appears to be ingestion, although there are reports of intravenous and inhalational administration [29]. DMAA can be insufflated, but is reported to be painful unless dissolved in water and administered using an atomiser [19]. Oral doses are in the range of 10–100 mg. Effects are reported to last between three and four hours [19,29].

Pharmacodynamics

Animal studies in the 1950s characterised DMAA as a sympathetic system agonist and

Dimethylamylamine (DMAA) or Methylhexanamine

User names: *Embrace, Go-E, Energy, Entropy, Dr Feelgood, BluE Hummer, Fast Layn, Nemesis, Pinkys, Crankd, Diablo, Molotov*

Molecular weight: 115.22 g/mol

Formula: $C_7H_{17}N$

Melting point: 120-130 °C

IUPAC* name:
4-methylhexan-2-amine

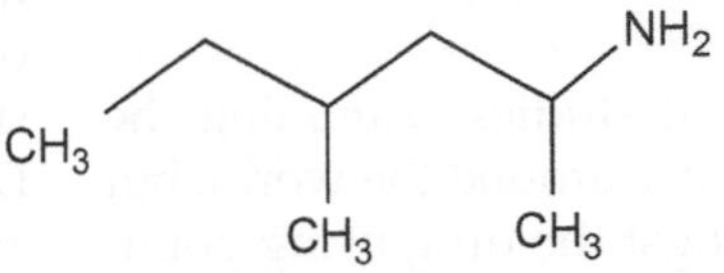

FIGURE 17.2 Chemical structure, physical properties and user names for DMAA. *(IUPAC – International Union of Pure and Applied Chemistry.)

vasoconstrictor. It has been postulated that DMAA acts either as a noradrenaline reuptake inhibitor or noradrenaline-releasing agent; similar to the action of monoamine compounds of which DMAA shares structural similarity [27].

A recent study examined the effects of DMAA in 10 healthy volunteers administered DMAA alone, caffeine alone or DMAA and caffeine. The study found DMAA acutely (maximum effect between 60–90 minutes) increased systolic BP (mean maximal increase 16 mmHg or 14% of baseline) and diastolic BP (mean maximal increase 8 mmHg or 8% of baseline) in a dose dependent manner, without any effect on heart rate. Circulating adrenaline and noradrenaline concentrations were measured and did not increase significantly, prompting the authors to suggest DMAA may exert effects via a different mechanism. However, catecholamine concentrations were only sampled twice during the study protocol [30].

User reports of DMAA's clinical effects suggest predominant sympathetic stimulant activity, as would be expected from its pharmacological activity [19,29].

POSITIVE OR DESIRED CLINICAL EFFECTS

No formal studies examining the behavioural effects of DMAA exist; however, some information can be obtained from published user reports. Users report variable effects; sympathomimetic effects appear to predominate over psychoactive effects. The majority of, but not all, users report euphoria and some report increased focus and productivity. Hallucinations are not reported. Variable effects on appetite are reported [19,29].

The potential benefits of DMAA contained with geranium extract obtained from the *Pelargonium graveolens* plant have attracted the attention of nutritionists and complementary health practitioners. Benefits are said to include weight loss, enhanced mood, increased exercise performance and benefits from the action of geranium extract as an antioxidant. Recently three studies were published in an open access journal, which all concluded DMAA offered potential benefits to humans as a dietary supplement and was safe to use [31–33]. It is difficult to draw conclusions from this series published by the same group, as there is no documentation of DMAA dose administered. There was a non-significant rise in mean systolic BP (13 mmHg) and mean diastolic BP (13 mmHg) in seven men receiving a dietary supplement allegedly containing DMAA [32].

Prevalence of Use

There are no formal studies examining the prevalence of DMAA use around the world, but currently its use amongst the drug using population appears relatively low. Use of DMAA as a psychoactive drug was first reported on an Internet user forum in 2006 [19]. DMAA use as a dietary supplement in the USA appears to becoming more prevalent [34,35]. In 2011 Canada removed DMAA from its list of permitted dietary supplements [36].

DMAA use became popular in New Zealand in 2008 following introduction of legislation restricting the availability of BZP [25]. In 2009 the New Zealand government banned the powdered form of DMAA and restricted the availability of other forms, but did not ban its use. Although there was no formal study of the prevalence of use within the New Zealand population, in 2009 an expert committee on drug use considered it was likely to be low [37]. However, a survey of BZP users in Auckland found 11.9% of respondents had used DMAA following the classification of BZP [29]. In addition legal-high distributors claimed to have sold 100 000 DMAA tablets in New Zealand in 2008–2009, and there were a number of subsequent reports of significant clinical toxicity [25,29,38].

Between 2009 and 2011 a number of high-level sportspersons from a number of different sports have allegedly tested positive for DMAA, suggesting a belief amongst the sporting community that DMAA enhances athletic performance [19].

In 2011 the National Advisory Committee on Drugs (NACD) in Ireland reported DMAA as an 'emerging psychoactive substance' (not identified prior to May 2010) and identified DMAA

within capsules purchased in Irish head shops (DMAA was contained in 13 of 38 products purchased and analyzed) [24]. DMAA was often combined with caffeine in these tablets. An anonymous on-line survey of legal high users in Ireland in 2011 found 1.9% of respondents had tried DMAA [24]. There are currently no reports of DMAA use on the Erowid website. The website Drugs-Forum reports products containing DMAA, geranium extract or *P. graveolens* as being available in the UK, Netherlands, USA, Australia, New Zealand and Europe. Drugs-Forum also contains a number (n = 84) of user reports dating from April 2007 [19]. Prevalence of DMAA use in the UK has not been reported using popular surveys such as the Mixmag/Guardian Global Drug Survey [13]. The EMCDDA EDND database reports seven seizures of DMAA within Europe during 2010–2011 [39].

Acute Toxicity

Animal Data

The LD_{50} for DMAA in mice is 39 mg/kg when injected intravenously and 185 mg/kg when administered into the intraperitoneal cavity [40].

Human Data

A number of case reports demonstrate the occurrence of haemorrhagic stroke in association with DMAA exposure. Case reports highlighting this adverse effect are consistent with limited human studies suggesting hypertension as a significant physiological response to DMAA.

In 2008 three patients aged between 17 and 30 years of age were reported to have presented to emergency departments in New Zealand with severe headache, nausea and vomiting following ingestion of 'party pills' containing DMAA. All three patients were hypertensive. Two required treatment with intravenous glyceryl trinitrate to control hypertension, while in the other case hypertension was controlled using benzodiazepines [29]. A 45-year-old male was reported to suffer a haemorrhagic stroke 18 hours after using DMAA in New Zealand [29]. A 25-year-old female was reported to have suffered a thrombotic stroke after using DMAA in France, although there are few published details regarding these cases [41].

A 21-year-old male presented to a New Zealand hospital after reportedly ingesting 150 mg of caffeine, ethanol and two tablets said to contain 99.9% pure DMAA. Within 30 minutes he developed severe headache and became confused. A friend took him home and he went to sleep. He was drowsy and had slurred speech the next day and, after not improving, presented to the local emergency department that evening. Examination revealed confusion and right-sided facial and limb weakness. He was normotensive and had a normal heart rate. Urgent CT scan of the brain demonstrated a large haemorrhage with mass effect adjacent to the left basal ganglia. No evidence of aneurysm was found on cerebral angiography. The patient was discharged on day 20 with residual impairment to memory, abstract reasoning and right hand function. Analysis of an identical pill obtained from the supplier found 278 mg of DMAA per capsule; however, there was no biological analytical confirmation of exposure [25].

A further three cases of DMAA-associated cerebral haemorrhage have been reported from the Canterbury region in New Zealand. A 36-year-old male purchased a legal party drug called Cocaine Party Powder and ingested one-quarter of the 'recommended' dose. The powder was documented to contain 50 mg of 1,3-dimethylamine HCl. Within 60 minutes he developed severe headache, slurred speech, ataxia and weakness of his right hand. He waited overnight to see whether his symptoms would resolve, but woke with a left facial droop and left-sided weakness. Assessment in the emergency department found normal values for heart rate, blood pressure and temperature.

Computerised tomography of the brain showed a large right external capsule cerebral haematoma ($64 \times 27 \times 23$ mm), but no evidence of an aneurysm. Angiography showed an arterial pattern consistent with a vasculitis, but immune and infectious markers for vasculitis were negative. DMAA was found in plasma at a concentration of 0.76 mg/L. DMAA, ethanol, nicotine and paracetamol were detected in the patient's urine. His symptoms improved over two days and he was discharged without any residual functional neurological deficit [38].

Thirty minutes following ingestion of two tablets labelled as 'Pure X-S' purchased as party pills, a 21-year-old female developed sudden severe frontal headache and associated dizziness. She vomited and reported that her limbs twitched involuntarily. The pills listed the main ingredient as *Pelargonium* (common name geranium) extract at a dose of 75 mg. At the time of ED presentation the patient was agitated, hypertensive (185/100 mmHg), and had heart rate 85 b.p.m., temperature 35.9°C, normal blood glucose concentration, GCS of 15/15, no focal neurological findings, dilated pupils (6 mm diameter) and normal oxygen saturation of 100%. Computerised tomography of the brain demonstrated a subarachnoid haemorrhage overlying the right frontal lobe and extending into the underlying sulcus, but with no mass effect. Subsequent cerebral angiography was unremarkable. She was discharged well three days post presentation. A plasma concentration in a sample obtained 100 minutes post exposure revealed a DMAA concentration of 1.09 mg/L. No other drugs were detected. Analysis of the tablet found: DMAA 66 mg, caffeine 84 mg, phenethylamine, and palmitic and steric acid [38].

The third case involved a 41-year-old male who collapsed 30 minutes after ingesting a white powder dissolved in water, which he had been offered as a stimulant. He complained of headache and vomited. In the emergency department he was agitated, confused, hypertensive (blood pressure 240/120mmHg), and had a heart rate of 54b.p.m.; he had a normal temperature and blood glucose concentration. He had no focal neurological signs, but his pupils were dilated (6mm) and he was diaphoretic. Computerised tomography of the brain revealed an $11 \times 6 \times 3$mm left basal ganglia haemorrhage. Blood pressure was controlled with labetalol. He made a full recovery over two days with no residual focal neurological deficit. DMAA at a concentration of 2.31mg/L was found in a blood sample taken two hours post ingestion [38].

The deaths of two US army servicemen who suffered 'heart attacks' while undertaking fitness exercises following DMAA exposure have been reported in the mainstream media. DMAA was reportedly detected in biological samples at the time of autopsy [34,42].

A publication in abstract form describes a previously well 24-year-old male who presented to the ED with headache, palpitations and chest pain one hour after ingesting a product allegedly containing caffeine, arginine alpha-ketoglutarate, DMAA, creatine, beta-alanine and schizandrol A. The patient reported a viral illness during the previous month. He was tachycardic (heart rate 130b.p.m.) and hypertensive (BP 180/100mmHg). He was administered intravenous fluid (the volume administered is not reported) and developed acute pulmonary oedema requiring intubation, nitroglycerine and diuretics. He was also administered antibiotics. Echocardiography demonstrated severe left ventricular hypertrophy, hypokinesis of basal segments and an ejection fraction of less than 20%. Magnetic resonance imaging did not find an infective or infiltrative cause for the cardiomyopathy and urine toxicological screening was negative for cocaine. Biological samples were not analyzed to determine the presence of DMAA. He was discharged on day 14-post admission with improved cardiovascular function [43].

Further toxicological information is available through published Internet user reports.

Adverse or unwanted effects reported by users include nausea, vomiting, diaphoresis, headache, drowsiness, palpitations, appetite depression and appetite stimulation, and the desire to use more DMAA. Users also report worse nausea when DMAA is used with ethanol [19,44].

Chronic Toxicity

Animal Data

There are no published animal reports examining the chronic toxicological effects of DMAA exposure.

Human Data

There are no published human reports examining the chronic toxicological effects of DMAA exposure.

Dependence and Abuse Potential

Animal Data

There are no published animal or *in vitro* studies investigating the dependence/abuse potential of DMAA.

Human Data

There are no published reports examining possible dependence to DMAA. There are user reports of feeling the desire to re-dose with more DMAA, which may increase the potential for dependence and abuse [44]. However, the Drugs-Forum website reports that users who experience euphoria on initial use of DMAA also report that additional doses seem to be associated with less euphoria, a pattern that is unlikely to lead to dependence [19].

SYNTHETIC COCAINES

Numerous synthetic analogues of cocaine have been created with the intention of utilising them for medicinal purposes. Dimethocaine has been available since the 1930s and was originally used as a local anaesthetic under the trade name Larocaine [24]. Dimethocaine and 4-fluorotropacocaine have been found in recreational products sold in head shops in Europe [24]. 4-fluorotropacocaine was reported to the EMCDDA early warning system by Finnish authorities in 2008, and dimethocaine was found in seized products in a number of European countries in 2010 and 2011 [45,46].

Pharmacology

Physical and Chemical Description

Dimethocaine and 4-fluorotropacocaine are benzoic acid esters. 4-fluorotropacocaine has a similar structure to cocaine, but has a fluorine atom in the phenyl ring and has lost a carboxyl group on the tropane ring (Fig. 17.3). Dimethocaine does not have a tropane ring, but shares a similar structure with the local anaesthetic procaine, which does not have psychoactive properties. 4-fluorotropacocaine and dimethocaine are found as white powders at room temperature [47,48].

Pharmacokinetics

There are no formal animal or human studies examining the pharmacokinetic properties of 4-fluorotropacocaine or dimethocaine. Cocaine and other synthetic cocaines with ester structures are hydrolyzed by esterases in the liver [49]. Dimethocaine would be expected to also undergo hydrolysis potentially producing 4-amino benzoic acid and 3-(diethylamino)-2-dimethylpropanol. Like cocaine, both dimethocaine and 4-fluorotropacocaine are unlikely to be significantly active orally because of metabolism by gastrointestinal esterases.

Although limited by reporting bias and the inability to confirm exact substance exposure, user reports provide some pharmacokinetic information [50,51,52]. Dimethocaine is typically insufflated with an onset of action of 10–30 minutes, peak effects at 60–120 minutes and duration of action with some 'after-effects' of 4–6 hours. Typical doses are 50–200 mg. Users report minimal effects following ingestion or smoking dimethocaine. Intravenous administration of dimethocaine is reported to produce rapid onset, but short duration (<10 minutes) of effects [24].

4-fluorotropacocaine users report insufflating doses of 50–200 mg with onset of effects after 30–60 minutes. Duration of effect is reported as between three and five hours. 4-fluorotropacocaine is reported by users to be 'substantially less potent than cocaine' [53].

Pharmacodynamics

There have been no pharmacodynamic studies of dimethocaine or 4-fluorotropacocaine in humans. In 1996 Rigon and Takahasi studied the physiological and behavioural effects of dimethocaine in mice [54]. They found positive 'locomotor stimulant' actions and behavioural effects similar to the action of cocaine seen in other animal models. They concluded dimethocaine had significant dopaminergic actions [54]. Further rodent studies suggest dimethocaine acts by inhibiting dopamine reuptake [55,56]. Evidence suggests dimethocaine is equivalent in terms of potency to cocaine as a local anaesthetic and roughly half the potency of cocaine as a stimulant [57]. However, animal studies suggest that in order to achieve dopamine active transporter (DAT) inhibition of >65%, which is associated with euphoria, 10 times the dose of dimethocaine is required compared to cocaine [54–56]. Dimethocaine has been demonstrated to substitute for cocaine-induced repetitive behaviour in rats [58,59]. In summary, it appears that both dimethocaine and 4-fluorotropacocaine have local anaesthetic actions and act as dopamine reuptake inhibitors in the central nervous system.

POSITIVE OR DESIRED CLINICAL EFFECTS

No human studies examining the behavioural or psychological effects of dimethocaine

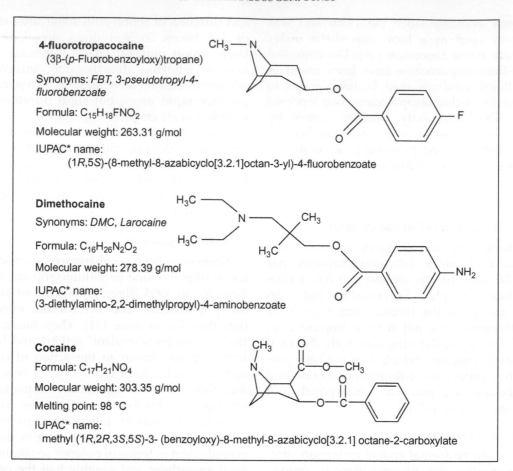

4-fluorotropacocaine
(3β-(p-Fluorobenzoyloxy)tropane)

Synonyms: *FBT, 3-pseudotropyl-4-fluorobenzoate*

Formula: $C_{15}H_{18}FNO_2$

Molecular weight: 263.31 g/mol

IUPAC* name:
(1R,5S)-(8-methyl-8-azabicyclo[3.2.1]octan-3-yl)-4-fluorobenzoate

Dimethocaine

Synonyms: *DMC, Larocaine*

Formula: $C_{16}H_{26}N_2O_2$

Molecular weight: 278.39 g/mol

IUPAC* name:
(3-diethylamino-2,2-dimethylpropyl)-4-aminobenzoate

Cocaine

Formula: $C_{17}H_{21}NO_4$

Molecular weight: 303.35 g/mol

Melting point: 98 °C

IUPAC* name:
methyl (1R,2R,3S,5S)-3- (benzoyloxy)-8-methyl-8-azabicyclo[3.2.1] octane-2-carboxylate

FIGURE 17.3   Chemical structure, physical properties and user names for 4-fluorotropacocaine, dimethocaine and cocaine. *(IUPAC – International Union of Pure and Applied Chemistry.)

or 4-fluorotropacocaine exist; however, information can be obtained from user reports published on-line and interviews with drug users. Internet user reports of exposure to dimethocaine obtained via Internet suppliers or head shops report mild stimulant effects. User or slang names for dimethocaine include: 'Mind melt', 'Amplify', 'Amplifier', 'Mint' and 'Mania'. Minimal euphoria is reported, with some users reporting none at all. Some users report increased clarity of thought, while others experience intellectual slowing following initial

stimulant effects and sedation lasting into the next day. Other desired or neutral reported effects include increased talkativeness, minimal effect on libido and the gastrointestinal system, and no residual 'coming-down' adverse effects [50,52,60,61].

On-line user reports of dimethocaine exposure contrast with interviews of drug users conducted by researchers from the National Advisory Committee on Drugs in Ireland. Analysis of drugs obtained from Internet suppliers and head shops in Ireland identified dimethocaine within two products known as 'Amplify' and 'Amplifier' [24,62]. Users who injected these substances report a short (three–six minute) 'crazy rush' (often rated better than effects induced by cocaine) followed by a few hours of feeling 'stoned'. Some users reported the drug as stronger than any they had ever used and said it provided a smooth, consistent experience. 'Amplify' and 'Amplifier' were also ingested and smoked in doses of up to 1000 mg at a time. Many of the users administered 'Amplify' or 'Amplifier' multiple times per day up to a total dose of 5–10 g [24]. A possible explanation for the incongruence between experiences reported by these two populations comes from a study published in the UK in 2010. Researchers analyzed 17 products purchased from UK based Internet suppliers. One product was labelled as containing dimethocaine, but in fact contained only caffeine and lignocaine [63]. This highlights the limitations of user reports in attributing behavioural and physiological effects to a particular substance. At present it is difficult to draw any definite conclusions regarding the positive effects of dimethocaine exposure.

There are very few user reports of 4-fluorotropacocaine exposure. Some users report short periods of cocaine-like effects, although the stimulant properties appear to predominate compared with any euphoric effects. Other positive reported effects include increased conversation and clarity of thought and a feeling of peace [64].

Prevalence of Use

Recreational use of synthetic-cocaine molecules is a relatively recent phenomenon, and the overall level of use in most countries appears low amongst both the NPS and illicit drug using populations. Dimethocaine was identified in a number of products obtained from head shops in Ireland in 2010 [24,62]. Two products named 'Amplify' and 'Amplifier' containing dimethocaine were reportedly abused by intravenous drug users in Dublin, Ireland in 2010–2011 [24]. A questionnaire of 329 'legal high' users in Ireland conducted by the National Advisory Committee on Drugs (NACD) in 2010 found 3% had used dimethocaine. Two other products previously found to contain dimethocaine were each reportedly used by 2.7% of respondents [24].

Prevalence of use of synthetic cocaines in the UK has not been reported using subpopulation surveys that have assessed use of some other NPSs such as the UK Mixmag/Guardian Global Drug Survey [13].

Seizures of substances containing dimethocaine in Sweden, Bulgaria, UK and Croatia were reported to the EMCDDA Early Warning System and recorded on the European Database on New Drugs (EDND) in 2011 [64]. There are only three user reports of dimethocaine use on the website www.erowid .org [50]. Other user websites report sporadic, but low volume dimethocaine use since 2008 [51,52,61]. Dimethocaine appears to be available in Australia, with at least one reported exposure documented on a user's forum [52].

4-Fluorotropacocaine was first reported to the EMCDDA in 2008 and has been found in a number of products including 'Star Dust' and 'Wack' (combined with desoxypipridol) sold in head shops in Ireland [24,65,66]. The Irish 2010 NACD questionnaire found 0.5% of respondents had used 4-fluorotropacocaine, while 7.5% had used 'Wack' and 5.9% 'Stardust' [24]. There are no user reports of 4-fluorotropacocaine use on the website www.erowid.org. Other user websites provide infrequent user reports [67].

Acute Toxicity

Animal Data

One study examining the effect of intracerebroventricular administration of dimethocaine in mice demonstrated impairment of learning and memory processes [68].

Human Data

There are no human studies examining human 4-fluorotropacocaine or dimethocaine toxicity. As dimethocaine is an anaesthetic-stimulant, it is likely to have effects similar to other such drugs. The sodium channel blocking properties of cocaine and other local anesthetics are known to produce cardiotoxicity. However, no reports of cardiotoxicity associated with dimethocaine have been published. Combined cocaine and ethanol exposure leads to genera-tion of cocaethylene, and there are some reports to suggest that this may be associated with a different pattern of toxicity [69,70]. There is no data to determine whether there may be any interaction between dimethocaine and ethanol.

Adverse or unwanted effects of dimethocaine reported on Internet user web-sites include peripheral vasoconstriction of limbs and male genitalia (not reported by all users), sedation following initial stimu-lant effects, tachycardia, diaphoresis, muscle twitching, nausea, vomiting, diaphoresis and in some cases the desire to re-dose. Users of dimethocaine report nose and throat numbness when the drug is insufflated, in keeping with a local anaesthetic effect [24,60,61].

Interviews conducted with intravenous drug users in Ireland exposed to the products 'Amplify' and 'Amplifier' previously found to contain dimethocaine provide a picture of more severe toxicity. Users who injected 'Amplify' or 'Amplifier' reported agitation, seizures and psychosis following short-term exposure. Reported effects when 'coming-down' from 'Amplify' or 'Amplified' include unpleasant hallucinations, agitation, depression, suicidal thoughts, loss of awareness, loss of motor con-trol and paranoia. These effects appeared to be worse in users with a pre-existing psychiatric illness. Effects observed and reported by carers of this group of drug users include weight loss, insomnia, aggression, abscesses and ulcers at injection sites [24].

In June of 2010 the Irish National Poisons Information Centre received 40 calls relating to adverse reactions related to the head shop product 'Whack', which had previously been analyzed and found to contain 4-fluorotropa-cocaine and desoxypripradol. Adverse effects included palpitations, hypertension, anxiety, increased respiratory rate and psychosis [24]. It is difficult to conclude to what degree either of the ingredients contributed to the adverse effects and/or whether there may have been other drugs in the 'whack' that may have contributed.

Internet user reports of 4-fluorotropacocaine exposure describe negative symptoms includ-ing extreme pain on insufflation or with rectal administration, sedation, gastrointestinal upset, abdominal pain, anxiety, reduced mood and the absence of euphoria [67].

Chronic Toxicity

Animal Data

There are no published animal reports examining the chronic toxicological effects of dimethocaine or 4-fluorotropacaine exposure.

Human Data

There is scarce published information regarding the chronic effects in humans of dimethocaine or 4-fluorotropacocaine expo-sure. Reports from users in Ireland who inject products possibly containing dimethocaine document soft tissue ulcers and abscesses, dependence and tolerance to the drug, and severe psychological disturbance when discon-tinuing regular use of dimethocaine [24].

Two patients are reported to have developed an acute psychotic state after using 'Whack', a product containing 4-fluorotropacocaine and desoxypipradrol. Both patients displayed sig-nificant anxiety and psychotic type symptoms, but recovered fully following treatment with atypical antipsychotic medications [71]. It is not possible to determine whether these features were due to 4-fluorotropacocaine or desoxypipradrol; however, as described in more detail in Chapter 10 on pipradrols, desoxypipradrol has been reported to be associated with acute psychiatric complications.

Dependence and Abuse Potential

Animal Data

Rodent studies suggest that dimethocaine is associated with inhibition of dopamine re-uptake and the compulsive need to re-dose, similar to effects that are observed with cocaine [54,55].

Human Data

There are no published reports examining possible dependence to dimethocaine or 4-fluorotropacocaine. Users of the products 'Amplify' and 'Amplifier', implicated as containing dimethocaine, report the occurrence of tolerance and the need to increase the amount of daily dosing [24].

REFERENCES

[1] Europol 2010 Annual report on the implementation of council decision 2005/387/JHA. Available: <http://www.emcdda.europa.eu/publications/implementation-reports/2010> [accessed 29.01.12].

[2] Mak SK, Chan MT, Bower WF, et al. Lower urinary tract changes in young adults using ketamine. J Urol 2011;186:610–4.

[3] Morris H (11-February-2011). Interview with a ketamine chemist: or to be more precise, an arylcyclohexylamine chemist. Vice Magazine. Available: <http://www.vice.com/read/interview-with-ketamine-chemist-704-v18n2> [accessed 29.01.12].

[4] Linders J, Furlano DC, Mattson MV, Jacobsen AE, Rice KC. Synthesis and preliminary biochemical evaluation of novel derivatives of PCP. Lett Drug Des Discov 2010;7:79–87.

[5] ACMD REPORT. Advisory council on the misuse of drugs (ACMD) methoxetamine report 2012. Available: <http://www.homeoffice.gov.uk/publications/agencies-public-bodies/acmd1/methoxetamine2012?view=Binary> [Last accessed 14.12.12].

[6] Erowid. Available <http://www.erowid.org/chemicals/methoxetamine/methoxetamine.shtml> [Accessed 29.01.12].

[7] Aroni F, Iacovidou N, Dontas I, Pourzitaki C, Xanthos T. Pharmacological aspects and potential new clinical applications of ketamine: reevaluation of an old drug. J Clin Pharmacol 2009;49(8):957–64.

[8] Dargan PI, Hudson S, Wood DM. Metabolites and potential metabolic pathways for the novel psychoactive substance methoxetamine. Clin Tox (Phila.) 2012;50:692–3.

[9] Rosenbaum CD, Carreiro SP, Babu KM. Here today, gone tomorrow … and back again. A review of herbal marijuana alternatives (K2, spice), synthetic cathinones (bath salts), kratom, salvia divinorum, methoxetamine, and piperazines. J Med Toxicol 2012;8:15–32.

[10] Wood DM, Davies S, Puchnarewicz M, Johnston A, Dargan PI. Acute toxicity associated with the recreational use of the ketamine derivative methoxetamine. Eur J Clin Pharmacol 2012;68:853–6.

[11] Blue Light. Available : <http://www.bluelight.ru/vb/search.php?searchid=369958>.

[12] DrugLink Nov/Dec 2011. Street drugs trend survey 2011. The ketamine zone. Available: <http://www.drugscope.org.uk/Resources/Drugscope/Documents/PDF/Publications/StreetDrugsSurvey2011.pdf> [accessed 29.01.12].

[13] 2012 Global Drug Survey. Available: <http://globaldrugsurvey.com/run-my-survey/2012-global-drug-survey>.

[14] Wood DM, Hunter L, Measham F, Dargan PI. Limited use of novel psychoactive substances in South London nightclubs. Q J Med 2012 Jun 19 [Epub ahead of print].

[15] European Information System and Database on New Drugs. Available: <http://ednd.emcdda.europa.eu/html.cfm/index7246EN.html?SUB_ID=149> [accessed 29.01.12].

[16] Ward J, Rhyee S, Plansky J, Boyer E. Methoxetamine: a novel ketamine analog and growing health-care concern. Clin Toxicol (Phila.) 2011;49:874–5.

[17] Hofer KE, Grager B, Müller DM, Rauber-Lüthy C, Kupferschmidt H, Rentsch KM, et al. Ketamine-like effects after recreational use of methoxetamine. Ann Emerg Med 2012;60:97–9.

[18] Shields JE, Dargan PI, Wood DM, Puchnarewicz M, Davies S, Waring S. Methoxetamine associated cerebellar toxicity: Three cases with analytical confirmation. Clin Tox 2012;50:438–40.

[19] Drugs Forum. Available: <http://www.drugs-forum.com/forum/showwiki.php?title=Methoxetamine> [accessed 29.01.12].

[20] Wood DM, Yew DT, Lam WP, Dargan PI. Chronic methoxetamine exposure in a mouse model

demonstrates that methoxetamine is not a 'bladder friendly' alternative to ketamine. Clin Tox (Phila.) 2012;50:694.

[21] Council on Pharmacy and Chemistry Methylhexamine Forthane (Lilly). JAMA 1950; 143:1155–1157.

[22] Ping Z, Jun Q, Qing L. A study on the chemical constituents of geranium oil. J Guizhou Inst Tech 1996;25:82–5.

[23] Cohen PA. DMAA as a dietary supplement ingredient. Arch Int Med 2012;172:1039–40.

[24] Kelleher C, Christie R, Lalor K, Fox J, O'Donnell C. An overview of new psychoactive substances and the outlets supplying them. : National Advisory Committee on Drugs (Ireland); 2011.

[25] Gee P, Jackson S, Easton J. Another bitter pill: a case of toxicity from DMAA party pills. N Z Med J 2010;123:124–7.

[26] World Anti Doping Agency. The world anti-doping code. The 2010 prohibited list. International standard. Available: <http://www.wada-ama.org/Documents/World_Anti-Doping_Program/WADP-Prohibited-list/WADA_Prohibited_List_2010_EN.pdf> [accessed 29.01.12].

[27] Miya TS, Edwards I. A pharmacological study of certain alkoxyalkylamines. J Pharm Sci 1953;42(2):107–10.

[28] Charlier R. Pharmacology of 2-amino-4-methylhexane. Arch Int Pharmacodyn Ther 1950;83(4):573–84.

[29] Advice to the expert advisory committee on drugs on: assessment of 1,dimethylamylamine(DMAA). Available: <http://ndp.govt.nz/moh.nsf/Files/ndp-advice-docs/$file/advice-to-eacd-dmaa-aug-2009.pdf> [accessed 29.01.12].

[30] Bloomer RJ, Harvey IC, Farney TM, Bell ZW, Canale RE. Effects of 1,3-dimethylamylamine and caffeine alone or in combination on heart rate and blood pressure in healthy men and women. Phys Sportsmed 2011 Sep;39:111–20.

[31] McCarthy CG, Canale RE, Alleman RJ, et al. Biochemical and anthropometric effects of a weight loss dietary supplement in healthy men and women. Nutr Metabolic Insights 2012;5:13–22. doi:10.4137/NMI.S8566. Available: <http://www.la-press.com/biochemical-and-anthropometric-effects-of-a-weight-loss-dietarysupple-article-a2946> [accessed 14.09.12].

[32] Farney TM, McCarthy CG, Canale RE, et al. Hemodynamic and hematologic profile of healthy adults ingesting dietary supplements containing 1,3-dimethylamylamine and caffeine. Nutr Metabolic Insights 2012;5:1–12. doi:10.4137/NMI.S8568. Available: <http://www.la-press.com/hemodynamic-and hematologic-profile-of-healthy-adults-ingesting-dietar-article-a2947> [accessed 14.09.12].

[33] McCarthy CG, Farney TM, Canale R, et al. A finished dietary supplement stimulates lipolysis and metabolic rate in young men and women. Nutr Metabolic Insights 2012;5:23–31. Available:<http://www.la-press.com/a-finished-dietary-supplement-stimulates-lipolysis-and-metabolic-rate--article-a2945> [accessed 14.09.12]..

[34] Lattman P, Singer N. Army studies workout supplements after deaths. New York Times. February 2, 2012. Available: <http://www.nytimes.com/2012/02/03/business/army-studies-workoutsupplements-after-2-deaths.html> [accessed 14.09.12].

[35] Vorce SP, Holler JM, Cawrse BM, et al. Dimethylamylamine: a drug causing positive immunoassay results for amphetamines. J Anal Toxicol 2011;35:183–7.

[36] Health Products and Food Branch, Health Canada. Canadian Natural Health Products ingredients database. August 16, 2011. Available: <http://webprod.hc-sc.gc.ca/nhpid-bdipsn/ atReq.do?atidwhats.quoi.2011.08.16&langeng#3> [accessed 14.09.12].

[37] Expert Advisory Committee on Drugs. Minutes of meeting of Thursday 6 August 2009. Available: <ndp.govt.nz/moh.nsf/pagescm/565/$File/eacd-minutes-6aug-09.doc> [accessed 29.01.12].

[38] Gee P, Tallon C, Long N, Moore G, Boet R, Jackson S. Use of recreational drug 1,3-Dimethylethylamine (DMAA) associated with cerebral hemorrhage. Ann Emerg Med 2012;60(4):431–4.

[39] EMCDDA. European information system and database on new drugs. Available: <http://ednd.emcdda.europa.eu/html.cfm/index7246EN.html?SUB_ID=124> [accessed 29.01.12].

[40] Merck index; an encyclopedia of chemicals, drugs, and biologicals, 11th ed. Rahway, NJ: Merck & Co.; 1989. vol 11: p. 957.

[41] Van Rosen F. Personal communication.

[42] Tritten TJ. Army probing connection between body building supplement, 2 deaths. Stars and Stripes. December 15, 2011.Available: <http://www.stripes.com/mobile/news/army-probingconnection-between-body-building-supplement-2-deaths-1.163652> [accessed 14.09.12].

[43] Salinger L, Daniels B, Sangalli B, et al. Recreational use of body-building supplement resulting in severe cardiotoxicity. Clin Toxicol 2011;49:573–4.

[44] Long J. Head shop drugs across Europe: Data from the EMCDDA. Presented at the national drugs task force 'legal highs' conference, 26 January 2010, Mullingar, Ireland.

[45] Europol 2010 Annual report on the implementation of council decision 2005/387/JHA. Available: <http://www.emcdda.europa.eu/publications/implementation-reports/2010> [accessed 29.01.12].

[46] European information system and database on new drugs. Available: <http://ednd.emcdda.europa.eu/

html.cfm/index7246EN.html?SUB_ID=123> [accessed 29 .01.12].

[47] Gibbons S. 'Legal highs' – novel and emerging psychoactive drugs: a chemical overview for the toxicologist. Clin Toxicol 2012;50:15–24.

[48] Singh S. Chemistry, design, and structure-activity relationship of cocaine antagonists. Chemical Reviews 2000;100:925–1024.

[49] Kolbrich EA, Barnes AJ, Gorelick DA, Boyd SJ, Cone EJ, Huestis MA. Major and minor metabolites of cocaine in human plasma following controlled subcutaneous cocaine administration. J Anal Toxicol 2006;30:501–10.

[50] Erowid. Erowid experience vaults. Available: <http://www.erowid.org/experiences/exp_front.shtml> [accessed 29.01.12].

[51] Dimethocaine. The user's guide. Available: <http://dimethocaine.org/> [accessed 29.01.12].

[52] Blue Light. Available: <http://www.bluelight.ru/vb/threads/450125-Dimethocaine-the-quot-larrikin-quot-relative-of-cocaine?highlight=dimethocaine> [accessed 29.01.12].

[53] Drugs-Forum. Available: <http://www.drugsforum.com/forum/showthread.php?t=65149> [accessed 29.01.12].

[54] Rigon AR, Takahashi RN. Stimulant activities of dimethocaine in mice: reinforcing and anxiogenic effects. Psychopharmacology (Berl.) 1996;127(4):323–7.

[55] Wilcox KM, Rowlett JK, Paul IA, Ordway GA, Woolverton WL. On the relationship between the dopamine transporter and the reinforcing effects of local anesthetics in rhesus monkeys: practical and theoretical concerns. Psychopharmacology 2000;153:139–47.

[56] Woodward JJ, Compton DM, Balster RL, Martin BR. In vitro and in vivo effects of cocaine and selected local anesthetics on the dopamine transporter. Eur J Pharmacol 1995;277:7–13.

[57] Wilcox KM, Kimmel HL, Lindsey KP, Votaw JR, Goodman MM, Howell LL. In vivo comparison of the reinforcing and dopamine transporter effects of local anesthetics in rhesus monkeys. Synapse 2005;58(4):220–8.

[58] Graham JH, Balster RL. Cocaine-like discriminative stimulus effects of procaine, dimethocaine and lidocaine in rats. Psychopharmacology 1993;110:287–94.

[59] Woolverton WL, Balster RL. Effects of local anesthetics on fixed interval responding in rhesus monkeys. Pharmacol Biochem Behav 1983;18:383–7.

[60] Wikipedia. Available: <http://en.wikipedia.org/wiki/Dimethocaine> [accessed 29.01.12].

[61] Drugs-forum. Available: <http://www.drugs-forum.com/forum/showthread.php?t=91881> [accessed 29.01.12].

[62] Kavanagh P, Sharma J, McNamara S, et al. Poster: 'Head shop 'legal highs' active constituents identification chart', June 2010. Available: <https://ednd-cma.emcdda.europa.eu/assets/upload/HS%20ID%20 Poster%20June%20Post-ban.pdf [accessed 29.01.12].

[63] Brandt SD, Sumnall HR, Measham F, Cole J. Second-generation mephedrone. The confusing case of NRG-1. BMJ 2010;341

[64] European information system and database on new drugs. Available: <http://ednd.emcdda.europa.eu/html.cfm/index7246EN.html?SUB_ID=123> [accessed 29.01.12].

[65] King LA. New drugs coming our way: what are they and how do we detect them? Presented at the EMCDDA conference: identifying Europe's information needs for effective drug policy. Lisbon, 6–8 May 2009.

[66] Kavanagh P, Angelov D, O'Brien J, et al. The syntheses and characterization 3β-(4-fluorobenzoyloxy) tropane (fluorotropacocaine) and its 3α isomer. Drug Test Anal 2011.

[67] Drugs Forum. Available: <http://www.drugs-forum.com/forum/showthread.php?t=66216> [accessed 29.01.12].

[68] Blatt SL, Takahashi RN. Memory-impairing effects of local anaesthetics in an elevated plus-maze test in mice. Braz J Med Biol Res 1998;31:555–9.

[69] Hayase T, Yamamoto Y, Yamamoto K. Role of cocaethylene in toxic symptoms due to repeated subcutaneous cocaine administration modified by oral doses of ethanol. J Toxicol Sci 1999;24:227–35.

[70] Wilson LD, Jeromin J, Garvey L, Dorbandt A. Cocaine, ethanol, and cocaethylene cardiotoxity in an animal model of cocaine and ethanol abuse. Acad Emerg Med 2001;8:211–22.

[71] El-Higaya E, Ahmed M, Hallahan B. Whack induced psychosis: a case series. Ir J Psychol Med 2011;28:S11–3.

This page intentionally left blank

Index

Printed and bound by CPI Group (UK) Ltd, Croydon, CR0 4YY

03/10/2024

01040326-0005